AMERICAN ARCTIC LICHENS

1. The Macrolichens

AMERICAN ARCTIC LICHENS

1. The Macrolichens

JOHN W. THOMSON

Illustrations by
BETHIA BREHMER

Supplementary drawings by
Lucy C. Taylor

COLUMBIA UNIVERSITY PRESS
New York
1984

Library of Congress Cataloging in Publication Data

Thomson, John Walter, 1913–
American Arctic lichens.

Bibliography: p.
Includes index.
Contents: v. 1. The macrolichens—
1. Lichens—Canada, Northern.   2. Lichens—Alaska.
3. Lichens—Arctic regions.   I. Brehmer, Bethia.
II. Title.
QK587.7.N67T46   1984   589.1′09719′9   84-5003
ISBN 0-231-05888-8 (v. 1. : alk. paper)

Columbia University Press
New York   Guildford, Surrey
Copyright © 1984 Columbia University Press
All rights reserved

Drawings © 1983 Board of Regents of the University of Wisconsin System,
Department of Botany, UW-Madison.

Printed in the United States of America

To Olive
And to the memory of our son
Douglas

Contents

Preface

MANY YEARS have gone into the production of this volume on the macrolichens, which is the first of a projected two-volume work on the lichens of the American Arctic. Currently lichenologists use an artificial separation of lichens into those of larger size, the macrolichens, which are generally foliose and fruticose in growth habit of the thallus, and those of smaller size, the microlichens, with squamulose and crustose growth habits. Some genera are difficult to place as their thallus may be somewhat intermediate in growth habit, and such must be placed somewhat arbitrarily in one or the other of the volumes. Additionally there are some lichens as in members of *Baeomyces* and *Cladonia* which may produce a crustose primary thallus which further develops into a fruticose thallus as the lichen fungus develops into fruiting stages. Such dual stage lichens are included in this volume.

Geographically the question of what species to include in such a volume is also a difficult matter of choice. The border line between the boreal forest and the treeless tundras is not everywhere a sharp one. There are inclusions of tundra within the northern portions of the boreal forest, and conversely, there are outlying stands of forest trees, especially black and white spruce and balsam poplar and willows, beyond the more continuous edges of the forest. The lichens of the northern edge of the boreal forest often accompany such isolated stands and although not primarily arctic in distribution will nevertheless be included. Many forest species also may extend their ranges beyond the trees into the tundras proper. Obviously there will be many lichens of the boreal forest which will not be occurring along the northern edge, and this manual should not be construed as covering all the lichens of the boreal forest. That would be an entirely different undertaking.

Inspiration for work on arctic lichens came in the form of a small grant from the Arctic Institute of North America to work during the summer of 1950 on the lichens of the vicinity of Churchill, Manitoba, where the then most accessible tundra could be reached by railroad. Later grants from the National Science Foundation during the years from 1958 to 1968 enabled me to reach a very diverse series of areas in the American Arctic, especially by use of float planes which by then had made far reaches of the Arctic more readily attainable. With the help of the Arctic Institute of North America and the National Science Foundation plus a grant from the Office of Naval Research, I was able to devote a summer of field work in 1958 at the Arctic Research Laboratory at Point Barrow, Alaska, accompanied by Dr. Sam Shushan. The volume *Lichens of the Alaskan Arctic Slope* (University of Toronto Press, 1979) was produced from the material collected on that expedition.

The Ninth International Congress at Montreal during the summer of 1959 had as one of the field trips an excursion well into the Arctic with stops at Great Whale River, Fort Chimo, Frobisher Bay on Baffin Island, Cambridge Bay on Victoria Island, Resolute on Cornwallis Island, Coral Harbor on Southampton Island, and at Shefferville, Labrador. There were excellent opportunities to gather material for study at all of those localities. Other sites in the Canadian Arctic visited in collaborative expeditions with James A. Larsen were Yellowknife, Fort Reliance, and Artillery Lake in the Great Slave Lake region during the summer of 1962, Coppermine during the summer of 1962, Inuvik and Canoe Lake and the Babbage River (Yukon) during 1964, Lake Ennadai during 1960, Lake Dubawnt and lakes just north of it during 1963, and an expedition from Churchill to Chesterfield Inlet and localities near the mouth of the Back River, led by Reid A. Bryson of the Department of Meteorology, University of Wisconsin, during 1959.

The knowledge that I was studying the arctic lichens also led many other investigators to submit material to me for identification. Especially important were the collections made by George W. Scotter of the Canadian Wildlife Service in many sectors of the Arctic and subarctic, by Barbara Murray in the Alaskan tundras, by Rudy Schuster on northern Ellesmere Island, by Mrs. Johanna Marr on Ungava, and by Paul Barrett on Devon Island. Very useful contributions of collections from the Arctic islands came from the meteorologist A. Innes-Taylor, and by Andreas Züst from the Carey Islands off Greenland. Dr. Ernst Abbe of the University of Minnesota made available to me the many tiny specimens of lichens collected by Margaret Oldenburg on her journeys to many unusual arctic localities. I am also indebted to Dr. Eilif Dahl for a very valuable set of duplicates of Greenland lichens collected by himself and by Bernt Lynge which provided most useful comparisons.

Through the years many other contributiors have sent specimens for checking and their locality records have all been incorporated on the maps. During the spring and summer of 1961 the grants from the National Science Foundation made possible study of the arctic materials in the European herbaria at Kew, London, Paris, Geneva, Oslo, Copenhagen, Lund, Stockholm, Uppsala, Helsinki, and Turku. Field work in the Alps with Eduard Frey, and in the Swedish Lapland area so well worked by A. H. Magnusson, the Torne Lappmark area, was also made possible by the grants. A most valuable experience which made possible months of work in the vast collections of the University of Helsinki and the University of Turku as well as field work in the Finnish Lapland area was afforded by an exchange professorship during January to June of 1965 to the University of Helsinki. A long expedition to Alaska and the Yukon during the summer of 1969 together with Dr. Teuvo Ahti enabled the collection of data on arctic-alpine sites accessible from the various highway systems. Further work in the field was made possible by Dr. C. D. Bird on an expedition to the upper reaches of the Arctic Red River near the Yukon boundary west of Great Bear Lake during the summer of 1978. An opportunity to see arctic lichens in the north of the USSR in the Kirovsk region came with an excursion of the Twelfth International Botanical Congress in Leningrad during 1975. In all I have made fourteen trips into tundra regions in search of the lichens.

Mapping the ranges of the species has depended upon the collation of data from a wide variety of sources. The foremost source was the large collection of specimens in the herbarium of the University of Wisconsin. This includes the specimens collected by myself and by many others on expeditions into the Arctic. This herbarium contains the largest collection of far northern lichens available in North America. Visits to the herbarium of the National Museums of Canada, the Farlow Herbarium, Harvard University, and the U.S. National Herbarium added data, as did the visits to the European herbaria, especially those at Copenhagen and Oslo where so many Greenland collections are stored. A major source of information on records from Greenland came from the papers by Lynge, Dahl, and K. and E. S. Hansen. Information on Icelandic records came from papers by Lynge and by Kristinsson. Valuable Alaskan records came from the papers by Krog (1962, 1968). Older literature reports have seldom been accepted for mapping because of the unreliability of determinations or differing concepts of species limits in taxonomy. Actually there is comparatively little literature on North American Arctic lichens prior to the extensive papers of Lynge on Greenland and his analysis of the lichens collected by Polunin.

In selecting the use of dot-mapping rather than the mapping of areas with overall shading such as may be gained by use of "zipotone," I am well aware of both the advantages and disadvantages of this method of expressing the ranges. The dots do cover square miles of terrain. But this is not too disconcerting in the Arctic where large areas may be covered by particular vegetational types. They do reflect the activity of collectors but this may be advantageous in indicating where more work may be necessary. Someone who has collected at a particular site may, and often does, become annoyed that that site is not represented by the dots on the maps. Actually one finds that such records may fill in lacunae on the maps but seldom extend the range of the species.

Perhaps such criticisms could have been avoided by representing the generalized ranges but my feeling was that it was better to represent the true data base for the range maps. At this time the collecting of the larger lichens has been sufficiently extensive to reflect the general patterns of distribution of the arctic species. At this time too, the collecting of the smaller lichens, the crustose and squamulose species, still needs to be much more complete and done by the more professional lichen students to give accurate suggestions of the ranges. Few of the general collectors bring back the rock inhabiting crustose species in sufficient amounts to represent the richness of these forms in a given locality. A sharp eye and a knowledge of the possible richness of the species to be expected are still necessary in working in new regions. The tiny dots on the maps which may puzzle some readers are not specimen records but are sites of settlements which make convenient orientation spots when plotting the specimen dots. The latter are the large dots. Some of the latter, as may be expected, may represent several overlapping collection records.

In addition to the field work with Sam Shushan, James A. Larsen, Charles D. Bird, and Teuvo Ahti I am indebted to many other fellow botanists and lichenologists for help and consultation, including Barbara Murray, Hildur Krog, Eilif Dahl, Eduard Frey, I. Mackenzie Lamb, Hannes Hertel, Irwin M. Brodo, William A. Weber, Ove Almborn, and Rolf Santesson.

In preparing this work I have relied very heavily, of course, on my field experiences for selection of the species and on my continued routine procedures of examining the specimens, using the microchemistry as well as the morphology. The chemistry has been especially helpful in dealing with the arctic specimens which are often ice-crystal blasted or otherwise altered in that harsh environment. When recent monographs have been available I have trusted these for many interpretations of the species but have also taken into account my personal prejudices and experiences. The user of this manual will find helpful, I hope, the lists of pertinent references for each genus. The current chaotic state of assignment and reassignment of genera to families led me to adopt the practice pioneered by Krog and Dahl and by Poelt in some of the European floras of using alphabetical generic arrangement in the book rather than a familial sequence. Puzzling perhaps may be the citation of synonyms. These are not intended to be comprehensive. That is left to the monographers. Included are the basionyms and the names which I have accepted. Added to these may be names current in the literature and such older names as may have appeared in the arctic literature.

The majority of the drawings of the species were prepared by Bethia Brehmer when she was staff artist in the Department of Botany, University of Wisconsin. Those prepared by Lucy C. Taylor include *Actinogyra mühlenbergii, Agyrophora scholanderi, Anaptychia kaspica, Cetraria commixta, Cladonia nipponica, Collema cristatum, Dermatocarpon arnoldianum, Fistulariella scoparia, Leprocaulon subalbicans, Leptochidium albociliatum, Leptogium lichenoides, L. minutissimum, L. palmatum, L. sinuatum, L. tenuissimum, Pilophorus cereolus, P. robustus, Psoroma hypnorum, Ramalina intermedia, R. obtusata, R. thrausta, Sticta weigelii,* and *Vestergrenopsis elaeina.* Four drawings prepared by Janet Mackenzie are *Thamnolia vermicularis, Xanthoria candelaria, X. elegans,* and *X. fallax.* Both Lucy C. Taylor and Janet Mackenzie prepared the drawings while staff artists at the University of Wisconsin. Carol Gubbins finished the drawings of *Fistulariella scoparia* and *Ramalina thrausta.* The scales in centimeters were added by Lucy Taylor, using the original specimens for the scale comparisons. The specimens which served as models are marked and filed in the herbarium of the University of Wisconsin. For those interested, the drawing technique involved using lithographic pencil on coquille paper. I am much indebted to the artists for their patient collaboration to portray the taxonomic characters as accurately as possible. Jean Arnold is to be thanked for preparing the final "clean copy" typing necessary for submission for publication.

Thanks are also owed to the University Research Committee for several summers of support giving time to work on the specimens and manuscripts.

Acknowledgment

THIS MATERIAL is based upon research supported by the National Science Foundation under Grants G4315, GB2702, GB7461, GB31115, and GB 3115-1 and 2 and is published with support of National Science Foundation Grant No. BSR-8304395.

The Foundation provides awards for research and education in the sciences. The awardee is wholly responsible for the conduct of such research and preparation of the results for the publication. The Foundation, therefore, does not assume responsibility for such findings or their interpretation.

Any opinions, findings, conclusions, or recommendations expressed in this publication are those of the author and do not necessarily reflect the views of the National Science Foundation.

AMERICAN ARCTIC LICHENS
1. The Macrolichens

Introduction

Geology

THE EASTERN Canadian Arctic, including two thirds of Canada, is underlain by the Precambrian rocks of the Canadian Shield. Over these rocks lies a belt of sedimentary rocks of Palaeozoic to Tertiary age, the belt running through the central Arctic islands and including the northern and western Arctic islands. The Precambrian rocks are mainly granites, gneisses, and schists. They extend westward toward the Mackenzie River valley but not including it. Outcrops of the Precambrian occur on Victoria Island in the western Arctic. The mountainous eastern parts of the chain of islands from Baffin Island northward through Devon Island and Ellesmere Island are composed of these rocks. This substrate strongly influences the presence of acidophilic lichens.

Around and to the west of the Canadian Shield lie complex mixtures of rocks, mainly sedimentaries such as limestones, shales, sandstones, dolomite, and quartzite, but with many intercalated basaltic lava flows, granites, and other rock types, some Precambrian, others ranging from Palaeozoic to Tertiary. It is difficult to summarize the complexity of the geology of this vast region and much interpretation and field study remains to be done. The distribution of calciphilic lichens is of course influenced by the occurrence of the basic rocks. This area of the interior plains of Canada forms a wedge west of the vast Canadian Shield and is broadest at its base which extends from western Manitoba to the Rocky Mountains in Alberta. It narrows northward until in the Arctic it extends only from the Mackenzie River to Cape Parry.

The Cordilleran Region, with very complex geology including many types of sedimentary as well as igneous rocks, extends northward into the Arctic west of the Mackenzie River from the Rocky Mountains through the Mackenzie Mountains, the Richardson Mountains, and bending around in the Yukon–Alaska region through the Brooks Range in Alaska. The many types of rocks yield many varied substrata for lichens and the high relief and deep dissection of this region give many microhabitats in infinite variety. The altitudes are such that alpine tundras extend the ranges of the tundra lichens far southward along the mountain peaks.

The availability of terrain for tundra lichens in North America was strongly influenced by the vast sheets of ice which covered the northern part of the continent during the Pleistocene. The heaviest glaciation centered in the Keewatin District west of Hudson Bay. From there the ice sheets spread in all directions but the montane glaciers along the Rocky Mountain chain were largely local and debouched eastward. Parts of the American Arctic, however, were not glaciated, notably western and northern portions of the arctic islands and the north slope of Alaska as well as parts of the Rocky Mountains and the British Mountains. Undoubtedly there were refugia for the arctic lichens from which they could spread rapidly following deglaciation.

Arctic Climates

The vast regions of Alaska and Canada in which the tundras form the predominant vegetational cover are influenced by a variety of geographical controls. Alaska and the Yukon, as well as parts of the Mackenzie River valley, fall within the influence of the western Cordillera. These mountain chains, including the Kenai, Alaska, Chugach, Wrangell, St. Elias, and Coastal ranges, as well as the Brooks and Richardson ranges and other montane heights, form a massive obstacle to moist cloudy air streams at low altitudes coming from the west, and these cause far-reaching perturbations over the entire continent (Bryson & Hare 1974). The plains to the east extending from the arctic coast to the prairies descend eastward from 1,300 m in Alberta to 250 m in Manitoba and present little obstacle to the flow of arctic air from the north. Farther east, centering on Hudson Bay, the Canadian Shield of granitic metamorphic rocks has somewhat elevated rims, especially to the east in Atlantic Canada, and this influences precipitation patterns but less so the flow of the air masses. In the Arctic Archipelago the high land and ice masses on Ellesmere, Baffin, and Devon Islands have strong effects locally but not so much on general airflow patterns. Greenland of course presents a strong barrier effect.

Permafrost underlies much of northern Canada and Alaska. It is continuous north of a line from the Seward Peninsula and the southern side of the Brooks Range to the eastern tips of Great Bear Lake, Great Slave Lake, Churchill, Manitoba, and northern Ungava. South of this line the permafrost is patchy from 500 to 1200 km southward into the boreal forest.

The Bering Sea, with its extensive winter ice cover and cold fogs in summer, keeps western coastal Alaska cloudy, cold, and moist. Hudson Bay has a similar influence, cooling eastern Canada and contributing to frequent snowfalls nearby. Along the Atlantic coast the cold Canadian and Labrador currents carrying pack ice and even bergs from Greenland may reach as far as Newfoundland, and with onshore winds can bring fog and clouds in summer. The Arctic Ocean is covered by pack ice 3 m thick in winter and summer and often chokes the channels between the islands of the Arctic Archipelago. Melting of the surface under the summer radiation alters the ice surface and stresses may break the pack but the climate remains cold and fogs are frequent coming off the ice pack.

Canada and Alaska are strongly influenced by the cyclones and anticyclones which travel in the moving waves of the circumpolar westerly current. The wavelengths occupy 70° to 30° longitude, with a trough moving eastward behind each cyclone center and a ridge following each anticyclone. Strong uplift of warmer air associated with the fronts at lower levels produces most of the dense clouds and precipitation associated with the fronts. The broad westerly flow may be divided into three conventionalized fronts along the edges of the air masses—Continental Arctic, Maritime Arctic, and Polar. In addition airstreams from the northern North Atlantic may penetrate eastern Canada and reach western Hudson Bay.

The bend of the mountain systems to the west in Alaska make the north highly accessible to airstreams and disturbances moving inland from the Bering and Chukchi seas. These are common in the warm season and produce the summer rains of the north, the Arctic Front lying across the north part of Alaska, the Pacific air moving across the Yukon, Mackenzie, Keewatin, and Arctic Archipelago. The maximum summer rainfall usually comes in August. In other seasons the region tends to lie within the circulation pattern of persistent lows centered over Baffin Island or south of it. Little weather change occurs and surface temperatures become very low, occasionally below −40° C.

The summers are short and warm, and the winters long and cold except where oceanic influences alter the regime. Mean air temperatures at 1.5 m in January range from −35° C in the area of the Boothia Peninsula to −25° C over most of the tundra regions. By April only the northern part of the Archipelago remains in the −25° C range and the major part of the tundra regions is within −5° to 20° C range. During July the tundra regions range from 5° C near the coasts to 15° C over most of the area, but by October the drop to 0° C or below has reached most of the area, ranging down to −20° C in the northernmost part of the Archipelago. The mean date of fall of daily air temperature to 0° C occurs from September 1 in the north to October 15 in the southern edge of the tundra zone. June is the month of thaw for the Arctic. The period of frost free days over the tundra is short, less than 50 for most of the Arctic, less than 40 over large sections of the north. The boreal forest may have 60 to 110 days (Bryson & Hare 1974).

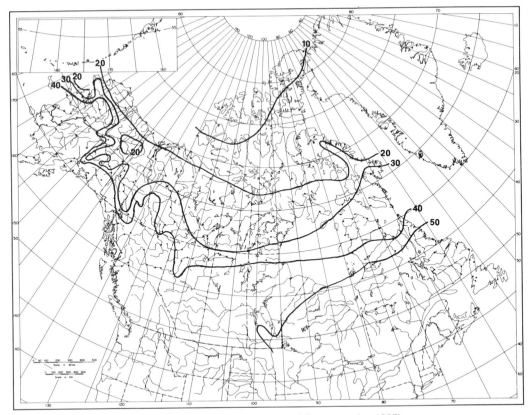

Mean annual precipitation in cm (after Canada Department of Transportation 1967).

Precipitation is very low in the American Arctic, the range varying from below 10 cm mean annual precipitation in the western Arctic Archipelago to 40 cm over most of the area, rising to 50 cm in a portion of the south.

The coastal plain south of Barrow, Alaska, has less than 15 cm precipitation yet is waterlogged with melt from the permafrost. The highest precipitation is in the high plateaus facing Baffin Bay. The main rainfall comes in late summer to early fall with August the wettest month. Winter snowfall is light in all areas, below 150 cm over much of the arctic, and on the Alaskan coastal plain and in the Queen Elizabeth Islands below 100 cm. Winds sweep the snow into drifts which gather in low spots and are important sites to seek unusual lichens.

Bryson (1966), in an excellent study of the relation of air masses and air streamlines to the boreal forest, has shown that a zone of rapid transition, the Arctic Frontal Zone, separating Arctic Air dominance from Pacific Air dominance, lies along the northern border of the boreal forest (the southern edge of the tundras) and suggests that summer air mass distribution might be an important causal factor for the distribution of forest versus tundra. Similar data for winter show a close coincidence of the winter position of the Arctic Frontal Zone with the southern border of the boreal forest.

REFERENCES ON CLIMATE

Bryson, R. A. 1966. Air masses, streamlines and the boreal forest. Geogr. Bull. 8:228–269.

Bryson, R. A. & F. K. Hare, eds. 1974. Climates of North America: World Survey of Climatology, 11:1–420. New York: Elsevier.

Hare, F. K. & M. K. Thomas. 1974. Climate Canada. Toronto: Wiley.

Sources for the History of American Arctic Lichenology

The lichen collections made in the American Arctic were for the most part small collections made incidentally to other activities of the early expeditions and by non-lichenologists. As they are already well summarized in previous publications it would be redundant to repeat them here. An excellent summary of the early collections from the Canadian eastern Arctic is available in Lynge (1947). The equally superficial early collections made in central and western Canada including the Northwest Territories and Yukon are summarized in Bird et al. (1980). More recent collections are also cited in that listing. A very useful summary of Greenland lichen research to 1950 is in Dahl (1950). Krog (1968) gives a summary of lichen exploration in Alaska. The reader should also consult the Arctic Bibliography volumes 1–16 (1953–1975) and the Recent Literature on Lichens listings published in *The Bryologist* from 1951 to 1978 by W. L. Culberson and subsequently by R. S. Egan.

REFERENCES

Bird, C. D., J. W. Thomson, A. H. Marsh, G. W. Scotter, & Pak Yau Wong. 1980. Lichens from the area drained by the Peel and Mackenzie Rivers, Yukon and Northwest Territories, Canada. 1. Macrolichens. Can. J. Bot. 58:1947–1985.

Dahl, E. 1950. Studies in the macrolichen flora of southwest Greenland. Medd. om Grønl. 150(2):1–176.

Krog, H. 1968. The macrolichens of Alaska. Norsk Polarinstitutt Skrift. 144:1–180.

Lynge, B. 1947. Lichenes in Botany of the Canadian Eastern Arctic. 2. Thallophyta and Bryophyta. Nat. Mus. Canada Bull. 97:298–369.

Geography and Distribution Patterns

Examination of the maps of currently known distribution of each of the species will show that they are highly individualistic. However the distributions do fall into about thirteen rather distinct patterns into which the species can be grouped. By far the greatest number of the lichens in the Arctic or reaching it are circumpolar if one groups the first three sets of patterns, the circumpolar arctic-alpine, the circumpolar low arctic and boreal, and the boreal species. The total of the three categories account for approximately 70 percent of the macrolichen flora. The balance fall into a larger number of categories which each includes a smaller proportion of the macrolichen flora.

1. Circumpolar arctic-alpine species. These have ranges which extend south in the mountains of the east, including the mountains of the Gaspé Peninsula, the Mt. Kataadn area in Maine, the White Mountains in New Hampshire, and the Adirondack Mountains in New York. In the west they reach southward in the Rocky Mountain and Cascade ranges as well sometimes in the Coastal ranges. Some extend their known ranges into Mexico and a few range southward into the southern hemisphere in the Andes chain. Some of the latter species have been classed as bipolar by Du Rietz (1926) although they do not really range into Antarctica.

There are 119 species that fall into this category and account for about 35 percent of the macrolichens: *Actinogyra polyrrhiza; Agyrophora lyngei, A. rigida; Alectoria ochroleuca, A. vexillifera; Arctomia delicatula, A. interfixa; Baeomyces carneus, B. placophyllus; Bryoria chalybeiformis, B. nitidula; Cetraria chlorophylla, C. commixta, C. cucullata, C. delisei, C. ericetorum, C. fastigiata, C. hepatizon, C. islandica, C. nigricans, C. nivalis, C. tilesii; Cladonia bellidiflora, C. cyanipes, C. ecmocyna, C. gracilis v. gracilis, C. macrophylla, C. macrophyllodes, C. mitis, C. phyllophora, C. pocillum, C. pleurota, C. pyxidata, C. rangiferina, C. stellaris, C. stricta, C. subcervicornis, C. subfurcata, C. uncialis; Collema ceraniscum, C. tenax, C. undulatum; Coriscium viride; Cornicularia aculeata, C. divergens, C. muricata; Dactylina arctica, D. ramulosa; Dermatocarpon cinereum, D. intestiniforme, D. lachneum; Ephebe lanata; Hypogymnia austerodes, H. oroarctica, H. physodes, H. subobscura, H. vittata; Leptogium lichenoides, L. minutissimum, L. saturninum; Massalongia carnosa; Nephroma arcticum, N. ex-*

pallidum; Omphalodiscus decussatus, O. krascheninnikovii, O. virginis; Pannaria pezizoides; Parmelia alpicola, P. disjuncta, P. exasperatula, P. fraudans, P. infumata, P. omphalodes, P. panniformis, P. saxatilis, P. stygia, P. substygia; Parmeliella arctophila; Polychidium muscicola; Pseudephebe minuscula, P. pubescens; Psoroma hypnorum; Siphula ceratites; Solorina bispora, S. crocea, S. octospora, S. spongiosa; Sphaerophorus fragilis, S. globosus; Stereocaulon alpinum, S. botryosum, S. capitellatum, S. coniophyllum, S. glareosum, S. incrustatum, S. rivulorum; Thamnolia subuliformis, T. vermicularis; Umbilicaria arctica, U. cylindrica, U. polyphylla, U. proboscidea, U. torrefacta; Xanthoparmelia centrifuga, X. incurva, X. separata.

2. Circumpolar low arctic and boreal species. It should be noted that among the species I have placed in this category as well as in the previous one are some which will be well known to workers as also being common in temperate regions. Perusal of the maps will reflect these ranges. For ranges below those parts of North America which are shown on the maps in this book an arrow at the southern border of the map suggests that one should consult Hale's *How To Know the Lichens.*

The 72 species listed in this category constitute about 21 percent of the arctic macrolichen flora: *Anaptychia kaspica; Baeomyces rufus; Bryoria simplicior; Cetraria pinastri, C. sepincola; Cladonia acuminata, C. amaurocraea, C. arbuscula, C. bacilliformis, C. cariosa, C. carneola, C. cenotea, C. chlorophaea, C. coccifera, C. cornuta, C. crispata, C. decorticata, C. deformis, C. digitata, C. fimbriata, C. gracilis ssp. turbinata, C. maxima, C. merochlorophaea, C. norrlinii, C. squamosa, C. sulphurina, C. symphycarpa; Collema flaccidum, C. glebulentum; Hypogymnia bitteri; Lobaria scrobiculata; Parmeliella praetermissa; Parmeliopsis ambigua, P. hyperopta; Peltigera aphthosa var. aphthosa, P. aphthosa var. leucophlebia, P. canina, P. canina var. rufescens, P. collina, P. lepidophora, P. malacea, P. polydactyla, P. scabrosa, P. venosa; Phaeophyscia endococcinea, P. sciastra; Physcia caesia, P. dubia; Physconia muscigena; Pilophorus robustus; Placynthium aspratile; Solorina saccata; Stereocaulon grande, S. paschale, S. subcoralloides, S. symphycheilum, S. tomentosum, S. vesuvianum; Umbilicaria deusta, U. vellea; Usnea cavernosa, U. dasypoga, U. glabrata, U. hirta, U. subfloridana, U. substerilis; Xanthoparmelia taractica, X. tasmanica; Xanthoria candelaria, X. elegans, X. fallax, X. polycarpa, X. sorediata.*

3. Circumpolar boreal species. These just barely reach into the southern edge of the tundra. I have placed 49 species or about 15 percent of the macrolichen flora in this category: *Bryoria capillaris, B. fuscescens, B. lanestris, B. nadvornikiana, B. tenuis; Cladonia botrytes, C. coniocraea, C. conista, C. glauca, C. macilenta, C. major, C. multiformis, C. pseudorangiformis, C. scabriuscula, C. subulata, C. turgida; Collema bachmanianum, C. crispum, C. cristatum, C. furfuraceum, C. limosum, C. polycarpon, C. tuniforme; Evernia mesomorpha; Leptogium sinuatum, L. tenuissimum; Nephroma bellum, N. helveticum, N. parile, N. resupinatum; Pannaria conoplea; Parmelia glabratula, P. olivacea, P. septentrionalis, P. sorediosa, P. squarrosa, P. subaurifera, P. sulcata; Parmeliella tryptophylla; Parmeliopsis aleurites; Peltigera horizontalis, P. membranacea; Physcia adscendens, P. aipolia, P. phaea; Platismatia glauca; Ramalina roesleri, R. pollinaria, R. thrausta.*

4. Amphi-Atlantic species. These species occur in the eastern American Arctic and also in Europe or Africa. This is a well-known pattern in the higher plants and has been the subject of much speculation. Within this group of lichens is a subgroup including *Coccocarpia erythroxyli, Baeomyces roseus, Sticta weigelii,* and *Umbilicaria caroliniana,* which have main distributions in the Appalachian regions and disjunct occurrences in Alaska.

Among the amphi-Atlantic lichens are 35 species including about 10 percent of the arctic macrolichens: *Actinogyra mühlenbergii; Agyrophora leiocarpa; Baeomyces roseus; Cetraria halei; Cladonia bacillaris; C. cryptochlorophaea, C. floerkeana, C. grayi; Coccocarpia erythroxyli; Dermatocarpon daedaleum; Ephebe hispidula; Lasallia pensylvanica; Pannaria hookeri; Parmelia elegantula; Phaeophyscia constipata; Physcia subobscura; Placynthium pannariellum; Pycnothelia papillaria; Ramalina intermedia, R. obtusata; Stereocaulon condensatum, S. dactylophyllum, S. groenlandicum, S. saxatile, S. spathuliferum, S. uliginosum; Sticta weigelii; Umbilicaria aprina, U. caroliniana* (in Asia), *U. cinereorufescens, U. havaasii, U. hirsuta, U. mammulata; Vestergrenopsis elaeina, V. isidiata.*

5. Amphi-Beringian species. These species occur in the western American Arctic and also in eastern Asia. There are 31 species, about 10 percent of the macrolichen arctic flora, that belong to this category. They appear to be of particular significance in attempts to account for the origin and distribution of the arctic lichens (Thomson 1972): *Asahinea chrysantha, A. scholanderi; Cetraria andrejevii, C. inermis, C. kamczatica, C. laevigata, C. nigricascens; Cladonia aberrans, C. granulans, C. kanewskii, C. metacorallifera, C. nipponica, C. pseudostellata; Cetrelia alaskana; Dactylina beringica; Evernia perfragilis; Fistulariella almquistii, F. scoparia; Leptogium crenatulum, L. parculum; Lobaria kurokawae, L. linita, L. pseudopulmonaria; Masonhalea richardsonii; Nephroma isidiosum; Parmelia almquistii; Stereocaulon apocalyptum, S. arenarium, S. intermedium, S. saviczii; Sticta arctica.*

6. Species confined to western North America. It is quite possible that these species are really amphi-Beringian but that counterpart collections have not yet been made in Asia. These 5 species comprise nearly 2 percent of the macrolichen flora: *Agyrophora scholanderi; Cladonia alaskana, C. thomsonii; Umbilicaria angulata, U. phaea.*

7. Species occurring in western North America and also in Europe. This is a well-known element in the flora of the western United States and Canada. Of those which reach the Arctic there are 8, comprising a little over 2 percent of the macrolichen flora: *Bryoria pseudofuscescens; Collema multipartitum; Cornicularia normoerica; Dactylina madreporiformis* (also Asia); *Dermatocarpon arnoldianum; Leptogium palmatum* (also Asia); *Parmelia olivaceoides; Usnea scabrata ssp. nylanderiana.*

8. Species in western North America and disjunct in Greenland. This is a highly interesting pattern. One can hypothesize that this is either a relict distribution pattern of great antiquity or else it is an anomalous long-distance dispersal difficult to account for in the light of the more continuous distributions of the majority of the arctic lichens. There are 8 species, representing about 2 percent of the macrolichens, that show this pattern: *Agyrophora rigida; Bryoria subdivergens; Collema callopismum; Dermatocarpon rivulorum; Endocarpon tortuosum; Leprocaulon subalbicans; Leptochidium albociliatum; Lobaria hallii.*

9. Species having their main distributions in the temperate regions but which extend their ranges to just reach into the arctic. There appear to be 9 species in this category, nearly three percent of the macrolichen flora: *Cladonia anomoea, C. verticillata; Dermatocarpon fluviatile, D. hepaticum, D. miniatum; Lasallia papulosa; Leptogium byssinum; Physconia detersa; Placynthium nigrum.*

10. Species in the Great Plains and in the Arctic. Of the macrolichens, only one species, reported by Ahti et al. from the Anderson River region, falls into this category: *Xanthoparmelia chlorochroa. Buellia elegans* also has this pattern among the microlichens (Thomson 1978 as *B. epigaea*).

11. Species known only from Greenland in the American Arctic. Two however, do also occur in Iceland. There are 3 in this group, 1 percent of the macrolichens: *Dermatocarpon lyngei; Ephebe multispora* (currently known only from Greenland); *Peltigera occidentalis* (recently added from the central American Arctic as well).

12. Truly bipolar species known from Antarctica and the American Arctic. This includes only two species: *Agyrophora leiocarpa; Neuropogon sulphureus.*

13. Species known only from the American Arctic. Of the macrolichens only one species falls into this category: *Teloschistes arcticus* (known from the western islands and the adjacent mainland).

Ecology

Knowledge of the ecology of the arctic lichens is exceedingly scant and presents a challenging situation for more intensive work. One of the commonest misconceptions of the tundra regions is that because it is treeless, somehow it is uniform across the vast expanses it occupies. On the contrary, there are many habitats within the tundras, reflecting not only the gradations of soil temperature, texture, pH, moisture content, and permafrost, but also the topography, the gradations of air moisture and precipitation, the length of growing season, the angle of the sun's rays at such high latitudes, the available radiation and its qualities, the presence and persistence of snow and the qualities of the snow, the intensity of the winds and related snow and ice crystal abrasion, the presence and activities of animals and of man, the influence of fires, natural or man-caused and, more recently, the problems of pollution which may be expected to cause changes in the vegetational cover.

One may select some vegetational communities for description but the gradations between are often very subtle and will require study using ordination techniques. Such were early suggested by Johnson et al. (1966) but few such studies other than those of Larsen (1965, 1971, 1973) and Richardson and Finegan (1977) have used such approaches. The majority of contributions on the ecology of the American Arctic lichens are based on discussions of the species rather than the communities. However, excellent discussions on the vegetation of Greenland may be consulted in papers by K. Hansen (1971) and E. S. Hansen (1978), and detailed discussions on Alaskan tundra may be found in Johnson et al. (1966), Williams et al. (1978), and Walker et al. (1980). Interior Canadian vegetation is best covered in papers by Polunin (1948), Larsen (1965, 1967, 1971a, 1971b, 1972, 1973), and by Zoltai and Johnson (1979). Barrett & Thomson (1974), Bird (1975) and Richardson & Finegan (1977) give recent information on the Arctic islands. Generalized summaries and species lists are also to be found in Munroe (1956), Thomson et al. (1978), and Thomson (1980).

A very conspicuous feature of the lichen vegetation in the ecotone between the boreal forest and the tundra is the dominance of yellow lichens in the lichen woodlands and the southern edge of the tundras. The hypothesis is suggested in Thomson (1982) that this may be related to large exposures to ultraviolet light in the continuous day regimes in the southern part of the Arctic. The usnic acid content of the yellow lichens as *Cladonia mitis, Cetraria cucullata, C. nivalis,* and the *Xanthoparmelia* species may absorb the ultraviolet rays and protect the algal layers from the detrimental effects of such rays. This may also give the lichens a slight competitive advantage as compared with vascular plants, which do not possess such a filtering mechanism. In addition, the usnic acid has allelopathic and antiherbivore effects which give these species further advantages in ecological competition situations. It is also suggested that in the far northern portions of the Arctic the darker lichens prevail in contributing a more somber appearance to the landscape. Large dark mats of *Cornicularia divergens, C. aculeata, Bryoria nitidula, Parmelia stygia,* and on the rocks *Actinogyra, Omphalodiscus, Umbilicaria,* dark species of *Cetraria,* and *Pseudephebe* contribute to the darker landscape. It is hypothesized that as the poles are approached the sun's rays have to pass through a greater depth of atmosphere and the ultraviolet is filtered out to a greater extent than in the south. Consequently less protection against such rays is necessary. To the contrary a greater advantage is conferred to such lichens as can extend the growing season and also attain higher operating temperatures with the dark colors which absorb heat more readily.

In considering the types of communities present in the arctic tundras it is necessary to take into account the availability of the species. Quite obviously the amphi-Beringian species will not ordinarily be available in the eastern Arctic, nor will amphi-Atlantic species ordinarily be expected to contribute to communities in Alaska.

Lichens of High Humus Soil Types

In the cold conditions and short active seasons of the Arctic the organic materials built up by the plants do not completely disintegrate and recycle. The organic materials gradually accumulate and, depending upon moisture conditions, may form more or less deep upper soil layers. These cause a grading off into a continuum of plant communities. On the high, dry, well-drained, often gravelly, soils with permafrost retreating deeply in the summer and oxidation of organic materials proceeding more rapidly, there will be a lichen-rich heath tundra on acid soil substrata. With increasing moisture in the slope, down slope, over seepages, or in places of greater melt of permafrost, the terrain may become more hummocky and irregular. Lichens are also abundant on these soil types which have much humus in them. Increasing moisture availability, especially in low wet spots as in swales or valleys or in the interior of low center polygons, will grade the hummocky types of tundra into tussock meadows. The latter are dominated usually by species of *Eriophorum* which form the high tussocks between which are organic soils, the whole making a community in which walking is very difficult. The permafrost may even be ascendent within the tussocks, maintaining a moist top which is easily unbalanced over the lubricated central ice core. These tussock meadows are usually poor in lichens although some species exist within the tussock tops mixed among the *Eriophorum* stems and leaves. A few lichens may exist on the low moist expanses between the tussocks. They include such moisture-loving species as *Cetraria andrejeviii, C. delisei,* and *Cetrelia alaskana,* the latter only in the western Arctic. Although there is a general abundance of lichens on such humus rich soils, sometimes over 50% of the ground cover consisting of lichens, there may be only a scattering of thalli on a given site. And then one may find an abundance of the same species in large swards over the next ridge with currently no clue as to the cause of the difference in abundance. Much work on the autecology of arctic lichens obviously remains to be done.

Characteristic of the humus rich soils is a long list of lichens from which one may expect (dependent on geographic distributions) such species as *Alectoria nigricans, A. ochroleuca, A. vexillifera* on drier sites; *Arctomia delicatula, A. interfixa; Asahinea chrysantha; Bacidia alpina, B. bagliettoana, B. obscurata, B. sphaeroides, B. trisepta; Biatorella fossarum; Bryoria nitidula; Buellia papillata; Caloplaca cinnamomea, C. jungermanniae, C. stillicidiorum, C. tetraspora, C. tiroliensis; Candelariella terrigena; Catillaria muscicola; Cetraria cucullata, C. elenkenii, C. ericetorum, C. islandica, C. kamczatica, C. nigricans, C. nivalis; Cetrelia alaskana* (western sector); *Cladonia aberrans, C. acuminata, C. alaskana* (western sector), *C. amaurocraea, C. anomaea, C. arbuscula, C. bacillaris, C. bacilliformis, C. bellidiflora, C. cariosa, C. carneola, C. cenotea, C. chlorophaea, C. coccifera, C. coniocraea* (only near forest border), *C.*

conista (rare), *C. cornuta, C. crispata, C. cyanipes, C. decorticata, C. deformis, C. digitata, C. ecmocyna, C. fimbriata, C. floerkeana* (only Greenland), *C. glauca, C. gracilis, C. kanewskii* (alpine in Alaska), *C. macrophylla, C. maxima, C. merochlorophaea, C. mitis, C. multiformis, C. nipponica, C. norrlinii, C. phyllophora, C. pocillum* (also on calcareous soils), *C. pseudorangiformis, C. pseudostellata* (Alaska), *C. rangiferina, C. scabriuscula, C. squamosa, C. stellaris, C. stricta, C. subcervicornis, C. subfurcata, C. subulata, C. sulphurina, C. turgida, C. uncialis, C. verticillata; Coccocarpia erythroxyli* (Alaska only); *Collema ceraniscum; Coriscium viride* (wet hummocks and tussocks); *Cornicularia aculeata, C. divergens, C. muricata; Dactylina arctica, D. beringica, D. madreporiformis, D. ramulosa; Hypogymnia austerodes, H. subobscura; Icmadophila ericetorum; Lecanora beringii, L. castanea, L. epibryon, L. leptacina; Lecidea assimilata, L. berengeriana, L. fusca, L. granulosa, L. luteovernalis, L. tornoensis, L. uliginosa, L. vernalis; Lecidella wulfenii; Leciophysma finmarkicum, L. furfurascens; Leptochidium albociliatum* (Greenland); *Leptogium arcticum, L. lichenoides, L. palmatum, L. sinuatum, L. tenuissimum; Lobaria kurokawae* (Alaska), *L. linita, L. pseudopulmonaria* (Alaska); *Lopadium coralloideum, L. fecundum, L. pezizoideum; Masonhalea richardsonii; Massalongia carnosa; Microglaena muscorum; Nephroma arcticum, N. expallidum, N. isidiosum; Ochrolechia androgyna, O. frigidum, O. grimmiae, O. inaequatula, O. upsaliensis; Pachyospora verrucosa; Pannaria conoplea, P. pezizoides; Parmelia omphalodes, P. saxatilis, P. sulcata; Parmeliella arctophila, P. praetermissa, P. tryptophylla; Peltigera aphthosa, P. canina, P. collina, P. lepidophora, P. malacea, P. occidentalis* (rare), *P. polydactyla, P. scabrosa; Pertusaria bryontha, P. coriacea, P. dactylina, P. glomerata, P. octomela, P. oculata, P. subdactylina, P. subobducens, P. trochisea; Phaeophyscia constipata; Physconia muscigena; Platismatia glauca; Polyblastia gelatinosa, P. gothica, P. sendtneri; Polychidium muscicola, P. umhausense; Psoroma hypnorum; Rinodina mniaraea, R. nimbosa, R. roscida, R. turfacea; Sphaerophorus globosus; Stereocaulon alpinum, S. grande, S. paschale; Sticta arctica, S. weigelii* (Alaska); *Thamnolia subuliformis,* and *T. vermicularis.*

In the lowest, wettest sites which still support some lichens may be found *Cetraria andrejevii, C. delisei; Cetrelia alaskana* (Alaska and vicinity); *Cladonia phyllophora,* and *C. stricta.*

Seepage areas below snowbanks and where spring water seeps out to give a summer supply of cold water may have a gradient of species with *Lecidea ramulosa* and *Siphula ceratites* within the coldest portions, *Stereocaulon glareosum* a little way below, and *Stereocaulon rivulorum; Cetraria andrejevii, C. delisei; Cladonia stricta,* and *Solorina crocea* still further below but still in very cool persistent water. Such seepages may have a higher percentage of mineral soil than the lichen heaths above.

A curious and characteristic high humus community in the arctic is to be found on the dung of lemmings, ptarmigan, and arctic hares. The dung apparently does not disintegrate for many decades and becomes populated with such species as *Buellia papillata; Caloplaca cinnamomea, C. jungermanniae, C. stillicidiorum, C. tetraspora, C. tirolensis; Candelariella terrigena; Catillaria muscicola; Lecanora epibryon; Lecidea assimilata; Rinodina mniaraea,* and *R. turfacea.*

The microhabitats presented by the long persistent snowbanks do not alter appreciably the species present but they do enable certain species to fruit more abundantly than in the more severe conditions in more open habitats. *Dactylina arctica, D. madreporiformis,* and *D. ramulosa* fruited almost exclusively just above the banks. But *Cetrarias* as *C. cucullata, C. nivalis,* and *C. islandica* also appeared to fruit better in such habitats. The snowbank areas are usually in declivities where the meager snows could accumulate in the lee of the winds. Often these areas in the Arctic could be noted from a distance by the presence of *Cassiope* which also had a preference for the upper parts of the slopes where the snow accumulated. It was quite apparent that the gradual meltdown of the snowbanks would sort the lichen communities according to the availability of moisture and the length of time the earthbank would be exposed by the recession of the snow. In some places the snowbanks are so long persistent that the growing season is too short at the bottom of the slope for any vegetation to become established and only bare earth is visible toward the end of the growing season. But downslope from this bare earth may be the seepages with the lichens mentioned above.

Lichens of Mineral Soils

The lichens growing on the mineral soils are strongly influenced by frost perturbations as well as by the pH of the soils. Early stages in succession are particularly well seen on the frost boils, which are common, especially on ground with a high clay content. The early stages of the frost boils may be pioneered by the primary thalli of several species of *Stereocaulon, S. glareosum,* and *S. rivulorum* being very common in such habitats. Several species of *Pertusaria* and *Lecanora epibryon* are also early colonizers. *Baeomyces carneus, B. roseus,* and *B. rufus* are also in these habitats. As stabilization proceeds, the lichens listed for calcareous soils may gain a foothold if the substratum has a high pH. If humus is accumulating on the frost boil then crustose and other species of the organic soil substrata may gain a foothold and become especially prominent on the humus mat surrounding the more open ''boil'' area. Gradually the boil may become obliterated as the humus mat becomes more consolidated and the general tundra vegetation with a rich assemblage is resumed.

On calcareous soils there is a rich flora characteristic of such habitats. The soils of the *Dryas* and *Saxifraga oppositifolia* ''barrens'' are good examples of such habitats. The lime-containing pebbles constituting in many cases a windblown ''desert pavement'' provide a habitat for the limestone rock lichens. The soils between are characteristically settled by such lichens as *Cetraria tilesii; Cladonia pocillum; Collema bachmanianum, C. glebulentum, C. tenax; Dactylina madreporiformis, D. ramulosa; Dermatocarpon cinereum, D. hepaticum; Endocarpon pusillum; Evernia perfragilis; Fulgensia bracteata, F. fulgens; Gyalecta foveolaris; Heppia lutosa; Lecanora lentigera; Lecidea subcandida; Leptogium lichenoides; Pannaria pezizoides; Peltigera lepidophora, P. venosa* (especially beside animal burrows); *Protoblastenia terricola; Psora decipiens, P. rubiformis; Solorina crocea* (if wet enough), *S. bispora, S. octospora, S. saccata, S. spongiosa; Thamnolia subuliformis, T. vermicularis; Toninia caeruleonigricans, T. lobulata,* and *T. tristis.*

If the soils are noncalcareous they may be dominated by such species as *Alectoria nigricans; Baeomyces carneus, B. placophyllus, B. roseus, B. rufus, Bryoria nitidula, B. tenuis; Buellia elegans; Cladonia cariosa, C. decorticata, C. pleurota, C. pyxidata, C. symphycarpa, C. thomsonii* (in Alaska), *C. turgida* (near forest border), *C. uncialis, C. verticillata; Cornicularia aculeata, C. divergens, C. muricata; Geisleria schnogonoides; Lecidea cuprea, L. ramulosa* (in seepages); *Lepidoma demissa; Mycoblastus alpinus, M. sanguinarius; Pseudephebe minuscula, P. pubescens; Psora globifera; Ramalina almquistii* (western), *R. pollinaria; Stereocaulon alpinum, S. glareosum, S. incrustatum, S. paschale, S. rivulorum, S. tomentosum; Thamnolia subuliformis,* and *T. vermicularis.*

Rock Lichen Communities

Lichens on the rocks depend upon whether the rocks are acidic or basic. The exposure and moisture relationships also strongly influence the lichen covers. Use of the rocks as bird perches will also strongly influence the lichen cover. As a result of glaciation the rock habitats occur not just as outcrops but as boulders or pebbles in the till and may project above the general level of the tundra heaths. Many isolated enormous boulders may stand in the tundra, left by the wastage of the ice sheets. Periglacial boulder streams may be covered by an abundance of lichens. The permafrost sorting of materials which develop the stone nets also give habitats soon populated by the appropriate epilithic lichens. Boulder fields and "fell fields" are common arctic phenomena and also become heavily lichen covered. Most of the lichens on rocks are crustose species but some foliose or fruticose species also commonly occur.

The limestone rock substrata may have any of a large number of possible characteristic species, dependent upon the exposure of the rocks, the shading, moisture relationships, and other factors. A very obvious indicator of calcareous rocks in the Arctic is *Cetraria tilesii* which grows on the soil between the rocks. Any of the following series of species may be found on calcareous rocks: *Acarospora badiofusca, A. glaucocarpa, A. scabrida, A. schleicheri, A. smaragdula, A. veronensis; Agyrophora rigida; Aspicilia annulata, A. anseris, A. caesiocinerea, A. cinerea, A. cinereorufescens, A. disserpens, A. fimbriata, A. heteroplaca, A. lesleyana, A. nikrapensis, A. pergibbosa, A. ryrkapiae, A. sublapponica; Bacidia coprodes; Blastenia exsecuta; Buellia alboatra, B. immersa, B. notabilis, B. vilis; Caloplaca cirrochroa, C. festiva, C. fraudans, C. granulosa, C. murorum; Candelariella athallina, C. aurella; Collema callopismum, C. crispum, C. cristatum, C. multipartitum, C. polycarpon, C. tuniforme, C. undulatum; Dermatocarpon intestiniforme, D. miniatum; Glypholecia scabra; Ionaspis epulotica, I. heteromorpha, I. ochraceella, I. reducta, I. schismatopsis; Lecania alpivaga; Lecanora annulata, L. atra, L. candida, L. crenulata, L. dispersa, L. flavida, L. intricata, L. microfusca, L. muralis, L. nordenskioldii, L. polychroma, L. polytropa, L. rupicola, L. torrida; Lecidea atromarginata, L. hypocrita, L. lapicida, L. lithophila, L. marginata, L. melaphanoides, L. speirea, L. umbonata, L. vorticosa; Lecidella carpathica, L. goniophila, L. stigmatea; Microglaena muscorum; Pertusaria subplicans; Placopsis gelida; Placynthium aspratile, P. nigrum; Polyblastia hyperborea, P. integrascens, P. sommerfeltii, P. thelodes; Porocyphus coccoides; Protoblastenia rupestris; Pyrenopsis pulvinata; Rhizocarpon alaxensis, R. chioneum, R. disporum, R. geographicum, R. inarense, R. intermediellum, R. umbilicatum; Rinodina bischoffii, R. occidentalis; Sarcogyne regularis; Staurothele clopima, S. perradiata; Thelidium acrotellum, T. aenovinosum, T. pyrenophorum; Tremolecia jurana; Verrucaria aethiobola, V. devergens, V. muralis, V. nigrescens, V. obnigrescens, V. rupestris; Xanthoria candelaria, X. elegans,* and *X. sorediata.*

Lichens on Bones and Antlers

Old bones and antlers scattered over the tundras eventually become covered with a number of lichens which have a preference for calcareous substrata. They may have species from the following list: *Bacidia subfuscula; Buellia punctata, B. stigmatea; Caloplaca festiva, C. stillicidiorum, C. tiroliensis; Candelariella aurella, C. vitellina; Lecanora beringii, L. crenulata, L. dispersa, L. epibryon; Lecidella stigmatea; Parmelia sulcata; Physcia caesia, P. dubia; Xanthoria elegans.* A lichenist will already have noted that several of these are species which ordinarily occur not on calcareous rocks but upon highly organic substrata as decaying mosses and other vegetation or on dung. The organic content of the bones and antlers may be an influence in such occurrences.

When considering the lichens that may be found on acid rocks such as granites and schists as well as sandstones, a number of factors in addition to the chemistry operate to restrict the presence of the species or may promote their presence. The direction of exposure, the amount of shading to sunlight or rain shadow, the rill chanels down which precipitation trickles, the use by birds or mammals as perches all may play a part in the ecology of such lichens.

In places where the rock streams or boulders are open to water running beneath the rocks there is a likelihood of finding *Haematomma lapponicum, Huilia flavocaerulescens, Lecidea lapicida,* and *Tremolecia atrata. Rhizocarpon chioneum* appears to grow best in shaded crevices.

The species of *Ionaspis* are best found on the boulders at the edge of tarns at or just above the water surface, and on boulders in temporary runoff channels.

    The dung of birds on the perch boulders which project above the tundra and are used as overlooks gives a characteristic nitrophile-calciphile flora which may include *Acarospora fuscata, A. peliocypha; Candelariella aurella, C. corallorhiza, C. vitellina; Lecanora badia, L. muralis; Lecidea atrobrunnea; Phaeophyscia sciastra; Physcia caesia, P. dubia, Parmelia sulcata; Rhizoplaca chrysoleuca, R. melanopthalma; Xanthoria candelaria, X. elegans, X. sorediata; Umbilicaria arctica, U. decussata,* and *U. hirsuta.*

    On the trickle surfaces one is most likely to find such species as *Acarospora pelioscypha, Leprocaulon subalbicans, Massalongia carnosa,* and *Umbilicaria deusta.* On the other hand, *Acarospora chlorophana* and *A. molybdina* are in the shade and dry conditions under overhangs.

    The list of species which grow on acid rocks is very long and it is best to check the species keys rather carefully in case one has happened to collect one of the rarer species. Some typical acidophiles are *Acarospora chlorophana, A. fuscata, A. veronensis; Agyrophora rigida, A. scholanderi; Amygdalaria peliobotrya; Asahinea scholanderi; Aspicilia cinerea, A. cingulata, A. composita, A. concinnum, A. elevata, A. perradiata, A. plicigera, A. rosulata, A. supertegens; Bryoria chlaybeiformis; Buellia malmei, B. moriopsis; Caloplaca festiva; Candelariella aurella; Catillaria chalybeia; Cetraria commixta, C. hepatizon; Dermatocarpon arnoldianum, D. fluviatile* (in or by water); *Dimelaena oreina; Diploschistes scruposus; Haematomma lapponicum; Huilia flavocaerulescens, H. macrocarpa; Hymenalia lacustris; Hypogymnia bitteri, H. oroarctica, H. subobscura; Ionaspis epulotica, I. suaveolens; Lasallia papulosa, L. pensylvanica; Lecanora aquatica, L. atrosulphurea, L. badia, L. caesiosulphurea, L. dispersa, L. frustulosa, L. granatina, L. intricata, L. polytropa, L. rupicola, L. stygioplaca; Lecidea aglaea, L. armeniaca, L. atrobrunnea, L. auriculata, L. brachyspora, L. carbonoidea, L. cinereorufa, L. confluens, L. diducens, L. glaucophaea, L. instrata, L. lapicida, L. leucophaea, L. lucida, L. melinodes, L. pallida, L. panaeola, L. pantherina, L. paupercula, L. picea, L. shushanii, L. tesselata; Lecidella carpathica, L. stigmatea; Nephroma isidiosum, N. parile; Omphalodiscus decussatus, O. krascheninnikovii, O. virginis; Parmelia almquistii, P. alpicola, P. exasperatula, P. fraudans, P. infumata, P. panniformis, P. stygia, P. sulcata; Pertusaria amara* var. *flotowiana, P. excludens, P. subplicans; Phaeophyscia endococccinea, P. sciastra; Physcia caesia, P. dubia; Pilophorus robustus; Platismatia glauca; Pseudephebe minuscula, P. pubescens; Ramalina intermedia; Rhizocarpon badioatrum, R. chioneum, R. copelandii, R. crystalligenum, R. disporum, R. eupetraeoides, R. eupetraeum, R. geographicum, R. grande, R. hochstetteri, R. polycarpum, R. riparium, R. superficiale; Rhizoplaca chrysoleuca, R. melanaspis, R. melanopthalma, R. peltata; Rhopalospora lugubris; Rinodina milvina; Sphaerophorus fragilis; Sporastatia polyspora, S. testudinea; Staurothele fissa; Stereocaulon botryosum, S. dactylophyllum, S. subcoralloides, S. symphycheilum; Tremolecia atrata; Umbilicaria angulata, U. arctica, U. caroliniana, U. cinereorufescens, U. cylindrica, U. deusta, U. hirsuta, U. hyperborea, U. proboscidea, U. torrefacta, U. vellea; Verrucaria devergescens, V. margacea; Vestergrenopsis elaeina, V. isidiata; Xanthoparmelia centrifuga, X. incurva, X. separata,* and *X. tasmanica.*

Epiphytic Lichens

    The barks of the trees of the forest border and the shrubs which occur in the tundras are inhabited by many lichens. Some of these are species which are common on the humus soils and extend their habitats up onto the bases of the birches, *Betula,* willows, *Salix,* alders, *Alnus,* and Labrador tea, *Ledum. Picea mariana,* the black spruce, is a favored habitat for a number of epiphytic lichens. It sometimes startles one to find tufts of ordinarily ground living species such as *Bryoria nitidula* or *Alectoria ochroleuca* epiphytic on the lower branches of spruce in protected situations. The most common of the lichens which occur on the bark and twigs of woody plants include: *Bacidia subincompta; Bryoria simplicior; Buellia disciformis, B. punctata; Caloplaca cerina, C. discolor, C. ferruginea, C. holocarpa, C. stillicidiorum, C. tiroliensis; Candelariella lutella, C. xanthostigma; Cetraria pinastri, C. sepincola; Evernia mesomorpha; Hypogymnia austerodes, H. bitteri, H. physodes; Lecanora coilocarpa, L. distans, L. epibryon, L. expallens, L. hageni, L. pulicaris, L. rugosella; Lecidea symmicta, L. vernalis; Lecidella elaeochroma, L. glomerulosa; Leptogium saturnium; Lobaria scrobiculata; Lopadium pezizoideum; Ochrolechia androgyna, O. frigida, O. upsaliensis; Parmelia exasperatula, P. septentrionalis, P. sulcata; Parmeliopsis ambigua, P. hyperopta; Pertusaria alpina, P. bryontha, P. dactylina, P. panyrga, P. sommerfeltii, P. velata; Physcia adscendens, P. aipolia; Ramalina pollinaria; Rinodina archaea, R. hyperborea, R. laevigata, R. septentrionalis, R. turfacea; Sticta weigelii* (Alaska); *Usnea glabrata, U. glabrescens, U. hirta, U. lapponica;* and *Xanthoria candelaria.*

Rotting wood, especially of spruce, reaches the Arctic via the down-rushing streams at melt time and is distributed along the shores. Wood also occurs in the isolated stands of spruce and willow along the southern edges of the tundra. On such wood may occur the array of lichens of humus soils, particularly on the more moist and rotten portions of the wood. Aside from such less expected species, those on the old wood of branches and logs include *Bryoria simplicior; Buellia disciformis, B. punctata, B. zahlbruckneri; Calicium abietinum, C. salicinum, C. trabinellum; Caloplaca discolor; Candelariella vitellina; Cetraria pinastri; Cladonia cenotea, C. cyanipes, C. deformis, C. fimbriata, C. sulphurina; Cyphelium caliciforme, C. inquinans, C. tigillare; Lecanora coilocarpa, L. vegae; Lecidea albohyalina, L. granulosa, L. hypopta, L. plebeja, L. punctella, L. symmicta, L. turgidula; Mycoblastus alpinus, M. sanguinarius; Mycocalicium subtile; Parmeliopsis aleurites, P. ambigua, P. hyperopta; Phaeocalicium populneum;* and *Rinodina lecideoides.*

One should also be aware of the possible presence of lichens which grow over other lichens such as *Buellia nivalis* on *Xanthoria* and *Caloplaca invadens* on *Lecanora* and *Placynthium, Epilichen scabrosa* on *Baeomyces, Lecidea vitellinaria* on *Candelariella,* and *Rhizocarpon pusillum* on *Sporastatia testudinea.*

The most conspicuous vagant lichen in the Arctic is *Masonhalea richardsonii,* which is never attached to the substratum but when dry rolls with the wind, then spreading and coming to rest when moistened. Other vagant forms may be formed by *Rhizoplaca chrysoleuca* becoming detached from the rocks on which it grows.

REFERENCES ON GEOGRAPHY AND ECOLOGY

Ahlstrup, V. 1977. Cryptograms on imported timber in West Greenland. Lichenologist 9:113–117.
—— 1982. The epiphytic lichens of Greenland. Bryologist 85:64–73.
Barrett, P. E. & J. W. Thomson. 1975. Lichens from a high arctic coastal lowland, Devon Island, N.W.T. Bryologist 78:160–167.
Bird, C. D. 1975. The lichen, bryophyte, and vascular plant flora and vegetation of the Landing Lake area, Prince Patrick Island, Arctic Canada. Can. J. Bot. 53:719–744.
Bird, C. D., J. W. Thomson, A. H. Marsh, G. W. Scotter, & P. Y. Wong. 1980–1981. Lichens from the area drained by the Peel and Mackenzie Rivers, Yukon and Northwest Territories, Canada. 1. Macrolichens. Can. J. Bot. 58:1947–1985. 2. Microlichens. Can. J. Bot. 59:1231–1252.
Johnson, A. W., L. A. Viereck, R. E. Johnson, & H. Melchior. 1966. Ch. 14. Vegetation and Flora. In N. J. Wilimovsky & J. N. Wolfe. Environment of the Cape Thompson region, Alaska. U. S. Atomic Energy Commission. p. 277–354.
Larsen, J. A. 1965. The vegetation of the Ennadai Lake area, N.W.T.: Studies in subarctic and arctic bioclimatology. Ecol. Monogr. 35:37–59.
—— 1971a. Vegetational relationships with air mass frequencies: Boreal forest and tundra. Arctic 24:177–193.
—— 1971b. Vegetation of Fort Reliance, Northwest Territories. Can. Field Nat. 85:147–178.
—— 1973. Plant communities north of the forest border, Keewatin, Northwest Territories. Can. Field Nat. 87:241–248.
Munroe, E. 1956. Canada as an environment for insect life. Can. Entomologist 88:372–476.
Richardson, D. H. S. & E. J. Finegan. 1977. Studies on the lichens of Truelove Lowland. In L. C. Bliss, ed. Truelove Lowland, Devon Island, Canada: A High Arctic Ecosystem.
Seward, M. R. D. 1977. Lichen Ecology. New York: Academic Press.

Thomson, J. W. 1972. Distribution patterns of American Arctic lichens. Can. J. Bot. 50:1135–1156.
—— 1982. Lichen vegetation and ecological patterns in the high arctic. J. Hattori Bot. Lab. 53:361–364.
Tieszen, L. L., ed. 1978. Vegetation and Production Ecology of an Alaskan Arctic Tundra. Ecol. Studies 29:1–686. Springer Verlag.
Walker, D. A., K. R. Everett, P. J. Webber, & J. Brown. 1980. Geobotanical Atlas of the Prudhoe Bay region, Alaska. Cold Regions Res. & Engr. Lab. Report 80-14:1–69. Hanover, N.H.
Williams, M. E., E. D. Rudolph, E. A. Schofield, & D. C. Prasher. 1978. The role of lichens in the structure, productivity, and mineral cycling in the wet coastal Alaskan tundra. Ch. 7 in Tieszen et al. 1978.
Zoltai, S. C. & J. D. Johnson. 1978. Vegetation-soil relationships in the Keewatin District. 1-95. Environmental-Social Program Northern Pipelines ESCOM No. AI-25. Environment Canada, Ottawa.

Key to the Genera

1. Thallus fruticose—2.
    2. Thallus hollow—3.
        3. Thallus bone white, smooth, pointed .... *Thamnolia*
        3. Thallus greenish yellow, yellow, or straw-colored—4.
            4. Upright podetia with internal cartilaginous layer; primary thallus present or absent—5.
                5. Outer cortex present or podetia with a superficial layer of tomentum; primary thallus squamulose or absent... *Cladonia*
                5. Outer cortex lacking, no tomentum present, outer medullary hyphae forming a persistent subcontinuous pseudocortex; primary thallus granulose to verrucose .. *Pycnothelia*
            4. Podetia lacking internal cartilaginous layer; primary thallus absent.............. *Dactylina*
    2. Thallus with either solid or at most partly fistulose center—6.
        6. With a solid central axis within a cottony medulla and an outer cortex; thallus pendent—7.
            7. Thallus evenly colored .... *Usnea*
            7. Thallus variegated black and yellow with black banding .... *Neuropogon*
        6. Thallus lacking the solid central axis strand—8.
            8. Thallus all white, tips rounded, sides ± fluted, growing in moist seepages......... *Siphula*
            8. Thallus greenish yellow, straw-colored, blue-gray or brown—9.
                9. Thallus and podetia or pseudopodetia blue-gray (sometimes with a brownish tinge)—10.
                    10. Thallus covered with granulose to squamulose phyllocladia, often with a tomentum between the phyllocladia—11.
                        11. With cephalodia containing *Nostoc* or *Stigonema* interspersed among the phyllocladia or on a primary thallus; soredia present or absent .... *Stereocaulon*
                        11. Lacking cephalodia; leprose sorediate .... *Leprocaulon*
                    10. Thallus lacking phyllocladia or tomentum, the surface either smooth or decorticate—12.
                        12. Thallus much branched, shrubby; primary thallus lacking or papillate isidiate; pseudopodetial surface smooth, usually shining, the center cottony; apothecia uncommon, becoming a mass of black, powdery, spherical spores .... *Sphaerophorus*
                        12. Pseudopodetia simple to few branched at tops; primary thallus persistent, granulose to areolate and with conspicuous pink cephalodia; center composed of longitudinal loose hyphae surrounded by dense mechanical tissue; apothecia usually present, smooth, shiny, spores ellipsoid .... *Pilophorus*
                9. Thallus not blue-gray, usually greenish yellow, straw-colored, white, or brown—13.
                    13. Thallus with a continuous primary crustose or squamulose thallus on which are distributed short, erect stipes bearing large capitate apothecia .... *Baeomyces*
                    13. Primary thallus lacking, the thallus composed of erect or pendent, branching podetia—14.

14. Longitudinal sections showing cartilaginous strands interspersed in a cottony medulla—15.
    15. Thallus fistulose and hollow with numerous lateral perforations . . . . *Fistulariella*
    15. Thallus not inflated or hollow, lacking numerous lateral perforations—16.
        16. Cross section of branches flattened . . . . *Ramalina*
        16. Cross section of branches round or angular . . . . *Evernia*
14. Longitudinal sections showing only a uniform medulla—17.
    17. Thallus flattened or channeled in cross section . . . . *Cetraria*
    17. Thallus cylindrical in cross section—18.
        18. Cortex of irregularly or radially oriented hyphae . . . . *Cornicularia*
        18. Cortex of longitudinally oriented hyphae—19.
            19. Algae of thallus bright green, thallus over 4 mm tall (except *Pseudephebe minuscula*)—20.
                20. Thallus lacking pseudocyphellae, dark brown to black, appressed and attached to rock or gravel; cortex with superficial cells differentiated; spores frequently present, 8 per ascus, hyaline, simple . . . . *Pseudephebe*
                20. Thallus often with pseudocyphellae, yellow-green or brown to black, erect or pendent, loose on substratum if on soil, or attached to trees or shrubs; cortical cells not differentiated; spores rare, 8 or 2–4 per ascus, hyaline or brown, simple—21.
                    21. Thallus on ground, rarely on lower tree branches, yellow-green or brown to black with paler base; pseudocyphellae always present, conspicuously raised; medullary hyphae knobby ornamented; containing either usnic or alectorialic acid; if spores are present 2–4 per ascus and brown. . . . . *Alectoria*
                    21. Thallus usually epiphytic on trees or shrubs but terrestrial in some species; pseudocyphellae present or not, adpressed to slightly raised, medullary hyphae not knobby ornamented; lacking usnic acid; spores rare to unknown, 8 per ascus, hyaline . . . . *Bryoria*
            19. Algae of thallus blue-green; thallus less than 4 mm tall, forming small filamentous dendroid masses—22.
                22. Lobes longer than 1 mm; spores 1–2 celled—23.
                    23. Growing on rocks and attached by holdfasts, ± erect; algae *Stigonema* . . . . *Ephebe*
                    23. Growing on soil or over mosses, lacking holdfasts, ± decumbent; algae *Nostoc* or *Scytonema*—24.
                        24. Hyphae of thallus longitudinally oriented in outer parts, paraplectenchymatous in older parts; apothecia with proper exciple only; spores 8, 1–2 celled, ellipsoid to spindle-shaped; algae *Nostoc* . . . . *Polychidium*
                        24. Hyphae of thallus forming irregular nets of angular cells; apothecial margin thalloid; spores 16–24, simple, oval; algae *Scytonema* . . *Zahlbrucknerella*
                22. Lobes shorter than 1 mm; medulla of irregular netted cells; spores 8, 4–11 celled, spindle-shaped . . . . *Arctomia*

1. Thallus foliose—25.
    25. Thallus umbilicate—26.
       26. Fruits composed of flask-like perithecia imbedded within the thallus and only the mouths showing on the upper surface—27.
          27. With algae present in hymenial layer . . . . *Endocarpon*
          27. Algae lacking in the hymenial layer . . . . *Dermatocarpon*
       26. Fruits absent or composed of disk-like or gyrose apothecia usually superficial or sessile on the thallus—28. (See supplementary key to sterile specimens).
          28. Thallus pustulate . . . . *Lasallia*
          28. Thallus not pustulate—29.
             29. Apothecia with a flat disk—30.
                30. Disk with a sterile secondary button or a fissure in the center . . *Omphalodiscus*
                30. Disk surface smooth, lacking the central fissure or button . . . . *Agyrophora*
             29. Apothecia convex, the disk with raised furrows—31.
                31. Furrows of the disk concentric, a proper margin present around the disk edge . . . . *Umbilicaria*
                31. Furrows of the disk radial, proper margin lacking . . . . *Actinogyra*
    25. Thallus non-umbilicate—32.
       32. Thallus of small cochleate, blue-green to dark green lobes with paler inrolled margins; upper cortex paraplectenchymatous; algae *Polycoccus* (do not mistake primary squamules of a *Cladonia* for this. They have *Trebouxia* algae) . . . . *Coriscium*
       32. Thallus lacking the above set of characters—33.
          33. Thallus gelatinous when wet, the algae blue-green and distributed through the thallus—34.
             34. Cross section showing a cortical layer 1 or more cells thick.—35.
                35. Thallus lobes with smal hyaline cellular hairs projecting especially from the margins and exciple; spores 2-celled . . . . *Leptochidium*
                35. Thallus lobes lacking small hyaline cellular hairs—36.
                    36. Spores few to many celled, muriform; underside with a cortex . . *Leptogium*
                    36. Spores only 2–3 celled; underside ecorticate, of closely interwoven hyphae . . . . *Massalongia*
              34. Cross-section lacking cortex—37.
                37. Spores with several cells, 1 to several septate to usually muriform . . . . *Collema*
                37. Spores single celled . . . . *Leciophysma*
           33. Thallus not gelatinous when wet; the algae in distinct layers in the thallus, usually in a single layer of several cells beneath the upper cortex—38.
             38. Algae of the thallus blue-green—39.
                39. Thallus lobes very small, less than 2 mm broad—40.
                    40. Thallus lobes longitudinally striated; spores 12–16 per ascus . . . . *Vestergrenopsis*
                    40. Marginal lobes not striated, at most with 1 groove; spores 8 per ascus—41.
                        41. Thallus dark brown to black, on rocks; algae *Rivularia* or *Scytonema* . . . . *Placynthium*
                        41. Thallus brown. on soil or mosses; algae *Nostoc* . . . . *Massalongia*
                39. Thallus lobes larger, more than 2 mm broad—42.
                42. Underside cyphellate . . . . *Sticta*
                42. Underside lacking cyphellae—43.
                  43. Underside densely clothed with bluish black rhizoids . . . . *Coccocarpia*
                  43. Underside with rhizoids but these not dense nor bluish black—44.
                    44. Apothecia borne on the lower surface of the thallus which is

then reflexed; lower surface pubescent or glabrous, sometimes
tubercles or papillate . . . . *Nephroma*

44. Apothecia borne on margins or upper surface of thallus—45.

    45. Spores simple—46.

        46. Apothecia with thalloid margin . . . . *Pannaria*

        46. Apothecia with proper margin . . . . *Parmeliella*

    45. Spores 2 or more celled—47.

        47. Lower surface corticate, at least partially—48.

            48. Apothecia adnate on upper surface; lower sur-
face with erect tomentum at least partially; spores
3-septate . . . . *Lobaria*

            48. Apothecia sunken in depressions in upper sur-
face; lower surface lacking erect tomentum; spores
2-celled, brown, with round tips . . . . *Solorina*

        47. Lower surface lacking cortex; apothecia borne erect
on margins; spores 4–8 celled, thin-walled, hyaline.
tips pointed . . . . *Peltigera*

38. Algae of the thallus green (but may have blue-green algae in external or internal
cephalodia)—49.

    49. Thallus of small granulose yellowish brown squamules with golden luster;
a thick thalloid exciple; often with cephalodia containing *Nostoc,* the main
alga *Myrmecia;* spores hyaline, 1-celled, the walls warty . . . . Psoroma

    49. Without the above set of characters—50.

        50. Thallus with internal cephalodia or warty cephalodia external on the
upper or lower surface of the thallus—51.

            51. Thallus with veins or veins and white interspaces below—52.

                52. Thallus orange below; the blue-green algae in a distinct layer
below the green algal layer . . . . *Solorina crocea*

                52. Thallus dark below with white interspaces; the external as small
gray lobes or galls on the thallus surface . . . . *Peltigera*

            51. Thallus lacking veins below—53.

                53. The cephalodia distributed in warts on either the upper or lower
surfaces of the thallus—54.

                    54. Apothecia on margins . . . . *Peltigera*

                    54. Apothecia in depressions in upper surface . . . . *Solorina*

                53. The cephalodial algae within the thallus—55.

                    55. Apothecia in depressions in upper surface of the thallus;
blue-green algae in a layer below the green algae . . . *Solorina*

                    55. Apothecia on upper surface or upturned lower surface; blue-
green algae in glomerules or clumps within the thallus
medulla—56.

                        56. Thallus ± scrobiculate . . . . *Lobaria*

                        56. Thallus not scrobiculate . . . . *Nephroma*

        50. Thallus lacking cephalodial algae—57.

            57. Fruits composed of perithecia and imbedded in thallus; thallus
squamulose to crustose—see choice 27.

            57. Fruits disk-like lecanorine; thallus foliose—58.

                58. Thallus bright yellow . . . . *Cetraria*

                58. Thallus gray, yellow-green, brown, or black—59.

                  59. Lobes small, narrower than 1 mm—60.

                    60. Cortex of interwoven hyphae; lobes with long mar-

ginal cilia . . . . *Anaptychia*

60. Cortex ± cellular in appearance; lobes usually lacking cilia on margins—61.

   61. Tips shining; medulla UV + (divaricatic acid); spores simple, hyaline . . . . *Parmeliopsis*

   61. Tips dull; medulla UV − (no divaricatic acid); spores 2-celled, brown—62.

      62. Cortex K + yellow, containing atranorin . . . . *Physcia*

      62. Cortex K − , lacking atranorin—63.

         63. Upper cortex paraplectenchymatous; lower cortex usually paraplectenchymatous; spores with angular lumina and smooth surface, less than 25 u long. . . . . *Phaeophyscia*

         63. Upper cortex scleroplectenchymatous; lower cortex prosoplectenchymatous; spores with uniform walls and minutely verrucose surface, more than 25 μ long. . . . . *Physconia*

59. Lobes large, more than 1 mm broad—6.

   64. Upper side with network of raised ridges around depressions (scrobiculate); underside with pale or dark tomentum around bald spots . . . . *Lobaria*

   64. Upper side lacking raised network of ridges (or if with network then lacking tomentum below); underside bare or with rhizinae but not with fine tomentum—65.

      65. Medulla hollow . . . . *Hypogymnia*

      65. Medulla well developed—66.

         66. Apothecia almost always present; disk dark brown; spores dark brown, 2-celled; upper cortex paraplectenchymatous—67.

           67. Cortex K + yellow (atranorin) . . . . *Physcia*

           67. Cortex K − (no atranorin) . . . *Phaeophyscia*

         66. Apothecia often absent; spores if present simple (if 2-celled, brown see *Physconia*)—68.

           68. Pycnidia internal, the mouths appearing as black spots distributed over the surface of the thallus—69.

              69. Upper cortex scleroplectenchymatous; upper surface usually pruinose . . . . *Physconia*

              69. Upper cortex paraplectenchymatous or prosoplectenchymatous—70.

                 70. Rhizinae present on lower surface—71.

                    71. Thallus very yellowish, containing usnic acid. . . . . *Xanthoparmelia*

                    71. Thallus gray-green or

brown, lacking usnic acid
.... *Parmelia*

70. Rhizinae lacking; upper cortex very thick; texture of thallus very horn-like .... *Masonhalea*

68. Pycnidia superficial, projecting from and mainly on margins of thallus or else lacking—72.

72. Plants of a horn-like consistency, free from substratum and rolling in winds................ *Masonhalea*

72. Plants softer, attached to substratum—73.

73. Cortex paraplectenchymatous (with the hyphae densely coherent but with the lumina large and a cellular appearance) ...
.... *Cetraria*

73. Cortex prosoplectenchymatous (the hyphae very thick walled and the lumina minute)—74.

74. Rhizinae absent; purple pigment usually produced in lower part of medulla, especially in dying parts
.... *Asahinea*

74. Rhizinae present; no purple pigment produced in medulla—75.

75. Caperatic acid present (feathery crystals in GE); upper cortex I−, medulla I+ blue; upper surface usually pseudocyphellate, soredia common, conidia without inflated ends; spores small, to 8.5 u, subspherical..
.... *Platismatia*

75. Caperatic acid absent; upper cortex often I+ blue, always pseudocyphellate, esorediate in our species; conidia with inflated ends; spores larger but unknown in our species ..... *Cetrelia*

Key to Sterile Specimens of the Arctic Umbilicariaceae
    The rock tripes, Umbilicariaceae, so often occur without apothecia that this supplementary
key to the species occurring in the arctic is offered as an aid to such specimens.
1. Thallus pustulate—2.
    2. Underside smooth to tiny papillose, light brown to darker but not black . . . . *Lasallia papulosa*
    2. Underside black charred, strongly verrucose areolate . . . . *Lasallia pensylvanica*
1. Thallus not pustulate—3.
    3. Thallus upper side with papillate, cylindrical or flattened isidia; the thallus paper thin, dull dark
       brown to black both above and below . . . . *Umbilicaria deusta*
    3. Thallus lacking isidia, sorediate or not—4.
        4. Thallus sorediate toward margins; underside brown, areolate, with numerous pale rhizinae
           . . . . *Umbilicaria hirsuta*
        4. Thallus lacking soredia—5.
            5. Thallus with vermiform ridges but the ridges not forming a network—6.
                6. Underside with black area around umbilicus, rest of underside dove gray . . . .
                   . . . . *Umbilicaria arctica*
                6. Underside entirely brown to clove brown or black—7.
                    7. Center portion over umbo reticulate and frosted, the reticulations fading margin-
                       ally to vermiform ridges; underside smooth, sooty black or with paler patches,
                       marginally pruinose granular. . . . . *Agyrophora lyngei*
                    7. Center portion over umbo also vermiform ridged, not frosted, similar to margins;
                       underside entirely black . . . . *Umbilicaria hyperborea*
            5. Upper side smooth or chinky areolate or reticulate rugose—8.
                8. Upper side smooth or chinky areolate—9.
                    9. Upper side smooth—10.
                        10. Underside lacking rhizinae, trabeculae, or lamellae—11.
                            11. Thallus monophyllus, light brown; underside dove gray to dark gray,
                                verrucose . . . . *Umbilicaria phaea*
                            11. Thallus polyphyllus, black-brown; underside black to black-brown,
                                smooth . . . . *Umbilicaria polyphylla*
                        10. Underside with rhizinae and/or trabeculae or lamellae—12.
                            12. Underside with rhizinae only, lacking trabeculae or lamellae—13.
                                13. Underside with a thick black nap of ball-tipped rhizinae—14.
                                    14. Upper side white to violet-brown, sometimes stained reddish, often
                                        pruinose; underside with both cylindrical and ball tipped rhi-
                                        zinae . . . . *Umbilicaria vellea*
                                    14. Upper side olive-brown to darker brown—15.
                                        15. Underside with black cylindrical, papillate rhizinae which are
                                            also ball tipped; under surface with coarse black verrucae;
                                            thallus unchanged when wet . . . . *Actinogyra polyrrhiza*
                                        15. Underside with only short ball-tipped rhizinae; smooth to
                                            coarsely papillate: thallus greener when wet . . . .
                                            . . . . *Umbilicaria mammulata*
                                13. Underside lacking ball-tipped rhizinae, the rhizinae attenuate—16.
                                    16. Underside pink or light brown—17.
                                        17. KC−, lacking lichen substances; thallus margins more or less
                                            perforated, often with rhizoid-like outgrowths; light to dark
                                            brown above; lower cortex smooth to finely granular, some-
                                            times finely pruinose to margin; thallus 450 μ thick . . . . . .
                                            . . . . *Umbilicaria cylindrica*
                                        17. KC+ red, with gyrophoric acid; margins irregular to torn;
                                            olive to yellowish brown above; lower side with well-devel-

oped buttresses around the umbilicus, smooth to granular, epruinose; thallus ca. 266 μ thick ..... *Omphalodiscus virginis*

16. Underside at least partly black, verrucose—18.

    18. Upper side brown to olive brown, umbo not frosty; underside black or reddish black, lacunose; rhizinae simple to cribrose or split .... *Umbilicaria caroliniana*

    18. Upper side dark brown to black, umbo frosty pruinose; underside black; rhizinae flat or terete attenuate .... *Agyrophora scholanderi*

12. Underside with trabeculae or lamellae with or without rhizinae—19.

  19. Underside with trabeculae or lamellae but no rhizinae—20.

    20. Margins perforate; thallus rigid to membraneous and fragile; trabeculae on underside sometimes becoming fimbriate, occasionally poorly developed; underside smooth to finely granulose papillose............................ *Umbilicaria torrefacta*

    20. Margins not perforate but may be cut or torn; thallus rigid; trabeculae form plate-like layers over underside; underside areolate .... *Actinogyra mühlenbergii*

  19. Underside with trabeculae or lamellae plus rhizinae—21.

    21. Rhizinae black; underside with a mat of black branched granulose rhizinae but *no* ball-tipped short rhizinae; underside black verrucose, the lamellae and trabeculae near the umbilicus .... *Umbilicaria angulata*

    21. Rhizinae pale; underside black verrucose with short ball-tipped rhizinae; umbilicus with central ray-like lamellae which become fibrillose........................ *Umbilicaria cinereorufescens*

9. Upper side chinky areolate, not reticulate ridged except at the umbo—22.

  22. Rhizinae absent—23.

    23. Underside papillose areolate, dark brown or brown with black patches .... *Agyrophora rigida*

    23. Underside smooth to subpapillose but not papillose areolate, sooty black .... *Agyrophora leiocarpa*

  22. With a mat of black, branched granulose rhizinae on the black verrucose underside and lamellae and trabeculae near the umbilicus .... *Umbilicaria angulata*

8. Upper side reticulate ridged, the region around the umbo also often pruinose frosted with crystalline material—24.

  24. Lower surface with rhizinae—25.

    25. Lower surface with sparse, short cylindrical rhizinae; underside entirely pale gray .... *Umbilicaria hyperborea* var. *radicicula*

    25. Lower surface with cylindrical to flattened, coarse, pale rhizinae in patches or over lower surface; underside sooty black with scattered gray patches .. .. *Umbilicaria aprina*

  24. Lower surface without rhizinae (note 3 choices next)—26.

    26. Underside paler in center, sooty black around, sometimes paler on outer edges; thallus very thin, 120–225 μ thick; medulla very compact.............. .... *Umbilicaria havaasii*

    26. Underside black all over or with occasional ochraceous patches; the margins may have clear pruinose granular band—27.

    27. Thallus 300 μ thick, medulla loose, underside without radiating strands around the flat umbilicus (apothecia of leiodisk type)... *Agyrophora lyngei*

    27. Thallus 600 μ thick; underside with paler band toward margin, medulla very loose; thallus usually more reticulate and perforate, often with small

INTRODUCTION

epithalline excrescences; underside with folds around umbilicus (apothecia of omphalodisk type, seldom present) ...... *Omphalodiscus decussata*

26. Underside all pale mouse gray or darker close to the umbilicus, smooth or smooth to finely chinky—28.

    28. Lower side pruinose, smooth to finely chinky; thallus 300–500 μ thick; apothecia omphalodisk............... *Omphalodiscus krascheninnikovii*

    28. Lower side with gray pruina only toward margins, smooth; thallus 120–230 μ thick; apothecia gyrodisk .... *Umbilicaria proboscidea*

Treatments of the Genera

ACTINOGYRA Schol.

Nyt. Mag. f. Naturvid. 75:28. 1936.

Thallus foliose, umbilicate; upper cortex paraplectenchymatous; heteromerous, medulla dense; lower cortex paraplectenchymatous to scleroplectenchymatous, under surface with well-developed rhizinae or with lamellae.

Apothecia adnate, radiate, and without a common exciple around the gyri as in Umbilicaria; hypothecium poorly developed; asci with 8, simple, hyaline spores.

Algae: *Trebouxia*.

Type species: *Actinogyra mühlenbergii* (Ach.) Schol.

REFERENCE

Llano, G. A. 1950. A monograph of the lichen family Umbilicariaceae in the Western Hemisphere. Navexos P-831, Office of Naval Research, Washington, D.C.

1. Underside with a dense nap of black, ball-tipped rhizines, apothecia rare . . . . 2. *A. polyrrhiza*
1. Underside without rhizines but with lamellae and/or trabeculae, apothecia common . . . .
. . . . 1. *A. mühlenbergii*

1. Actinogyra mühlenbergii (Ach.) Schol.

Nyt. Mag. f. Naturvid. 75:28. 1936. *Gyrophora mühlenbergii* Ach., Lich. Univ. 227. 1810. *Lecidea mühlenbergii* (Ach.) Spreng., Syst. Veg. 4:262. 1827. *Umbilicaria mühlenbergii* (Ach.) Tuck., Enum. North Amer. Lichens 55. 1845.

Thallus to 20 cm broad, usually less, monophyllus, stiff, with folds or pleats centrally, the edges torn or deeply cut; upper surface smooth, dull, olive-brown; underside olive-brown, darker toward the umbilicus, with papillate lamellae and trabeculae which form plate-like layers over the underside; lower cortex areolate.

Thallus 210–280 μ thick, cortex thin, brown, medulla about 33 μ, lower cortex thick 130–180 μ.

Apothecia common, adnate, in indentations in the upper side, convex, of free gyri, branching dichotomously, lacking a common exciple around the edge; parathecium black externally, brown internally; hypothecium indefinite, hymenium 85–99 μ, paraphyses unbranched, septate, 3.5 μ; asci with 8 spores, spores simple, hyaline, 8.5–13 × 3.5–5 μ.

Reactions: K−, C+ red, KC+ red, P−.

Contents: gyrophoric acid.

This species grows on acid rocks, including sandstones. It has a disjunct Asian–eastern North America range. In the latter it occurs from the Great Bear Lake region east to Newfoundland, south to South Carolina. Llano (1950) reported a specimen "between Point Barrow and Mackenzie River, Capt. Pullen, 8/1849." It has not subsequently been collected this far to the northwest of the rest of its range. Neither has this species been found on Greenland.

2. Actinogyra polyrrhiza (L.) Schol.

Nyt. Mag. f. Naturvid. 75:28. 1936. *Lichen polyrrhizos* L., Sp. Pl. 1151. 1753—*Umbilicaria polyrrhiza* (L.) Ach., Vetensk.-Akad. Nya. Handl. 15:92. 1794—*Gyrophora polyrrhiza* (L.) Körb., Parerg. Lich. 41. 1859. *Gyrophora diabolica* Zahlbr. in Herre, Proc. Wash. Acad. Sci. 7:366. 1906.

Thallus to 10 cm broad, mono- or polyphyllus, with folds in center, upper surface smooth, dark brown, dull or shining; underside dark brown or black, the lower cortex areolate with coarse black verrucae and with black cylindrical rhizines which are papillate and also ball tipped.

Thallus about 420 μ thick, upper cortex 25 μ, medulla and lower cortex not differentiated, 180–280 μ thick.

Apothecia infrequent, to 5 mm broad, adnate, convex, gyri free, without common exciple; parathecium black; hymenium 50 μ, spores 8, simple, hyaline, 8.5–9 × 3.4–6 μ.

Reactions: K−, C+ red, KC+ red, P−.

Actinogyra mühlenbergii (Ach.) Schol.

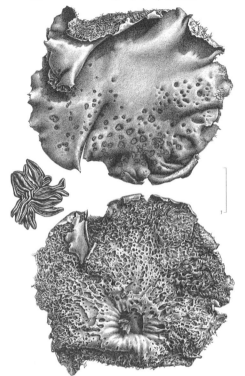

Contents: gyrophoric acid (± umbili-caric acid, Krog 1973).

This is a species which grows on acid rocks. It is in the North and South Temperate zones; in North America it is western from British Columbia south to California but east in Cook County, Minnesota. Llano (1950) lists and maps a specimen from northern Quebec "Hudson Strait, Foul Bay, J. Macoun No. 27 13/3/1914(CAN)." But Macoun had retired to Vancouver Island in 1912 and undoubtedly the specimen came from there. There is a Foul Bay just south of Victoria but none on Hudson Straits. Macoun habitually used poor labeling habits!

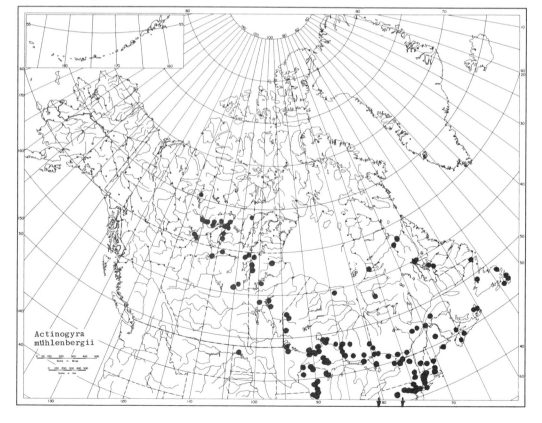

Actinogyra
mühlenbergii

Actinogyra polyrrhiza (Ach.) Schol.

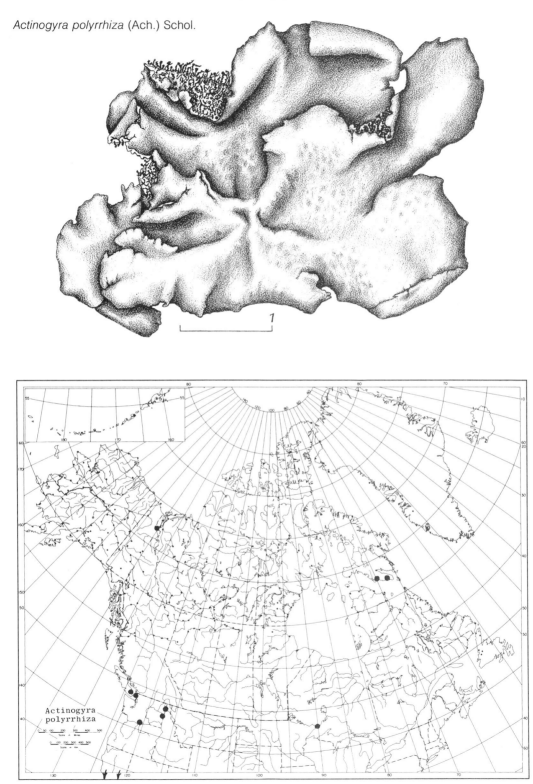

AGYROPHORA Nyl. em. Llano

Lich. Environs Paris 43. 1896.

Foliose, monophyllous or polyphyllous, attached by a central umbilicus; upper surface weakly ribbed or folded; lower surface smooth, areolate-papillate to verrucose, with or without rhizinae.

Apothecia common or rare, lecideine, sessile or stipitate; with proper margin, the disk entire or with superficial cracks, black; spores 8, hyaline, simple.

Algae: *Trebouxia*.

Type species: *Agyrophora atropruinosa* (Schaer.) Nyl.

REFERENCES

Llano, G. A. 1950. A monograph of the lichen family Umbilicariaceae in the western hemisphere. Navexos P-831, Office of Naval Research, Washington, D.C.

Llano, G. A. 1956. New Umbilicariaceae from the western hemisphere, with a key to genera. J. Wash. Acad. Sci. 46:183–185.

1. Rhizinae absent on the underside.
   2. Upper side chinky-areolate, not reticulately ridged except in center near umbo.
      3. Underside papillose-areolate .... 3. *A. rigida*.
      3. Underside smooth to subpapillose but not papillose-areolate .... 1. *A. leiocarpa*
   2. Upper side reticulate ridged centrally and vermiform ridged marginally, underside smooth, sooty black .... 2. *A. lyngei*
1. Rhizinae present on underside; upper side smooth except for slight reticulation near the umbo; underside black verrucose .... 4. *A. scholanderi*

1. Agyrophora leiocarpa (DC.) Gyeln.

Ann. Mycol. 30:444. 1932. *Umbilicaria leiocarpa* DC. in Lam. & DC., Flore Française 3d ed. 2:410. 1805.

Thallus umbilicate, mono- or polyphyllus, usually deeply incised and folded; margins fragile and slightly perforate; upper side rough-ribbed becoming chinky areolate toward the margins, the center pruinose over the umbo, dull, dark brown to brown-black; lower surface smooth to subpapillose, sooty roughened, the margins paler, tiny areolate.

Thallus ca. 250 μ thick; upper cortex 33 μ, medulla 33 μ, lower cortex about 83 μ.

Apothecia stipitate, to 2 mm broad, disk flat, black, margin even, dark; hypothecium black; hymenium 50 μ, brownish; paraphyses unbranched, septate, the tips capitate, to 2.4 μ; spores simple, hyaline, middle dark stained, 4–4.8 × 12–15 μ.

Reactions: not checked.

Contents: unknown.

This is a rare bipolar species known only from western Greenland (Inglefield Land in the north and Igdlut in the south central part) and northern Somerset Island in North America. In Europe it occurs in

Norway, Switzerland, and Hungary. I have not seen this species.

2. Agyrophora lyngei (Schol.) Llano

Monograph of Umbilicariaceae 57. 1950. *Umbilicaria lyngei* Schol., Nyt. Mag. Naturv. 75:19–22. 1936.

Thallus umbilicate, to 3.5 cm broad, monophyllus, the margins thin, torn, or incised, umbo reticulate, the reticulations fading marginally; upper side dull, dark brown to brown-black with frosty pruina over the umbo; underside smooth, sooty black or with paler patches, marginally with a pruinose-granular band.

Thallus 300 μ thick, upper cortex about 40 μ, medulla about 83 μ, lower cortex about 66 μ.

Apothecia rare, stipitate, to 2 mm broad; disk black, flat; paraphyses unbranched, septate, tips capitate, to 3.4 μ; spores simple, hyaline, 4.9–7 × 9.8–14 μ.

Reactions: K−, C−, KC−, P−.

Contents: Hale (1956) reported gyrophoric acid. This was not confirmed by Krog (1973) who reported traces of norstictic acid only

Agyrophora leiocarpa (DC.) Gyel.

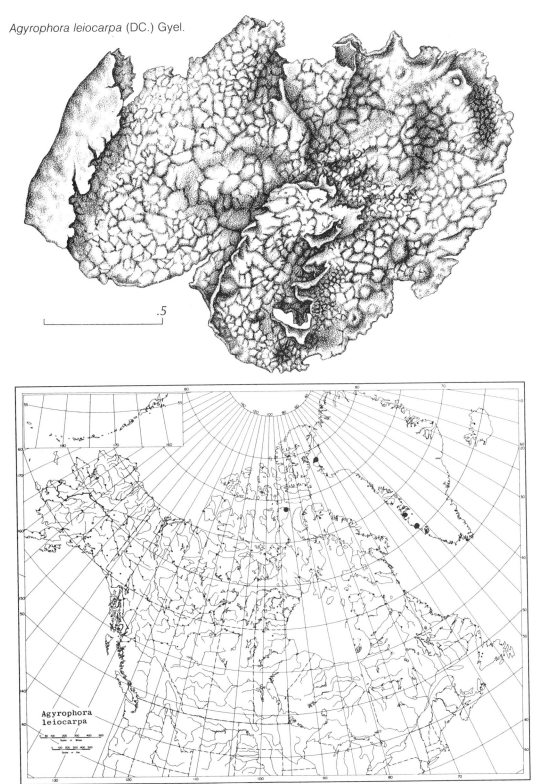

Agyrophora lyngei (Schol.) Llano

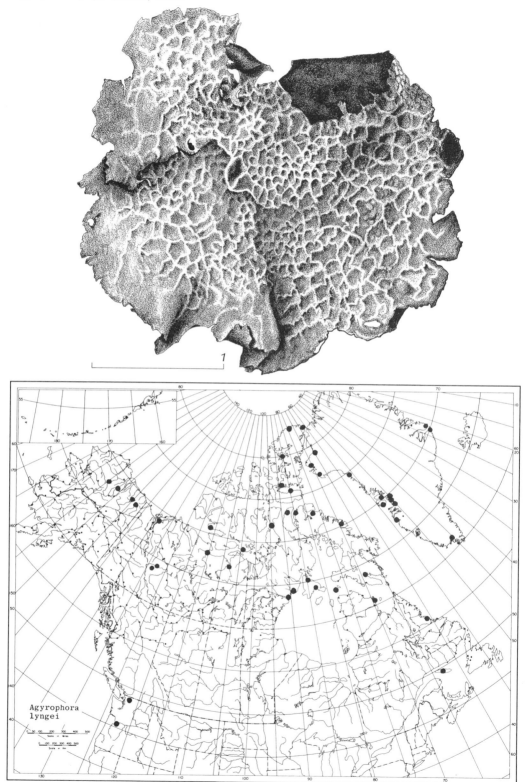

1

Agyrophora rigida (DR.) Llano

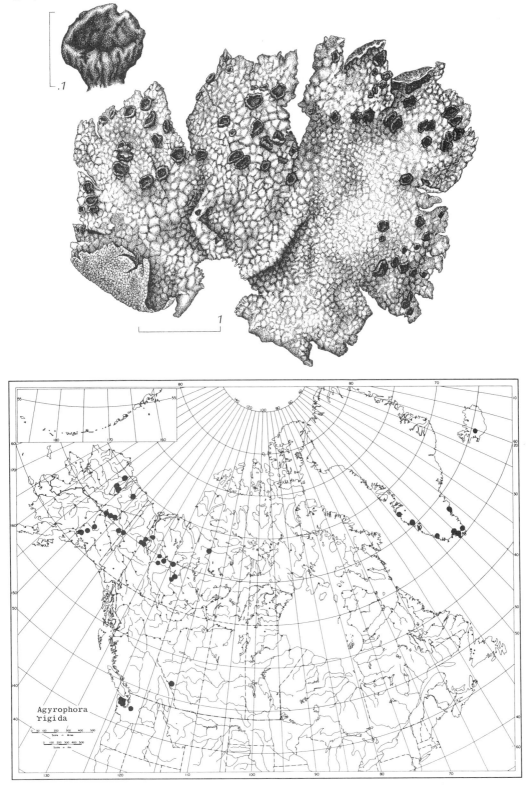

Agyrophora scholanderi Llano

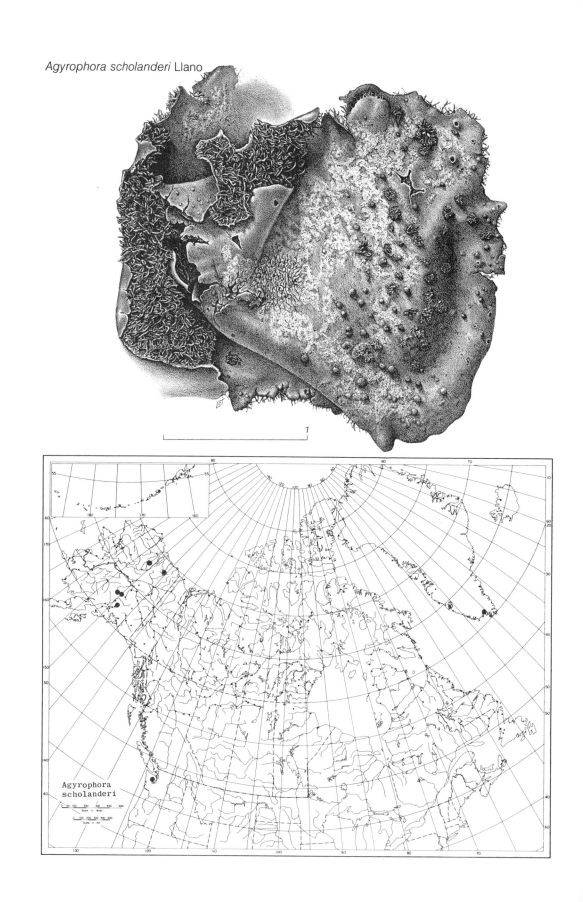

1

with TLC. I did not find any gyrophoric acid in my material.

This species is reported to be not associated with bird droppings. It is high arctic-alpine but not in the European Alps. In North America its range is high northern but with outliers into Oregon and on the island of Anticosti in the Gulf of St. Lawrence.

3. Agyrophora rigida (DR.) Llano

A monograph of the lichen family Umbilicariaceae 58. 1950. *Gyrophora rigida* DR., Ark. f. Bot. 19:3–4. 1926. *Umbilicaria rigida* (DR.) Frey, Hedwigia 71:117. 1931. *Umbilicaria coriacea* Imsh., Bryologist 60:257. 1957.

Thallus foliose, mono- or polyphyllus, rigid and leathery, folded, the margins usually incised or perforate, the umbo becoming chinky areolate or dissected and with a frosty bloom; upper side dull, buff to dark brown or black-brown, chinky-areolate; underside smooth to dimpled and with the lower cortex in a pattern of small areolae, or papillose-areolate to verrucose, dark brown or in brown and black patches, usually marginally darker.

Thallus 180–200 μ thick; upper cortex of palisadeplectenchyma, 50 μ thick; algal layer discontinuous; medulla ca. 50 μ thick; lower cortex about 70 μ, the cell walls thick and dark.

Apothecia common, stipitate, 1–2 mm broad, the disk black, flat, the margins slightly shiny, parathecium of isodiametric cells; hypothecium brown; hymenium 100 μ, brownish; paraphyses septate, branching toward the tips, 2–4 μ; spores 8, simple, hyaline, 9.6–15 × 2.4–9 μ.

Reactions: K − , C − , KC − , P − .
Contents: Gyrophoric acid reported by Hale (1956) but I am unable to confirm this in GE or GAQ reagents or by the negative color reactions above. Krog (1968) reported no lichen substances in Alaskan material but one Norwegian specimen contained traces of norstictic acid.

This is a species of acid rocks and is reported to like bird perches. It is arctic-alpine, circumpolar in distribution in all probability, as in North America it ranges across the far north in Alaska and Canada, south to Washington, and is in Greenland but absent from eastern North America.

4. Agyrophora scholanderi Llano

J. Wash. Acad. Sci. 46:183. 1956.

Thallus small, to 2 cm broad, mono- or polyphyllus, the margins thin, slightly incised; the umbo raised, raised ridged and frosty pruinose; upper side smooth, slightly reticulate veined, with minute papillae, dark brown to black; underside verrucose, black, with flat or rounded rhizines, which are attenuate.

Apothecia few to numerous, adnate, to 1 mm broad; disk black, flat, the margin irregular; paraphyses unbranched, capitate, to 2.5 μ; spores 8, simple, hyaline, 5–7.6 × 11–14 μ.

Reactions: K − , C + red medulla, KC + red, P − .

Contents: contains gyrophoric acid (Krog 1968), unlike the other species of *Agyrophora*.

A species of acid rocks, this appears to be a species of the western mountains. The type locality is near Lake Peters on the north slope of Alaska and it ranges south into the mountains of Washington. Alstrup (1979) recently reported it from Narssaq, Greenland.

ALECTORIA Ach.

In Luyken, Tent. Hist. Lich. 95. 1809. *Evernia* Sect. *Alectoria* (Ach.) Fr., Syst. Orb. 1:237. 1825. *Cornicularia* Sect. *Alectoria* (Ach.) Duby, Bot. Gall. 2:616. 1830. *Parmelia* Sect. *Alectoria* (Ach.) Spreng., Fl. Halen. 2d ed. 2:521. 1832. *Bryopogon* Link, Grund. Kraüt. 3:164. 1833. *Alectoria* subgen. *Eualectoria* Th. Fr., Lich. Scand. 1:19. 1871. *Eualectoria* Gyel., Ann. Mus. Nat. Hung. 28:283. 1934. *Alectoria* sect. *Phaesporae* Hue, Nouv. Arch. Mus. Paris ser. 4, 1:93. 1899. *Ceratocladia* Del. in Schwendener in Naegeli, Beitr. Bot. 2:149. 1860. *Alectoria* sect. *Ochroleuca* Gyel., Feddes Repert. 38:243. 1935.

Thallus fruticose, erect, decumbent or pendent; branching terete to compressed and angular or foveolate; greenish-yellow to yellow or brownish to black and paler at the base in one species; pseudocyphellae present, usually abundant, fusiform, white, raised; isidia lacking, soredia rare; cortex of periculinal hyphae in a thick matrix; medullary hyphae ornamented with projections.

Apothecia lateral, with thalloid margin of same color as the thallus; disk brown to black; asci clavate, thick walled; spores 2–4, simple, ellipsoid, with distinct hyaline epispore, brown at maturity; pycnidia rare.

Type species: *Alectoria sarmentosa* (Ach.) Ach.

REFERENCES

Ahti, T. & D. L. Hawksworth. 1974. Notes on the lichens of Newfoundland. 4. *Alectoria*. Ann. Bot. Fenn. 11:189–196.
Brodo, I. M. & D. L. Hawksworth. 1977. *Alectoria* and allied genera in North America. Opera Botanica 42:1–164.
DuRietz, G. E. 1924. Lichenologiska Fragment VI. Skandinaviska *Alectoria*-arten. Svensk Bot. Tidsk. 18:141–155.
DuRietz, G. E. 1926. Vorarbeiten zu einer "Synopsis Lichenum." 1. Die Gattungen *Alectoria, Oropogon* und *Cornicularia*. Ark. f. Bot. 20A. 11:1–43.
Hakulinen, R. 1965. Über die Verbreitung und das Vorkommen einiger Nordlichen Erd.- und Steinflechten in Ostfennoskandien. Aquilo, Ser. Bot. 3:22–66. 1965.
Hawksworth, D. L. 1972. Regional studies in *Alectoria* (Lichenes). 2. The British species. Lichenologist 5:181–261.
Laundon, J. R. 1970. Lichens new to the British flora. 4. Lichenologist 4:297–308.
Motyka, J. 1964. The North American species of *Alectoria*. Bryologist 67:1–44.

1. Thallus yellow or yellow-green; cortex KC+ yellow, containing usnic acid.
   2. Thallus erect, main stems terete, usually less than 2 mm diameter; medulla KC− or KC+ yellowish, not red, containing diffractaic acid . . . . 2. *A. ochroleuca*
   2. Thallus prostrate to decumbent, main stems flattened, expanded, foveolate, usually broader than 2 mm (to 4 cm wide); medulla KC+ red, UV+, containing alectoronic acid . . . .
   . . . . 3. *A. vexillifera*
1. Thallus pale brownish at the base, the tips dark brown to blackish; cortex C+ red, P+ yellow, containing alectorialic acid . . . . 1. *A. nigricans*

1. Alectoria nigricans (Ach.) Nyl.

Lich. Scand. 71. 1861. *Cornicularia ochroleuca* var. *nigricans* Ach., Lich. Univ. 615. 1810. *Alectoria boryana* Del. in Th. Fr., Nova Acta Reg. Soc. Sci. Upsal. 3:28. 1860 (nom. inval.). *Alectoria ochroleuca* var. *nigricans* (Ach.) Körb. Pareg. Lich. 5. 1859; *Alectoria ochroleuca* (ssp.) *thulensis* Th. Fr. *Alectoria nigricans* var. *thulensis* (Th. Fr.) Vain., Ark. f. Bot. 8:14. 1909; *Alectoria nigricans* var. *tschuctschorum* Vain. *Alectoria nigricans* f. *goemoerensis* Gyeln., Nyt. Mag. Naturv. 70:44. 1932. *Bryopogon nigricans* f. *goemoerensis* (Gyeln.) Gyeln., Feddes Repert. 38:239. 1935; *Bryopogon nigricans* f. *thulensis* (Th. Fr.) Gyeln. *Alectoria nigricans* var. *nipponica* Asah., J. Jap. Bot. 13:696. 1936.

Thallus fruticose forming large entangled mats, podetia to 10 cm tall, usually somewhat less; branches anisotomic dichotomous, the angles usually acute, branches terete toward tips, becoming flattened and slightly thickened toward the base which commonly dies; branches dull, the base pale brown or whitish or pinkish, the upper parts chestnut brown to black, discoloring herbarium packets on long standing in them, pseudocyphellae abundant, slightly raised, fissural, to 1 mm long; soredia have been described from Greenland (f. *sorediata* Dahl).

Apothecia rare, lateral but appearing ge-

niculate, margin concolorous with the thallus, becoming excluded; disk concave to flat then convex, yellowish brown to brown 2–9 mm broad; spores 2, hyaline to pale brownish, ellipsoid, with hyaline epispore, 20–41 × 14–28 u. Pycnidia unknown.

Reactions: Cortex and medulla K+ faintly yellow, to reddish, C+ reddish, KC+ red, P+ yellowish.

Contents: alectorialic acid, traces of barbatolic acid and an unknown K+ red substance.

This species grows in large swards over soil, humus, and gravels. It is especially common in heath tundras. It may also grow on the lower branches of spruce. In boulder trains it is often common between the boulders. Its range is circumpolar, arctic-alpine, extending slightly into the northern edge of the boreal forest. It ranges south to the Gaspé Peninsula in the east, in the west to New Mexico and Washington, but is also bipolar, ranging in New Zealand, the southern tip of South America, and on islands in the southern hemisphere.

Alectoria nigricans (Ach.) Nyl.

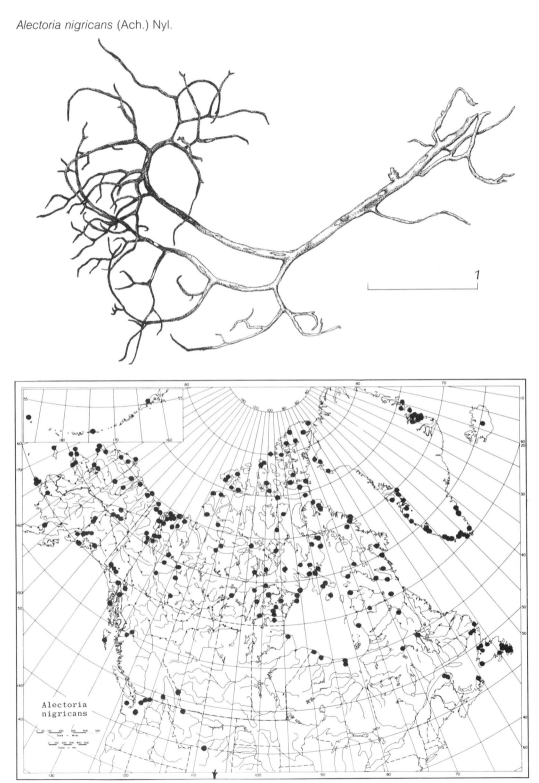

2. Alectoria ochroleuca (Hoffm.) Mass.

Sched. Crit. Lich. Ital. 47. 1855. *Usnea ochroleuca* Hoffm., Descr. Adumb. Plant. Crypt. 2:7. 1791. Details of the nomenclature of this species may be sought in Laundon (1970) and Hawksworth (1972).

Thallus fruticose, erect, stiff, to 15 cm tall but usually much less, sometimes decumbent; branching anisotomic dichotomous, the branches terete, to 3 mm broad, the apices sometimes drooping, smooth to slightly ridged or tuberculate, with numerous longitudinally oriented, raised, whitish pseudocyphellae to 1 mm long; lower parts yellow to yellow-green, often blackened in parts, apices curved, pointed, black to blue-black.

Apothecia rare, lateral but sometimes geniculate; margin thalloid and of same color as the thallus; disk flat to concave, orange-yellow to reddish brown or black, to 7 mm diameter; spores 2 (–4) per ascus, simple, ellipsoid with hyaline epispore, becoming dark brown when mature 28–46 × 14–28 μ.

Reactions: Cortex K−, C−, KC+ yellow, P−; medulla K−, C− or rarely C+ red, KC−, P−.

Contents: Usnic acid with diffractaic acid or alectoronic acid (KC+ red strain). Acid deficient strains are also known. The diffractaic acid strain is the common one.

Growing on soil or humus especially in heath tundras, often in sheltered places between frost-riven boulders in fell fields, this species also occurs on the lower branches of spruce and willow.

It is a circumpolar, arctic-alpine species, ranging far southward in South America in the Andes. In North America it is a common species in the Arctic and southward into the northern fringes of the boreal forest. It also occurs in the mountains in Mexico.

3. Alectoria vexillifera (Nyl.) Stiz.

Annals Naturhist. Hofmus. Wien 7:122. 1892. *Evernia ochroleuca* var.·*cincinnata* Fr., Lich. Eur. Ref. 22. 1831. *Cornicularia ochroleuca* var. cincinnata (Fr.) Schaer., Enum. Crit. Lich. Eur. 5. 1850. *Alectoria ochroleuca* var. *cincinnata* (Fr.) Nyl., Syn. Lich. 1:282. 1858–60. *Alectoria ochroleuca* f. *cincinnata* (Fr.) Fellm., Lich. Arct. 53. 1863. *Alectoria sarmentosa* var. *cincinnata* (Fr.) Nyl., Flora 52:444. 1869. *Bryopogon sarmentosus* var. *cincinnatus* (Fr.) Müll. Arg., Flora 72:362. 1889. *Alectoria ochroleuca* f. *tenuior* Cromb., J. Bot. Lond. 10:232. 1872. *Alectoria ochroleuca* ssp. *vexillifera* Nyl. in Kihlman, Medd. Soc. Fauna Flora Fenn. 18:48. 1891. *Alectoria cincinnata* f. *fasciata* Sern., Svensk Bot. Tidsk. 1:178. 1907. *Alectoria cincinnata* var. vexillifera (Nyl.) Räs., Annals Acad. Sci. Fenn. 34:22. 1931. *Alectoria cincinnata* f. *vexillifera* (Nyl.) Savicz, Lichenotheca Ross. 4: no. 33. 1935. *Bryopogon ochroleucus* var. *tenuior* (Cromb.) Gyel., Feddes Repert. 38:248. 1935. *Alectoria sarmentosa* var. *cincinnata* f. *vexillifera* (Nyl.) Magn., Forteckn. ö″ Skand. Vaxt. 4:68. 1936. *Alectoria sarmentosa* ssp. *vexillifera* (Nyl.) Hawksw., Taxon 19:241. 1970.

Thallus fruticose, prostrate, rigid except at the apices, to 10 cm long; branching anisotomic dichotomous at the base, isotomic dichotomous toward the apices, rather sparse, the main branches weakly to strongly compressed, the sides ridged or striated or foveolated, greenish yellow to yellow, the apices of same color or becoming black.

Apothecia not seen.

Reactions: cortex K−, KC+ yellow, C−, P−; medulla K−, KC− or KC+ red, P−.

Contents: usnic acid with accessory atranorin in the cortex, and accessory alectoronic acid in the medulla. Hawksworth (1972) reports also possibly squamatic acid and α-collatolic acid in the medulla in some.

Growing in open, more or less windswept gravelly areas and on old frost boils, sometimes mixed in heath vegetation, this species shares the same chemistry and variability of chemistry as *A. sarmentosa* with which many lichenologists classify it, e.g., Brodo & Hawksworth (1977). It does occur farther north and in more exposed ecological conditions as well as being decumbent instead of pendent on trees as in *A. sarmentosa*. Ahti and Hawksworth (1974) mention finding intermediates on Newfoundland and in Europe, but that these two populations seem to be quite different over most of North America. Possibly these are ecads instead of species but in view of the differing habits and ranges I have taken the course of retaining them as separate species until experimentation decides this question. The range of *A. vexillifera* is circumpolar arctic-alpine with the North American range extending south to Newfoundland, Alberta, and Washington. Its close relative, *A. sarmentosa*, is confined to the boreal forest.

Alectoria ochroleuca (Hoffm.) Mass.

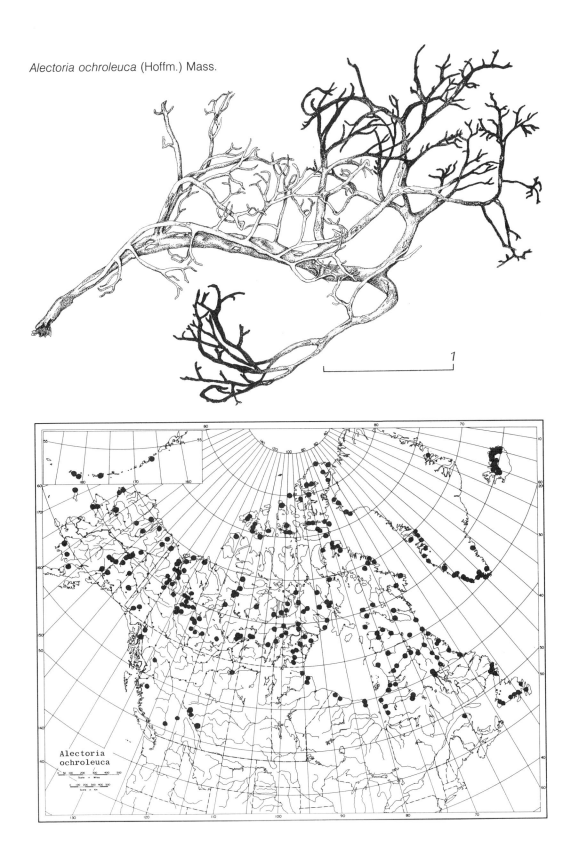

1

Alectoria
ochroleuca

Alectoria vexillifera (Nyl.) Stiz.

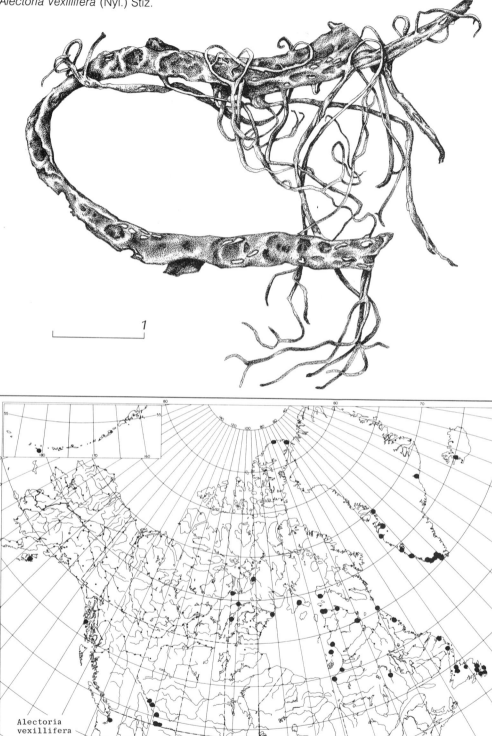

1

ANAPTYCHIA Körb

Grundr. Krypt.-Kunde 197. 1848.

Foliose to subfruticose; lobes dorsiventral, branching dichotomously or irregularly, corticate above with interwoven hyphae oriented more or less longitudinally, lower cortex present or absent; margins ciliate in the arctic species; rhizines scant.

Apothecia sessile or on short stalks, lecanorine; epithecium pale brownish; hypothecium pale and not sharply limited from the exciple; hymenium hyaline, I + blue; paraphyses septate, sparsely branched toward the tips, the tips brown and capitate; asci cylindrical, spores 8, brown, 2-celled.

Algae: *Trebouxia*.

Type species: *Anaptychia ciliaris* (L.) Körb.

REFERENCE
Kurokawa, S. 1962. A monograph of the genus *Anaptychia*. Beih. Nova Hedwigia 6:1–115.

1. Anaptychia kaspica Gyeln.

Ann. Crypt. Exot. 4:166. 1932. *Parmelia ciliaris* var. *angustata* Tuck., Amer. Acad. Arts. Sci. 1:224. 1848. *Physcia ciliaris* var. *angustata* (Tuck.) Müll. Arg., Flora 65:318. 1882. *Anaptychia ciliaris* β *angustata* (Tuck.) Mass., Mem. Lichenogr. 35. 1853.

Thallus foliose to subfruticose, brownish gray to brown, the margins with long cilia which also at intervals attach the thallus loosely to the substratum; mainly dichotomously branched, the lobes narrow, 0.2–1.5 mm broad, dorsiventral; the upper side rounded and with very fine pubescence, the edges rolled over to make the underside canaliculate; the lower cortex lacking on underside and white loose hyphae showing.

Lobes 200–300 μ thick; upper cortex 30–120 μ thick, algal layer not continuous, medulla 140–180 μ.

Apothecia toward the tips, more or less stalked, 1–4 mm broad; disk dark brown, sometimes white pruinose; margins entire, the back spinulose; hymenium 150–170 μ, hyaline, spores 8, brown, 1-septate, the walls of uniform thickness, thin, 30–42 × 12–18 μ.

Reactions: K−, C−, KC−, P−.

Contents: No lichen substances demonstrated.

This species grows on barks (*Picea* and *Thuja*) and commonly on rocks. It appears to be a circumpolar boreal species which is not commonly collected. The similar *Anaptychia ciliaris* (L.) Körb. with which it has been confused has broader lobes, more than 2 mm broad, and lacks the spinules on the apothecia. *A. kaspica* in North America ranges from Alaska east to Labrador and Newfoundland, south to Minnesota, Michigan, and New England. This is the only species of *Anaptychia* to reach the Arctic. Although present in Iceland and in North America it has apparently not yet been collected in Greenland.

ARCTOMIA Th. Fr.

Lichenes Arctoi Europ. Groenl. hact. cogn. 287. 1860.

Thallus granular crustose or short-lobed rosette forming, red-brown or olive, attached by rhizines; with a cortex of one or partly several celled brown pigmented cells; medulla of a netted or irregular group of cells, the cells short, cylindrical or spherical.

Apothecia red-brown, without thalloid exciple, the proper exciple present, convex to spherical, often conglomerate; paraphyses net-like interwoven and anastomosing; the tips clavate and pigmented brown; hypothecium and hymenium hyaline to brownish, the asci broadly clavate; spores 8, hyaline, several celled. Pycnidia partly immersed; conidia short, cndobasidial.

Anaptychia kaspica Gyeln.

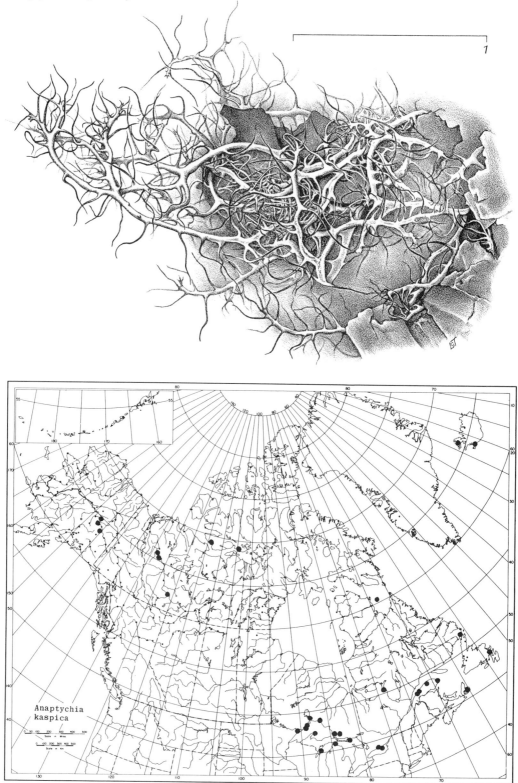

1

Algae: *Nostoc.*

Type species: *Arctomia delicatula* Th. Fr.

REFERENCE
Henssen, A. 1969. Eine studien über die Gattung *Arctomia*. Sv. Bot. Tids. 63:126–138.

1. Thallus red-brown to olive; lobes shorter than 0.2 mm; spores 7–11 celled, mostly over 50 u
   long . . . . 1. *A. delicatula*
1. Thallus rust-brown, lobes 0.2–0.6 mm long; spores 4–7 celled, shorter than 50 u . . . . . 2. *A. interfixa*

1. Arctomia delicatula Th. Fr.

Lich. Arct. Europ. Groenl. 287. 1860. *Arctomia delicatula* ssp. *cisalpina* Hult., Bot. Not. 1891:84. 1891.

Thallus crustose-gelatinous to coarsely granular, dispersed or in a crustose mass, rust-brown to rarely olive, the lobes less than 0.2 mm broad; cortex well marked; medulla of loose network of cells.

Medullary hyphae 2–3 μ thick when young, thickening to 5–11 μ.

Apothecia red-brown, often shining, strongly convex to spherical, to 0.5 mm broad, frequently conglomerate; hymenium 95–110 μ, the upper gelatin brown; subhymenium 30–95 μ; exciple 10–80 μ broad, radiate or in old apothecia paraplectenchymatous; paraphyses 1–2 μ broad, the tips capitate, 5–8 μ, pigmented brown; asci 70–95 × 12–24 μ; spores 8, spindle-shaped, 7–10 celled, 35–75 × 4–7 μ.

This species grows on mosses, over other lichens, on earth and humus, and old wood. It is a circumpolar species known from Greenland, Iceland, Spitzbergen, Novaya Zemlya, Siberia, Fennoscandia, and the British Isles. In North America it has been collected on Ellesmere Island. It is probably usually overlooked because of its small size.

2. Arctomia interfixa (Nyl.) Vain.

Ark. f. Bot. 8(4):98. 1909. *Pannularia interfixa* Nyl., Flora 68:446. 1885. *Arctomia delicatula* ssp. *andraearum* Th. Fr., Bot. Not. 1866:151. 1866.

Thallus forming small rosettes or dispersed, the lobes 0.2–0.6 mm broad, somewhat imbricated in the rosette forms, rust-brown; cortex brown.

Cortex not as well defined as in *A. delicatula;* medullary hyphae 5–11 u thick, or irregularly grouped ellipsoid or spherical cells.

Apothecia red-brown, convex, to 0.5 mm broad; hymenium 70–110 μ; subhymenium 80–245 μ, paraphyses 1–2 μ, the tips capitate, 4–8 μ, and brown; asci 50–90 × 8–21 μ; spores 8, 4–7 celled, spindle-shaped 26–46 × 3.5–6 μ.

This species grows over mosses and other lichens. It is undoubtedly circumpolar in distribution. It has been reported from Siberia, Norway, Sweden, Finland, Novaya Zemlya, Spitzbergen, and Bear Island. It appears to be rarely found in North America, specimens being known from Baffin Island, N.W.T., and the Olympic Mountains in Washington.

ASAHINEA W. L. & C. F. Culb.

Brittonia 17:183. 1965.

Thallus foliose, the lobes broad, yellowish or grayish; upper side pseudocyphellate in one species, isidiate in another; the margin not ciliate, commonly with blackish spots and edges; underside black to the center, brown marginally, lacking any rhizinae; upper cortex prosoplectenchymatous (with thick walls and tiny lumina).

Apothecia rare, marginal, imperforate; the spores hyaline, simple, ellipsoid, 8–13 × 5–8 μ; pycnidia rare, marginal, immersed; conidia straight, 5–10 μ.

Algae: *Trebouxia.*

Type species: *Asahinea chrysantha* (Tuck.) W. L. & C. F. Culb.

Arctomia delicatula Th. Fr.

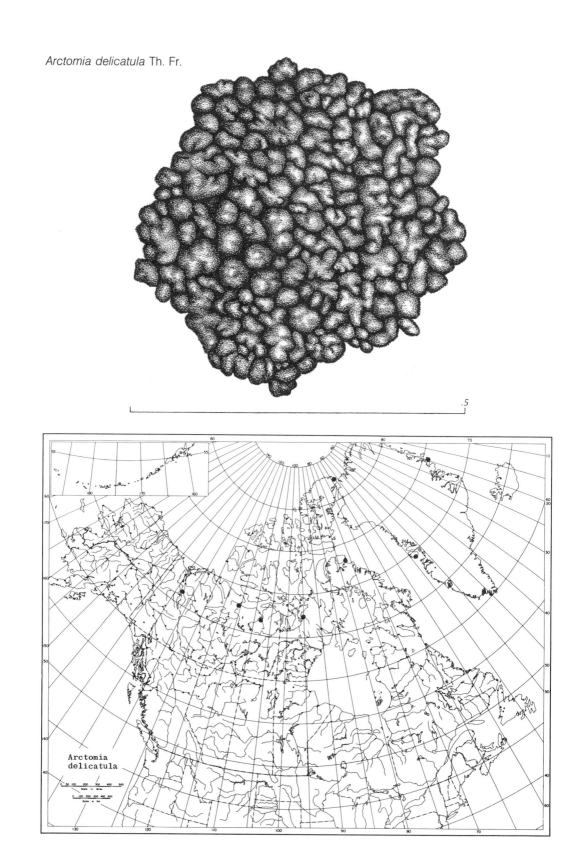

.5

Arctomia
delicatula

REFERENCES

Culberson, W. L. & C. F. Culberson. 1965. *Asahinea*, a new genus in the Parmeliaceae. Brittonia 17:182–190.

Hakulinen R. & T. Ulvinen. 1966. *Asahinea chrysantha* (Tuck.) W. L. Culb. & C. F. Culb. in Fennoskandien. Ann. Univ. Turku 36:101–105.

1. Thallus yellow, mottled or edged with black, lacking isidia but pseudocyphellate above, usnic acid present . . . . *1. A. chrysantha*
1. Thallus whitish to tan, mottled with black and becoming mostly black; upper surface isidiate, non-pseudocyphellate (but scars of isidia may resemble pseudocyphellae), usnic acid absent . . . .
. . . . *2. A. scholanderi*

1. Asahinea chrysantha (Tuck.) W. L. & C. F. Culb.

Brittonia 17:184. 1965. *Cetraria chrysantha* Tuck., Am. J. Arts. Sci. II. 15:423. 1858. *Platysma septentrionale* Nyl., Syn. Lich. 1:315. 1860. *Cetraria septentrionalis* (Nyl.) Almq., Lich. Kusten Beringsmeeres 523. 1887 (comb. illeg., sine ref.). *Cetraria chrysantha* f. *cinerascens* Asah., J. Jap. Bot. 10:481. 1934. *Platysma chrysanthum* (Tuck.) Dahl., Rev. Bryol. Lich. 21:129. 1952.

Thallus medium to large, the lobes to 3 cm broad, rounded; upper side with large and small reticulate pitting, pseudocyphellate, the pores few or many, mainly on the ridges, yellow to yellowish-tan with the edges black; lower surface jet black, smooth or wrinkled, brown toward the margins, lacking rhizines.

Upper cortex 16–26 μ thick, densely filled with minute brownish crystals; medulla white or partly lavender especially in older parts, 93–248 μ thick; lower cortex 12–22 μ thick.

Apothecia rare, described by Culbersons from Japanese material, submarginal to laminal, to 2.4 cm., disk brownish black; hy-

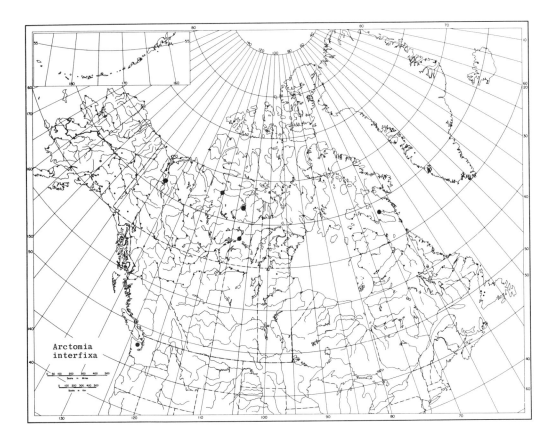

Arctomia
interfixa

Asahinea chrysantha (Tuck.) W. L. & C. F. Culb.

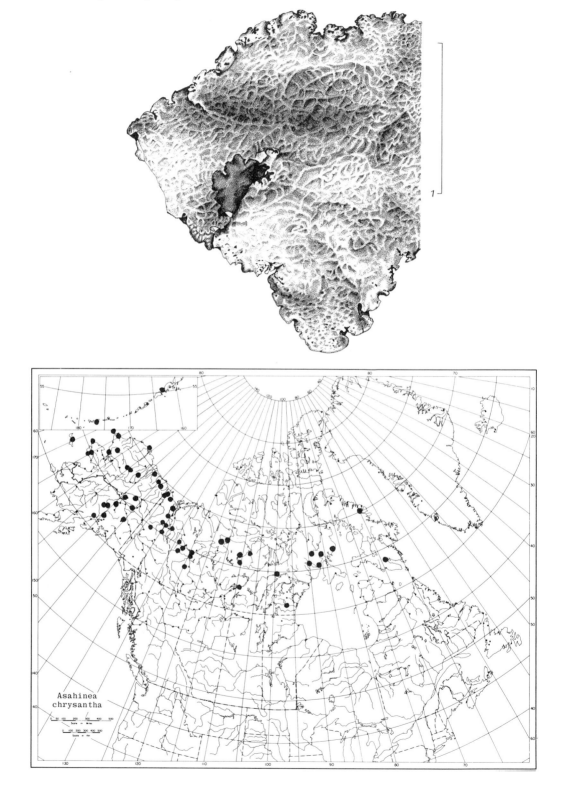

Asahinea
chrysantha

Asahinea scholanderi (Llano) W.L. & C.F. Culb.

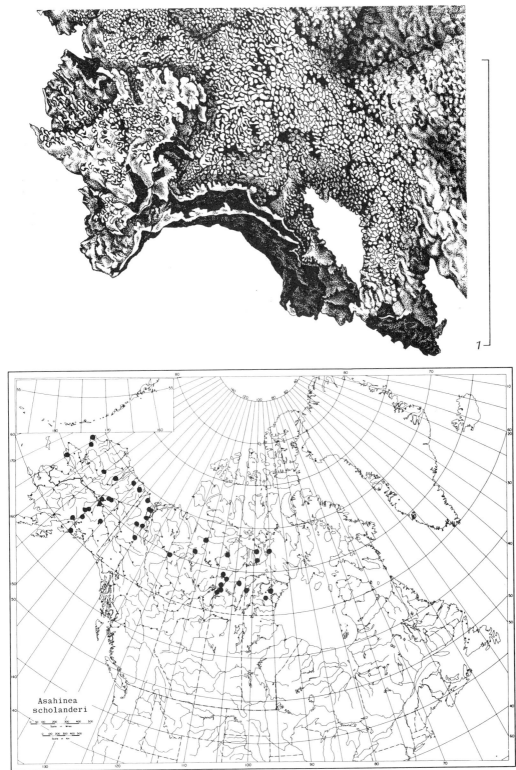

1

menium hyaline 52–70 μ, hypothecium I+ lavender, proper exciple I−; spores 8, ellipsoid, 8–13×5–8 μ; pycnidia rare, marginal; conidia 5–6.5×1 μ.

Reactions: medulla K−, C−, KC+ red or pink, P−.

Contents: usnic acid and atranorin in upper cortex, alectoronic acid ± α-collatolic acid in medulla.

This species grows over boulders and plant debris and humus, occasionally on soil in tundras. A member of the Beringian Element in the Arctic, it ranges into Finland in Europe and in North America east to Baffin Island. It does not appear to range southward in the Rocky Mountains, remaining arctic only.

2. Asahinea scholanderi (Llano) W. L. & C. F. Culb.

Brittonia 17:187. 1965. *Cetraria scholanderi* Llano., J. Wash. Acad. 41:197. 1951. *Cetraria saviczii* Oxn. & Rass., Not. Syst. Leningrad 13:6. 1960.

Thallus medium-sized, the lobes to 1.5 cm broad, forming rosettes on the substratum; the lobes imbricated and massed; upper side whitish or bluish gray to tan, mottled with black or mainly black, lacking pseudocyphellae, with numerous erect unbranched or rarely slightly branched cylindrical isidia, colored like the surface or tipped with black or all black; lower surface jet black, brown toward the margins, smooth or minutely pitted, lacking rhizines.

Upper cortex 18–26 μ; medulla white or partly lavender, 65–240 μ; lower cortex 16–26 μ thick.

Apothecia rare, seen only once (Culberson 1965), 3 mm broad, exciple isidiate, the isidia with pyncidia in the tips; spores 8, subglobose to ovoid, 10–13×5–8 μ; pycnidia small, immersed in the tips of isidia; conidia straight or slightly curved, 5–10 μ long.

Reactions: medulla K−, C−, KC+ red or pink, P−.

Contents: atranorin in upper cortex, alectoronic and α-colatollic acids in the medulla.

This species grows over boulders and less commonly on humus in the tundras. A member of the Beringian Element in the Arctic, its range encompasses Siberia and in North America east to the area of Baker Lake and the mouth of the Back River.

BAEOMYCES Pers.

Usteri, Neue Ann. der Bot. 1:19. 1794.

Thallus crustose, granulose, squamulose, or marginally foliose, attached by medullary hyphae or rhizines; cortex with one or more paraplectenchymatous layers or composed of interwoven hyphae more or less parallel to the upper surface or lacking; algal layer continuous.

Apothecia round, finally swollen, often multiparted, on more or less distinct stipes of dense or loose strands, often containing algae, from the base partly or entirely overgrown by an algal layer with a cortex similar to that of the thallus; hypothecium and exciple not distinct from the interior of the stipe, hyaline and lacking algae; paraphyses simple or the upper part sparingly branched, slender; asci cylindrical; spores 8, fusiform or ellipsoid, hyaline, 1–4 celled; pycnidia immersed in warts in the thallus, fulcra endobasidial; conidia short, bacilliform.

Algae: *Cystococcus*.

Type species: *Baeomyces roseus* Pers.

REFERENCE

Thomson, J. W. 1967. The lichen genus *Baeomyces* in North America. Bryologist 70:285–298.

1. Apothecia pink, soon swollen and emarginate; thallus, stipes, and apothecia containing baeomycic acid . . . . 3.*B*. roseus

1. Apothecia brownish, concave to flat and marginate at first, later swollen and with reflexed margin; thallus, stipes, and apothecia containing stictic or norstictic acid.
  2. Thallus margin distinctly foliose-squamulose, the marginal squamules usually several mm long, the center also squamulose; containing stictic acid . . . . 2. *B. placophyllus*
  2. Thallus partly sorediate, partly of small squamules less than 1 mm long, the squamules sometimes raised and attached by the margins.
    3. Thallus K+ yellow, containing stictic acid, squamules small . . . . 4. *B. rufus*
    3. Thallus K+ yellow becoming red, containing norstictic acid; squamules to 2 mm broad . .
    . . . . 1. *B. carneus*

1. Baeomyces carneus (Retz.) Flörke
Deutschl. Lich. 8:16. 1821. *Lichen ericetorum* a. *carneum* Retz., Flor. Scand. Prodr. 224. 1779.

Thallus squamulose or partly areolate-squamulose, the squamules 0.2–2 mm long, 0.2–0.7 mm broad, flat or partly convex, ascending or adnate, more or less sorediate with labriform soralia, whitish or ashy, the margin crenulate, underside whitish, occasionally rhizinose.

Apothecia to 2 mm broad, solitary or clustered, sessile or subsessile to long-stipitate, the stipes to 5 mm tall and furrowed; disk flat to more commonly convex, the edges sometimes reflexed, reddish brown, sometimes slightly pruinose, the pale margin disappearing; cortex of thallus variable, partly of vertical paraplectenchyma, partly of hyphae fanning out from between the algal glomerules and becoming interwoven, and partly forming a vertical decomposed layer, the hyphae leptodermatous, 1.7–2.2 μ in diameter; stipes solid, the hyphae pachydermatous, 4 μ in diameter, algae present to below the apothecium; epithecium brownish; exciple poorly developed, thin, pale; hypothecium to 100 μ thick, brownish; hymenium hyaline or brownish, 70–100 μ; paraphyses lax, slender, 1–3 μ thick, slightly branched at the tips, the tips scarcely thickened; asci cylindrical, 60–75 × 5–8 μ; spores 8, uniserate or the apical ones biserate, spindle shaped, simple or 1-septate, hyaline, 3–11 × 2–2.75 μ.

Reactions: K+ yellow turning red, P+ yellow to orange.

Contents: norstictic acid.

This species grows on soils high in clays, especially on frost boils in the Arctic. It is not only in the coniferous forest but also in the tundras further north.

This species is circumpolar and boreal to arctic but is also found in the West Indies, the Philippines, Hawaii, and New Zealand.

2. Baeomyces placophyllus Ach.
Meth. Lich. 323. 1803.

Thallus forming rosettes, the center squamulose, the margins lobate foliose, the marginal lobes close to the substratum and raised only at the tips; main lobes to 5 mm broad, gray-green, when dry whitish or ochraceous, somewhat pruinose; the margins with pale whitish isidia or papillae 0.2 mm wide, broadening into squamules; podetia large, to 6 mm tall and 2 mm broad, deeply fissured, naked or covered with squamules.

Apothecia to 4 mm broad, one to a few per podetium; disk soon swollen, the margin often reflexed, light ochraceous brown to brownish red; cortex 10–20 μ thick, the hyphae more or less parallel to the surface, the algal layer glomerulate and the hyphae between the glomerules rising and fanning over the surface, the hyphae leptodermatous 2.5–3 μ diameter; hyphae of the podetium gelatinous, dense, 2.5–3 μ in diameter with numerous elongate interstices containing algae; hymenium 90–150 μ, I−; paraphyses slender, simple or branched toward the tips; asci cylindrical, 90–130 × 6–8 μ; spores 8, fusiform, 1-celled, 8–14 × 2–4 μ. Conidia unknown.

Reactions: K+ yellow, P+ orange.

Contents: stictic acid.

This species grows on sand, clay soil, and raw humus. It is circumpolar in the boreal forest but has also been collected in Java. In North America it ranges

Baeomyces carneus (Retz.) Flörke

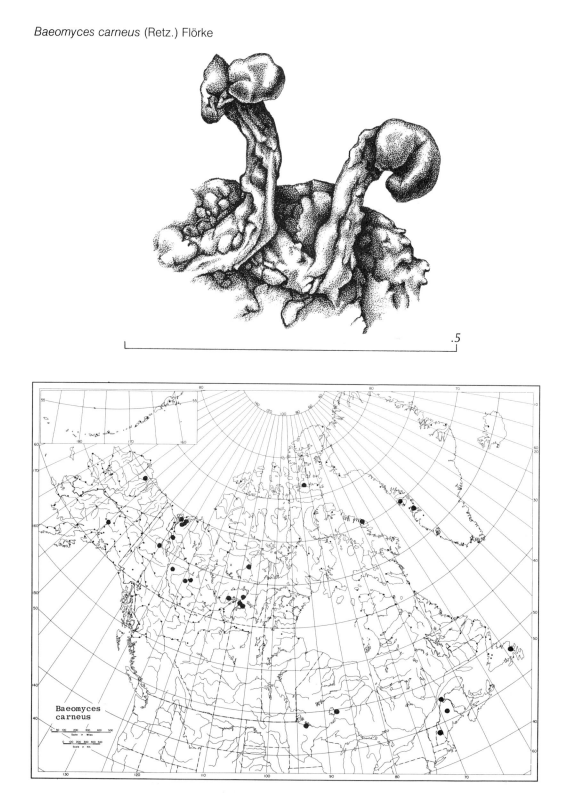

.5

Baeomyces placophyllus Ach.

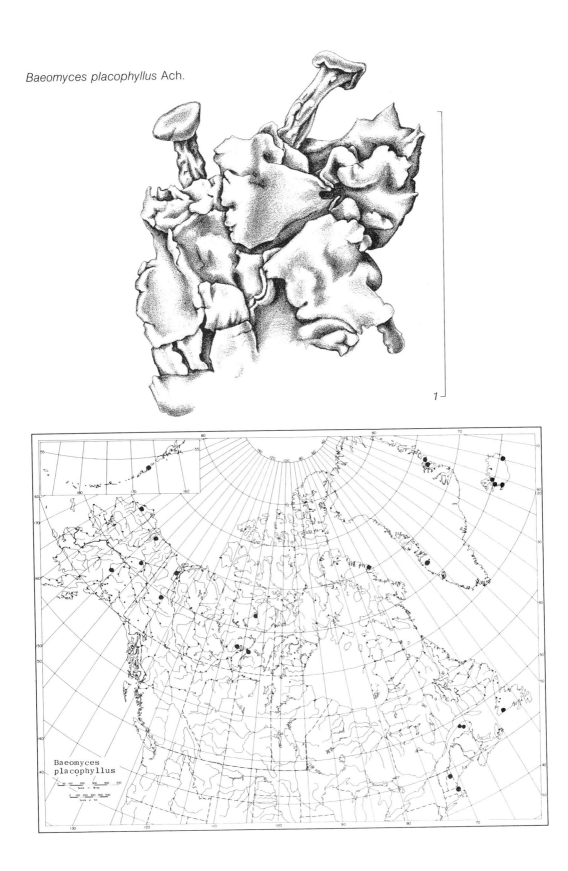

1

Baeomyces
placophyllus

from Newfoundland to Alaska and south to New Hampshire and Massachusetts.

3. Baeomyces roseus Pers.
Usteri, Neue Ann. der Bot. 1:19. 1794.

Thallus gray, white, or roseate tinged, crustose, smooth or powdery, covered with spherical or flattened verruculose warts with a narrowed base, in fertile plants the warts 0.1–0.3 mm broad, in sterile ones to 1 mm; podetia 2–6 mm tall, 1 mm thick, smooth or fissured, whitish.

Apothecia pink, 1–4 mm broad, spherical, emarginate or with a very narrow margin; cortex of the thallus 50 μ thick, of vertically oriented hyphae with thin cell walls; algal layer with the hyphae also more or less vertically oriented; medulla very loose, in parts forming a paraplectenchyma; cortex of the warts composed of interwoven hyphae parallel to the surface; interior of the podetium of strands of thin-walled hyphae 2.5–3 μ in diameter, upper part fistulose, the hyphae under the hymenium thicker, to 5 μ diameter, vertically oriented and not sharply separated from the subhymenium; subhymenium ca. 50 μ thick, yellowish; hymenium 100 μ, I + pale blue (color soon disappearing); paraphyses 2–2.55 μ, simple, little thickened above; asci cylindrical, 85 × 6 μ, thin walled; spores fusiform, mainly 1-celled, occasionally indistinctly 2-celled, 12–26 × 2.5–3 μ.

Reactions: thallus and podetia K + yellow, PD + yellow.

Contents: baeomycic acid, squamatic acid, and unidentified substance (Dey 1977).

This species grows on soils with much clay content, in the Arctic occurring on frost boils. It is circumpolar temperate in distribution. In North America it is common east of the Mississippi River in the eastern states south to Alabama and Georgia. In Canada it is in the eastern provinces to Newfoundland and is rare across the Northwest Territories to Alaska.

4. Baeomyces rufus (Huds.) Rebent.
Prodr. Flor. Neomarch. 315. 1804. *Lichen rufus* Huds., Flora Anglica 443. 1762.

Thallus crustose, powdery sorediate, greenish or whitish gray; partly squamulose, the squamules to 1 mm broad, somewhat raised, sometimes imbricate, with a warty to smooth cortex; podetia short, rarely to 6 mm tall, the lower part or to just below the apothecium covered with an algal zone, the stalk flattened or cylindrical, almost always fissured.

Apothecia concave or flat, becoming slightly convex, the disk red-brown, seldom reddish or pink; cortex lacking in the sorediate part of the thallus, of three types in the flat or squamulose part: 1) mainly in more or less vertical paraplectenchyma 2–3 cells thick; 2) of hyphae spreading out from between the algae and becoming more or less flat-interwoven; 3) forming a more or less vertical but decomposed layer, macroscopically whitish and pruinose. Inner layer of podetia of parallel thick-walled hyphae and containing many algal cells; hypothecium distinct, at edges gradually merging into a palisade plectenchyma; hymenium 90–105 μ, I−; paraphyses slender, 1–1.8 μ, the tips scarcely thickened; asci cylindrical, 75–100 × 7–9 μ; spores 8, often indistinctly 2-celled, ellipsoid, 8–13 × 2.5–4.5 μ. Conidia ellipsoid, 4–5 × 1 u.

Reactions: K + yellow, P + orange.

Contents: stictic acid, plus traces of other substances (Dey 1977).

This species grows on soil and also over rocks in very moist places, especially seepages. It is especially characteristic of the boreal coniferous forests and is circumpolar. Its northern border of distribution coincides roughly with that of the coniferous forest, including the outlier along the Thelon River in the Northwest Territories. The southern limits reach California in the west, New Mexico, and in the Appalachian Mountains in the east.

Baeomyces roseus Pers.

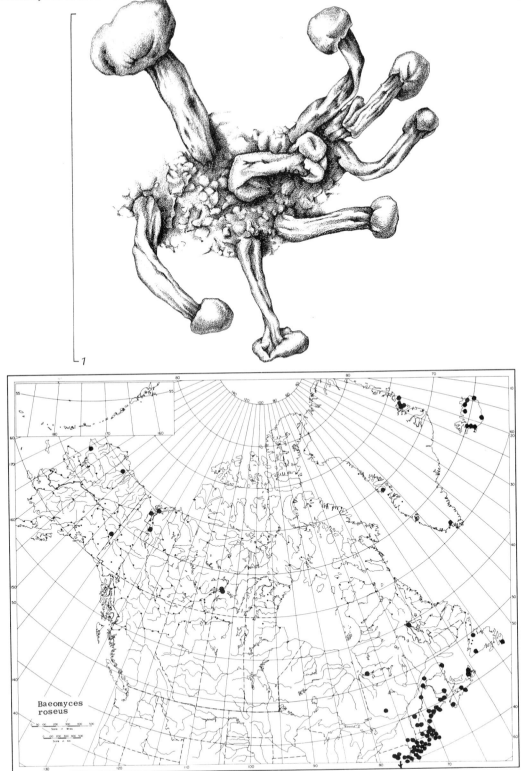

1

Baeomyces
roseus

Baeomyces rufus (Huds.) Rebent.

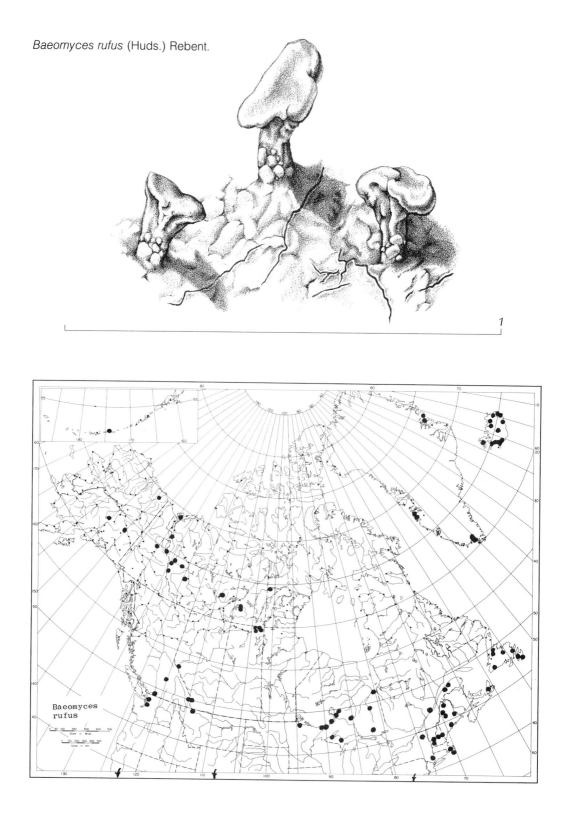

1

Baeomyces
rufus

BRYORIA Brodo & Hawksw.

Opera Botanica 42:78. 1977. *Setaria* Michx., Flora Bor. Amer. 2:331. 1803 (name rejected by conservation of same name in grasses). *Bryopogon* Th. Fr., Nova Acta Reg. Soc. Sci. Upsala ser. 3, 3:25. 1860. Alectoria subgen. *Bryopogon* Th. Fr., Lich. Scand. 1:23. 1871. *Alectoria* sect. *Hyalosporae* Hue, Nouv. Arch. Mus. Paris ser. 4, 1:86. 1899. *Alectoria* sect. *Bryopogon* (Th.Fr.) Zahlbr., Cat. Lich. Univ. 6:375. 1930.

Thallus fruticose, erect, decumbent, or usually pendent; branching terete to slightly compressed, sometimes angular, or foveolate, or twisted isotomic or anisotomic dichotomous; greenish gray or brown to black, some species with lateral spinules with constricted bases, isidia lacking but some species with isidia-like spinules borne in soralia; soralia present or absent, pseudocyphellae present or absent, fusiform, to tuberculate; of longitudinally oriented hyphae, the outer cortex of periclinal hyphae with little matrix; hyphae of medulla not ornamented.

Apothecia lateral to appearing geniculate, rare to unknown in some speices; margin thalloid, concolorous with the thallus, ciliate in a few species; disk reddish brown to dark brown, sometimes yellow-pruinose; epithecium brown or yellowish, hypothecium hyaline or brownish; paraphyses little branched, coherent; asci clavate, thick-walled; spores 8, ellipsoid, lacking epispore, hyaline, simple. Pycnidia rare.

Type species: *Bryoria trichodes* (Michx.) Brodo and Hawksw.

REFERENCES
Ahti, T. & D. L. Hawksworth. 1974. Notes on the lichens of Newfoundland. 4. *Alectoria*. Ann. Bot. Fenn. 11:189–196.
Brodo, I. M. & V. Alstrup. 1981. The lichen *Bryoria subdivergens* (Dahl) Brodo & D. Hawksw. in Greenland and North America. Bryologist 84:229–235.
Brodo, I. M. & D. L. Hawksworth. 1977. *Alectoria* and allied genera in North America. Opera Botanica 42:1–164.
DuRietz, G. E. 1926. Vorarbeiten zu einer "Synopsis Lichenum." 1. Die Gattungen *Alectoria, Oropogon* und *Cornicularia*. Ark. f. Bot. 20A: 11:1–43.
Hawksworth, D. L. 1972. Regional studies in *Alectoria* (Lichenes). 2. The British species. Lichenologist 5:181–261.
Howe, R. Heber, Jr. 1911. American species of *Alectoria* occurring north of the fifteenth parallel. Mycologia 3:106–150.
Motyka, J. 1964. The North American species of *Alectoria*. Bryologist 67:1–44.

In this genus the color tests should be carried out on filter paper as per instructions in Brodo & Hawksworth (1977).

Key to the Species in the Arctic
1. Thallus K + red, P + yellow, containing norstictic acid; dark brown to black, shining, pseudocyphellae fissural, elongate to 1 mm . . . . 8. *B. pseudofuscescens*
1. Thallus K − or K + yellow, lacking norstictic acid, pseudocyphellae short.
  2. Soralia present.
    3. Thallus persistently K + bright yellow, KC + red, P + orange-yellow, containing barbatolic, alectorialic and fumarprotocetraric acids; basal branches blackish, apical grayish or brownish, lateral branches at right angles . . . . 6. *B. nadvornikiana*
    3. Thallus not persistently K + yellow, branches uniformly colored, lateral branches usually at acute angles.
      4. Soralia P −, greenish black to white; plants short, more or less erect, tufted; brown to dark brown, shining; medulla and cortex P − . . . . 9. *B. simplicior*
      4. Soralia P + red, containing fumarprotocetraric acid.
        5. On rocks and soil, main branches much larger than the secondary, to 0.5 mm thick, often foveolate or flattened; soralia usually sparse . . . . 2. *B. chalybeiformis*
        5. Epiphytic; main branches only slightly thicker than the secondary, to 0.4 mm thick, nearly terete.
          6. Medulla P −.
            7. Branches spirally twisted and contorted to foveolate, shining, olive-brown to black . . . . 13. *B. vrangiana*

7. Branches not contorted or foveolate, fuscous to dark brown.
   8. Branches even in diameter, not brittle, over 0.2 mm diameter.
      9. Branches shining, olive to olive-brown, angles of main branches obtuse and rounded; soralia fissural .... 4. *B. glabra*
      9. Branches dull, grayish to fuscous but not olivaceous, angles acute to obtuse, soralia tuberculate or fissural .... 3. *B. fuscescens*
   8. Branches uneven in diameter.
      10. Pseudocyphellae conspicuous; branches cervine brown; medulla usually P+ but may be P− .... 12. *B. trichodes*
      10. Pseudocyphellae lacking, brown-black to olive brown or black, brittle .... 5. *B. lanestris*
  6. Medulla P+ red.
    11. Pseudocyphellae present, thallus brown; branches uneven in diameter, brittle .... 12. *B. trichodes*
    11. Pseudocyphellae lacking; thallus fuscous brown to brown; branches very even in diameter, not brittle .... 3. *B. fuscescens.*
2. Soralia lacking.
  12. Thallus K+ bright yellow, containing barbatolic ± alectorialic acid; thallus pendent on trees; grayish to fuscous brown or rarely darker .... 1. *B. capillaris*
  12. Thallus K−, erect, on ground, dark brown to red-brown or blackish.
    13. Main branches uniformly red-brown, conspicuously longitudinally foveolate and flattened; medulla P−; known only from SW Greenland .... 10. *B. subdivergens*
    13. Main branches not red-brown and foveolate; medulla P+ red.
      14. Thallus uniformly black or brown; branching angles acute, tertiary branches common, branches coarse, 0.5–0.8 mm diameter; common and abundant species on ground... 7. *B. nitidula*
      14. Thallus black at base, pale brown or brown toward the apices; secondary branches at right angles to the primary, tertiary branches rare and at acute angles, branches slender, 0.3–0.5 mm diameter; a rare species usually on rock faces and boulders .... 11. *B. tenuis*

1. Bryoria capillaris (Ach.) Brodo & Hawksw.

Opera Botanica 42:115. 1977. *Parmelia jubata* var. *capillaris* Ach., Meth. Lich. 273. 1803. *Alectoria capillaris* (Ach.) Cromb., J. Bot., London 9:177. 1871. A further very large series of synonyms, none of which seem to have been used in arctic papers, can be sought in Hawksworth (1972).

Thallus pendent, to 10–15 (–30) cm long; the branching isotomic to anisotomic dichotomous, the angles usually acute, the branches terete, slender, less than 0.5 mm diameter, twisted or compressed especially the larger main branches, greenish gray to gray, sometimes brown to dark brown, dull; lateral spinules lacking, pseudocyphellae present, fusiform, white very short; soredia absent.

Apothecia rare in North America, reported by Brodo and Hawksworth to be lateral, to 1.5 mm broad, the disks convex, the margin soon disappearing, disk brown, epruinose; spores 8, hyaline, simple, ellipsoid to subglobose, 5.3–6.8 × 4 − 4.5 μ.

Reactions: cortex, medulla, and soralia K+ bright yellow, C+ pink or C−, KC+ red, P+ yellow.

Contents: barbatolic acid ± alectoralic acid.

This is a common species on conifers in the boreal forest, especially in the more humid regions. Although more characteristic of the southern part of the boreal forest it has been collected in the Arctic in the spruce forest at the edge of the tundras at Artillery Lake in central Canada.

2. **Bryoria chalybeiformis** (L.) Brodo & Hawksw.

Opera Botanica 42:81. 1977. *Lichen chalybeiformis* L., Sp. Pl. 2:1153. 1753. *Lichen jubatus* var. *chalybeiformis* (L.) Web., Spicil. Fl. Gött. 229. 1778. *Usnea chalybeiformis* (L.) Baumg., Fl. Lips. 591. 1790. *Parmelia jubata* var. *chalybeiformis* (L.) Ach., Lich. Univ. 593. 1810. *Cornicularia jubata* var *chalybeiformis* (L.) DC., in Lam. & DC., Fl. Fr. ed. 3, 2:332. 1805. *Alectoria jubata* var. *chalybeiformis* (L.) Ach., Lich. Univ. 593. 1810. *Alectoria chalybeiformis* (L.) Gray, Nat. arr. Brit. Pl. 1:408. 1821. *Evernia jubata* var. *chalybeiformis* (L.) Fr., Lich. Eur. Ref.: 20. 1831. *Bryopogon chalybeiformis* (L.) Link., Grund. Kräut. 3:164. 1833. *Bryopogon jubatus* var. *chalybeiformis* (L.) Rabenh., Deutsch. Krypt.-Fl.2: 120. 1845. *Evernia bicolor* var. *chalybeiformis* (L.) Fr., Summ. Veg. Scand.: 103. 1846. *Alectoria jubata* f. *chalybeiformis* (L.) Lindsay, Proc. Linn. Soc. Bot. 9:375. 1867. *Alectoria jubata* ssp. *chalybeiformis* (L.) Th. Fr., Lich. Scand. 1:25. 1871. *Alectoria prolixa* var. *chalybeiformis* (L.) Stiz., Annals Naturhist. Hofmus. Wien 7:129. 1892. *Setaria jubata* var. *chalybeiformis* (L.) Samp., Lich. Port. Exs. 3, no. 269. 1923.

Thallus fruticose, prostrate on rocks or partly hanging, to 10–15 cm long; very dark, olive-black or gray-black, not uniformly brown, dull rarely shining, irregularly submonopodial or dichotomous; the angles of branching broad, narrower in hanging plants, main branches to 0.5 mm thick, irregularly curved, usually irregularly deformed, slightly flattened or cylindrical, longitudinally wrinkled or slightly sulcate; soredia irregularly distributed over the branches, becoming tuberculate then emptying and appearing more fissure-like; medulla white.

Apothecia unknown.

Reactions: cortex and medulla K −, C −, KC −, P − or exceptionally the medulla P + red in parts; soredia P + red.

Contents: fumarprotocetraric acid in the soralia.

This species grows on earth, particularly at the edges of old frost boils, over and among tundra vegetation and over the edges of rocks as well as on rock faces. It is occasionally on wood and the lower branches of trees. It is a circumpolar and bipolar arctic-alpine species widely distributed north of the tree line in the Arctic and appearing to be absent in the

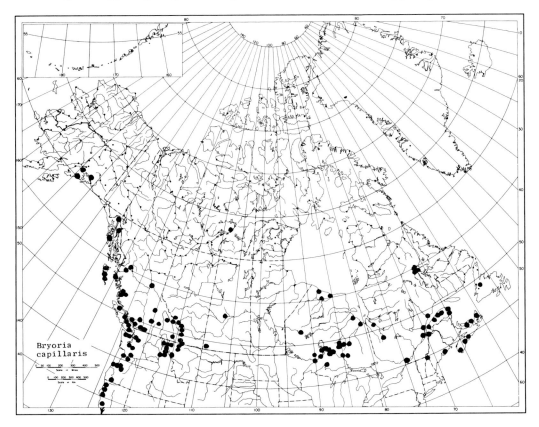

Bryoria capillaris

Bryoria chalybeiformis (L.) Brodo & Hawksw.

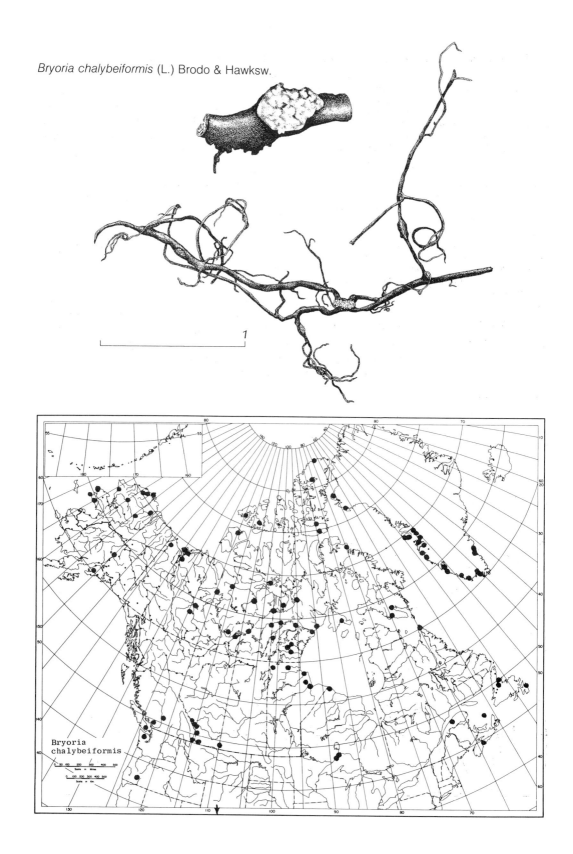

1

Bryoria
chalybeiformis

boreal forest but reappearing south of the boreal forest in Newfoundland in exposed places, in Nova Scotia and the Gaspé Peninsula, and the north shore of Lake Superior.

3. Bryoria fuscescens (Gyeln.) Brodo & Hawksw.

Opera Botanica 42:83. 1977. *Alectoria fuscescens* Gyeln., Nyt. Mag. Naturvid. 70:55. 1932. *Alectoria jubata* f. *sorediata* Harm., Bull. Soc. Sci. Nancy, ser. 2, 31:205. 1897. *Alectoria jubata* var. *sorediata* (Harm.) Oliv., Mem. Acad. Cienc. Art. Barcelona ser. 3, 16:446. 1921. *Alectoria jubata* f. *sorediifera* Anders, Str. Laubflecht. Mitteleur. 181. 1928. *Alectoria fuscescens* var. *albosorediosa* Gyeln., Nyt. Mag. Naturvid. 70:56. 1932. *Bryopogon fuscescens* var. *albosorediosus* (Gyeln.) Gyeln., Acta Geobot. Hungar. 2:164. 1937. *Bryopogon fuscescens* f. *breviusculus* Gyeln. ibid. *Bryopogon fuscescens* (Gyeln.) Gyeln., ibid. p. 137. *Bryopogon pacificus* Gyeln., Acta Geobot. Hung. 2:166. 1939. *Alectoria fuscescens* f. *opaca* Mot. ex Bystr., Annals Univ. Mariae Curie-Sklodowska, sect. C, 18:415. 1963.

Thallus pendent or prostrate, 5–15 (–45) cm; branching isotomic or anisotomic at the base, becoming anisotomic toward the apices, branches terete to twisted and foveolate, even or uneven in diameter, 0.3–0.4 (–0.6) mm diameter, the angles acute to obtuse, lateral spinulose branches sometimes present; pale fuscous to brown or blackish, the bases paler than the apices; pseudocyphellae absent, soralia sparse or abundant, fissural or tuberculate, sometimes becoming spinulose.

Apothecia described by Hawksworth (1972) from Scotland, to 1.5 mm diameter, lateral; the margin concolorous with the thallus; disk brown to dark brown; spores 8, hyaline, simple, ellipsoid, $6–9 \times 5 \mu$.

Reactions: cortex and medulla both K−, KC−, C−, P+ red at least in part or medulla P−; soralia P+ red.

Contents: fumarprotocetraric acid with accessory chloratranorin.

This species grows on conifers, *Picea, Abies, Pinus,* and *Betula*. It is a boreal forest species circumpolar in the northern hemisphere and is also in Africa. In North America it reaches the Arctic on the outliers of spruce and other trees at the edge of the

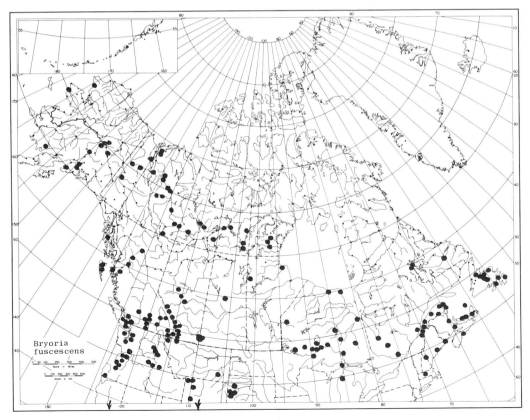

Bryoria
fuscescens

tundras. Hawksworth (1972) gives an analysis of differences between this species and *B. chalybeiformis* pointing out the thinner main branches, the paler color, the paler bases, the more fissural soralia, and the cortex and medulla being more commonly P+ red than in *B. chalybeiformis*.

4. Bryoria glabra (Mot.) Brodo & Hawksw.

Opera Botanica 42:86. 1977. *Alectoria glabra* Mot., Fragm. Florist. Geobot. 6(3):448. 1960.

Thallus fruticose, hanging to 10 cm long; olive-brown, brown, smooth, shining; the angles with the branches obtuse, the main branches dichotomous, slender, to 0.3–0.4 mm diameter; lacking pseudocyphellae or short lateral branches; sorediate, the soralia fissural, bearing such masses of soredia that they become capitate, the soredia then falling and leaving crater-like fissures.

Apothecia unknown.

Reactions: cortex and medulla K −, C −, KC −, P −; the soralia P+ red.

Contents: fumarprotocetraric acid.

This species grows on conifers, particularly *Picea* and *Larix*. It is a western North American endemic boreal and montane species which reaches the arctic in the vicinity of the mouth of the Mackenzie River. It has been reported from Newfoundland in the east by Ahti and Hawksworth (1974).

5. Bryoria lanestris (Ach.) Brodo & Hawksw.

Opera Botanica 42:88. 1977. *Alectoria jubata* var. *lanestris* Ach., Lich. Univ. 593. 1810. *Cornicularia jubata* var. *lanestris* (Ach.) Duby, Bot. Gall. 2:616. 1830. *Alectoria jubata* f. *lanestris* (Ach.) Hue, Nouv. Arch. Mus. Paris, ser. 3, 2:277. 1890. *Alectoria prolixa* var. *chalybeiformis* f. *lanestris* (Ach.) Stiz., Annals Naturhist. Hofmus. Wien 7:129. 1892. *Alectoria jubata* var. *chalybeiformis* f. *tenerrima* Cromb., Monogr. Brit. Lich. 1:212. 1894. *Alectoria jubata* var. *chalybeiformis* f. *lanestris* (Ach.) Harm., Lich. Fr. 3:433. 1907. *Bryopogon lanestris* (Ach.) Gyeln., Feddes Repert 38:227. 1935. *Bryopogon jubatus* f.

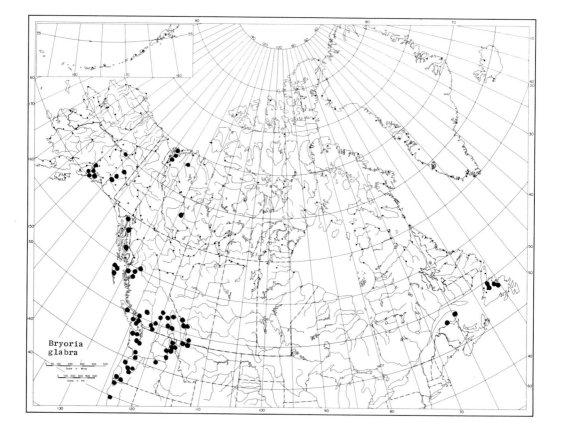

Bryoria
glabra

Bryoria lanestris (Ach.) Brodo & Hawksw.

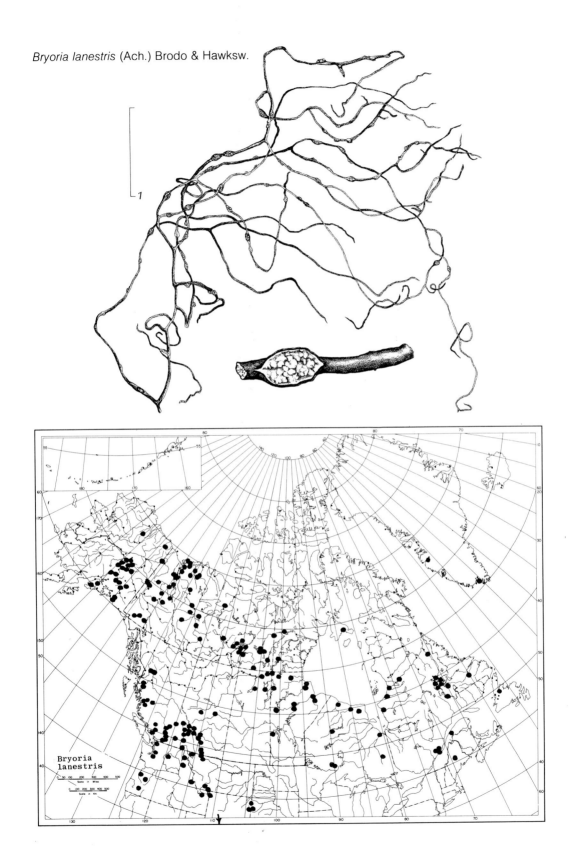

Bryoria
lanestris

lanestris (Ach.) Oksn., Viznachnik Lish. URSR: 276:1937. *Bryopogon negativus* Gyeln., Acta Geobot. Hung. 2:164. 1937. *Bryopogon chalybeiformis* f. *lanestris* (Ach.) Choisy, Bull. Mens. Soc. Linn. Lyon 22:184. 1953. *Alectoria tenerrima* Mot., Bryologist 67:31. 1964. *Bryopogon tenerrimus* (Mot.) Bystr., Ann. Univ. Mariae Curie-Sklodowska C, 26:274. 1971.

Thallus pendent, to 15 cm long; branching irregular, isotomic to anisotomic dichotomous, branching angles acute; branches very uneven in diameter, straight, sometimes flattened toward the base, very brittle and fragmenting in the herbarium packets, to 0.3 mm diameter; black, brown-black or olive blackish, dull, rarely shining; lateral spinules lacking; pseudocyphellae lacking; soralia sparse to abundant, fissural, white, short.

Apothecia unknown.

Reactions: cortex and medulla K −, C −, KC −, P −; soralia P + red.

Contents: fumarprotocetraric acid.

This species grows especially on conifers but also on *Salix* and *Betula* as well as on old wood. It also grows on rocks and soil, especially where its range extends north of the forest. It is a boreal forest species, circumpolar although apparently absent in Asia according to Hawksworth (1972). It is one of the very common species in the boreal forest across North America. Its very brittle branches of very uneven diameter and very dark color aid its differentiation from *B. fuscescens*.

6. Bryoria nadvornikiana (Gyeln.) Brodo & Hawksw.

Opera Botanica 42:122. 1977. *Alectoria nadvornikiana* Gyeln., Acta Fauna Fl. Univ. ser. 2, 1:6 1932. *Bryopogon implexus* var. *nadvornikianus* (Gyeln.) Gyeln., Feddes Repert. 38:242. 1935. *Bryopogon altaicus* Gyeln., Acta Geobot. Hung. 2:166. 1937. *Alectoria altaica* (Gyeln.) Räs., Ann. Bot. Soc. Zool.-Bot. Fenn. Vanamo 12:34. 1939. *Alectoria altaica* var. *spinulosa* Räs., Ann. Bot. Soc. Zool.-Bot. Fenn. Vanamo 12:34. 1939. *Bryopogon nadvornikianus* (Gyeln.) Gyeln., Ann. Mus. Nat. Hung., Bot. 32:154. 1939. *Alectoria karelica* Räs., Ann. Bot. Soc. Zool.-Bot. Fenn. Vanamo 12(1)34:1939. *Alectoria implexa* var. *nadvornikiana* (Gyeln.) Zahlbr., Cat. Lich. Univ. 10:555. 1940. *Alectoria curta* f. *pallidior* Ostm. in Hasselr., Ark. Bot. 30A(13):3. 1943. *Alectoria nadvornikiana* var. *extensa* Mot., Fragm. Florist. Geobot. 4:236. 1958. *Alectoria nadvornikiana* var. *eciliata* Mot., ibid. 237. 1958. *Bryopogon eciliatus* (Mot.) Bystr., Ann. Univ. Mariae Curie-Sklodowska C, 26:271. 1971.

Thallus caespitose to pendent, to 9 cm. long; branching anisotomic dichotomous, the angles toward the base broad, the angles toward the apices acute; branches terete, even, straight, sometimes slightly twisted; the basal branches dark brown to black, the apical ones gray-green to pale or olive brown, paler than the base, with lateral spinulose branches; soralia abundant or sparse, tuberculate or fissural, white; pseudocyphellae short, varying in abundance.

Apothecia not seen, a single one noted by Ahlner in Norwegian material was not dissected by him.

Reactions: Cortex, medulla and soralia K + yellow, C − or C + pink, KC + red, P + deep orange to reddish.

Contents: barbatolic acid, small amounts alectorialic acid, fumarprotocetraric acid, traces of atranorin, chloratranorin.

This species may grow on trees as *Betula, Picea, Abies* as well as on vertical rock faces and boulders. It is in the boreal forest in Europe and North America, in the latter ranging from Newfoundland to southern Alaska and British Columbia and is at the southern edge of the tundra only at tree line in central Canada. Here it may extend beyond the trees on rocky outcrops, moist and partly shaded.

7. Bryoria nitidula (Th.Fr.) Brodo & Hawksw.

Opera Botanica 42:107. 1977. *Bryopogon jubatum* var. *nitidulum* Th. Fr., Nova. acta Reg. Soc. Upsal. ser. 3, 3:25. 1860. *Alectoria jubata* var. *nitidula* (Th. Fr.) Linds., Trans. Bot. Soc. Edinb. 10:58. 1870. *Alectoria nitidula* (Th. Fr.) Vain., Medd. Soc. Fauna Fl. Fenn. 6:116. 1881. *Bryopogon nitudulum* (Th. Fr.) Elenk. & Savicz, Trav. Mus. Bot. Acad. Imp. St. Petersb. 8:26. 1911. *Bryopogon nitidulum* f. *caespitosa* Savicz, Trud. Student. Nautsch. Fis. Mat. Fak. Univ. Jurjew 3:40. 1911. *Alectoria bicolor* var. *nitidula* (Th. Fr.) DR., Ark. Bot. 20A(11):17. 1926. *Alectoria lanea* f. *caespitosa* (Savicz) Gyeln., Nyt. Mag. Naturv. 70:60. 1932. *Alectoria nitidula* f. *caespitosa* (Savicz) Zahlbr. Cat. Lich. Univ. 6:396. 1930. *Bryopogon bicolor* var. *nitidulum* (Th. Fr.) Gyeln., Feddes Repert. 38:237. 1935. *Bryopogon bicolor* var. *nitidulum* f. *caespitosus* (Savicz) Gyeln., ibid. *Alectoria irvingi* Llano, J. Wash. Acad. Sci. 41: *Bryopogon irvingii* (Llano) Bystr., Ann. Univ. Mariae Curie-Sklodowska C, 26:275. 1971.

Thallus erect to decumbent, forming intricately entangled mats from 2 or 3 to 10 cm

Bryoria nadvornikiana (Gyel.) Brodo & Hawksw.

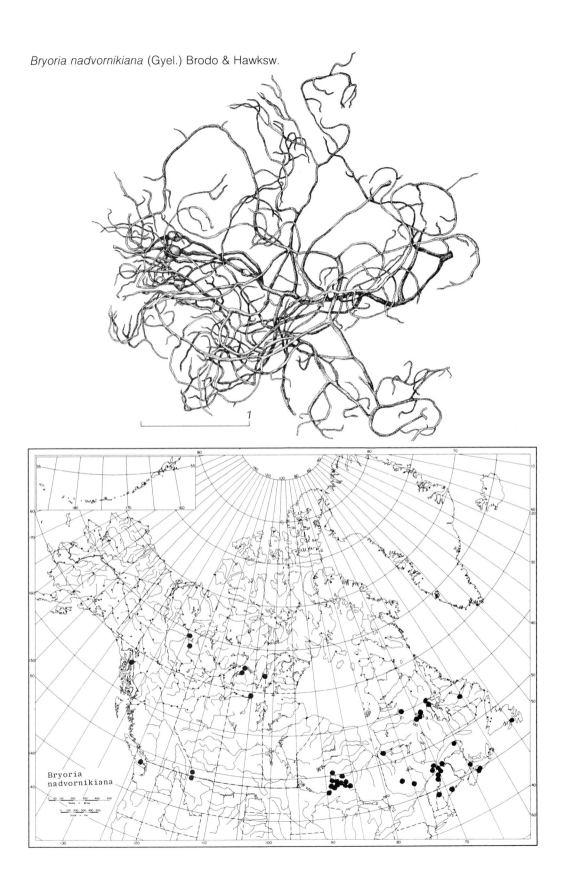

Bryoria nitidula (Th. Fr.) Brodo & Hawksw.

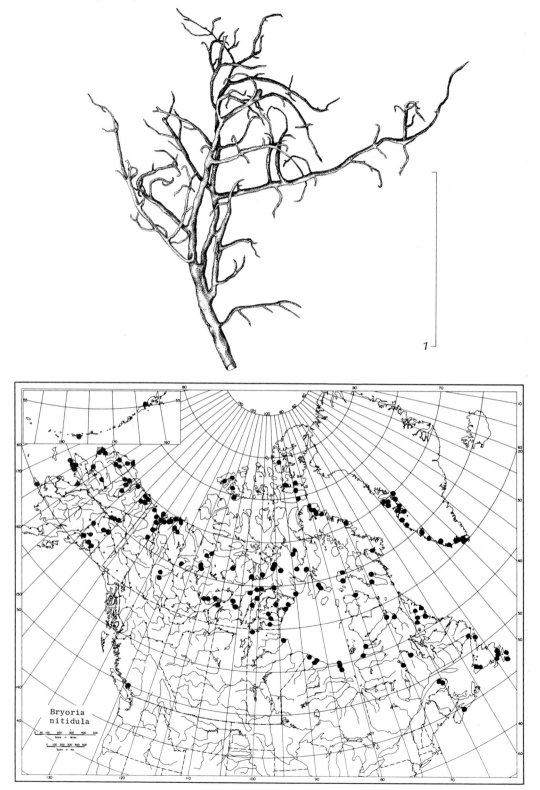

tall; anisotomically dichotomous in branching, the angles wide, the tips tapering gradually, cylindrical, occasionally slightly flattened, the branches 0.5–0.7 mm diameter, the main branches 1–2 mm at most, occasionally deformed in part, the branches sometimes forming loops; dark brown or blackish at the base, the branches olive-brown to dark-brown, shining, with conspicuous brown pseudocyphellae which are flat to slightly raised; soredia lacking.

Apothecia not known.

Reactions: Cortex and medulla K −, C −, KC −, the medulla P + strongly red.

Contents: fumarprotocetraric acid.

This species grows commonly on gravel and heath tundras in well-drained situations, on both humus and bare soils, sometimes on boulders along with *B. chalybeiformis*. It is often intermixed thoroughly with *Cornicularia divergens* and *Alectoria ochroleuca*. It is a circumpolar arctic-alpine species, common in the far north but also ranging into the forest in open areas as far south as Nova Scotia and British Columbia.

8. **Bryoria pseudofuscescens** (Gyeln.) Brodo & Hawksw.

Opera Botanica 42:127. 1977. *Alectoria pseudofuscescens* Gyeln., Ann. Mus. Nat. Hung., Bot. 28:283. 1934. *Alectoria norstictica* Mot., Bryologist 67:33. 1964. *Alectoria subtilis* Mot. Bryologist 67:32. 1964. *Bryopogon norsticticus* (Mot.) Bystr., Ann. Univ. Mariae Curie-Sklodwska 26:274. 1971. *Bryopogon subtilis* (Mot.) Bystr. ibid. Additional synonyms in Hawksworth 1972; 252.

Thallus pendent, to 15 cm long; branching isotomic dichotomous but becoming anisotomic toward the apices at times, angles of the main dichotomies varying, rarely with short perpendicular lateral branches; branches even or uneven in diameter, to 0.3 mm diameter, pale brown to grayish, darkening to dark brown or black, dull or shiny; soralia lacking, pseudocyphellae abundant, white, short to elongate fusi-

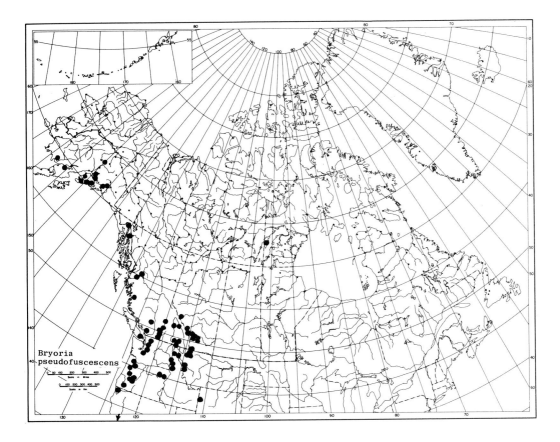

Bryoria
pseudofuscescens

form, to 1 mm long, sometimes spiraling around the branches.

Apothecia rare, to 0.5 mm broad, margin becoming excluded; disk reddish brown, concave to convex; hymenium 36–40 μ; spores 8, hyaline, simple, broadly ellipsoid to subspherical, 5.4–7.5 × 4–4.5 μ.

Reactions: cortex and medulla K + yellow turning red (rarely seen), C –, P + yellow.

Contents: norstictic acid ± traces of connorstictic acid.

This species occurs on conifers. It is in Europe and also is fairly common in the western U.S. and Canada to Alaska. An outlier of this species is in the Dubawnt Lake area in central northern Canada, unaccountably far from the main North American range.

9. Bryoria simplicior (Vain.) Brodo & Hawksw.

Opera Botanica 42:109. 1977. *Alectoria nidulifera* f. *simplicior* Vain., Medd. Soc. Fauna Fl. Fenn. 6:115. 1881. *Bryopogon nitidulum* f. *caespitosus* Savicz, Trudy Student. Nautsch. Fis. Mat. Fak. Univ. Jurjew 3:40. 1911. *Alectoria simplicior* (Vain.) Lynge, Norske Vid.-Akad. Oslo, Mat.-nat. Kl. 1921(7):212. 1921. *Alectoria nitidula* f. *caespitosa* (Savicz) Zahlbr., Cat. Lich. Univ. 6:396. 1930. *Alectoria lanea* f. *caespitosa* (Savicz) Gyeln., Nyt. Mag. Naturvid. 70:60. 1932. *Alectoria simplicior* var. *lapponica* Gyeln., Nyt. Mag. Naturvid. 70:62. 1932. *Bryopogon simplicior* (Vain.) Gyeln., Feddes Repert. 38:233. 1935. *Bryopogon simplicior* f. *lapponicus* (Gyeln.) Gyeln., Feddes Repert. 38:234. 1935. *Bryopogon bicolor* var. *nitidulum* f. *caespitosus* (Savicz) Gyeln., Feddes Repert. 38:237. 1935. *Bryopogon simplicior* f. *albidosorediosus* Gyeln., Acta Geobot. Hung. 2:166. 1937. *Alectoria simplicior* f. *subintricans* Vain., Ann. Univ. Turku., ser. A, 7(1):8. 1940. *Alectoria simplicior* f. *typica* Vain., ibid. *Alectoria nana* Mot., Bryologist 67:16. 1964. *Bryopogon nanus* (Mot.) Bystr., Ann. Univ. Mariae Curie-Skolodowska C., 26:371. 1971.

Thallus short fruticose, to 5 cm long, forming small, rather dense tufts to slightly hanging, sometimes attached to the substrate at the tips as well as the base; brownish black to black, dull or slightly shining, isotomically dichotomous or nearly so, the angles between the branches acute, making the plants appear brush-shaped, the main branches to 0.3 mm thick, the tips thinner, straight or slightly curved, round or slightly deformed; sorediate, the soralia fissural, sharply delimited, round or oval, young soredial masses convex, the soralia becoming crateriform as the soredia are shed, lacking isidial spinules among the soredia.

Apothecia not known.

Reactions: cortex, medulla and soredia all K –, C –, KC –, P –.

Contents: no lichen substances present.

This species grows on conifers *(Picea, Larix, Pinus)* and on *Betula,* occasionally on old wood, rarely on rocks. It is usually mixed with other species of *Bryoria* and is characteristic of the northern part of the boreal forest across North America and Europe and Asia. In North America it reaches the southern edge of the tundra and is also on isolated conifers north of the main timberline in central Canada and near the mouth of the Mackenzie River where it reaches its northernmost stands. Its occurrence in Greenland as reported by Dahl is on earth and between mosses.

10. Bryoria subdivergens (Dahl) Brodo & Hawksw.

Opera Botanica 42:153. 1977. *Alectoria subdivergens* Dahl, Medd. om Grønl. 150(2):145. 1950.

Thallus erect or decumbent, to 10 cm tall, more or less caespitose; the branching isotomic and anisotomic dichotomous, the main branches to 2 mm diameter, irregular, with long foveolae but no foramina, lacking pseudocyphellae, lacking soredia, brown to black-brown.

Apothecia lateral, to 3 mm broad, emarginate, brown; spores 8, hyaline, simple, subglobose, 6–8 μ.

Reactions: K –, C –, KC –, P –.

Contents: no lichen substances known.

This species is known in the arctic from the type specimen collected by Dahl on the ground among mosses at Quagssimiut, the Julianehaab, Holsteinsborg, and Sukkertoppen districts in southwestern Greenland and specimens reported from near Godthab by Alstrup (1979). Other reports, including those by Dahl and by the author have proved to be errors based on the close similarity to *Cornicularia divergens.* From that species it is to be distinguished by the C – instead of C + red reaction of the medulla and by the jigsaw puzzle-like cells of the cortical hyphae, the lack of pseudocyphellae, and the more dull appearance. Brodo and Alstrup (1981) have shown that this species, like *Leptochidium albociliatum* has a disjunct western North America (Montana) and Greenland distribution pattern.

Bryoria simplicior (Vain.) Brodo & Hawksw.

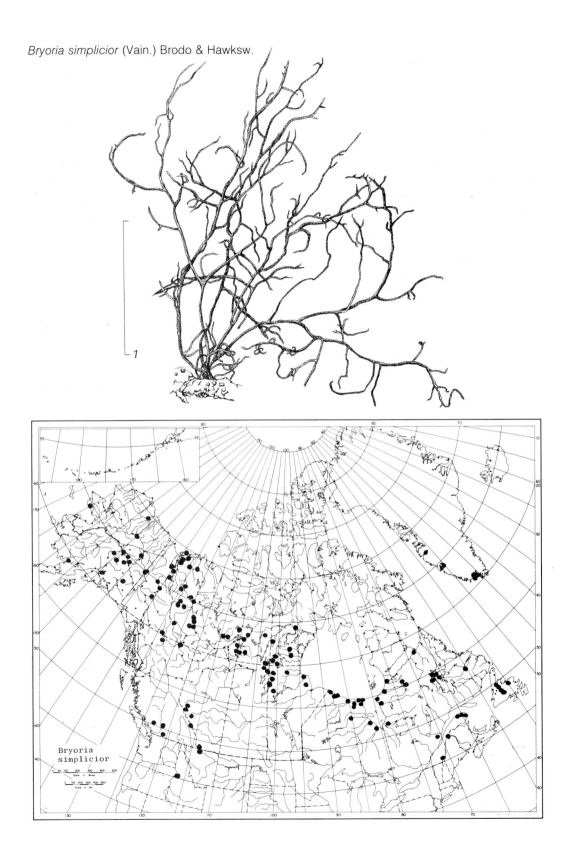

11. Bryoria tenuis (Dahl) Brodo & Hawksw.
Opera Botanica 42:112. 1977. *Alectoria tenuis* Dahl, Meddl.
om Grønl. 150(2):144. 1950. *Alectoria bicolor* var *subbicolor*
Mot., Bryologist 67:9. 1964. *Bryopogon tenuis* (Dahl) Bystr.,
Ann. Univ. Mariae Curie-Sklodowska C., 26:271. 1971.

Thallus erect to decumbent or subpendent, 4–6 cm long; the branches isotomic dichotomous toward the base, becoming anisotomic dichotomous toward the apices, the angles between the branches usually acute; branches terete, even in diameter, sometimes slightly compressed toward the base, to 0.4 mm diameter, the base becoming black, the apical branches paler than the base, pale brown to brown, some lateral spinules at right angles to the main stems; pseudocyphellae sparse to abundant, fissural, usually dark; soralia lacking.

Apothecia known only from one collection from Alaska; lateral, margin concolorous with the thallus, becoming excluded, concave to becoming convex, disk yellowish brown to reddish brown, to 1.5 mm broad; spores 8, hyaline, simple, subglobose, $7–9.5 \times 5–7 \mu$.

Reactions: inner cortex and medulla K−, C−, KC−, P+ red at least in part.
Contents: fumarprotocetraric acid.

Growing on mosses on trees and on rocks, also in rock crevices, this species is reported as being oceanic in Europe, Greenland, and North America where it ranges south into the Appalachian Mountains, in Alaska and British Columbia, and inland in the District of Keewatin. It is a poorly known, seldom collected species.

12. Bryoria trichodes (Michx.) Brodo & Hawksw.
Opera Botanica 42:92. 1977. *Setaria trichodes* Michx., Fl.
Boreal.-Amer. 2:331. 1803. *Alectoria americana* Mot., Fragm.
Florist. & Geobot. 6(3):449. 1960. *Alectoria canadensis* Mot.,
Bryologist 67:35. 1964.

Thallus pendent, to 11 cm long; the branches even or uneven in diameter, brittle, is-

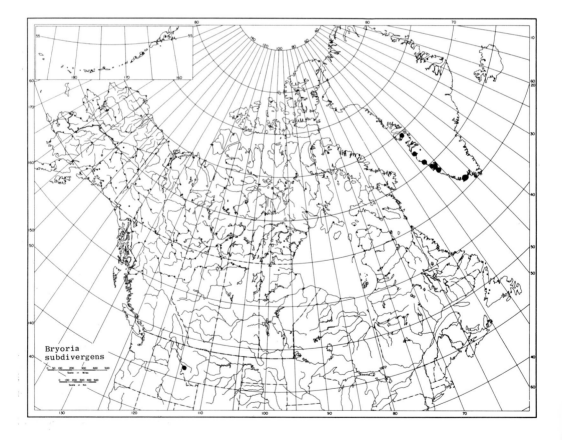

Bryoria
subdivergens

otomically dichotomous, pseudocyphellate, fuscus to cervine brown; soredia, when present, fissural, white to dark, or soredia absent or sparse.

Apothecia rare, lateral, disk to 1.5 mm broad, flat to convex, light brown, dull, spores 8, simple, hyaline, ellipsoid $7 \times 5 \ \mu$.

This species grows on conifers and on old wood. It is in the boreal forest of North America and Japan. It reaches the Arctic in the timber stands at the edge of the forest in the Great Slave Lake area.

13. **Bryoria vrangiana** (Gyeln.) Brodo & Hawksw.

Opera Botanica 42:97. 1977. *Alectoria jubata* var. *stricta* Ach., Lich. Univ. 592. 1810. *Alectoria vrangiana* Gyeln., Magy. Bot. Lapok. 31:46. 1932. *Alectoria ostrobotniae* var. *taboronsis* Gyeln., Acta Fauna Fl. Univ. Ser. 2, 1:7. 1932. *Bryopogon canus* f. *taboronsis* (Gyeln.) Gyeln., Feddes Repert. 38:241. 1935. *Bryopogon vrangianus* (Gyeln.) Gyeln., Feddes Repert. 38:232. 1935. *Bryopogon pacificus* Gyeln., Acta Geobot. Hung. 2:4. 1937. *Bryopogon taboronsis* (Gyeln.) Gyeln., Annals Mus. Nat. Hung. 32:156. 1939. *Alectoria implexa* f. *taboronsis* (Gyeln.) Zahlbr., Cat. Lich. Univ. 10:555. 1940. *Bry-*

opogon jubatus var. *vrangianus* (Gyeln.) Dombr., Nov. Sist Niz. Rast. 7:286. 1970.

Thallus pendent, to 12 cm long; the branching isotomic dichotomous at the base but anisotomic dichotomous toward the apices, angles of the branches mainly acute, branches becoming spirally twisted and contorted to foveolate, the main branches to 1 mm diameter, often with irregular spinules which are not constricted at the base; thallus dark olive-brown to black, dull; pseudocyphellae lacking; soralia usually sparse, mainly tuberculate, some fissural, white, to 1 mm long.

Apothecia unknown.

Reactions: cortex and medulla K −, C −, KC −, P −; soralia P + red.

Contents: fumarprotocetraric acid.

This species grows on conifers and on wood. It appears to be in montane forest in Europe and in North America but with strangely disjunct stations reaching the edge of the Arctic in the Dubawnt Lake area and the west coast of James Bay.

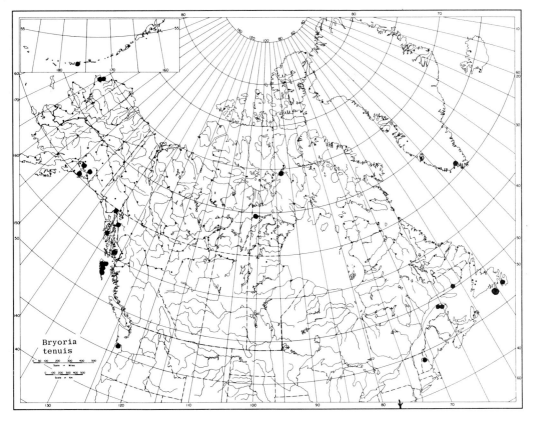

Bryoria
tenuis

Bryoria trichodes (Michx.) Brodo & Hawksw.

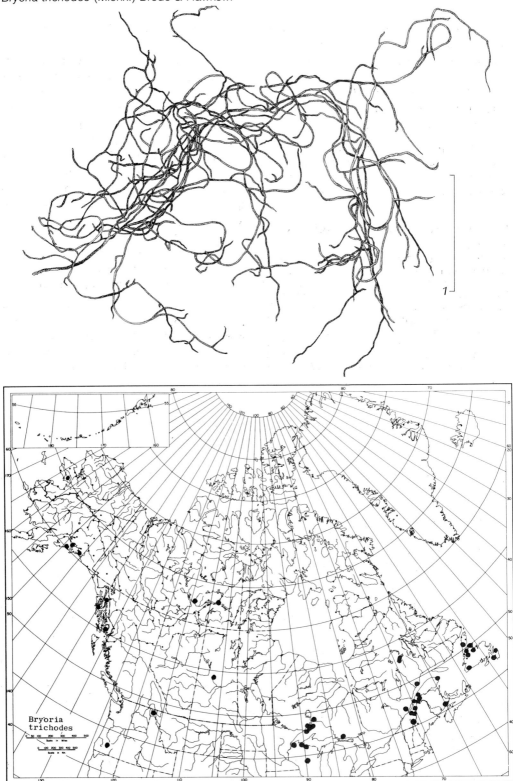

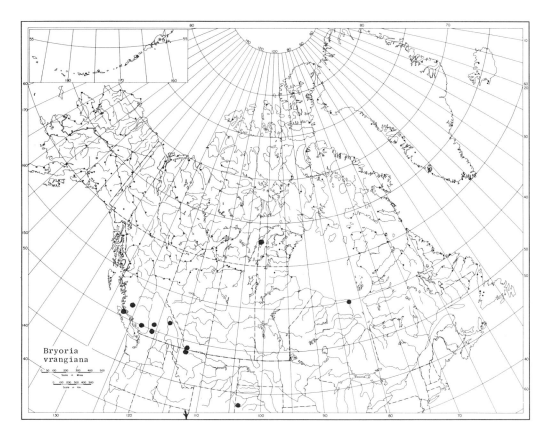

Bryoria
vrangiana

CETRARIA Ach.

Meth. Lich. 292. 1803.

Thallus foliose to fruticose and erect, more or less dorsiventral, attached to the substratum by the base or by scattered rhizinae; upper and lower sides corticate with prosoplectenchymatous or paraplectenchymatous cortex several cells thick; marginal cilia present in some species along the margins or on lamellae in some species.

Apothecia on the margins, sometimes appearing terminal, usually oriented toward the upper side of the thallus; margin thalloid; hypothecium hyaline; paraphyses unbranched to slightly branched, tips slightly thickened; spores 8, hyaline, simple, ellipsoid or spherical, thin walled. Pycnidia marginal and immersed in papillae or spinules, brown or black; fulcra endobasidial; conidia straight, ellipsoid, sometimes constricted in the middle.

Algae: *Chlorococcus*.

Type species: *Cetraria islandica* (L.) Ach.

REFERENCES

Bird, C. D. & A. H. Marsh. 1973. Phytogeography and ecology of the lichen family Parmeliaceae in southwestern Alberta. Can. J. Bot. 51:261–288.

Culberson, W. L. & C. F. Culberson. 1967. A new taxonomy for the *Cetraria ciliaris* group. Bryologist 70:158–166.

Karnefelt, I. 1977. Three new species of brown fruticose *Cetraria*. Bot. Not. 130:125–129.

—— 1979. The brown fruticose species of *Cetraria*. Opera Botanica 46:1–146.

Kristinsson, H. Chemical and morphological variation in the *Cetraria islandica* complex in Iceland. Bryologist 72:344–357. 1969.

Krog, H. 1973. *Cetraria inermis* (Nyl.) Krog, a new lichen species in the amphi-Beringian flora element. Bryologist 76:299–300.

Rassadina, K. A. 1948. On systematics and geography of the genus *Cetraria* in USSR. Botanicheskii Zhurnal 33:13–24.
Rassadina, K. A. & A. N. Oxner. 1960. Ad genus *Cetraria* ex USSR novitates. Bot. Materials Acad Nauk USSF 13:5–14.
Savicz, V. P. 1923. De lichene *Cetraria richardsonii* Hook. notula. Not. Syst. Inst. Crypt. Horti Bot. Petrop. 2:189–191.
Weber, W. A. & S. Shushan. 1955. The lichen flora of Colorado: *Cetraria, Cornicularia Dactylina* and *Thamnolia*. Univ. of Colo. Studies Biol. Ser. 3:115–134.

1. Thallus erect, fruticose, not strongly dorsiventral.
   2. Thallus yellow, containing usnic acid.
      3. Thallus straw yellow, leathery, not brittle, medulla white; on acid soils.
         4. Lobes cucullate, smooth above, base turning red when dying; containing usnic and pro-tolichesterinic acids . . . . 4. *C. cucullata*
         4. Lobes flattened, more reticulate above, base turning yellow when dying; containing usnic acid . . . . 16. *C. nivalis*
      3. Thallus golden yellow, brittle or leathery, medulla yellow; on twigs or on calcareous soils.
         5. Thallus brittle, esorediate, on calcareous soils; containing usnic, rangiformic and pinas-tric acids . . . . 19. *C. tilesii*
         5. Thallus leathery, not brittle, margins sorediate, on twigs; containing usnic, vulpinic and pinastric acids . . . . 17. *C. pinastri*
   2. Thallus brown, not containing usnic acid.
      6. Medulla UV + white; thallus tough, horny in texture, dark brown, becoming green when moistened, clearly dorsiventral but curled up when dry, underside with large pruinose patches, medulla KC + pink or red; containing alectoronic acid . . . . *Masonhalea richardsonii*
      6. Medulla UV −, thallus thinner, not horny in texture, dark brown, not becoming green when moistened, not clearly dorsiventral, underside not pruinose.
         7. Medulla I + blue, thallus pale to dark brown, base reddish or blackish.
            8. Thallus margins with cilia but no pycnidia, pseudocyphellae poorly developed; containing rangiformic and norrangiformic acids . . . . 15. *C. nigricascens*
            8. Thallus not ciliate, margins with pycnidia on spinules, pseudocyphellae present.
               9. Thallus dark brown; medulla K −.
                  10. Thallus P + red (fumarprotocetraric acid ± rangiformic and protolichesterinic acids), lobes usually broad, pseudocyphellae broad, both marginal and laminal . . . . 11. *C. islandica*
                  10. Thallus P − (fumarprotocetraric acid absent, protolichesterinic acid present), lobes narrow, cucullate; pseudocyphellae mainly marginal, narrow . . . . 6. *C. ericetorum*
               9. Thallus pale brown with white mottling; medulla K + yellow or red, P + red (fu-marprotocetraric acid) . . . . 13. *C. laevigata*
         7. Medulla I −.
            11. Medulla C + rose, containing gyrophoric and hiascinic acids,
               12. Lobes flattish, pale brown with yellowish base, dull, the branching bi-morphic with broad lower lobes and much tufting of the upper flat and markedly smaller branches; the tips acute, laminal and marginal pseudocy-phellae usually distinct . . . . 5. *C. delisei*
               12. Lobes subtubular and cucullate, not markedly bimorphic; the tips obtuse; laminal and marginal pseudocyphellae usually poorly developed. . . . . 7. *C. fastigiata*
            11. Medulla C −, lacking gyrophoric and hiascinic acids; lobes not bimorphic ex-cept sometimes in *C. andrejevii*.
               13. Thallus dorsiventral although not markedly so, paler below: margin not pseudocyphellate; epithecium K + violet; containing rangiformic and nor-rangiformic acids . . . . 15. *C. nigricascens*
               13. Thallus erect, cucullate (channeled above); epithecium K −; marginally pseudocyphellate.

14. Color dark, dark brown or black-brown, shining, growing on ground.
    15. Thallus tips broadly rounded, not tufted; branching sympodial, occasionally bimorphic and with the lower parts broad, the upper narrow; containing rangiformic and norrangiformic acids . . . .
    . . . . 1. *C. andrejevii*
    15. Thallus tips narrow; branching dichotomous, spinules rare; no laminal pseudocyphellae, those on margins few, narrow, inconspicuous; containing protolichesterinic acid and rangiformic acid. . . .
    . . . . 12. *C. kamczatica*
14. Color pale brown and olive-brown mottled, dull; thallus very irregular; containing protolichesterinic acid; mainly corticolous on subalpine shrubs . . . . 10. *C. inermis*
1. Thallus foliose, spreading, obviously dorsiventral.
    16. Thallus bright yellow, sorediate, on twigs and bark; containing usnic, vulpinic and pinastric acids . . . . 17. *C. pinastri*
    16. Thallus gray, brown, olive-brown, or black.
        17. Lobes sorediate on the edges, greenish brown, C −; containing protolichesterinic and rangiformic acids . . . . 2. *C. chlorophylla*
        17. Lobes lacking soredia.
            18. On bark or twigs, thallus brown, lobes short and broad.
                19. Medulla KC+ red, UV +, containing alectoronic acid (GE test) and sometimes α-collatolic acid, thallus gray-brown. . . . . 8. *C. halei*
                19. Medulla KC −, UV −, lacking alectoronic acid.
                    20. Pale brown, slightly taller and broader; epithecium K+ violet; containing rangiformic and norrangiformic acids . . . . 15. *C. nigricascens*
                    20. Dark brown, epithecium K −; containing protolichesterinic acid . . . . . .
                    . . . . 18. *C. sepincola*
        18. On soil or rocks; thallus deep brown to black.
            21. Blowing loose over the tundra, tough and horny in texture, dark brown when dry, greenish when moist; underside with large pruinose patches; medulla KC+ red or pink; containing alectoronic acid . . . . *Masonhalea richardsonii*
            21. Attached to soil or rocks, softer in texture, not markedly greenish when moist; not pruinose below; medulla KC −, lacking alectoronic acid but with other constituents.
                22. On soil; lobes marginally ciliate and spinulose, medulla I+ blue: containing protolichesterinic acid. . . . . 14. *C. nigricans*
                22. On rocks; lobes not marginally ciliate.
                    23. Conidia with the center narrow, the ends swollen; underside of thallus dark; medulla K+ yellow; containing stictic acid . . . 9. *C. hepatizon*
                    23. Conidia cylindrical, the ends not swollen; underside of thallus pale; medulla K −; containing α-colatollic acid . . . . 3. *C. commixta*

1. Cetraria andrejevii Oxn.

J. Bot. Acad. Sci. Ukraine 1(3–4):44. 1940.

Thallus fruticose, forming small mats and tufts, erect, intertwined and fusing in the groups, irregularly branched, the branch tips markedly enlarged and rounded, dilating upwards, the margins partly inrolled and the thalli therefore partly canaliculate; both sides deep shining brown to chestnut brown, the dying parts paler yellowish brown; with weak pseudocyphellae; the margins lacking isidia or cilia, the small spinules with pycnidia very scant.

Upper cortex prosoplectenchymatous,

Cetraria andrejevii Oxn.

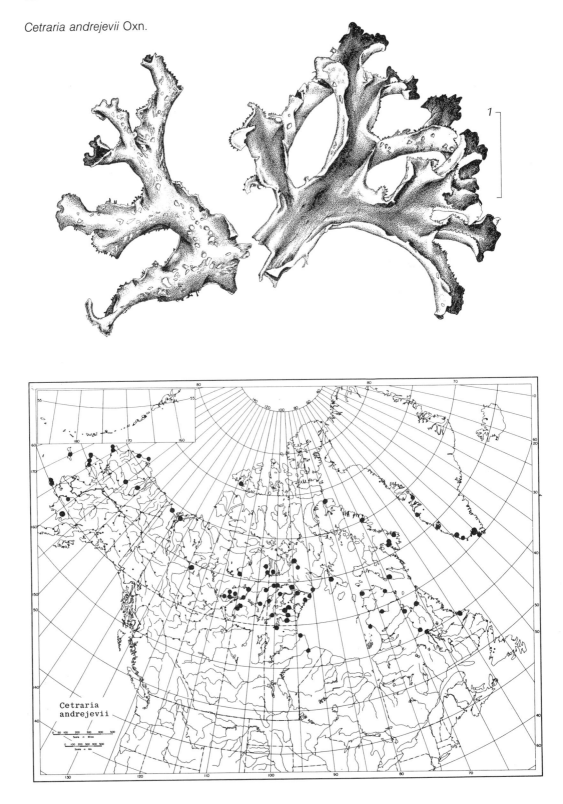

25–40 μ, the outer 7–10 μ brown and the hyphae oriented parallel with the surface; medulla fairly dense, 80–125 μ, the algae in scattered glomerules; lower cortex like the upper.

Apothecia rare, at the tips of lobes, the reverse wrinkled; epithecium yellowish brown; hypothecium pale, asci clavate; spores 8, hyaline, simple, ellipsoid, 8–10 × 4–5 μ.

Reactions: medulla K –, C –, KC –, I –, P –.

Contents: rangiformic and norrangiformic acids.

Growing on the ground in very wet places, especially with cold seepage as below late snow banks and along stream banks where there is springtime overflow, sometimes in very moist places in bogs. This species was described from Siberia and ranges across Alaska and the Northwest Territories to Greenland. It may possibly be characterized as amphi-Beringian broad ranging. In North America it is distinctly Arctic.

2. Cetraria chlorophylla (Willd.) Vain.

Acta Soc. Fauna Flora Fenn. 13:7. 1896. *Lichen chlorophyllus* Willd. in Humb., Florae Frib. Specim. 20. 1793. *Cetraria scutata* Auct.

Thallus small, the lobes to 3 cm long, 10 mm broad, usually less, the lobes forming rosettes, deeply cut, flat to more or less channeled, irregularly branched; upper side brown or greenish brown, dull to shining, in places slightly sulcate; sorediate along the margins with white to dark gray soredia, few soredia formed on the upper surface, occasionally with small isidia which break out into soredia; underside paler than the upper, pale brown or whitish, lacking pseudocyphellae, smooth or with rhizinae, the rhizinae pale, dispersed or in groups.

Cortex prosoplectenchymatous, 25–30 μ thick, the outer 3–5 μ brownish; medulla loose, the algae scattered in glomerules; lower cortex like the upper.

Apothecia rare, at the tips of lobes, cir-

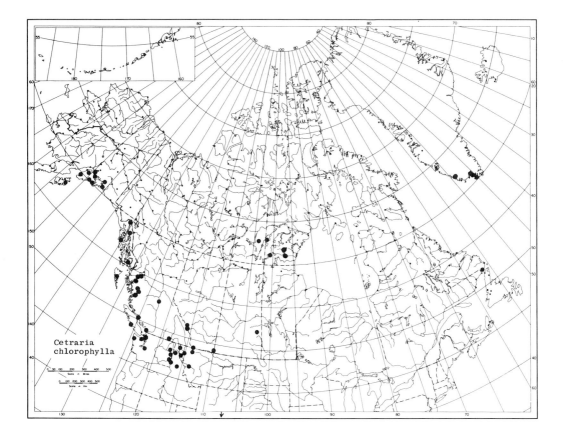

Cetraria
chlorophylla

cular, to 5.5 mm broad; margins sorediate, disk chestnut brown to light brown, flat or reflexed, dull to slightly shining, epruinose; hypothecium hyaline, hymenium hyaline, the upper part yellowish brown, 50 μ thick; asci 37×7–8 μ; paraphyses slender, to 2 μ thick; spores subspherical 8×5 μ.

Reactions: K−, C−, KC−, P−, I−.

Contents: Protolichesterinic and rangiformic acids. Atranorin has been reported several times from European specimens.

Growing on the twigs of spruce, alder, fir, larch, etc., this is a circumpolar boreal and temperate species, in North America fairly common in the west from Alaska to California, rather rare on the interior of the continent and in Labrador and Greenland.

3. Cetraria commixta (Nyl.) Th. Fr.

Lich. Scand. 1:109. 1871. *Lichen fahlunensis* L., Sp. Pl. 1143. 1953. *Platysma commixtum* Nyl., Syn. Lich. 1:310. 1860. *Platysma fahlunense* (L.) Vain., Medd. Soc. Fauna Fl. Fenn. 14:4. 1886 (non Nyl. 1860). *Cetraria fahlunensis* (L.) Vain., Medd. om Grønl. 30:127. 1905 (non (Schaer. 1833).

Thallus foliose, forming rosettes on rock surfaces, the lobes narrow, to 2 mm broad, slightly channeled as the previous species; upper side bronzy brown to black brown, shining or dull; underside pale or becoming brown, with scattered black rhizinae.

Apothecia at the tips of short lobes, more or less central on the thallus, to 7 mm broad; margin thin, inflexed, irregularly thickened, disk of the same color as the thallus, slightly shining, smooth, concave; epithecium brownish, hypothecium pale; hymenium 45–55 μ; spores 8, hyaline, simple, ellipsoid, 8–10×4.5 μ; spores 8, hyaline, simple, ellipsoid, 8–10×4.5 μ; Conidia cylindrical or the center thickest, not the ends, 3–4×1 μ.

Reactions: K−, C−, KC−, P−.

Contents: α-collatolic acid.

Growing on rocks and gravel in open dry places, this is a circumpolar, arctic-alpine species, in North America ranging from Labrador to Alaska, south to New York, Colorado, and Washington.

4. Cetraria cucullata (Bell.) Ach.

Meth. Lich. 293. 1803. *Lichen cucullatus* Bell., Osservaz. Bot. 54. 1788.

Thallus erect, fruticose, to 8 cm tall, scattered or in masses, the lobes to 4 mm broad, cucullate (channeled) in shape, the edges crisped, the tips somewhat reflexed, smooth; yellow or straw colored, sometimes greenish-yellow, the base turning red upon dying; the edges with dark, short-spinulose pycnidia; underside with narrow whitish pseudocyphellae along the edges; both sides corticate.

Cortex prosoplectenchymatous with cells filled with granular pigmented inclusions, 16–75 μ thick; medulla 37–175 μ, white, dense, the algae scattered within, the hyphae to 10 μ thick.

Apothecia infrequent, mainly formed in late snow areas; circular to elliptical, to 8 mm broad; margins concolorous with the thallus, sometimes slightly crenulate, mainly entire; disk reddish brown, pale brownish, or yellowish brown, epruinose, smooth; epithecium pale brownish; hymenium 35–45 μ, hyaline; hypothecium hyaline; spores hyaline, simple, ellipsoid, 7.5–8.5×3–4 μ.

Reactions: K−, C−, KC+ yellow, P−.

Contents: Usnic and protolichesterinic acids.

This species grows in many tundra habitats, scattered in windy sites, forming larger tufts and colonies in sheltered spots, growing well in late snow areas and fruiting there. A circumpolar arctic and boreal species, it ranges south to New England in the east and to New Mexico in the west in the mountains.

5. Cetraria delisei (Bory) Th. Fr.

Kgl. Svensk. Vetens.-Akad. Handl. 7:11. 1867. *Cetraria islandica* var. *delisei* Bory ex Schaer., Enum. Crit. Lich. Europ. 16. 1850. *Cetraria aculeata* var. *hiascens* Fr., Lich. Europ. Ref. 36. 1831. *Cetraria hiascens* (Fr.) Th. Fr., Lich. Scand. 1:99. 1871. *Cetraria hiascens* var. *delisei* (Bory) Vain., Medd. om Grønland 30:126. 1909.

Thallus fruticose, erect, forming tufts or mats in low places; the branches markedly bimorphic with the basal part of broad lobes and the outer tip parts very narrow and densely branched; the branches pale to dark brown on both surfaces, the tip parts usually paler, flattened, pseudocyphellae lacking on the upper side, scattered on the lower side, and lacking spi-

Cetraria commixta (Nyl.) Th. Fr.

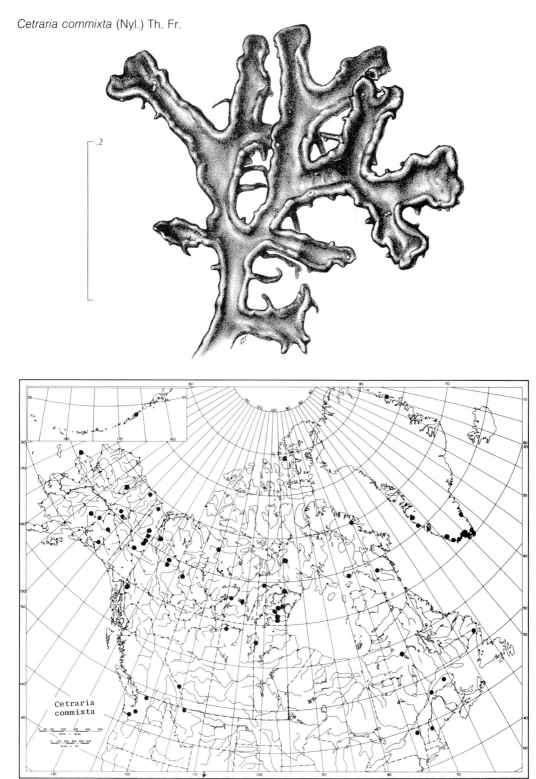

Cetraria
commixta

Cetraria cucullata (Bell.) Ach.

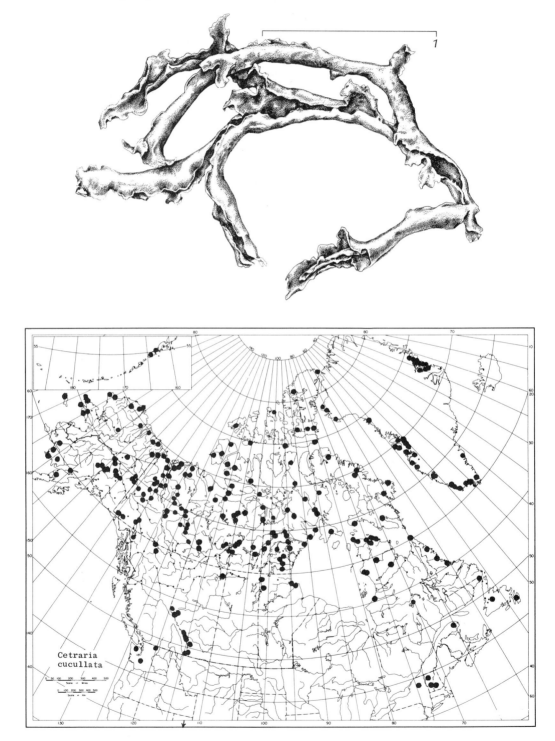

Cetraria delisei (Bory) Th. Fr.

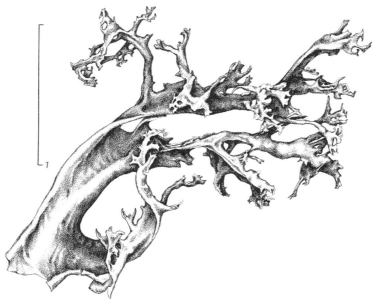

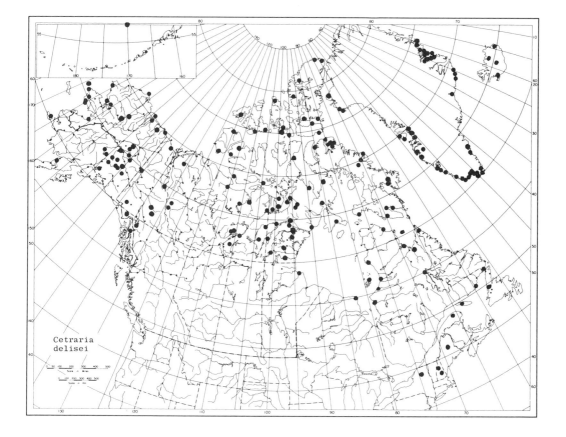

Cetraria
delisei

nules along the margins, the dying base yellowish.

Cortex prosoplectenchymatous, 75–175 μ thick; medulla loose, hyphae very irregular, to 10 μ in diameter; algae scattered.

Apothecia not seen.

Reactions: K −, C+ rose, KC+ red, P −.

Contents: hiascinic and gyrophoric acids.

This species grows in low, wet swales particularly where water flows from late snowbanks or in spring overflow areas along riverbanks. *Cetraria andrejevii* is a regular associate. It is circumpolar high arctic and alpine and occurs in the European Alps. In North America it ranges across Canada and Alaska, south to the White Mountains in the east in New Hampshire and into British Columbia in the west but not into Alberta.

During the summer of 1959 I noted that the Eskimos at the mouth of the Back River, Northwest Territories, were gathering this lichen for use as fuel in cooking. It burned with a resinous flame.

6. Cetraria ericetorum Opiz
Seznam rostlin Kveteny Ceské. 173. 1852.

Thallus fruticose, erect, to 9 cm tall; more or less dicotomously or irregularly branched, the margins inrolled to make the lobes more or less channeled, usually narrower than in *C. islandica,* the tips partly reflexed; the upper side from dark olive brown to brown or reddish brown or blackish brown, smooth, shining; the underside with elongate whitish pseudocyphellae along the margin, rarely over the midsurface which is of the same color as the upper surface or slightly paler; the margins with projecting thorn-like processes containing the pycnidia and also with occasional cilia.

Cortex about 25 μ thick, the outer 5–8 μ brown pigmented, prosoplectenchymatous; medulla dense, the algae distributed through it, 50–65 μ thick, the hyphae 3–5 μ in diameter.

Apothecia rare, at the tips of enlarged lobes, the margin inflexed, the disk red-brown, smooth; epithecium brown; hymenium 40–60 μ; hypothecium pale; asci clavate; spores 8, hyaline, simple, ellipsoid, 7.5–10 × 3.5–5 μ.

Reactions: K −, C −, KC −, P −.

Contents: protolichesterinic acid.

This species grows on soils and humus, especially among mosses, in dry or moist habitats, in Iceland shown by Kristinsson to be more continental in habitat preference than *C. islandica* (P+) type thalli. It is a circumpolar boreal and arctic-alpine species with variations. See the discussion under *C. islandica*.

7. Cetraria fastigiata (Del. ex Nyl. in Norrl.) Kärnef.
Bot. Not. 130:128. 1977. *Cetraria delisei* ssp. *fastigiata* (*C. fastigiata*) Del. ex Nyl. in Norrl., Not. Sällsk. Fauna Fl. Fenn. Förh. 13:323. 1873. *Cetraria hiascens* var. *fastigiata* (Del. ex Nyl. in Norrl.) Vain., Medd. Soc. Fauna Fl. Fenn. 6:119. 1881.

Thallus fruticose, dichotomously branched, the lobes 2–5 (−10) mm broad, canaliculate or subtubular, with short lateral branches, the tips obtuse; underside dark brown, gray-brown, or yellow brown, smooth and pitted, usually glossy; upper side of same color, more pitted, dull or glossy; marginal pseudocyphellae very narrow and scarcely discernible, laminal pseudocyphellae sparse and weak; marginal projections to 1 mm long, sparse or absent, rarely on lobe tips, the tips obtuse.

Apothecia marginal on lobe ends, disk to 10 mm broad, brown, margin undulate; spores 8, hyaline, ellipsoidal, 2.4–3.6 × 6–8.4 μ. Pycnidia on marginal projections, conidia cylindrical, 0.5 × 5 μ.

Reactions: medulla I −, K −, C+ weak reddish to red, KC −, P −.

Contents: gyrophoric and hiascinic acids.

This species grows in open moist sites in depressions in the tundra and bogs. It is apparently circumpolar and in North America occurs in Newfoundland and trans-Canada to Alaska. The old records from ''Newfoundland'' by Despreaux were unspecified as to locality and could conceivably have come from the coast of Labrador.

8. Cetraria halei W. L. & C. F. Culb.
Bryologist 70:161. 1967. *Nephroma americana* Spreng., Kongl. Vetensk. Akad. Handl. 1820:49. 1820. *Peltigera americana* (Spreng.) Spreng., Syst. Veg. 4(1):306. 1827, non *Cetraria americana* (Gyeln.) Sato in Nakai & Honda, Nov. Fl. Jap., Parmeliales 1:26. 1939.

Cetraria ericetorum Opiz

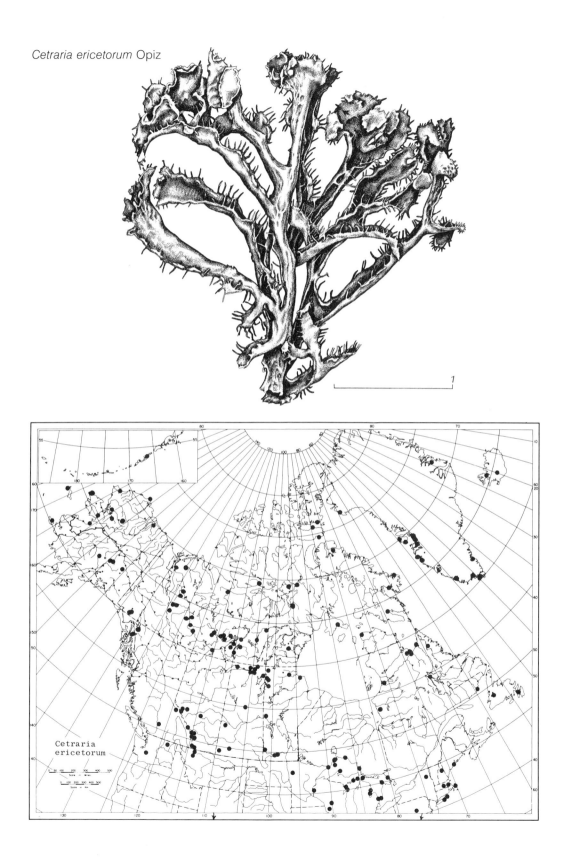

1

Cetraria
ericetorum

Thallus small, foliose, lobes to 1–3 mm broad, forming rosettes, flat; the upper surface more or less wrinkled, the margins with short cilia and also pycnidia; dull or shining, greenish brown to brown or olive-brown above, the underside white to brown with scattered rhizinae.

Upper cortex 12–50 μ thick, algal layer very uneven, the medulla very loose; lower cortex like the upper.

Apothecia common, marginal, pedicellate, to 3 mm broad, the margin concolorous with the thallus, inflexed, irregular, the disk brown, dull, epruinose; hypothecium hyaline, upper part of the hymenium and the epithecium brown, hymenium 50 μ; asci subcylindric; spores 8, simple, hyaline, nearly globose, 5–7 μ.

Reactions: medulla K + yellow or K –, C –, KC + red, P –. UV +.

Contents: alectoronic acid, with accessory atranorin and/or α-collatolic acid.

This species grows on conifers and old conifer wood as well as *Populus, Betula* and *Alnus*. In North America it is common in the eastern deciduous forest but ranges westward into British Columbia and Alaska. It reaches the Arctic along the northern edge of the tree line and on isolated tree stands. This species is also reported from Finland. The other chemical variants of the *C. ciliaris* species complex, as *C. ciliaris* with C + red reaction (containing olivetoric acid) and *C. orbata* with C –, KC – reaction (containing protolichesterinic acid) do not reach so far north.

9. Cetraria hepatizon (Ach.) Vain.

Termeszetr. Füzetek 22:278. 1899. *Lichen hepatizon* Ach., Lich. Suec. Prodrom. 110. 1798. *Cetraria fahlunensis* Schaer., Lich. Helv. Spicil. 5:255. 1833. *Parmelia fahlunensis* var. *vulgaris* Schaer., Enum. Crit. Lich. 48. 1850. *Platysma fahlunense* Nyl., Syn. Lich. 1:309. 1860.

Thallus foliose, loosely attached to rocks or gravel by rhizines, forming large rosettes of irregularly branched lobes; upper side brown to blackish brown, shining, the margins thickened and forming a channeled upper side, with embedded dark pycnidia (in the similar appear-

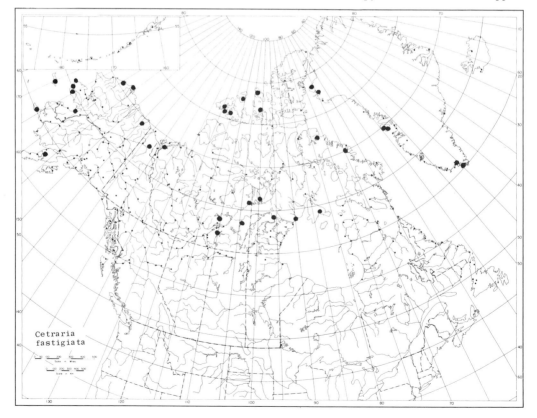

Cetraria
fastigiata

Cetraria halei W. L. & C. F. Culb.

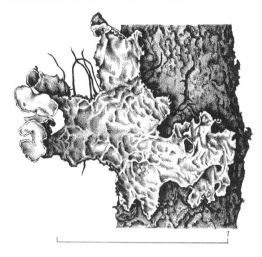

ing *Parmelia stygia* the pycnidia are distributed over the surface); underside dark brown to black, rhizinate; medulla white.

Apothecia at the tips of lobes but appearing toward the center of the thallus; to 5 mm broad; margin entire to much crenulate, brown; disk dark brown, almost the same color as the thallus, shining or dull; epithecium brown; hypothecium yellowsih; hymenium 45–50 μ; spores 8, hyaline, simple, ellipsoid, 6.5–10×5–7 μ; conidia with the ends thickened, 4–6×1μ.

Reactions: medulla K+ yellowish becoming reddish, C−, KC−, P+ red-orange.

Contents: stictic acid; my previous report of rangiformic acid in this species is incorrect.

This species is commonly growing on exposed dry rocks and gravel eskers in arctic and boreal regions. It is circumpolar and in North America ranges southward to New York in the east, to the Lake Superior region in the center, and to Arizona in the western mountains.

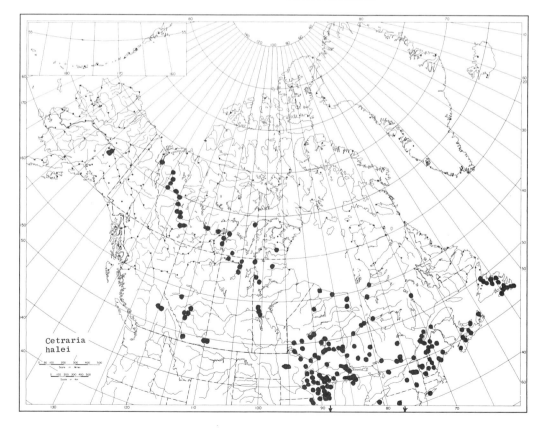

Cetraria
halei

Cetraria hepatizon (Ach.) Vain.

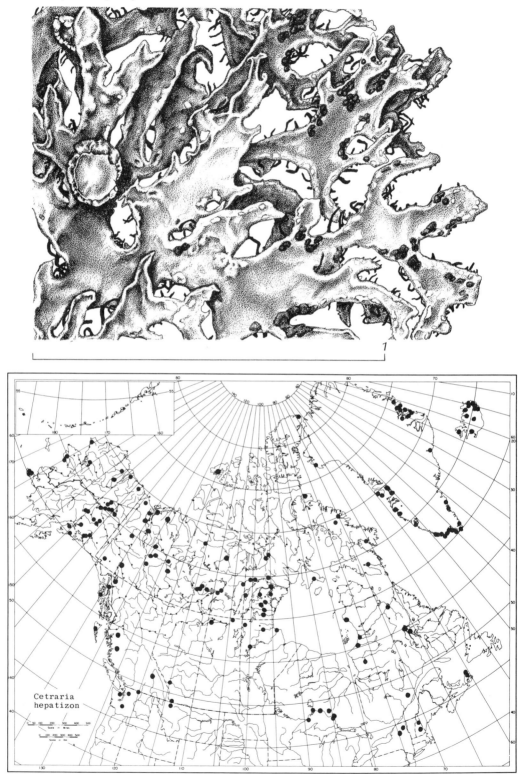

10. Cetraria inermis (Nyl.) Krog
Bryologist 76:229. 1973. *Cetraria crispa* f. *inermis* Nyl., Bull.
Soc. Linn. Normandie Ser. 4. 1:214. 1887. *Cetraria tenuifolia*
var. *inermis* (Nyl.) Räs., Kuopion Luonnon Yst. Yhdist. julk. B.
2:45. 1952.

Thallus fruticose, weakly dichotomously branched, lobes flat or slightly canaliculate, to 3 mm broad; underside pale brown to
white, smooth and glossy; upper side dark brown,
olive-green, or yellowish green, smooth or rarely
with transverse ridging; dull, marginal pseudocyphellae very distinct on outer parts of lower
surface; laminal pseudocyphellae absent; marginal projections few and inconspicuous, to 0.5
mm long.

Upper cortex paraplectenchymatous, 30–
40 μ, medulla 70–110 μ, lower cortex paraplectenchymatous, 20–40 μ.

Apothecia lateral marginal, substipitate,
to 4 mm broad, disk brown, margin slightly
crenulate, hymenium 30–60 μ, asci clavate to
cylindrical, spores 8, round to subglobose, 3.5–
6 × 3.5–5 μ.

Reactions: medulla I−, K−, C−,
KC−, P−.

Contents: lichesterinic and protolichesterinic acids.

This species grows on the twigs of *Betula* and
Salix, occasionally on the ground. It is an amphi-
Beringian species with a very restricted range in eastern Siberia and Alaska. The related *C. subalpina* Imsh.
has indistinct marginal pseudocyphellae on the underside, is smaller, and contains protolichesterinic acid
and two unknown substances. It is more southerly in
distribution along the Pacific coast and a Point Barrow report could not be substantiated by Kärnefelt
(1979).

11. Cetraria islandica (L.) Ach.
Meth. Lich. 293. 1803. *Lichen islandicus* L., Sp. Plant. 1145.
1753. See Zahlbr. Cat. Lich. Univ. 6:329 et seq. for synonyms
and described forms and variants.

Thallus fruticose, erect, to 10 cm tall;
more or less dichotomously branched or irregularly branched, the margins inrolled to make

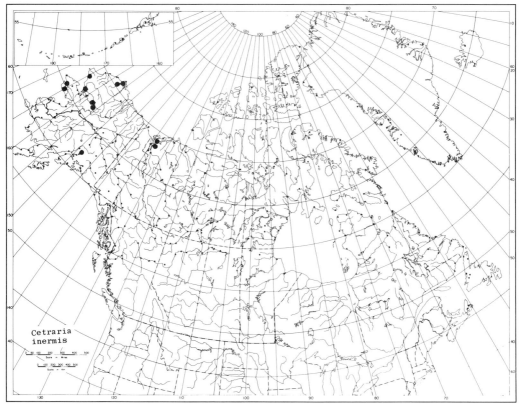

Cetraria
inermis

Cetraria islandica (L.) Ach.

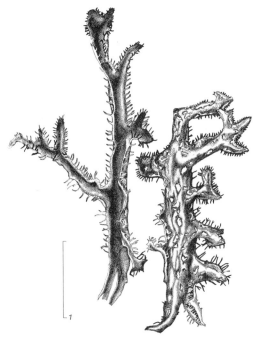

1

the lobes more or less channeled, the tips partly reflexed; upper side olive-brown to dark brown or blackish brown, smooth; the underside paler than the upper side, with white pseudocyphellae scattered over the surface and more elongate ones common along the edges; the margins with cilia and also with abundant pycnidia on projecting thorn-like processes; the dying base often turns red; medulla white.

Cortex prosoplectenchymatous, 25–30 μ thick, the outer 5–8 μ brown pigmented; the medulla white, dense, the algae glomerulate and scattered throughout, to 75 μ thick, the hyphae 3.5 μ in diameter; lower cortex like the upper.

Apothecia rare, on broadened thallus tips, to 18 mm broad; margin poorly developed, partly inflexed, the outer reverse side roughened; disk light or dark reddish brown, dull, epruinose; epithecium yellowish brown; hypothecium yellowish; hymenium pale yellowish; spores 8, hyaline, ellipsoid to ovoid, simple, 8–9.5 × 3–5 μ.

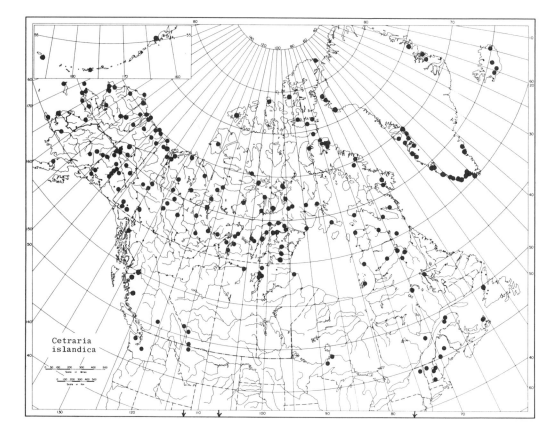

Cetraria
islandica

Reactions: K −, C −, KC −, P + red.

Contents: fumarprotocetraric acid plus accessory protolichesterinic and rangiformic acids.

Growing on moist or dry tundras among mosses or in the open, occasionally on old wood or twigs of spruce in the part protected by snow in winter, this is a circumpolar arctic and boreal lichen with many variations of size, branching, and shape. In North America it ranges south to New England in the east and to New Mexico in the west.

The tremendous variability of the *Cetraria islandica*–*C. ericetorum* group gives great difficulties in the treatment of the arctic material. Kristinsson (1969) in studying the islandic material suggested that hybridization is the source of some of the difficulties. Unlike the situation in Europe where the morphological characters and the presence or absence of fumarprotocetraric acid correlated well to treat the populations as belonging to two distinct species, *C. islandica* and *C. ericetorum,* he found confusing crossing of the characters with the gamut of the morphological characters within each of the PD + and PD − strains. A similar confusing situation exists with respect to the North American material. Kärnefelt (1979) has also attempted to bring order out of the chaos within this group. He has very interesting results. Like Kristinsson he found that the morphological and chemical characters criss-cross and in his solution he deemed the presence or absence of the fumarprotocetraric acid was not as important as the morphology, although the chemistry is used in his keys, the species appear under each of the PD reactions. PD + material with continuous marginal pseudocyphellae he segregated as *C. laevigata;* specimens with both marginal and laminal pseudocyphellae and a smooth surface he classed as *C. islandica* ssp. *islandica;* specimens with the lobes usually lacking marginal pseudocyphellae and the lobes with the surface usually ridged or pitted and the lobe shape more cucullate than in ssp. *islandica* he placed as ssp. *crispiformis.* This latter subspecies might also have PD − populations according to Kärnefelt. All three of the above taxa are common in the American arctic.

Within the PD − populations Kärnefelt treated two ssp. of *C. ericetorum* and again the same two ssp. of *C. islandica.* If both marginal and laminal pseudocyphellae were present and the lobes were flat, these were treated as *C. islandica* ssp. *islandica.* Those with only marginal pseudocyphellae and rather canaliculate lobes were treated as *C. ericetorum* ssp. *ericetorum.* The specimens lacking or with indistinct pseudocyphellae, usually more or less canaliculate, fell into three categories: one with the surface smooth or only weakly ridged and with marginal pseudocyphellae only on the upper part of the thallus, *C. ericetorum* ssp. *ericetorum,* and two with the surface ridged and pitted, one with short lobes, 2–3 cm tall and with the surface prominently reticulate ridged, *C. ericetorum* ssp. *reticulata* (Räs.) Kärnef., the other with taller lobes, 3–6 cm tall and pitted or ridged but not strongly reticulated, *C. islandica* ssp. *crispiformis* (Räs.) Kärnef.

The maps and statements by Kärnefelt indicate that he considers *C. islandica* ssp. *islandica* and ssp. *crispiformis* common across the American arctic but the latter as being North American and western European instead of circumpolar as the former. *C. ericetorum* ssp. *ericetorum* he considered as primarily northern European and only rarely in North America in Newfoundland and Quebec and also occurring in Greenland and Iceland. *C. ericetorum* ssp. *reticulata* replaced it westward in the tundras and in alpine western North America as well as around the north of Lake Superior. From the former *C. ericetorum* group Kärnefelt has also segregated populations in the southern margin of the range as *C. arenaria* Kärnef., a PD − species characterized by usually flattened lobes with marginal pseudocyphellae, a more grayish brown color, and acicular marginal projections. Its range encompassed roughly the area just beyond and just within the maximum of the Wisconsin glaciation, from the Atlantic coastal sands across the Great Lakes and to but not on the Rocky Mountains. A disjunct locality in Colombia was also reported.

Although downplaying the value of fumarprotocetraric acid in solving the puzzling situation in this group, Kärnefelt emphasized the value of the aliphatic compounds in aiding in the taxonomy, *C. ericetorum* contained lichesterinic but not protolichesterinic acid and also two unknown substances, named A and B. *C. islandica* usually contained both lichesterinic and protolichesterinic and rarely any A or B. *C. arenaria* contained lichesterinic and protolichesterinic acids and no A or B. Of course TLC testing is necessary in checking for these substances and a chart of the characteristic spots was given by Kärnefelt.

12. Cetraria kamczatica Savicz

Bull. Jard. Imp. Bot. Pierre le Grand 14:119. 1914.

Thallus fruticose, forming small mats with intertangled lobes; dichotomously branched, the angles of branching wider than in *C. islan-*

Cetraria kamczatica Savicz

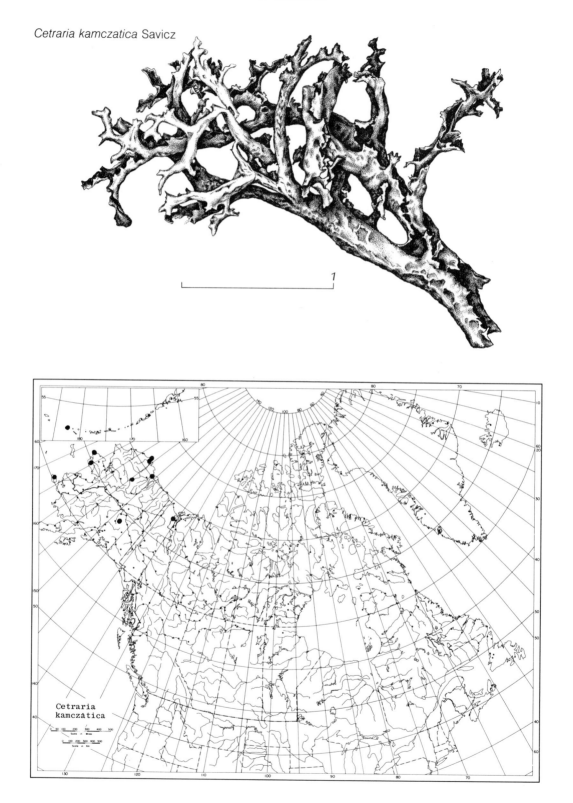

1

Cetraria
kamczatica

dica, the tips becoming turned back, the edges inrolled and not only canaliculate but occasionally the edges fusing to become tubular; the surfaces chestnut brown, shining, the margins wavy, occasionally dentate, with rare spinules which are poorly developed; lacking pseudocyphellae or with few narrow and tiny pseudocyphellae on lower margins.

Upper cortex prosoplectenchymatous, 25 μ thick, the outer 5–7 μ brown, the inner part hyaline; medulla white, dense, the algae in scattered glomerules 25–75 μ thick; lower cortex as the upper.

Apothecia not seen.

Reactions: medulla K –, C –, KC –, P –, I –.

Contents: protolichesterinic and rangiformic acids.

Growing among mosses in dry tundras, this is an amphi-Beringian species with a narrow range, being known from eastern Siberia and in the Aleutian Islands and Alaska Range as far east as Canoe Lake near the mouth of the Mackenzie River.

13. Cetraria laevigata Rass.
Bot. Materialy 5:133. 1945.

Thallus fruticose, erect, to 11 cm tall, the lobes 2–10 mm broad; irregularly branched, somewhat sympodial, the branches inrolled and canaliculate as in *C. ericetorum* and *C. islandica,* the tips slightly reflexed; upper side pale brown; underside paler, the base very pale pinkish but not red; the whole giving a much paler looking plant than the previous two species; underside with marginal pseudocyphellae which are wider and more continuous than in *C. ericetorum.*

Cortex prosoplectenchymatous, 40–75 μ thick; medulla loose, to 260 μ thick.

Apothecia not seen.

Reactions: K + yellow to red, KC + yellow to red, P + red.

Contents: fumarprotocetraric acid.

Growing in moister habitats than the previous two species, it is possibly a broad-ranging amphi-Beringian arctic species, in North America ranging from Alaska eastward through upper Canada. In Europe it extends only into the European USSR.

14. Cetraria nigricans (Retz.) Nyl.
Herb. Musei Fenn. 109. 1859. *Lichen islandicus* var. *nigricans* Retz., Flor. Scand. Prodrom. 227. 1779. *Cetraria islandica* var. *erinacea* Schaer., Enum. Crit. Lich. Eur. 16. 1850. *Cetraria islandica* var. *islandica* subvar. *crispa erinacea* (Schaer.) Boist., Nouv Flor. Lich. 2:47. 1903. *Cetraria crispa* var. *erinacea* (Schaer.) Oliv., Mem. Soc. Nation. Sci. Natur. Cherbourg 36:167. 1907. *Cetraria odontella* var. *nigricans* (Retz.) Lynge, Bergens Mus. Aarb. 9:79. 1910.

Thallus foliose, forming tufts and small mats on the substrate, the lobes narrow, to 1 mm broad, slightly channeled, the margins with cilia which may be simple or branched; upper side dark brown or olive-brown; lower side pale brown to tan, with sparse short and inconspicuous pseudocyphellae.

Apothecia but no spores were present in North America in specimens from Franklin Mts., McConnell Range, N.W.T. Bird 27188. In Scandinavian material apothecia were to 3 mm broad, the margins spinulose, the disk convex, reddish-brown, shining; hypothecium hyaline; hymenium with the upper part yellowish brown, 65–75 μ; spores 8, simple, hyaline, ellipsoid, 9–10 × 4–5 μ.

Reactions: medulla K –, C –, KC –, P –, I + blue.

Contents: protolichesterinic and rangiformic acids.

This species grows on rocks and soil. It is circumpolar and low arctic, ranging in North America from Labrador and Newfoundland to Alaska. Its southern areas are in the alpine or in very exposed sites.

15. Cetraria nigricascens (Nyl. in Kihlm.) Elenk.
Mem. Acad. Imper. Scienc. St. Petersb. Ser. 8: 27:14. 1909. *Platysma nigricascens* Nyl. in Kihlm., Medd. Soc. Fauna Fl. Fenn. 18:50. 1891. *Cetraria hiascens* var. *rhizophora* Vain., Ark. Bot. 8:23. 1909. *Cetraria nigricascens* Elenk., Acta Horti Petropol. 32:85. 1912 (nom. illeg.) *Cetraria arctica* Magn., Svensk Bot. Tids. 30:251. 1936 (non *C. arctica* (Richards) Tuck. 1881) *Cetraria sibirica* Magn., ibid. 253. 1936. *Cetraria rhizophora* (Vain.) Rass., Bot. Materially, Notul. System. Sect. Crypt. Inst. Bot. Nominie V. L. Komarovii Acad. Sci. URSS 6:13: 1949. *Cetraria rhizophora* f. *arboricola* Rass., ibid., 6:14. 1949. *Cetraria nigricascens* var. *tominii* Rass., ibid., 6:16. 1949. *Ce-*

Cetraria laevigata Rass.

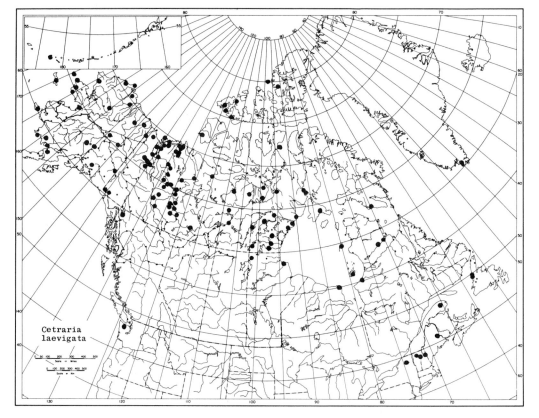

Cetraria
laevigata

Cetraria nigricans (Retz.) Nyl

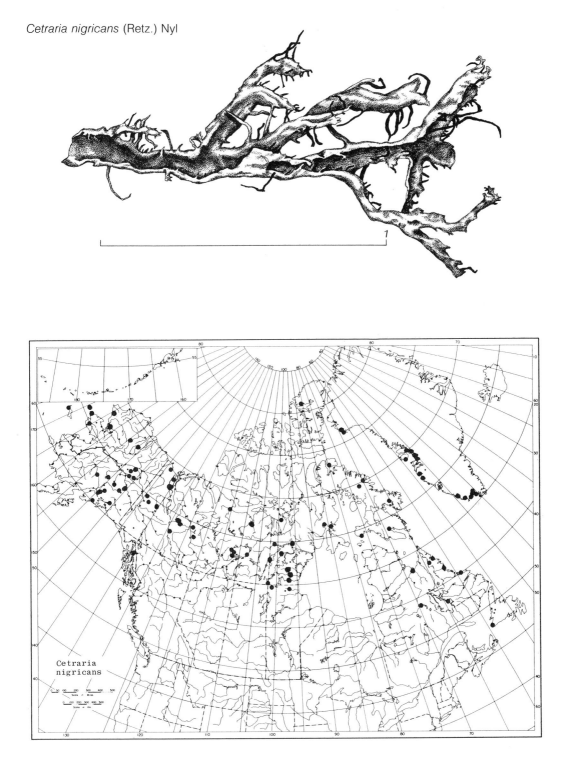

Cetraria nigricascens (Nyl. in Kihlm.) Elenk.

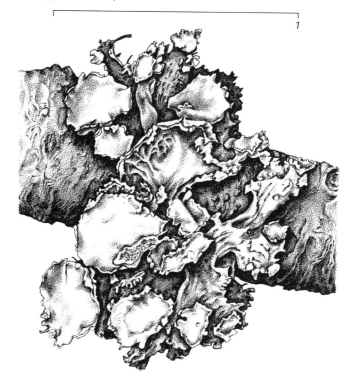

1

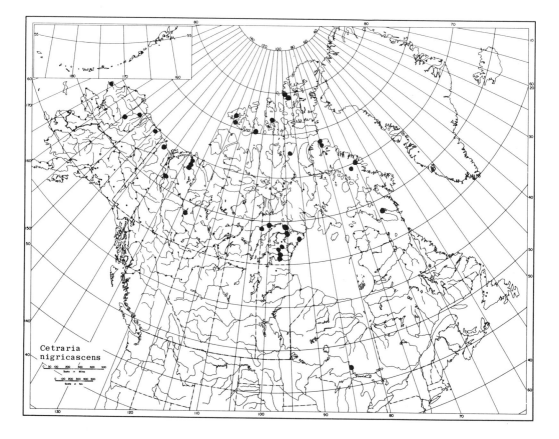

Cetraria
nigricascens

traria magnussonii Llano, J. Wash. Acad. Sci. 41:198. 1951.
Cetraria elenkinii Krog, Norsk Polarinst. Skr. 144:199. 1968.
Cetraria elenkinii var. *tominii* (Rass.) Rass., in I. I. Abramov, Handbook of the Lichens of the USSR 1:383. 1971.

Thallus dorsiventral but loose on the substratum and rising almost fruticose, lobes to 2 mm broad and to 4 cm long, slightly channeled above; upper side dark olive-brown to blackish brown, smooth; underside nearly the same color, slightly dented in spots; pseudocyphellae few or absent; margins with cilia which are simple or sparingly branched; tips not rounded, pointed; pycnidia nearly lacking. Cortex prosoplectenchymatous.

Apothecia not known.

Reactions: K −, C −, KC −, P −.

Contents: rangiformic and norrangiformic acids.

Growing on soils and the twigs of *Betula*, this species appears to be a rather rare member of the Beringian element in the arctic, ranging in the rare collections as far east as Baffin Island in the American Arctic.

16. Cetraria nivalis (L.) Ach.
Meth. Lich. 294. 1803. *Lichen nivalis* L., Species Plant. 1145. 1753.

Thallus erect to spreading erect, single lobes or forming tussocks; irregularly branched with crisped tips; upper side straw-yellow, reticulate sulcate, dull; underside of same color with scattered white pseudocyphellae; base turning yellow to brown when dying; both sides corticate.

Cortex prosoplectenchymatous 15–30 μ thick, the outer cells inspersed with pigmented granules; medulla loose, 80–150 μ thick, the algae scattered in the medulla; lower cortex similar to the upper.

Apothecia infrequent, at the sides or the tips of the lobes, circular, the margins crenulate, concolorous with the thallus, the reverse netted sulcate; disk yellow-brown to brown; epithecium pale brownish; hymenium pale; hypothecium hyaline; asci clavate; spores 8, hyaline, simple, ellipsoid, or with one end thicker, 5–8 × 3–5 μ.

Reactions: K −, C −, KC + yellow, P −.
Contents: usnic acid.

Growing on soil or between mosses on soil, rarely on twigs where snow covers the lower part of the plants, this species is arctic-alpine and boreal with a wide range in North America, south in the mountains in the east to New England and in the west to New Mexico.

17. Cetraria pinastri (Scop.) S. Gray
A Natur. Arrang. Brit. Plants 1:432. 1821. *Lichen pinastri* Scop., Flora Carn. 2:382. 1772. *Cetraria juniperina* var. pinastri (Scop.) Ach., Meth. Lich. 298. 1803. *Cetraria caperata* Vain., Acta Soc. Fauna Flora Fenn. 13:7. 1896.

Thallus foliose, crisped lobate, the lobes rounded, overlapping; upper side yellow to grayish green, smooth, dull; underside pale yellow to brownish yellow with scattered whitish rhizinae; margins with elongate soralia forming masses of yellow soredia.

Upper cortex paraplectenchymatous, 25–50 μ thick, the cells heavily pigmented with yellow granular inclusions; medulla very loose, the hyphae 4–5 μ diameter, much incrusted with yellow pigment granules, making the medulla yellow; lower cortex with the cells clearer than those of the upper cortex, with few granular inclusions, paraplectenchymatous, 25–50 μ thick.

Apothecia rare, marginal toward the tips of the lobes, the margin confluent with the upper side of the thallus, sorediate, the disk 2.5 mm broad, red-brown, flat, epruinose; epithecium brownish; hypothecium hyaline; hymenium 70 μ, hyaline; spores 8, hyaline simple, ellipsoid, 6–8 × 6 μ.

Reactions: K −, C −, KC −, P −.
Contents: usnic, pinastric and vulpinic acids.

This species grows on a variety of broad-leaved and coniferous trees and shrubs, common on old wood, occasionally on humus, soils, and rocks. It is a circumpolar boreal and temperate species which in North America ranges south in New England, the Lakes states, and in the west to New Mexico.

18. Cetraria sepincola (Ehrh.) Ach.
Meth. Lich. 297. 1803. *Lichen sepincola* Ehrh., Phytophyll. Ehrhart, No. 90. 1780.

Thallus of small lobes radiating up from the substrate, rounded, to 10 mm tall but usually less, to 2–3 mm broad; upper side smooth,

Cetraria nivalis (L.) Ach.

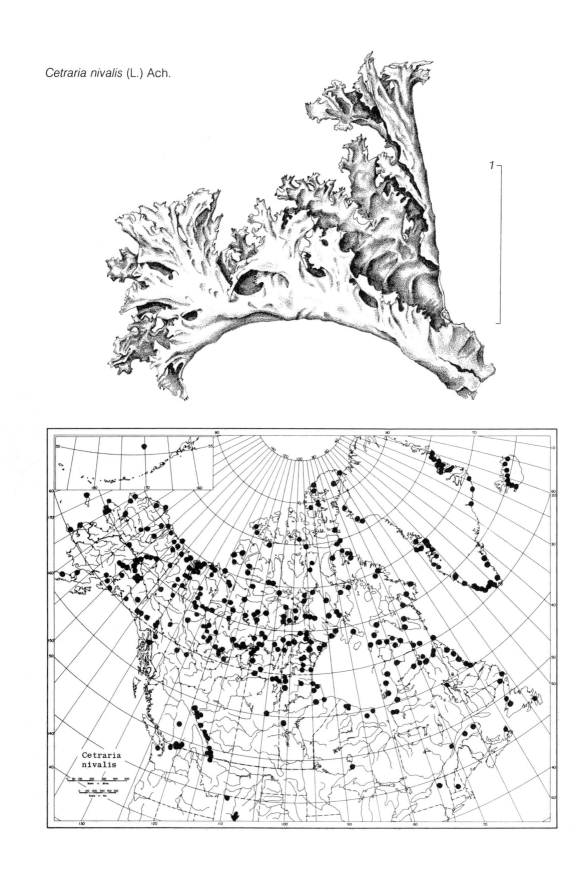

Cetraria pinastri (Scop.) S. Gray

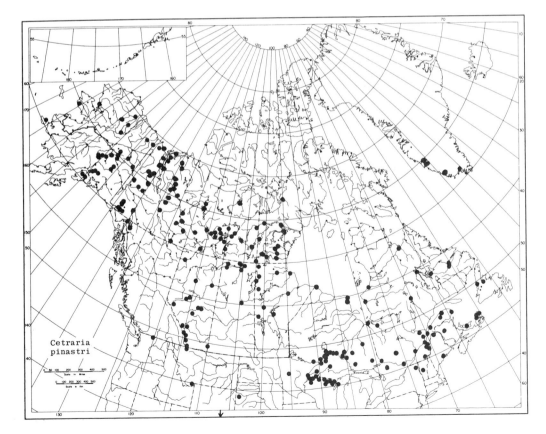

Cetraria
pinastri

CETRARIA

Cetraria sepincola (Ehrh.) Ach.

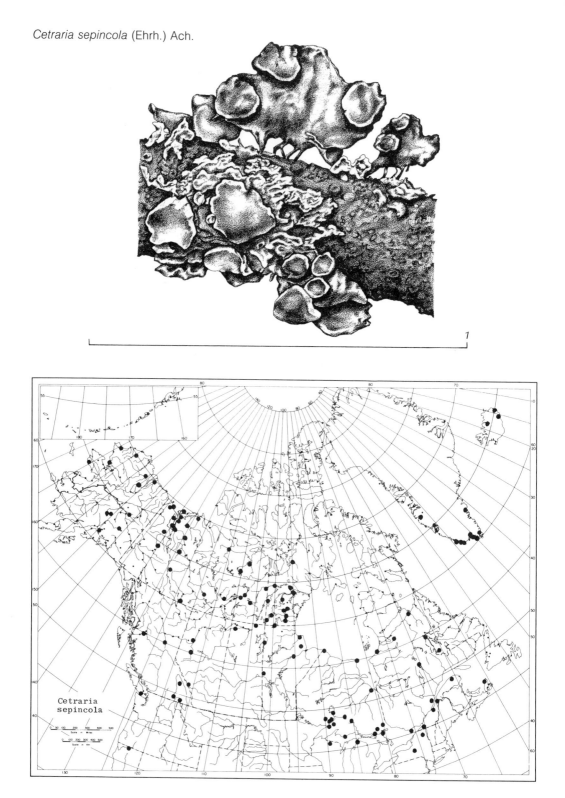

1

Cetraria
sepincola

Cetraria tilesii Ach.

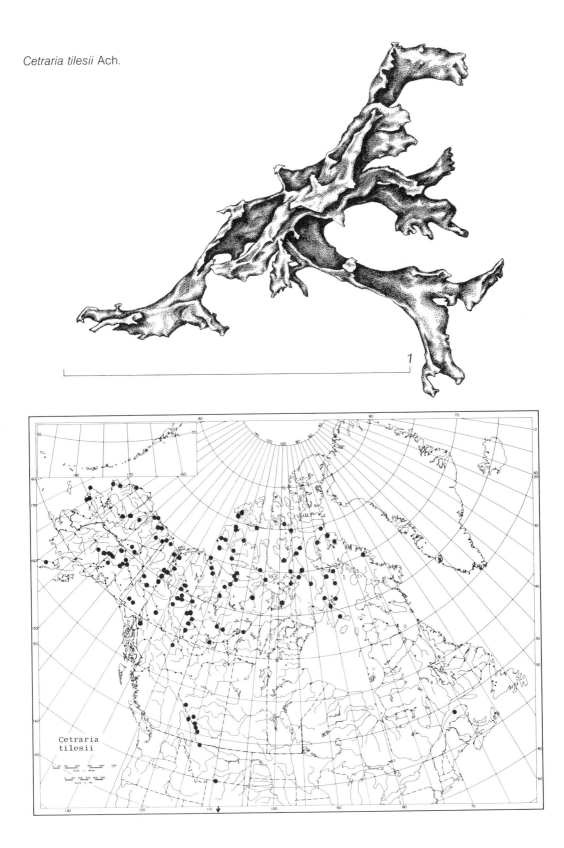

1

Cetraria
tilesii

dull or shining, brown, olive-brown, sometimes ashy-brown; lower surface whitish or pale brown, with scattered simple or slightly branched rhizinae.

Upper cortex prosoplectenchymatous, to 25 μ thick; medulla loose, lower cortex like the upper.

Apothecia very numerous, often obscuring the thallus; margin thick, smooth to crenulate, concolorous with the thallus; disk reddish brown, shining, epruinose, to 3 mm broad; epihymenium and upper part of the hymenium brown, paraphyses conglutinate; hymenium 50 μ, asci broadly clavate; spores 8, hyaline, simple, ellipsoid, $7-10 \times 4.5-6.5$ μ.

Reactions: K$-$, C$-$, KC$-$, P$-$.

Contents: protolichesterinic acid.

This species grows especially abundantly on the twigs of *Betula*, particularly the dwarf birches, but also occurs on the twigs and dead wood of other shrubs as *Ledum, Rhododendron, Alnus,* and *Vaccinium*. It is a circumpolar species of the boreal forest and of the lower part of the arctic, growing in the tundras in low moist areas on the shrubs.

19. Cetraria tilesii Ach.

Synopsis Lich. 228. 1814. *Cetraria juniperina* var. *terrestris* Schaer., Lich. Helvet. Spec. 10. 1823. *Platysma juniperinum* var. *terrestre* (Schaer.) Nyl., Act. Soc. Linn. Bordeaux 21:295. 1856. *Platysma tilesii* (Ach.) Nyl., Bull. Soc. Linn. Normand 1:257. 1887. *Cetraria juniperina* var. *tilesii* (Ach.) Th. Fr. Nova Acta Reg. Scient. Upsala. 3:138. 1861. *Cetraria nivalis* var. *tilesii* (Ach.) Boist., Nouve.Flore Lich. part 2:48. 1903. *Platysma juniperinum* var. *tilesii* (Ach.) Oliv., Mem. Soc. National. Scienc. Natur. Cherbourg 36:172. 1907. *Cetraria terrestris* (Schaer.) Fink, Mycologia 11:299. 1919.

Thallus erect, foliose, brittle, golden yellow, on calcareous soils; branching irregular with narrow lobes to 3–4 mm broad, tips somewhat dentate; upper side smooth to slightly sulcate; underside smooth, of same color as above; medulla yellow, lacking rhizinae.

Cortex very irregular in thickness, varying from 8–37 μ in section even in same thallus, prosoplectenchymatous but with a few paraplectenchymatous groups of cells here and there; medulla yellow, very loose, of irregular shaped hyphae with much pigmentary material over them and with much crystaline mineral inclusions between, about 75 μ thick, the hyphae to about 7 μ.

Apothecia not seen.

Growing on calcareous soils and between calcareous gravels, this is a circumpolar arctic-alpine species originally described from Kamtchatka. In North America it has been collected south to the Gaspé Peninsula in the east and to New Mexico in the west (Egan 1971). It has apparently not yet been collected in Greenland.

The exceedingly loose medulla and the thinness of the cortex in spots probably contribute to the brittleness of this species, a characteristic quite unlike the toughness of *C. juniperina* with which it has often been associated. Arctic reports of *C. juniperina* usually refer to this species.

CETRELIA W. L. Culb. & C. F. Culb.

Contr. U.S. Nat. Herb. 34:490. 1968.

Thallus foliose, large to medium, the lobes to 2.5 cm broad; upper surface ashy white to greenish white, pseudocyphellate; in extra territorial species can be isidiate or sorediate; under surface black, brown at the margins, rhizinate; upper cortex prosoplectenchymatous (with thick walls and narrow lumina).

Apothecia and pycnidia not known in the sole species reaching the arctic.

Type species: *Cetrelia alaskana* (C. Culb. & W. Culb.) W. Culb & C. Culb.

REFERENCE
Culberson W. L. & C. F. Culberson. 1968. The lichen genera *Cetrelia* and *Platismatia* (Parmeliaceae). Contr. U.S. Nat. Herb. 34:449–558.

This genus differs from *Parmelia* in having marginal rather than laminal pycnidia. It differs from *Asahinea* in having rhizines on the underside. It differs from the true Cetrarias in having a prosoplectenchymatous instead of paraplectenchymatous upper cortex according to Dahl (1952) and the Culbersons (1968).

Only the following species is known from the American Arctic.

1. Cetrelia alaskana (C. Culb. & W. Culb.) W. Culb. & C. Culb.

Contr. U.S. Nat. Herb. 34:492. 1968. *Cetraria alaskana* C. Culb. & W. Culb., Bryologist 69:200. 1966.

Thallus medium-sized, to 7 cm broad, the lobes to 1.5 cm broad, pale yellowish tan or brownish in the herbarium, greenish gray in the field, often mottled with black; smooth, esorediate, lacking isidia, pseudocyphellate with minute pores; lower surface jet black centrally, the margins chestnut brown; rhizines few, short, papilliform or longer; upper cortex 16–29 u, medulla 86–135 u, lower cortex 16–23 u.

Apothecia and pycnidia unknown.

Reactions: cortex K+ yellow; medulla K−, KC− or KC+ pale pink, C−, P−.

Contents: atranorin in the cortex, imbricaric acid in the medulla, plus an unknown substance giving the KC+ pink reaction.

This species grows in rather moist type of tundra on the tussocks and between the tussocks of *Eriophorum*. It appears to be a Beringian radiant species although known presently only from Alaska and as far east as Coppermine, N.W.T. It will undoubtedly be found in eastern Siberia.

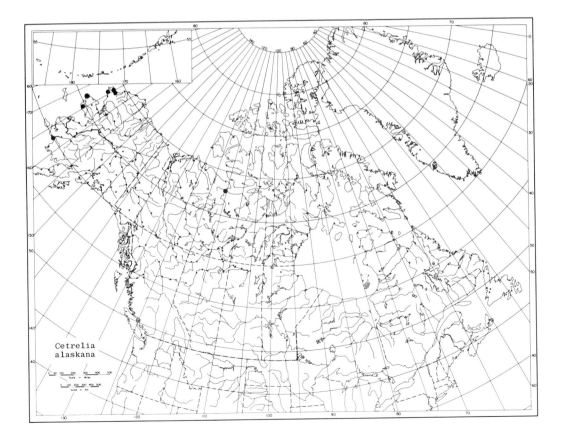

Cetrelia
alaskana

CLADONIA [Hill] Hill

Hist. Plant. 2 ed. 91. 1773. *(nom. cons.)*. *Cladonia* Hill, Hist. Plant. 91. 1751. *Pyxidium* Hill, Hist. Plant. 94. 1751. *Cladona* Adans., Fam. Plant. 2:6. 1763.

Primary thallus persistent or disappearing, crustose or squamulose to foliose; the hyphae usually loose, rarely dense or conglutinate; the algae green, conglomerate in an irregular layer in the upper part of the medulla; the upper cortex usually present and dense, or more or less vertical conglutinate hyphae; the lower side usually lacking a cortex. Podetia, the vertical developments of the stipe of the apothecia, arising from the upper side of the primary thallus, subulate or obtuse or cup-forming or becoming branched; solid when young but soon becoming hollow, with an outer cortex which may persist or disintegrate, with a medullary layer which may be dense or loose, and containing algae in the outer part; the base persistent or dying and disappearing, growth proceeding at the apices; the apices sterile or with apothecia. In one group of species the outer part becomes arachnoid, the inner part of the medullary layer becomes hard and horny (chondroid) entirely or in strands.

Apothecia at the tips of the podetia or sessile on the primary squamules, lecideine, red or pale brown or darker brown, rarely yellow or orange, with a narrow margin or more usually becoming immarginate and convex; the margin of the same color as the disk or paler; the exciple radiate; the apothecium usually larger than the diameter of the supporting podetium; the hypothecium pale to reddish, the hyphae conglutinate; the hymenium narrow, 30–70 μ thick, hyaline or pale, becoming more or less blue with iodine; the paraphyses slender, 2–5 μ thick; the apices either not thickened or becoming clavate or capitate, scarcely or not at all moniliform, simple or sparingly branched; the asci cylindrical to clavate, quite narrow, 7–14 μ thick, thickened or not at the apices; spores usually 8, irregularly biseriate, fusiform oblong or ovoid, 6–24 × 2–4.5 μ, simple, hyaline.

Pycnidia at the apices of the podetia, the margins of the cups, the sides of the podetia or the edges or upper surface of the primary squamules; sessile or short-stalked, variable in shape, cylindrical, ovoid, top-shaped or capitate; the base constricted or not, the ostiole small or dilated; black, brown, ashy, or red, containing a hyaline or reddish jelly; the sterigmata variously dichotomously, trichotomously, polychotomously, or irregularly branched, producing the conidia at the apices or rarely at the axils; the conidia cylindrical, filiform, 5–14 × 0.5–1 μ, straight or curved. Algae *Trebouxia*.

The genus Cladonia is notoriously variable among lichens and the taxonomy is consequently difficult. In the Arctic the abrasive action of windborne ice as well as the harsh climatic conditions make determination of specimens more difficult than in milder climes. One must rely very heavily therefore upon chemical characters which are very useful in this genus. Summaries with pictures of the crystal reaction products may be found in Hale (1969) and Thomson (1968). Thin layer chromatographic procedures are best consulted in the two papers by Chicita Culberson (1970, 1972).

REFERENCES

Ahti, T. 1961. Taxonomic studies on reindeer lichens *(Cladonia,* subgenus *Cladina)*. Annal. Bot. Soc. Zool. Bot. Fenn. "Vanamo" 32(1):1–160.
—— 1964. Macrolichens and their zonal distribution in boreal and arctic Ontario, Canada. Annal. Bot. Fenn. 1:1–35.
—— 1966. Correlation of the chemical and morphological characters in *Cladonia chlorophaea* and allied lichens. Annal. Bot. Fenn. 3:380–390.
—— 1973. Taxonomic notes on some species of *Cladonia*, subsect. Unciales. Annal. Bot. Fenn. 10:163–184.
—— 1974. The identity of *Cladonia theiophila* and *C. vulcani*. Annal. Bot. Fenn. 11:223–224.
—— 1978a. Nomenclatural and taxonomic remarks on European species of *Cladonia*. Ann. Bot. Fenn. 15:7–14.
—— 1978b. Two new species of *Cladonia* from western North America. Bryologist 81:334–338.
—— 1980. Taxonomic revision of *Cladonia gracilis* and its allies. Ann. Bot. Fennici 17:195–243.
Ahti, T. & I. M. Brodo. 1981. *Cladonia labradorica* sp. nov., and *C.* kanewskii in Canada. Bryologist 84:238–241.

Bowler, P. A. 1972. The distribution of four chemical races of *Cladonia chlorophaea* in North America. Bryologist 75:350–354.

Culberson, Chicita F. 1972. Improved conditions and new data for the identification of lichen products by a thin-layer chromatographic method. J. Chromatog. 72:113–125.

Culberson, Chicita F. & H. Kristinsson. 1970. A standardized method for the identification of lichen products. J. Chromatog. 46:85–93.

—— 1969. Studies on the *Cladonia chlorophaea* group: a new species, a new *meta*-depside, and the identity of "novochlorophaeic acid." Bryologist 72:431–443.

Culberson, W. L. 1969. The chemistry and systematics of some species of the *Cladonia cariosa* group in North America. Bryologist 72:377–386.

Hale, Mason E. 1969. The Lichens. Dubuque, Iowa: Wm. C. Brown.

Kristinsson, H. 1971. Morphological and chemical correlations in the *Cladonia chlorophaea* complex. Bryologist 74:13–17.

Thomson, J. W. 1967 (1968). The lichen genus *Cladonia* in North America. Toronto: Univ. Toronto Press.

Tønsberg, T. 1975. *Cladonia metacorallifera* new to Europe. Norweg. J. Bot. 22:129–132.

Wetherbee, R. 1969. Population studies in the chemical species of the *Cladonia chlorophaea* group. Mich. Bot. 8:170–174.

Yoshimura, I. 1968. Lichenological notes I. Some species of *Cladonia* with taxonomic problems. J. Hattori Bot. Lab. 31:198–204.

Key to the Arctic Species of Cladonia

1. Podetia much branched and interwoven.
    2. Surface of podetia decorticate and arachnoid, squamules lacking, cups lacking.
        3. Podetia P −, lacking fumarprotocetraric acid.
            4. Podetia forming rounded heads, tips of branches divergent, the branches in isotomic dichotomies and polytomies around open axils; pycnidial jelly red; plant containing usnic acid ± accessory perlatolic and pseudonorrangiformic acids . . . . 52. *C. stellaris*
            4. Podetia not forming equal branching systems, having a main branch with smaller side branches; pycnidial jelly colorless; containing usnic acid and accessory rangiformic acid . . . . 39. *C. mitis*
        3. Podetia P + red (fumarprotocetraric acid) or P + yellow (psoromic acid).
            5. Thallus forming rounded heads; branching in equal dichotomies or polytomies; P + yellow, containing psoromic acid . . . . 1. *C. aberrans*
            5. Thallus not forming rounded heads; branching unequal; P + red, containing fumarprotocetraric acid.
                6. Podetia K + yellow; containing atranorin, lacking usnic acid (KC −); podetia gray or whitish gray, branching mainly tetrachotomous . . . . 49. *C. rangiferina*
                6. Podetia K −, lacking atranorin, containing usnic acid (KC + yellow); podetia yellowish gray to glaucescent greenish, branching mainly trichotomous or tetrachotomous . . . . 6. *C. arbuscula*
    2. Surface of podetia corticate, at least in part, and not arachnoid; squamules sometimes present; cups present or lacking.
        7. Plant K −, lacking atranorin, usnic acid present.
            8. Plant P + red or P + yellow.
                9. Plant P + red (fumarprotocetraric acid); branching dichotomous or in whorls of 3–4; podetia with branches nearly vertical; surface slightly dull arachnoid, areolate in places; inner wall nearly continuous with bulges but no strands lining the inner wall; "diterpene" crystals lacking at the tips of the podetia . . . . 3. *C. alaskana*
                9. Plant P + yellow at tips of branches (psoromic acid); branching mainly polytomous and divaricate; outer surface smooth; inner wall with loosely reticulate shallow network of strands; tiny "diterpene" crystals at the extreme tips of the podetia . . . . 41. *C. nipponica*
        8. Plant P −.
            10. Often with distinct but small cups, the cups denticulate; cortex smooth or verruculose; containing usnic and barbatic acids . . . . 4. *C. amaurocraea*
            10. Lacking cups; branch tips somewhat spinescent, outer surface smooth; containing usnic acid (with accessory squamatic acid in *C. uncialis*).
                11. Inner surface of the podetium smooth, the cartilaginous layer continuous; tips of podetia lacking tiny "diterpene" crystals.

    12. Containing usnic + or − squamatic acid; circumpolar and common . . . .

. . . . 61. *C. uncialis*

    12. Containing usnic and hypothamnolic acids (S-shaped crystals in GE); Alaska and east Asia . . . . 47. *C. pseudostellata*

  11. Inner surface of podetium with flat embedded cartilaginous strands; tips of podetia with tiny "diterpene" crystals; containing only usnic acid; in Alaska and east Asia . . . . 32. *C. kanewskii*

7. Plant K+ yellow, containing atranorin, lacking usnic acid; podetia branching dichotomously to tetrachotomously, pale gray.

  13. Plant P+ red, containing fumarprotocetraric acid; primary thallus present, of very large squamules (5–25 × 1–7 mm); surface not arachnoid . . . . 60. *C. turgida*

  13. Plant P−, primary thallus usually lacking; surface of scattered areoles over ± arachnoid cortex.

    14. Thallus containing atranorin + unknowns (in TLC) . . . . 59. *C. thomsonii*

    14. Thallus containing atranorin, merochlorophaeic acid and psoromic acid (but P+ in extract only) . . . . 46. *C. pseudorangiformis*

1. Podetia little or not branched, awl-shaped, cup-shaped, or with few branches.

  15. Podetia forming cups.

    16. Apothecia and pycnidia red.

      17. Podetia sorediate.

        18. K−, P−; primary squamules small to medium.

          19. Podetia slender, tall, broadening at tip into shallow cups; farinose sorediate; containing usnic acid and bellidiflorin.

            20. Containing zeorin, UV−; podetia regular, little split . . . . . . 23. *C. deformis*

            20. Containing squamatic acid, medulla UV+; podetia with split and torn sides . . . . 57. *C. sulphurina*

          19. Cups short, flaring from near the base, granulose-sorediate.

            21. Finely sorediate without coarsely isidiate granules; containing usnic acid and zeorin . . . . 44. *C. pleurota*

            21. Coarsely isidiate-granular and small squamulose; containing didymic acid with usnic acid and squamatic acid (UV+) and ± bellidiflorin . . . .

. . . . 38. *C. metacorallifera*

        18. K+ yellow, P+ orange; primary squamules large; containing thamnolic acid, substance G, and accessory bellidiflorin . . . . 24. *C. digitata*

      17. Podetia lacking soredia.

        22. Cups short, regular, little squamulose; dying bases not strong yellow.

          23. Podetia smooth to with broad, low verruculae; UV−, containing usnic and barbatic acids . . . . 15. *C. coccifera*

          23. Podetia very rough with smaller, more protuberant verruculae; UV+, containing squamatic and usnic acids ± bellidiflorin.

            25. Containing didymic acid . . . . 38. *C. metacorallifera*

            25. Lacking didymic acid . . . . 30. *C. granulans*

        22. Cups taller, very squamulose, dying bases very yellow; containing usnic and squamatic acids and bellidiflorin. . . . . 9. *C. bellidiflora*

    16. Apothecia and pycnidia brown.

      26. Podetia sorediate.

        27. Cups with the edges inrolled, opening to the interior of the podetium, farinose sorediate; containing only squamatic acid (UV+) . . . . 13. *C. cenotea*

        27. Cup edges not inrolled, interior not opening into the podetium.

          28. Podetia with the soredia in rounded patches toward the upper part, a large part of the base corticate, the cups small on top of very long cylindrical podetia . . . . 18. *C. cornuta*

          28. Podetia mostly covered with soredia, only a small part of the base corticate.

29. Podetia very slender and tall, the cups small, poorly developed; podetia brownish white to whitish, P+ red (fumarprotocetraric acid); soredia very finely farinose . . . . 56. *C. subulata*
29. Podetia stouter, shorter, the cups well developed, goblet-shaped; podetia yellowish, greenish, greenish gray, P+ or P−; soredia coarser, granulose or farinose.
    30. Cups trombone-shaped with a short, quickly flaring cup on top of a slender podetium; soredia farinose . . . . 26. *C. fimbriata*
    30. Cups goblet-shaped, gradually flaring from the base; soredia granular.
        31. Podetia P+ red or P−, KC− or KC+ red but not yellow, lacking usnic acid.
            32. Cups tall and stout, 20–40 mm tall, stalk to 3 mm thick . . . . 35. *C. major*
            32. Cups short and stout, to 10 mm tall, 1 mm thick. *C. chloro-phaea* group. Chemistry is necessary for separation of these species.
                33. Containing grayanic acid (±fumarprotocetraric acid) dark brown to gray, roughly corticate at base; sorediate with white to green-brown or gray soredia . . . . 31. *C. grayi*
                33. Lacking grayanic acid.
                    34. Containing merochlorophaeic acid (±sekikaic acid, ±fumarprotocetraric acid); K+ red, KC+ red on active parts; cups broad, surface rough, dark brown or even blackish, apothecia on flattened stalks . . . . 37. *C. merochlorophaea*
                    34. Lacking merochlorophaeic acid.
                        35. Containing cryptochlorophaeic acid, C+ red, K+ red, KC+ red in acetone extracts; podetia pale green, very coarsely verruculose stalks . . . . 20. *C. cryptochlorophaea*
                        35. Lacking cryptochlorophaeic acid.
                            36. With bourgeanic acid (substance H); color greenish, soredia farinose, base smooth . . . . 17. *C. conista*
                            36. Lacking bourgeanic acid (substance H); containing only fumarprotocetraric acid or no acids; with coarse soredia, light greenish gray, stalks smooth . . . . 14. *C. chlorophaea*
        31. Podetia P−, KC+ yellow; containing usnic acid, lacking fumar-protocetraric acid; inner layers of podetia translucent when soredia are shed; apothecia pale brown . . . . 12. *C. carneola*
26. Podetia lacking soredia.
    37. Podetia proliferating from the centers of the cups.
        38. Podetia K−, P+ red; containing fumarprotocetraric acid . . . . 62. *C. verticillata*
        38. Podetia K+ yellow, P+ red; containing atranorin and fumarprotocetraric acid.
            39. Podetia with broad cups, cortex smooth, basal squamules well developed.
                40. Basal squamules long, somewhat grayish on the underside, podetia brown and cracked . . . . 54. *C. subcervicornis*
                40. Basal squamules broad, purely white on underside; podetia gray and entire . . . . 34. *C. macrophyllodes*
            39. Podetia with cups narrow when compared with length, basal squamules poorly developed; podetia with black spots at base . . . . 53. *C. stricta*
    37. Podetia proliferating from the margins of the cups.

41. Podetia K+ yellow, containing atranorin. . . . . 25. *C. ecmocyna*
41. Podetia K−, lacking atranorin.
   42. Cups opening widely into the inside of the podetia.
      43. Podetia becoming decorticate and thickly squamulose but not sorediate . . . .
      . . . . 51. *C. squamosa*
      43. Podetia with persistent cortex . . . . 19. *C. crispata*
   42. Cups closed on the interior or with a sieve-like membrane across the interior.
      44. Cups with interior partly closed by a lacerate or punctured membrane, cups
      sometimes proliferate. . . . . 40. *C. multiformis*
      44. Cups closed to the interior.
         45. Cups much split and irregular . . . . 43. *C. phyllophora*
         45. Cups regular, not split.
            46. Podetia tall, with cups flaring rapidly toward tips of podetia; podetia
            olive-green . . . . 29. *C. gracilis*
            46. Podetia short, cups flaring gradually from the base; podetia blue-green
            or gray-green.
               47. Primary squamules thin, not rosette forming . . . . 48. *C. pyxidata*
               47. Primary squamules thick, leathery, forming rosettes around the
               podetia . . . . 45. *C. pocillum*
15. Podetia not forming cups, awl-shaped with pointed tips or tipped with apothecia which form
heads on the stalk-like podetia.
   48. Podetia with red apothecia.
      49. Podetia and primary squamules sorediate; base not yellow; UV−; containing bar-
      batic acid plus accessory usnic and didymic acids and substance F.
         50. Podetia more or less corticate at least in patches throughout entire length; base
         entirely corticate; containing accessory usnic, didymic acids and substance 2 . .
         . . . . 27. *C. floerkeana*
         50. Podetia entirely corticate; containing didymic acid, usnic acid, and accessory
         substance F . . . . 7. *C. bacillaris*
      49. Podetia and primary squamules esorediate but very squamulose on podetia, base yel-
      low; UV+, containing squamatic and usnic acids and bellidiflorin . . . . . 9. *C. bellidiflora*
   48. Podetia with brown apothecia or apothecia lacking.
      51. Podetia K+ yellow or yellow turning red.
         52. Sides of podetia torn and fissured.
            53. Podetia K+ yellow, P−, containing only atranorin . . . . 11. *C. cariosa*
            53. Podetia K+ yellow turning red, P+ orange-red; containing atranorin and
            norstictic acid . . . . 58. *C. symphycarpa*
         52. Sides of podetia not torn and fissured, with or without apothecia.
            54. Esorediate, cortex continuous; P+ red, K+ yellow; containing fumarpro-
            tocetraric acid and atranorin. . . . . 25. *C. ecmocyna*
            54. Cortex sorediate-granulose
               55. P+ orange-red, K+ yellow; containing atranorin and norstictic acid
               . . . . 2. *C. acuminata*
               55. P+ bright yellow, K+ yellow, containing atranorin and psoromic
               acid . . . . 42. *C. norrlinii.*
      51. Podetia K−, lacking atranorin.
         56. Podetia P+ yellow, containing psoromic acid; with peltate squamules on sur-
         face . . . . 33. *C. macrophylla*
         56. Podetia P− or P+ red, lacking psoromic acid.
            57. Podetia sorediate.
               58. Podetia P+ red, UV−, containing fumarprotocetraric acid.
                  59. Soredia granular; podetia slender, decorticate areas pellucid; pri-
                  mary squamules narrow, finely divided . . . . 5. *C. anomaea*

59. Soredia farinose.
   60. Podetia very slender and tall, 30–100 mm tall; unbranched or with tiny branches, brownish white to white; primary squamules very tiny to lacking . . . . 56. *C. subulata*
   60. Podetia shorter or commonly branched and long with elongate branches, greenish or bluish gray.
      61. Podetia short, to 30 mm tall, unbranched, decorticate, and sorediate over most of the podetium, with few squamules on the podetium . . . . 16. *C. coniocraea*
      61. Podetia tall, 30–110 mm, dichotomously branched; upper part sorediate or isidiose-sorediate, usually with many squamules . . . . 50. *C. scabriuscula*
58. Podetia P−, lacking fumarprotocetraric acid; medulla UV+ or UV−.
   62. Podetia gray in color, lacking usnic acid, KC−, UV+.
      63. Containing perlatolic acid (tufts of long slender needles in GE) . . . . 22. *C. decorticata*
      63. Containing squamatic acid (short asymmetrical hexagonal crystals in GEd); brown dendritic crystals at edge of coverslip in K_2CO_3) . . . . 28. *C. glauca*
   62. Podetia yellow in color, containing usnic (except *C. bacillaris*) and barbatic acids, KC+ yellow, UV−.
      64. Podetia with acute tips.
         65. Podetia tall and slender, 20–80 mm tall . . . . . 21. *C. cyanipes*
         65. Podetia short, less than 20 mm tall . . . . . . 8. *C. bacilliformis*
      64. Podetia with blunt, rounded tips; containing barbatic acid with accessory didymic and usnic acids . . . . 7. *C. bacillaris*
57. Podetia lacking soredia.
   66. Podetia P+ red, containing fumarprotocetraric acid.
      67. 50–100 mm tall, to 2.5 mm thick . . . . 36. *C. maxima*
      67. 30–80 mm tall, to 1 mm thick . . . . 29. *C. gracilis* var. *gracilis*
   66. Podetia P−, lacking fumarprotocetraric acid (note 3 choices in 68).
      68. Podetia yellowish, KC+ yellow; containing usnic and barbatic acids (latter forming rhombic crystals in GE, no crystals in K_2CO_3).
         69. Podetia tiny, usually less than 10 mm tall, tipped with pale brown apothecia, on rotting wood . . . . 7. *C. bacillaris*
         69. Podetia taller than 10 mm, usually acute tipped and lacking apothecia, on soil and humus . . . . 4. *C. amaurocraea*
      68. Podetia olive-green, corticate, KC−, lacking usnic acid; containing squamatic acid (thick elongate hexagons in GE, brown dendrites at edge of coverslip in K_2CO_3), often with divaricate branchlets . . . . 55. *C. subfurcata*
      68. Podetia gray-green, with abundant squamules, decorticate between, UV+, containing squamatic acid . . . . 51. *C. squamosa*

1. Cladonia aberrans (Abb.) Stuck.

Acad. Nauk C.C.C.P. 11:12. 1956. *Cladonia alpestris* f. *aberrans* Abb., Bull. Soc. Sci. Bretagne 16:93. 1939. *Cladina aberrans* (Abb.) Hale & W. Culb. Bryologist 73:510. 1970.

This species is extremely similar to and perhaps conspecific with *C. stellaris* so that the description of that species may be consulted for the morphology. Slight trends suggested by Ahti are that the youngest branches are frequently very thick and obtuse and the outer medulla thicker than in *C. stellaris*. Many authors place this as the psoromic acid strain of *C. stellaris*.

Reactions: K−, KC+ yellow, P+ yellow.

Contents: l-usnic acid, perlatolic acid, and psoromic acid.

This species grows on soil and among mosses, sometimes over thin soil on rocks. It is probably circumpolar through approximately the same range as *C. stellaris* but is comparatively rarely collected. As comparison of the maps shows, there are large areas from which *C. stellaris* has been collected but from which this species has not. Obviously more study of this species is necessary.

2. Cladonia acuminata (Ach.) Norrl.

in Nyl. Norrl., Herb. Lich. Fenn. 57a. 1875. *Cenomyce pityrea* b. *acuminata* Ach., Syn. Lich. 254. 1814.

Primary squamules persistent, up to 7 mm long and 3 mm broad, narrowly elongate, sinuate-edged, becoming crenate, involute-concave; upper side glaucescent; underside white; the margins and underside of the primary squamules sometimes becoming granulose sorediate. Podetia 20–50 mm tall and up to 2.5 mm in diameter, cylindrical, cupless, the upper part branching irregularly or subfastigiately or sim-

ple; the sterile apices blunt and the fertile ones becoming dilated upwards; sides becoming chinky and sulcate, or subcontinuous corticate, sparsely granulose sorediate or esorediate, verruculose areolate or becoming decorticate, densely or loosely covered with squamules similar to the primary squamules; dull, not transparent where decorticate, the decorticate part chalky white, the granules and verruculae whitish to ashy glaucescent.

Apothecia reddish brown, large, commonly perforate or sublobate.

Reactions: K+ yellow to orange or red, KC−, P+ yellow.

Contents: atranorin and norstictic acid.

This species grows on soil rich in humus. It appears to be circumpolar boreal, and low arctic. Its North American range is ill defined because it has so seldom been collected. It ranges south to Maine, Vermont, and Pennsylvania. It is very closely related to *C. norrlinii*, from which it differs in the presence of norstictic acid instead of psoromic acid.

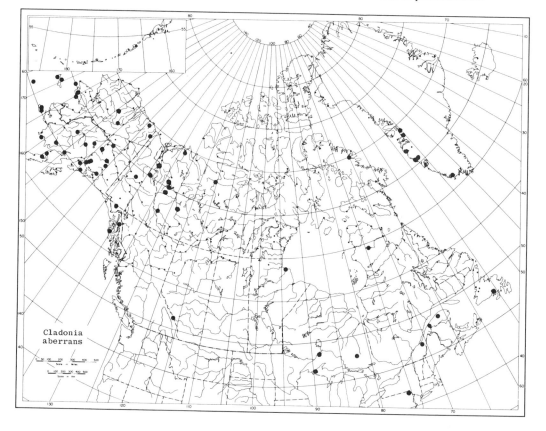

Cladonia aberrans

Cladonia acuminata (Ach.) Norrl.

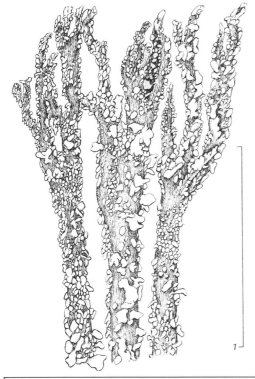

1

3. **Cladonia alaskana** Evans
Rhodora 50:35. 1947.

Primary thallus unknown. Podetia forming compact tufts; dying at the base and growing from the apices; branching irregularly by dichotomies or in whorls of 3 or 4, the branches ascending closely to each other; axils open or closed; podetial wall often split; 40–80 mm tall and up to 2 mm in diameter; the surface not at all glossy, somewhat arachnoid under a lens; pale gray tinged with yellowish or brownish tints, darker in older parts; the cartilaginous layer forming a continuous hollow cylinder around the central canal and forming a series of irregular bulges projecting into this canal; this layer sharply delimited from the outer medullary layer in which the algal clusters are located near the periphery in a single discontinous layer covered by the cortex; few small squamules on the podetia, up to 1 mm long and undivided or little lobed, rounded at apex; branches bearing the apothecia larger than the sterile branches and irregularly

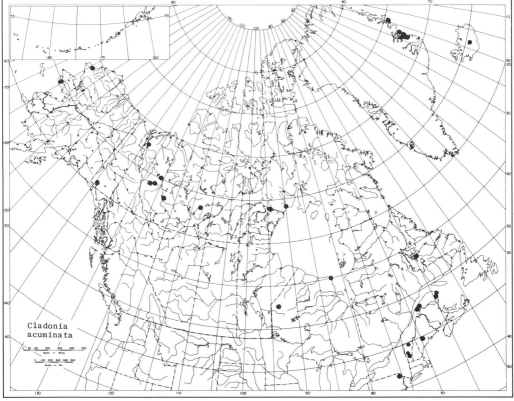

Cladonia
acuminata

Cladonia alaskana Evans

divided and split in the upper part, thus perforate at the apices but without actual cups.

Apothecia 1–2 mm in diameter; clustered or confluent around the apical perforations; pale to dark brown. Pycnidia at the apices of the podetia.

Reactions: K+ vaguely yellowish turning to gray, KC+ yellow, P+ red.

Contents: usnic and fumarprotocetraric acids plus accessory ursolic acid.

This species grows in lichen heath tundras, tussock tundras, and over boulders in tundras. It is an American endemic as far as currently known although it may actually be an amphi-Beringian element member as Krog (1968) mentions a Siberian specimen. It ranges from Alaska and the Yukon eastward in the tundras to the Thelon River and Churchill areas.

4. Cladonia amaurocraea (Flörke) Schaer.

Lich. Helvet. Spic. 34. 1823. *Capitularia amaurocraea* Flörke in Weber & Morh, Beitr. Naturk. 334. 1810.

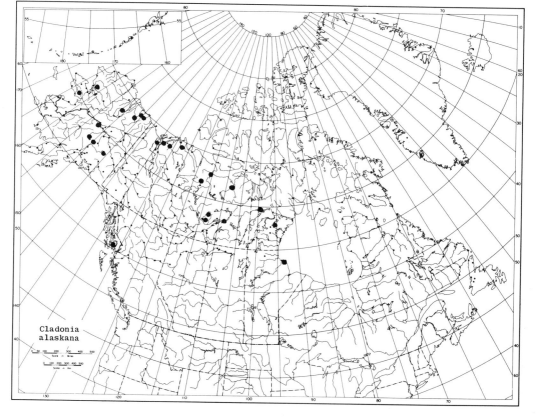

Cladonia
alaskana

Cladonia amaurocraea (Flörke) Schaer.

Cladonia amaurocraea f. *celotea* Ach. A form with cups.

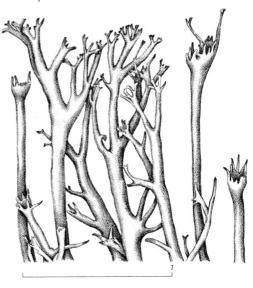

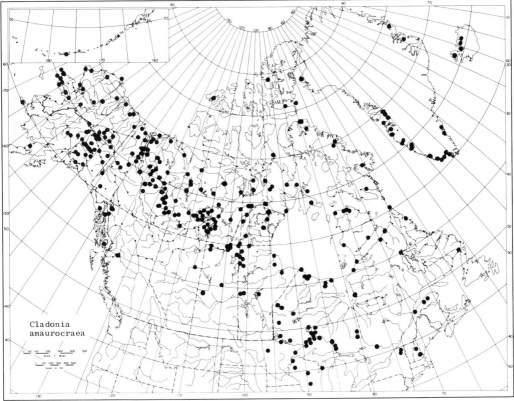

Cladonia
amaurocraea

Primary thallus rarely seen, usually disappearing, of small squamules up to 1.7 mm broad and long; squamules crenate or digitate incised, ascending, flat, sparse or in tufts; upper side glaucescent; underside white; esorediate. Podetia from the older dead parts or from fragments of old podetia lying on the ground, rarely from the primary squamules; in groups or in mats, the marginal podetia often more or less decumbent and the tips then curved upwards; 15–120 mm tall and up to 1.5 (–3.5) mm in diameter; dichotomously or sympodially branched, occasionally branching irregularly or from the margins of the cups in fascicles; often cupless and cylindrical; sometimes with narrow cups with flare rapidly; the interior membrane closed, or open into the interior by a single or several small holes; the margins of the cups with tapering spine-like proliferations and the tips of the sterile branches also tapering to spinescent tips; the axils are closed or slightly perforated, esorediate; the cortex continuous or more commonly dispersed areolate, the areolae little raised to rarely verruculose; shining or dull; impellucid; yellowish or yellowish glaucescent to glaucescent or olive-green; inner cartilaginous part of the medulla continuous, not sharply delimited from the outer part of the medullary layer.

Apothecia solitary or clustered at the tips of the podetia; middle-sized, up to 2 (–3.5) mm in diameter; brown or yellowish brown or reddish brown. Pycnidia on the margins of the cups or the tips of the podetia.

Reactions: K–, KC+ yellowish, P–.
Contents: usnic and barbatic acids.

This species grows on soil rich in humus, in tundras, and in bogs among mosses and other Cladonias. On the rock outcrops in the boreal forest it sometimes forms spectacular "polsters." It is a northern, circumpolar, arctic, and boreal species which in North America ranges from Alaska to Greenland and south to New York, Michigan, Minnesota, and Wisconsin.

Although well-developed plants usually possess cups, there are many plants which do not. Young stages with simple to only slightly branched podetia are sometimes difficult to determine. The color may vary also from quite gray-green to pale yellowish green plants resembling young plants of *C. uncialis* but differing in the chemistry.

5. Cladonia anomaea (Ach.) Ahti & P. James
Lichenologist 12:128. 1980. *Baeomyces anomaeus* Ach., Meth. Lich. 349, 1803. *Cladonia pityrea* (Flörke) Fr. Nov. Sched. Crit. 21. 1826. *Capitularia pityrea* Flörke, Berl. Magaz. 2:135. 1808.

Primary squamules persistent or disappearing, narrower and more delicate than in the other North American Thallostelides, 1–3 mm long and 1 mm broad; subdigitately lobed, sublinear or somewhat wedge-shaped or incised-lobed, involute or concave or flat, ascending; upper side glaucescent or rarely olivaceous; underside white; esorediate or granulose on the margins, or sometimes the entire primary squamules breaking up into masses of soredia. Podetia from the upper side of the primary squamules, 3–55 mm tall and up to 4 mm broad; cylindrical or top-shaped; with cups or cupless; the cups shallow and corticate on the inner surface, expanding gradually or abruptly, usually oblique and irregular with the margins dentate or proliferous; the cupless podetia simple, or irregular and sparingly branched, tapering or cylindrical; the cortex continuous, or areolate, or becoming decorticate and bearing coarsely granular soredia in more or less abundance; whitish glaucescent or ashy to olive-glaucescent; the inner cartilaginous layer well developed and giving a translucent appearance to the podetia after the soredia are dispersed; squamulose or lacking squamules.

Apothecia small to middle-sized; on the apices of the podetia or rarely on the margins of the cups; brown or reddish brown, rarely pale. Pycnidia on the tips of proliferations from the cups, or on the apices of the podetia or the upper side of the primary squamules.

Reactions: K– or K+ yellow, KC–, P+ red.

Contents: fumarprotocetraric acid plus, rarely, accessory atranorin in North American specimens. Asiatic specimens also contain homosekikaic acid.

This species grows on earth, tree bases, and rotten wood. It is almost cosmopolitan but its North American range is not well defined as it has been poorly collected. It is mainly temperate with some occurrences in the southern part of the boreal forest but with one disjunct occurrence in northern Yukon in tundra.

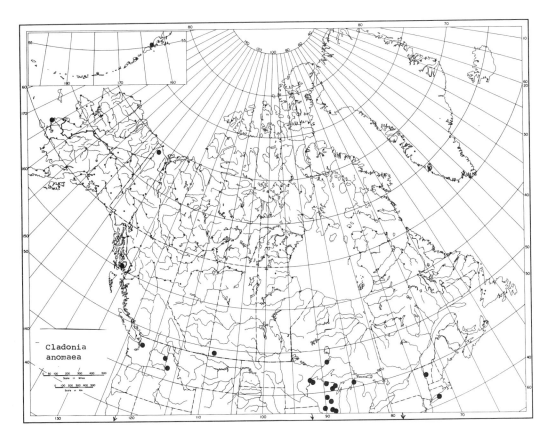

Cladonia
anomaea

6. Cladonia arbuscula (Wallr.) Rabenh.

Deutschl. Krypt. Flora 2(1):110. 1845. *Patellaria foliacea* var. m. *Arbuscula* Wallr., Naturges. Säul. Flecht. 169. 1829. *Cladonia arbuscula ssp. beringiana* Ahti, Ann. Bot. Soc. Zool. Bot. "Vanamo" 32:109. 1961. *Cladina arbuscula* (Wallr.) Hale, Bryologist 73:510. 1970.

Primary thallus usually disappearing and seldom seen, crustose, of subglobular verruculae 0.12–0.48 mm in diameter, pale yellow. Podetia forming tufts or in larger mats, robust, 50–100 (–150) mm tall and 1–2 (–4) mm in diameter; the branching typically anisotomic, tetrachotomous, or trichotomous, rarely dichotomous, the whorls of unequal branching so that the podetium forms a sympodium with unilateral growths, especially in the lower part, the upper part showing more equal branching, the uppermost branchlets terminated in 2–3 (–5) short, stout points which are all recurved toward one side, the fertile branchlets in a corymbose arrangement; the axils mainly open, sometimes widely so; pale yellow, becoming greenish glaucescent in more moist and shady habitats; the surface dull, little arachnoid-tomentose, smooth or rarely verruculose, lacking soredia (except in a rare form), impellucid.

Apothecia small; solitary or grouped at the tips of corymbose branchlets; dark brown. Pycnidia at the tips of the branchlets; containing colorless jelly.

Reactions: K−, KC+ yellow, P+ red.

Contents: usnic and fumarprotocetraric acids plus accessory ursolic acid and substance A.

This species grows on sandy soil, over earth on rock outcrops, and in bogs. It is circumpolar, boreal, montane, less common than *C. rangiferina* or *C. mitis*. It also occupies less extreme habitats than these species. In North America it ranges south into Maryland, Wisconsin, and Washington.

There has been controversy about the name of this species, some authors, including myself, preferring the usage of *C. sylvatica* (L.) Hoffm. for it. The differences of opinion rest upon interpretation of the usage of the epithet *sylvatica* by Linnaeus. Currently the argument of Ahti and of Santesson that the

Cladonia arbuscula (Wallr.) Rabenh.

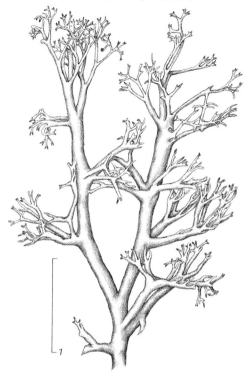

epithet *sylvatica* is an illegitimate name for typical *C. rangiferina* has gained acceptance. Subspecies *beringiana* Ahti differs in being more sparsely branched, the tips more erect, the color very yellow rather than yellow-gray, and the branches usually slender but varying to robust. It is common in the western American Arctic and is amphi-Beringian whereas ssp. *arbuscula* is in eastern Canada in two chemical strains, P+ and P−, according to Ahti. The P− strain, lacking fumarprotocetraric acid, is distinguished with difficulty from *C. mitis*.

7. Cladonia bacillaris (Ach.) Nyl.

Not. Sällsk. F. et Fl. Fenn. Förh. 8:179. 1866 (separate pre-print). *Baeomyces bacillaris* Ach., Meth. Lich. 329. 1803. *Cladonia bacillaris* var. *pacifica* Asahina, J. Jap. Bot. 15:666. 1939.

Primary squamules persistent, small to middle-sized, up to 3 mm long, crenate to lobed; upper side glaucescent to olive-glaucescent; underside white to darker at the base; esorediate or more commonly sorediate at the margin and on the underside. Podetia growing from the up-

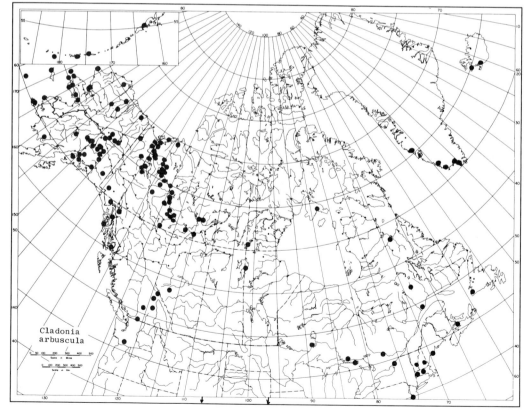

Cladonia
arbuscula

per side of the primary squamules; varying in height up to 50 mm but usually smaller, up to 10–20 mm; commonly broadening slightly to the blunt tip, producing a club-shaped podetium, rarely sharp-pointed; usually covered with greenish farinose soredia, the base with only a very short corticate patch; white throughout after dispersal of soredia although in one form with yellowish spots which turn violet with K.

Apothecia scarlet, small to medium-sized, up 2 mm; often imbedded in the tips of the podetia. Pycnidia on the surface and margins of the primary squamules.

Reactions: K−, KC− or KC+ yellowish, P−.

Contents; barbatic acid with accessory usnic and didymic acids and F.

This species grows on old logs, tree bases, earthen banks, and humus. It is cosmopolitan, in North America being low arctic boreal and temperate, ranging over most of the United States.

8. Cladonia bacilliformis (Nyl.) DT. & Sarnth.

Flecht-Tirol, Vorarlberg und Liechstenstein. 1902. *Cladonia carneola* var. *bacilliformis* Nyl., in Nyl. et Saelan, Herb. Musei Fenn. 79. 1859.

Primary squamules usually persistent; small, up to 3 mm long and 0.3 mm broad, subentire or crenate to incised-crenate, more or less concave, ascending or appressed; upper side yellowish; underside yellowish or yellowish white; farinose sorediate. Podetia short, 5–20 mm tall, up to 2 mm thick, unbranched, truncate or with narrow or obsolete cups, the margins of the cups with slender proliferations; thickly sorediate with farinose soredia, yellowish or strongly yellow; esquamulose.

Apothecia small, up to 1 mm; pale yellowish to pale brown. Pycnidia on the upper side and margins of the primary squamules and also on the tips of the podetia.

Reactions: K−, KC+ yellow, P−.

Contents: usnic and barbatic acids.

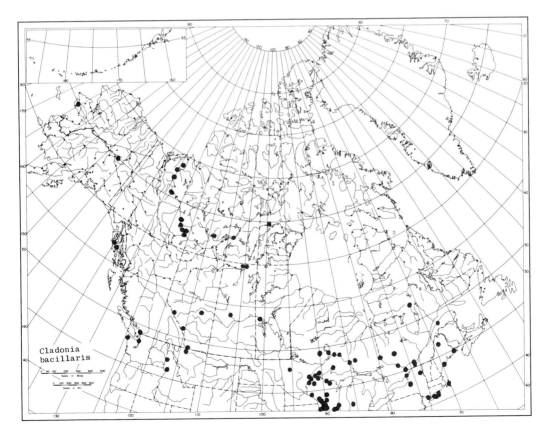

Cladonia bacillaris

Cladonia bacilliformis (Nyl.) Vain.

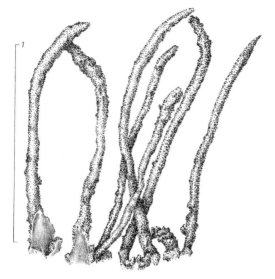

This species grows on mosses, on rocks, and on rotten wood. It is circumpolar arctic and boreal but rarely collected. It ranges southward in North America in the mountains to Colorado.

9. Cladonia bellidiflora (Ach.) Schaer.

Lich. Helv. Spic. 21. 1823. *Lichen bellidiflorus* Ach., Lich. Suec. Prodr. 194. 1798. *Cladonia hookeri* Tuck., Proc. Amer. Acad. Arts. & Sci. 1:247. 1848. *Cladonia bellidiflora* var. *hookeri* (Tuck.) Nyl. Synops. Lich. 1:221. 1858.

Primary squamules commonly disappearing, small or middle-sized. Podetia 20–50 mm tall, the base commonly dying; squamulose, the squamules small to middle-sized, 2–5 mm long, crenate or laciniate, ascending or erect, flat or involute, sparse to abundant; upper side yellowish or pale to glaucescent; underside white, becoming brown to black toward the base. Podetia cupless or with short and narrow cups which are quite regular or oblique, margins of cups subentire or dentate to radiate proliferate, eso-

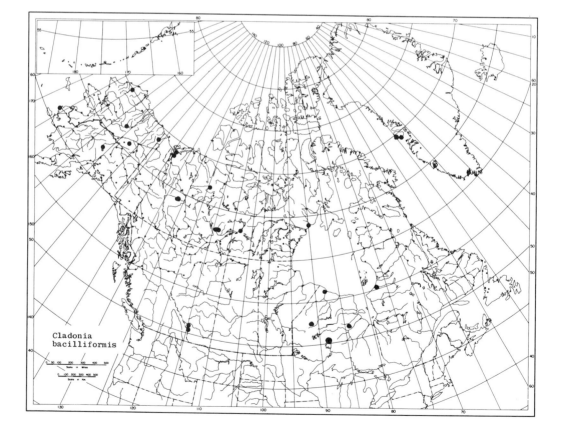

Cladonia
bacilliformis

Cladonia bellidiflora (Ach.) Schaer.

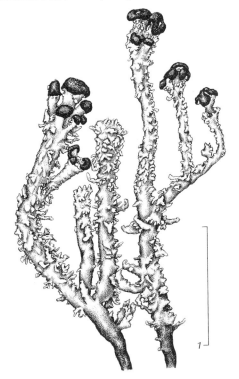

rediate; cortex continuous to chinky or areolate discontinuous, smooth or verruculose, squamulose, the squamules splitting off the cortex, the cortex yellowish or yellowish-glaucescent, the decorticate part white to faintly yellowish, opaque.

Apothecia scarlet, on proliferations or on the edges of the cups. Pycnidia on the apices of proliferations or the margins of the cups.

Reactions: K−, KC+ yellow, P−.

Contents: usnic and squamatic acids and bellidiflorin.

This species grows on soil and humus, often intermixed with mosses. It is circumpolar, arctic, and is also in the southern hemisphere. In North America it ranges into Newfoundland and Quebec in the east and southward along the west coast to California. The western distribution pattern appears rather anomalous for an arctic species; it should also be found along the mountain ranges.

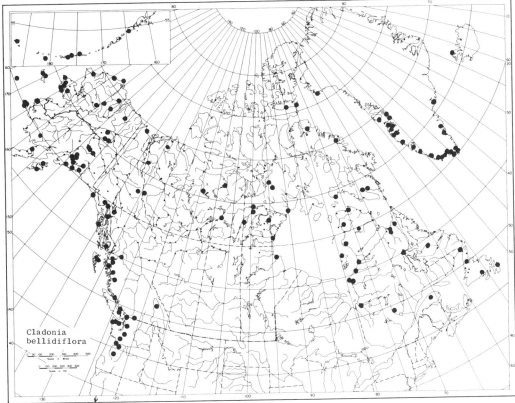

Cladonia
bellidiflora

Cladonia botrytes (Heg.) Willd.

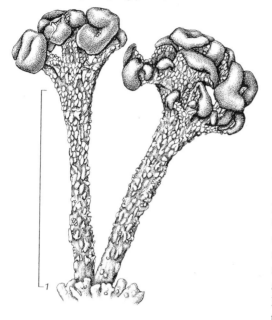

10. **Cladonia botrytes** (Hag.) Willd.

Fl. Berol. 365. 1787. *Lichen botrytes* Hag., Tent. Hist. Lich. 121. 1782.

Primary squamules persistent or disappearing, very small, not over 1.5 mm long, crenate to incised crenate or subdigitate laciniate, flat or involute or convex, ascending or appressed, rarely in little tufts; upper side yellowish or yellowish glaucescent, rarely olivaceous glaucescent; underside white, esorediate, or sparsely granulose sorediate. Podetia from the upper side of the primary squamules, usually rather short and less than 10 mm high, but sometimes up to 20 mm, cupless or rarely with pseudocups on the branches, sparingly branched above, the branches diverging widely, the axils closed or becoming split; the cortex verruculose areolate or diffract areolate, the areoles quite small, esorediate, and lacking squamules or with a few squamules at the base; opaque, dull, yellowish or yellowish glaucescent, the decorticate portion white, yellowish white or yellowish.

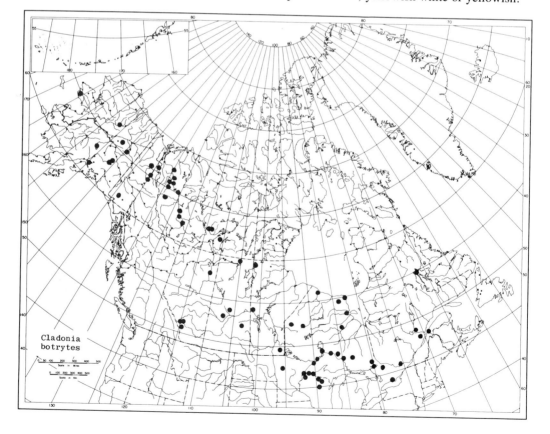

Cladonia
botrytes

Apothecia small to middle-sized, entire to lobate or perforated, solitary or grouped in small clusters at the tips of branches or on the main part of the podetium, pale brownish or waxy to brownish. Pycnidia on the upper side of the primary squamules or on the sides of the podetia.

Reactions: K−, KC+ yellowish, P−.
Contents: usnic and barbatic acids.

This is a very characteristic species of the coniferous forests, growing on rotting wood and occasionally on earth rich in humus. It is a circumpolar, boreal, species which reaches the tundras at the edge of the boreal forest trans-Canada and in Alaska.

11. Cladonia cariosa (Ach.) Spreng.
In Linn., Syst. Veg. 16 ed. 4:272. 1827. *Lichen cariosus* Ach., Lich. Suec. Prodr. 198. 1798.

Primary squamules persistent, small to middle-sized, 1–7 mm long and 1–2 mm broad, rarely larger, a little coarser and larger than those of C. *capitata,* irregularly divided, crenate, incised or rounded, flat or concave, sometimes reflexed, confluent or sometimes caespitose; upper side white or bluish green or olive-green; underside white or darkening toward the base; esorediate. Podetia from the margins or the upper surface of the primary squamules, 7–26 mm tall, up to 2 mm in diameter, cylindrical; cupless, terminated by apothecia, somewhat branched toward the top, the sides markedly fissured; cortex continuous at first but soon becoming verruculose areolate and fissured, the decorticate parts exposing the inner medullary layer which also becomes fissured and torn, the whole appearance suggesting half-dead podetia, this resemblance heightened by the pale white to bluish green or olive-green color.

Apothecia on the tips of the podetia and branches, dark brown and quite large, larger than in C. *capitata.* Pycnidia on the primary squamules.

Reactions: K+ yellow, KC−, P−, or rarely P+ red.
Contents: atranorin.

This species grows on soils, especially those rich in humus. It is circumpolar, arctic to temperate, but lacking in the high arctic. Its southern extent of range in North America is to Connecticut, Wisconsin, Arizona, and New Mexico.

12. Cladonia carneola (Fr.) Fr.
Lich. Eur. Ref. 233. 1831. *Cenomyce carneola* Fr. Sched. Crit. 23. 1824.

Primary squamules persistent or disappearing; middle-sized to large, up to 13 mm long and 10 mm wide but usually much less, broadly lobed or narrowly laciniate, crenate, flat or convolute, ascending; upper side yellowish glaucescent or yellowish to olive-green; underside white or yellowish, darkening toward the base; esorediate or with the underside sparsely granulose. Podetia arising from the primary squamules, up to 40 mm tall but usually much less; cup-bearing, the cups tapering from the base much like those of C. *deformis* or more abruptly flaring, entire margined or with proliferations from the margin, the interior of the cup not opening into the interior of the podetium; the base or varying heights of the podetium corticate, the cortex subcontinuous to areolate; the rest of the podetium yellowish farinose sorediate or rarely granulose sorediate, usually esquamulose or with a few squamules at the base; yellowish or yellowish glaucescent.

Apothecia middle-sized to large, up to 6 mm broad, on stipes or sessile around the margins or from the centers of the cups; pale, waxy yellowish or pale brown. Pycnidia from the margins of the cups or on the short stipes around the edges of the cups.

Reactions: K−, KC+ yellow, P−.
Contents: usnic acid ± zeorin.

This species grows on soil rich in humus and on rotting wood. It is a northern circumboreal and arctic species with the North American range mainly in Canada and Alaska and in the west into the northern states south to Oregon.

13. Cladonia cenotea (Ach.) Schaer.
Lich. Helvet. Spic. 35. 1823. *Baeomyces cenoteus* Ach., Meth. Lich. 345. 1803.

Primary squamules persistent or disappearing; small, 1–3 (−5) mm long and 1 mm

Cladonia cariosa (Ach.) Spreng.

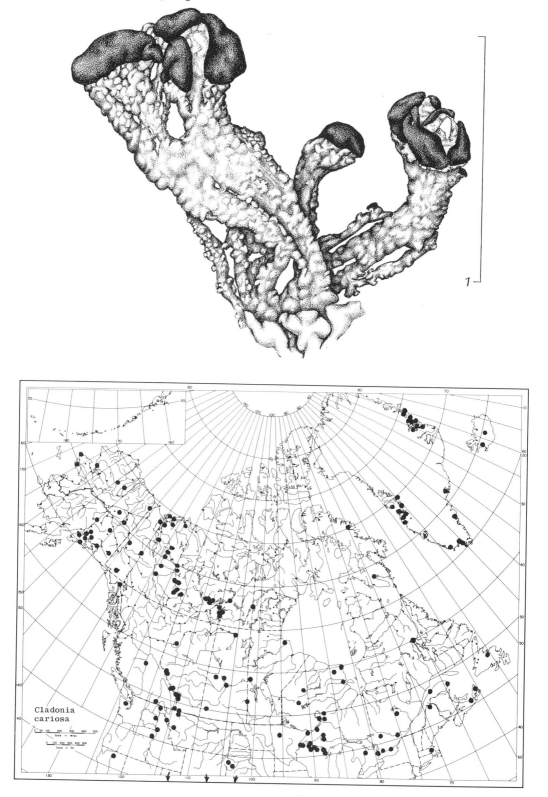

1

Cladonia
cariosa

Cladonia carneola (Fr.) Fr.

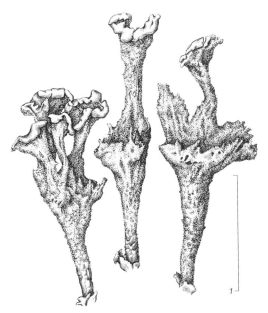

broad; thick; irregularly or subdigitately divided, the lobes crenate, ascending, flat or involute, scattered or tufted; the upper side glaucescent or pale to olive-green or brownish; the underside white; esorediate or the underside becoming sparsely sorediate. Podetia arising from the upper side of the primary squamules; the base persistent or dying; 5–100 mm tall and up to 5 mm in diameter; usually cup-bearing, the cups flaring gradually or abruptly, distinctly perforate and the margins inrolled into the perforation, the margins usually oblique, proliferate, with the proliferations acute or bearing small secondary cups, the podetia not usually branched below the cups; the cortex disintegrating over most of the upper part of the podetium, becoming farinose sorediate; whitish or variegated or brownish or greenish with the fine soredia; the lower portion corticate and more or less squamulose; the dying parts becoming black.

Apothecia small, livid brown or waxy pale brown, on the margins of the cups or the

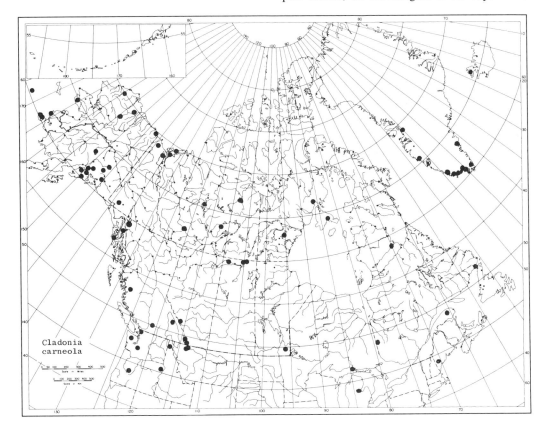

Cladonia
carneola

Cladonia cenotea (Ach.) Schaer.

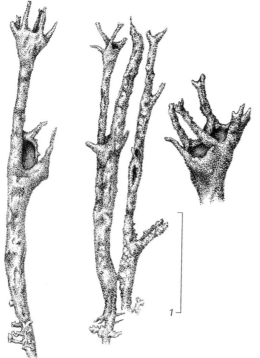

tips of the irregular proliferations, subsolitary, rarely conglomerate. Pycnidia on the margins of the cups.

          Reactions: K−, KC−, P−.
          Contents: squamatic acid.

    This species grows on soil rich in humus, on rotting wood, on stumps, occasionally on thin humus over rocks. It is circumpolar in the northern hemisphere and also occurs in Australia. It is boreal and arctic but not high arctic, in North America ranging south to California in the west and southward in the Appalachian Mountains in the east.

14. **Cladonia chlorophaea** (Flörke) Spreng.

in Linn., Syst. Veg. 16th ed. 4:273. 1827. *Cenomyce chlorophaea* Flörke ex Sommerf., Suppl. Fl. Lappon. 130. 1826. *Cladonia pyxidata* var. *chlorophaea* Flörke, Clad. Comm. 70. 1828.

    Primary squamules persistent or disappearing, small to middle-sized, 4–7 (–15) mm long, almost as broad, incised crenate or laci-

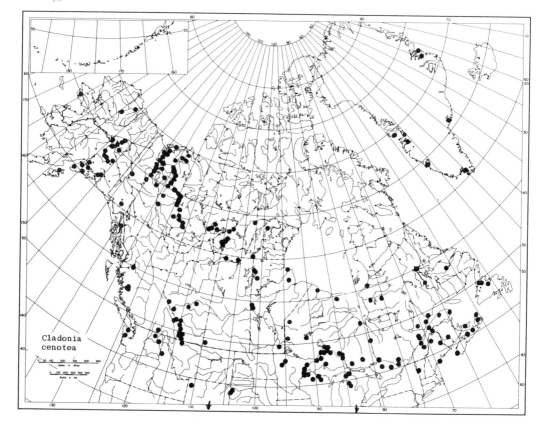

Cladonia
cenotea

Cladonia chlorophaea (Flörke ex Sommerf.) Spreng.

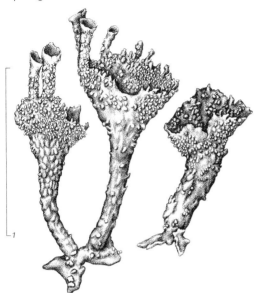

niately lobed; margin crenate, ascending, more or less involute-concave; upper side glaucescent or olive glaucescent or pale glaucescent, dull to rarely shining; underside white, darkening toward the base, esorediate or with sparse granulose soredia on the underside. Podetia cup-bearing; the cups flaring gradually, goblet-shaped, regular or irregular and with proliferations from the edges; if sterile the margins of the cups usually entire; the base of the podetium corticate; the cortex areolate or verruculose; podetia glaucescent or ashy to olivaceous; the upper part of the podetium and the inner surface of the closed cups decorticate and granulose sorediate; the decorticate areas becoming white.

Apothecia sessile, or stipitate from the margins of the cups; brown. Pycnidia on the margins of the cups or on the primary squamules.

Reactions: K−, or rarely K+ yellow, KC−, P+ red.

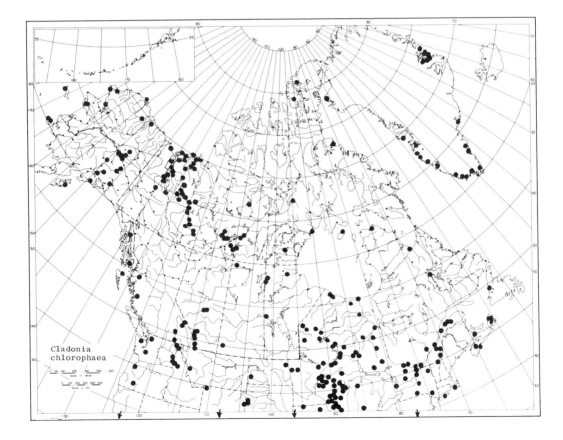

Cladonia chlorophaea

Contents: fumarprotocetraric acid, rarely with atranorin.

This species grows on soil, tree bases, rotting wood, on thin soil over rock outcrops, rarely on thick humus. It is a cosmopolitan species and in North America occurs from the low arctic through the boreal forest to temperate regions, including the southern states where it becomes rarer than in the north. This group has long been a source of difficulty in taxonomy with the morphological characters only weakly expressed. Within the recent past there have been a number of papers clarifying in part the intricate chemistry of the *C. chlorophaea* complex. Strong reliance has been placed on the chemistry and various segregates have been sorted based upon the particular substances which they contain. These segregates have been accepted as of species rank based upon the data in the papers in the bibliography and my own experience. There are apparent trends of morphology, ecology, and distribution patterns as well as the more easily determined chemical characters on which to base this decision. As treated here *C. chlorophaea* therefore becomes the residue of the sorediate goblet-shaped Cladonias which do not contain more than fumarprotocetraric acid, the latter not being deemed of taxonomic value in this group of lichens.

15. Cladonia coccifera (L.) Willd.
Fl. Berolin. 361. 1787. *Lichen cocciferus* L., Sp. Pl. 1151. 1753.

Primary squamules persistent or disappearing; small, or becoming as large as 12 mm × 5 mm, irregularly crenately incised or lobed; upper side yellowish to glaucescent or olive-colored, underside white or rarely yellowish, the base becoming orange to blackish-brown; esorediate. Podetia becoming up to 50 mm tall but usually shorter; the cups goblet shaped, flaring gradually, often oblique at the mouth, simple or with marginal proliferations which bear apothecia or more rarely cups in a second tier; corticate, the cortex yellowish to glaucescent, subcontinuous at the base and becoming verruculose areolate above, the verruculae contiguous or dispersed; esorediate, with or without squamules, the decorticate part opaque, white, or yellowish white.

Apothecia large, scarlet. Pycnidia occurring especially on the margins of the cups but also on the tips of proliferations from the margins.

Reactions: K−, KC+ yellow to orange, P−.

Contents: usnic acid, barbatic acid.

This species grows on soil, over rocks, and on rotten logs; it is abundant in muskegs on humus. Its range is circumpolar in the Arctic and boreal forest. It is also in the southern hemisphere and in the higher ranges in South America. In North America it ranges south into the northeastern states and into Colorado in the west.

16. Cladonia coniocraea (Flörke) Spreng.
in Linn., Syst. Veg. ed. 16. 4:272. 1827. *Cenomyce coniocraea* Flörke, Deutsch. Lich. 7:11. 1821.

Primary squamules persistent, larger than those of the previous few species, middle-sized to fairly long, thick, crenate or entire margined, convex or concave; upper side olive-green or glaucescent to brownish; underside white, becoming granulose sorediate. Podetia from the upper side of the primary squamules; up to 30 mm tall, usually less, 1–2 mm thick; markedly tapering, subulate, or with tiny cups at the tips; the interior of the cups sorediate; unbranched or slightly branched; only a small part of the base, or the part immediately below the apothecia corticate with subcontinuous or areolate to verruculose cortex, the major part decorticate and farinose sorediate, forming a whitish or yellowish or greenish powder; with or without podetial squamules.

Apothecia brown, growing from the margins of the cups or on the tips of the podetia. Pycnidia on the tips of the podetia or occasionally on the primary squamules.

Reactions: K− or K+ faintly brown, KC−, P+ red.

Contents: fumarprotocetraric acid.

This species grows on humus, rotting logs, tree bases, and over rocks in moist places. It is in the southern as well as in the northern hemisphere where it is circumpolar, boreal, to south temperate, reaching the Arctic only along the northern edge of the boreal forest and in Greenland.

Cladonia coccifera (L.) Willd.

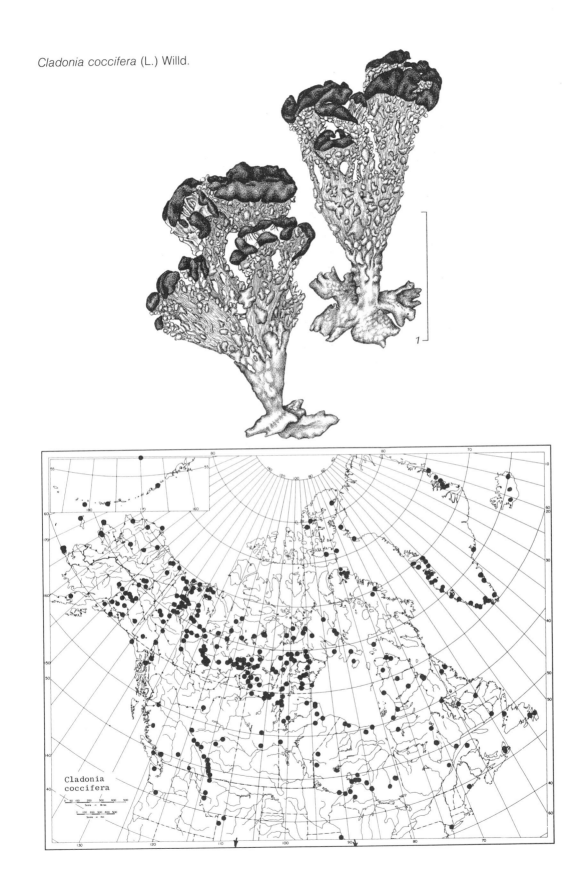

Cladonia
coccifera

17. Cladonia conista (Nyl.) Robb.

in Allen, Rhodora 32:92. 1930. *Cladonia fimbriata* f. *conista*
Nyl., Ann. Sci. Nat. Bot., Sér. 4, 15:370.1861. *Scyphophora
conista* (Ach.) S. F. Gray, Nat. Arrang. Brit. Clad. 421. 1821.

Podetia goblet-shaped as in *C. chlorophaea*, the base more smoothly corticate than
that of the other species in the group. The soredia are farinose rather than granular and the
color a lighter green than in most members of
the group. Occasionally the soredia may be
granular, but Kristinsson has shown that the
majority are less than 100 mu instead of over
100 mu with a peak at 50 mu vs 100 mu (1971).

Reactions: K−, C−. KC−. P+ red.

Contents: bourgeanic acid (substance H)
and fumarprotocetraric acid.

This species grows on soil and rotting logs.
The range is not clearly known at present. It is known
from Europe, Asia, and New Zealand as well as North
America where it ranges from Nova Scotia to southern Alaska, south to Virginia and Colorado. It is
probably circumpolar boreal and temperate. Outliers
of the range into the tundra occur in the Northwest
Territories at Ennadai Lake and the Gordon Lake area
near Great Slave Lake.

18. Cladonia cornuta (L.) Hoffm.

Plant. Lich. 1: pl. 25. 1791. *Lichen cornutus* L., Sp. Pl. 1152.
1753.

Primary squamules usually disappearing, middle-sized, 3–8 mm long, 1–4 mm broad,
irregularly lobed, sinuate or crenate or incised-
crenate, ascending; upper side glaucescent or
olive-green; underside white, esorediate or rarely
with the underside sparsely granulose. Podetia
arising from the upper side of the primary squamules, tall and slender, 20–120 (−160) mm tall
and up to 2.5 (−5) mm in diameter, usually
cupless and cylindrical or tapering; cups, when
present, very narrow, 2–4 mm broad, gradually
or abruptly flaring, the interior usually closed,
the margins proliferating or simple, the tips tapering; the cortex continuous to subareolate with

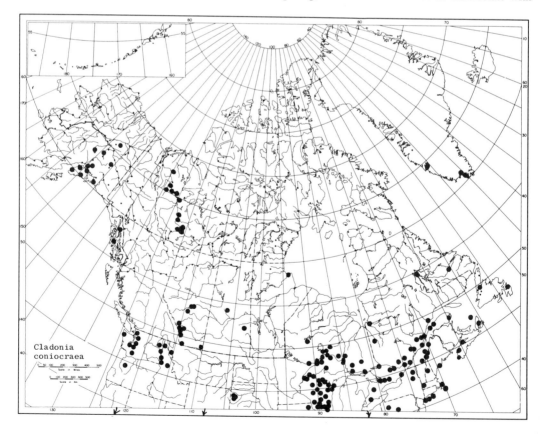

Cladonia
coniocraea

Cladonia conista (Nyl.) Robb.

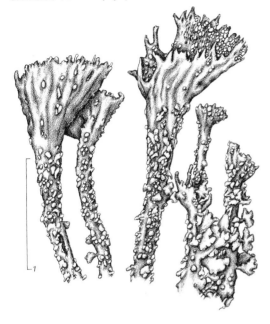

smooth areoles; the upper part of the podetium becoming sorediate in definite patches with farinose soredia or the entire top becoming covered with soredia, usually without squamules but in one form with squamules similar to the primary squamules on the podetia.

Apothecia on the margins of the cups or rarely on the apices of podetia, brown or reddish brown or pale, middle-sized, up to 4 mm broad, subentire or perforated or conglomerate. Pycnidia on the margins of the cups.

Reactions: K−, KC−, P+ red or orange-red; young parts and sorediate areas K+ yellow from abundance of fumarprotocetraric acid.

Contents: fumarprotocetraric acid.

This species grows on soil rich in humus, on sandy soil, sometimes on rotting wood. It is circumpolar, boreal, and arctic but not high arctic. It is also in the southern hemisphere. In North America it ranges southward into the northern United States and to Colorado in the west.

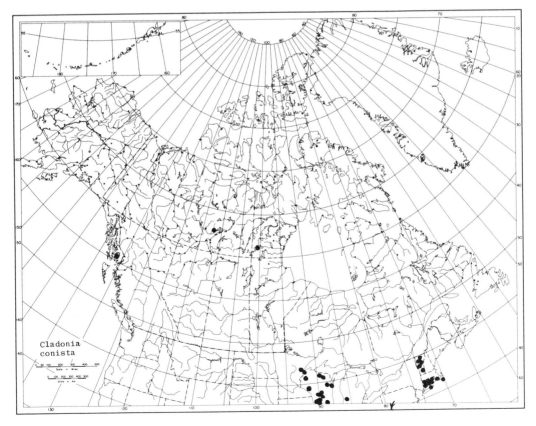

Cladonia
conista

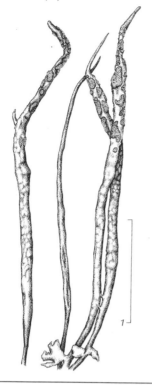

19. Cladonia crispata (Ach.) Flot.

In Wendt, Therm. Warmbr. 96.1839. *Baeomyces turbinatus ζ B. crispatus* Ach., Meth. Lich. 341. 1803. *Bo[sic]eomyces crispatus* Wahlenb., Flora Lapp. 456. 1812.

Primary squamules persistent or disappearing, middle-sized, 1–4 mm long, 0.5 mm broad, digitate-laciniate, crenate, ascending, flat or involute, scattered or in tufts, rarely forming cushions; the upper side glaucescent to olive brownish, rarely whitish glaucescent; underside white or the base browning or reddish brown, esorediate. Podetia arising from the upper side of the primary squamules; the base persistent or dying and the growth continuing from the apices; 10-110 mm tall, up to 5 mm in diameter; cupless or bearing cups, the cups perforated and open into the interior of the podetium, flaring abruptly; or cupless and the branches radiately or sympodially developed; the margins of the cups becoming repeatedly proliferate, the axils of the branches commonly perforate, the tips of the proliferations bearing cups, or blunt, or sub-

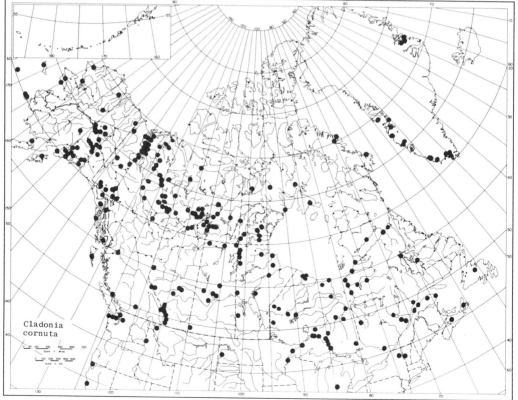

Cladonia
cornuta

Cladonia crispata (Ach.) Flot.

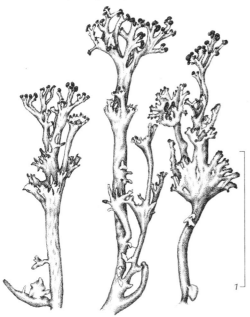

ulate, or bearing apothecia; esorediate; the cortex continuous or subcontinuous, smooth or with slightly developed areoles; shining or dull, impellucid; glaucescent or pale or olive-green or brownish to reddish brown, the dying parts becoming black.

Apothecia small, at the apices of the branches or of short subcorymbose branchlets; brown or reddish brown. Pycnidia on the margins of the cups or the apices of the branchlets or on the upper side of the primary squamules.

Reactions: K−, KC−, P−.
Contents: squamatic acid.

This species grows on earth, rotting logs, old stumps, and thin soil over rocks. It is circumpolar, arctic to temperate, and also in South America. In North America it is not in the high arctic but is common in the low arctic and in the boreal forest. It occurs in the New England states, Wisconsin and Minnesota in the central states, in the Rocky Mountains, and in California.

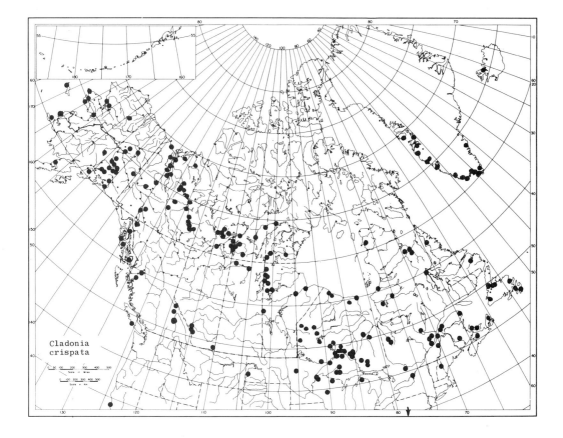

Cladonia
crispata

20. Cladonia cryptochlorophaea Asah.
J. Jap. Bot. 16:711. 1940.

Podetia goblet-shaped as in *C. chloro-phaea;* the cortex of the stalk is more coarsely verrucose than the other species of the *C. chlorophaea* complex and the soredia are coarser (Kristinsson 1971). The color is often a lighter greenish shade than that of the other species in the group.

Reactions: made in extracts only, K+ red, C+ red, KC+ red.

Contents: cryptochlorophaeic acid with accessory fumarprotocetraric acid.

This species grows on soil and the bases of trees. It is apparently circumpolar, boreal, and temperate, and also occurs in New Zealand. In North America it is common in the eastern states and eastern Canada, occurring north of the tree line in the Thelon River and Anderson river regions of Northwest Territories and south to Georgia and Texas.

21. Cladonia cyanipes (Somm.) Nyl.
Mem. Soc. Sci. Nat. Cherbourg 5:95. 1857. *Cenomyce cyanipes* Somm., Kong. Norske Videns. Skrift. 2:62. 1826.

Primary squamules persistent or disappearing; small or middle-sized, up to 6.5 mm long and 1 mm broad, usually narrowly laciniate, crenate, flat or involute concave, ascending; upper side esorediate, yellowish or yellowish glaucescent, the margins and the underside farinose or with granular soredia; the underside white or pale yellowish. Podetia 20–80 mm tall, slender, cupless; simple or sparingly branched, apices subulate or with rudimentary cups; distinctly yellowish or with a bluish tinge toward the base; the base slightly corticate or the podetium usually entirely decorticate and covered with farinose soredia.

Apothecia at the tips of the podetia; small, pale brownish or pale yellowish. Pycnidia at the tips of the podetia.

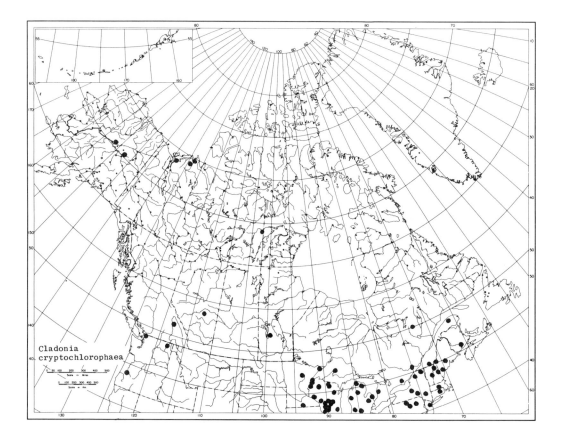

Cladonia
cryptochlorophaea

Cladonia cyanipes (Sommerf.) Nyl.

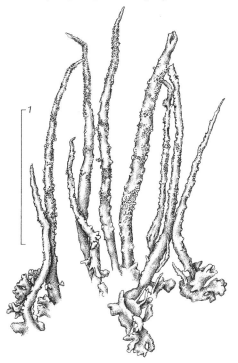

Reactions: K−, KC+ yellowish, P−.
Contents: usnic and barbatic acids.

This species grows on humus, soil rich in humus, and on rotten wood. It is circumpolar, arctic, and boreal, not in the high arctic. In North America it ranges south into the northern border states in the mountains and to Colorado and Washington.

22. Cladonia decorticata (Flörke) Spreng.
in Linn., Syst. Veg. 4:271. 1827. *Capitularia decorticata* Flörke, Beschr. Braunfr. Becherfl. 297. 1810.

Primary squamules usually persistent; small, up to 4 mm by 2 mm, irregularly or palmately divided, crenate, usually becoming concave; upper side glaucescent or olive-green or whitish glaucescent; underside white, darkening toward the base; esorediate or sparingly granulose. Podetia from the upper side of the primary squamules, 10–40 mm tall, up to 2.5 mm in diameter, cylindrical, cupless, simple or subsimple, dichotomous to trichotomous; the tips

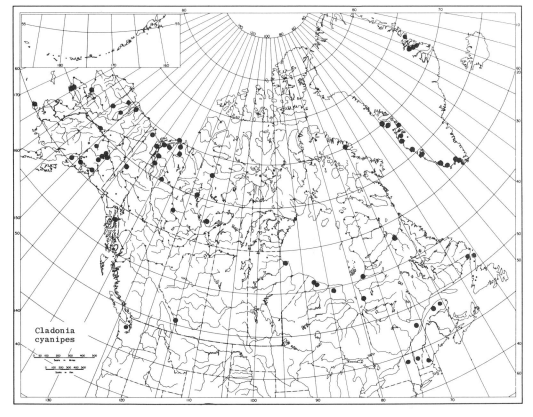

Cladonia
cyanipes

Cladonia decorticata (Flörke) Spreng.

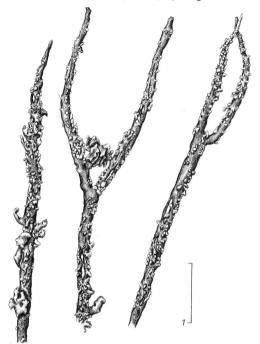

which bear apothecia dilated, those which are sterile blunt or subulate; the sides entire or fissured, sparingly granulose sorediate; cortex verruculose, the verruculae dispersed, partly developing into small squamules; the upper part of the podetium with reflexed, squarrose squamules similar to the primary squamules, the lower part becoming entirely squamulose, the part between the squamules decorticate; dull and not translucent, white or ashy or ashy brown, the verruculae and squamules white or ashy or olive-glaucescent.

Apothecia middle-sized to large or often small, confluent or lobate-fissured, dark brown or red-brown. Pycnidia on the upper side of the primary squamules and the base of the podetia.

Reactions: K−, KC−, P−.

Contents: perlatolic acid.

This species grows on soils which are sandy as well as on those rich in humus. It is circumpolar, arctic, and boreal, but not high arctic. In North

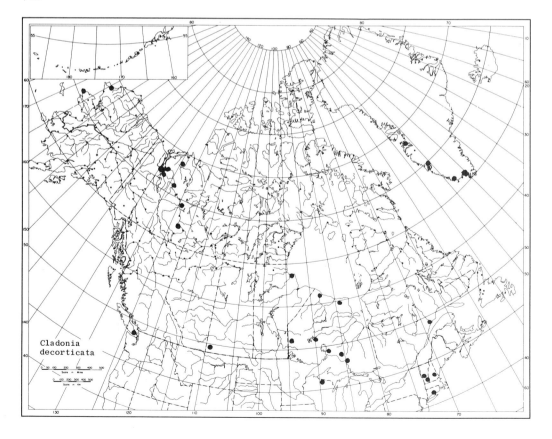

Cladonia
decorticata

Cladonia deformis (L.) Hoffm.

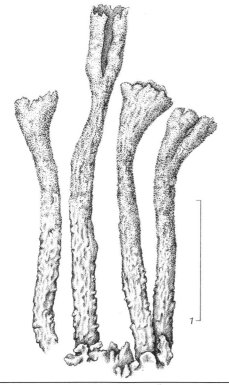

1

America it ranges south to Connecticut, New York, Wisconsin, and Colorado.

23. **Cladonia deformis** (L.) Hoffm.

Deutschl. Fl. 2:120. 1796. *Lichen deformis* L., Sp. Plant. 1152. 1753.

Primary squamules usually disappearing, small to middle-sized, 2–4 mm long, incised crenate or lobate, flat to convex or involute; upper side glaucescent to yellowish; underside white to pale brown to dark; esorediate or the underside sparingly sorediate. Podetia arising from the upper side of the squamules, 25–85 mm tall and up to 5 mm thick on the stalk; the cups gradually flaring, moderately or little expanded, the margin of the cups entire or dentate to proliferate with blunt or cup-bearing proliferations, the cups usually imperforate; the major portion of the podetium, except for the base, densely covered with fine, farinose soredia, the soredia quite yellowish; the basal cor-

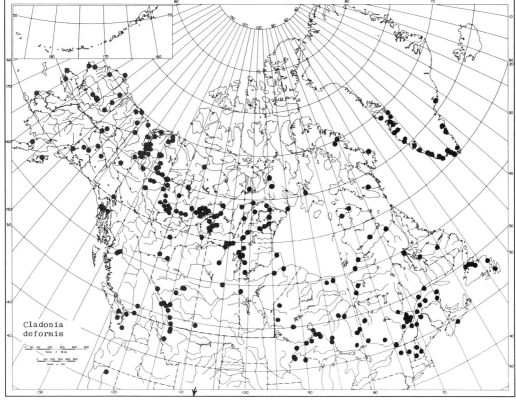

Cladonia
deformis

ticate portion continuous to chinky, yellowish glaucescent, the decorticate portion opaque.

Apothecia scarlet or pale, on the margins of the cups or on short marginal proliferations. Pycnidia on the margins of the cups or on the apices of the marginal proliferations.

Reactions: K−, KC+ yellow, P−.

Contents: usnic acid, zeorin, plus accessory bellidiflorin.

This species grows on soil rich in humus, among mosses, on rotten logs, on tree bases, and over rocks with a little humus. It is circumpolar in the Arctic and the boreal forest but is lacking in the high arctic. Southward it ranges to West Virginia in the Appalachian Mountains, and to Colorado in the Rocky Mountains.

24. Cladonia digitata (L.) Hoffm.
Deutschl. Flora 2:124. 1796. *Lichen digitatus* L., Sp. Pl. 1152. 1753.

Primary squamules largest of those of the *Cocciferae,* up to 15 mm broad, lobed and incised; upper side gray-green to olive-green; underside whitish to yellowish, deepening to brownish-orange toward base, finely sorediate especially along the margin. Podetia forming distinct cups which broaden gradually from the base, the major part of the podetium finely sorediate with whitish or yellowish or greenish soredia. Pycnidia and scarlet apothecia on the margins of the cups, many podetia sterile.

Reactions: K+ yellow, KC+ yellow, P+ deep orange to red.

Contents: Thamnolic acid and substance G, plus accessory bellidiflorin.

This species grows on soil rich in humus, on old logs and on bases of trees. Its range is circumpolar boreal and temporate but there are a few stations farther north in the tundra which could possibly represent relics of a former advance of the forest during the hypsithermal.

25. Cladonia ecmocyna Leight.
Ann. Mag. Nat. Hist. Ser. 3, 18:406. 1866. (*Non Cenomyce ecmocyna* Ach., Lich. Univ. 549. 1810).

Primary squamules disappearing. Podetia with the base dying and growth continuing from the top, with occasional branching to form large tufts; 50–100 mm tall or more; up to 2.5–5 mm in diameter; subsimple or with submonopodial branching; dull or nitid, glaucescent or whitish or pale to olive-green, or brownish in strong light; cupless or with narrow cups, with marginal or with occasional central proliferations which are long extended; the cortex continuous or subcontinuous, or areolate with smooth areoles, Apothecia brown, on the margins of the cups. Pycnidia on the margins of the cups.

Reactions: K+ yellow, KC−, P+ red.

Contents: atranorin and fumarprotocetraric acid.

This species resembles *C. gracilis* but prefers moister habitats, occurring beside late snowbanks and in bogs on soil high in humus. It is circumpolar, arctic alpine, and ranges south in the mountains in North America to New Hampshire, Maine, and Colorado. It is much more common in the western part of the continent than in the east. It tends to be much more squamulose than *C. maxima* with which it might otherwise be confused.

26. Cladonia fimbriata (L.) Fr.
Lich. Eur. Ref. 222. 1831. *Lichen fimbriatus* L., Sp. Pl. 1152. 1753.

Primary squamules persistent or disappearing; middle-sized, 2–10 mm long, digitately or irregularly lobed, crenate to sinuate or incised on the margins, flat to involute or concave above; the upper side glaucescent to olive-green or pale glaucescent; the underside white; esorediate or sparsely granulose sorediate below. Podetia from the upper side of the primary squamules; 10–20 mm tall, the stalk 1–2 mm thick; cup-bearing, the cups flaring rapidly to form trumpet shapes, entire margined or rarely proliferate, the interior closed; the surface sometimes slightly corticate at the base but mainly decorticate and farinose sorediate.

Apothecia brown; on the margins of the cups or on short stipes from the margins. Pycnidia from the margins of the cups.

Reactions: K−, or the young parts K+ brownish, KC−, P+ red.

Contents: fumarprotecetraric acid.

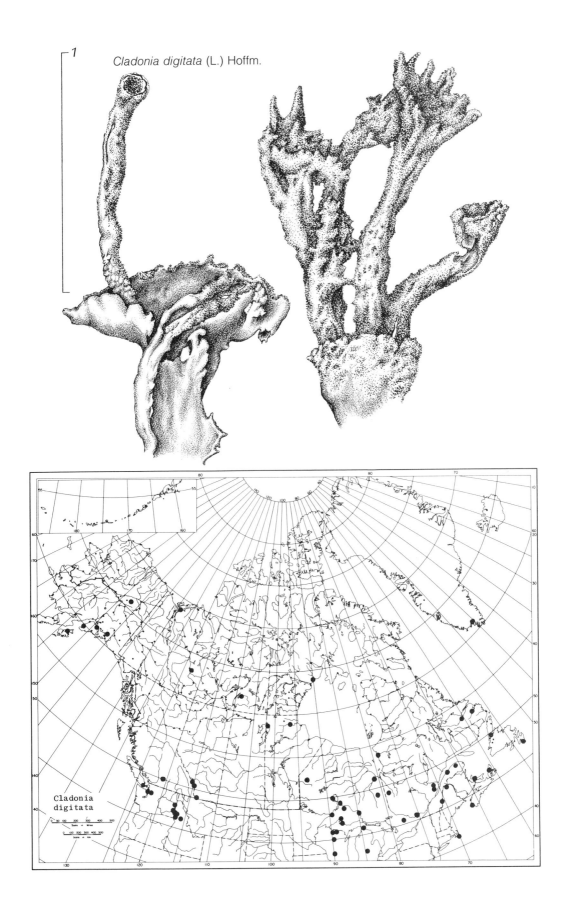

Cladonia digitata (L.) Hoffm.

Cladonia
digitata

Cladonia ecmocyna Leight. ssp. *ecmocyna*   *Cladonia ecmocyna* ssp. *intermedia* (Robb.) Ahti

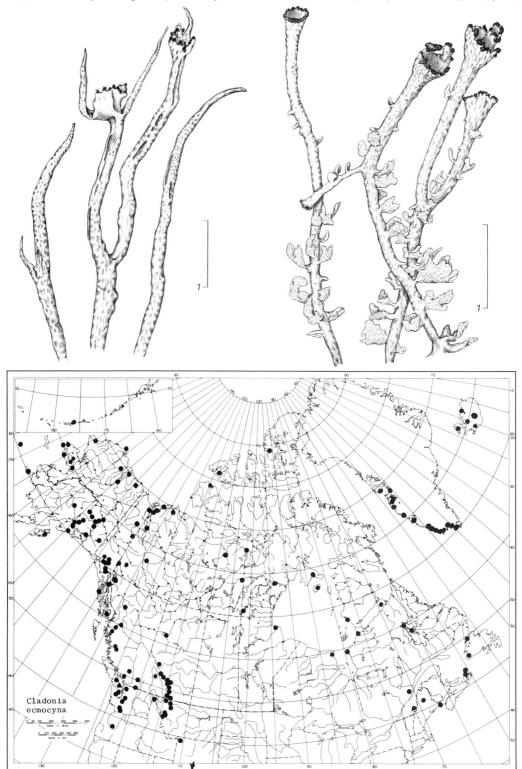

Cladonia
ecmocyna

Cladonia fimbriata (L.) Fr.

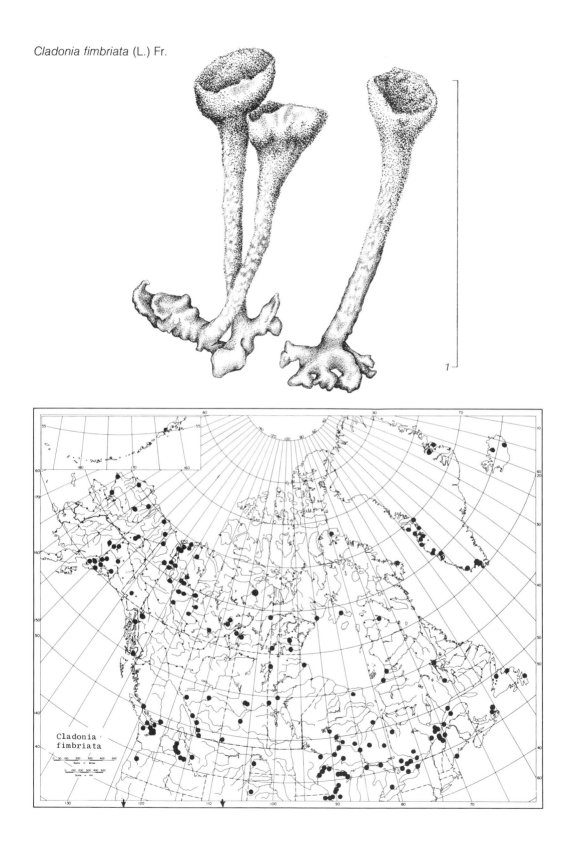

This species grows on earth, rotting logs, tree bases, and on thin soil over rock edges. It is circumpolar, boreal, and low arctic, ranging south in North America to North Carolina, Iowa, Colorado, and California.

27. Cladonia floerkeana (Fr.) Flörke

Clad. Comm. 99. 1828. *Cenomyce floerkeana* E. Fr., Lich. Suec. Exsicc. 82. 1824. *Cladonia leptopoda* Nyl., Synops. Lich. 226. 1858.

Primary squamules persistent, growing scattered or in depressed mats; small to middle-sized, up to 2 mm long, undivided to crenate to slightly lobed; upper side glaucous to olivaceous; underside white to dark at point of attachement; esorediate or the apices slightly granular sorediate. Podetia growing from the upper surface of the primary squamules; slender, cylindrical, 4–10 mm or rarely up to 45 mm tall, unbranched or sparingly branched in the upper part; the cortex variable, from smooth and unbroken to indistinct areolate, some podetia with the areolae separated by narrow bands which may be opaque or more or less translucent; most podetia with granular soredia intermingled with the areolae and more abundant on the upper part; podetial squamules present on some forms and resembling the primary squamules.

Apothecia scarlet and tipping the podetia. Pycnidia on the upper side and margins of the primary squamules, rarely also on the podetia.

Reactions: K−, KC+ orange, P−.

Contents: barbatic acid with accessory usnic and didymic acids and F.

This species grows on rotten wood, logs, on earth with humus, and on rocks in fissures containing soil rich in humus. It occurs in both the northern and southern hemispheres and in Europe and Asia as well as North America where it occurs in the east from Gaspé Peninsula west to Minnesota and south to Florida and Mississippi. It appears to reach the Arctic only in Greenland (Dahl 1950).

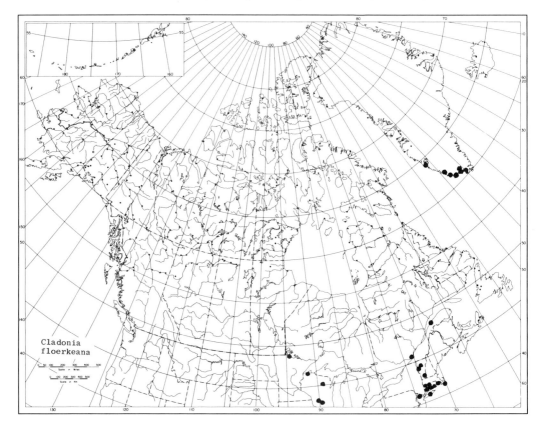

Cladonia floerkeana

Cladonia glauca Flörke

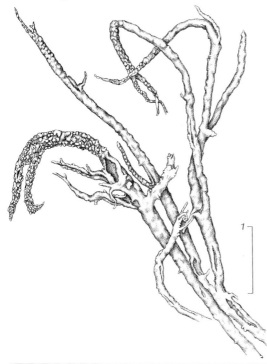

28. **Cladonia glauca** Flörke
Clad. Comm. 140. 1828.

Primary squamules persistent or disappearing; middle-sized, or small to large, 1–5 mm long and 1 mm broad; irregularly or subdigitately lobed or partly incised, the lobes crenate-incised, ascending, flat or involute, sparse or tufted; the upper side glaucescent or pale greenish white; the underside white; esorediate or the underside sparsely sorediate. Podetia arising from the upper side of the primary squamules, sometimes in small tufted groups; the base persistent or dying; 25–80 (–100) mm tall and up to 2.5 mm in diameter, cylindrical, cupless or rarely with minute cups up to 3 mm broad at the apices of the usually simple podetia, the rare branches suberect, the apices attenuate or blunt, the axils perforated or closed; entirely decorticate and farinose sorediate, or rarely with a small corticate area at the base, opaque, impellucid; whitish or ashy glaucescent or rarely ashy brown or

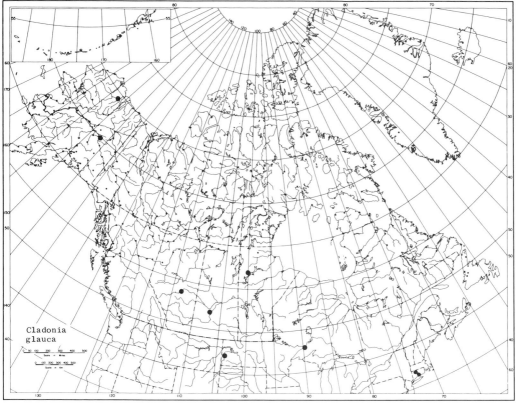

Cladonia
glauca

Cladonia gracilis ssp. *turbinata* (Ach.) Ahti

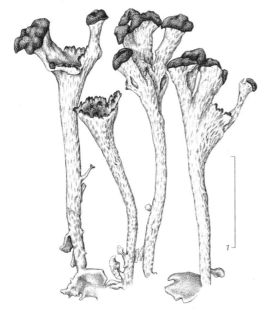

variegated; lacking squamules or with a few at the base.

Apothecia small; sometimes aggregated; at the tips of the podetia; commonly peltate and perforate; brown or yellowish-brown. Pycnidia on the podetia, at the edges of the axils and the cups, and also on the sides and apices of the podetia.

Reactions: K−, KC−, P−.
Contents: squamatic acid.

This species grows on soils rich in humus, bogs, and in heaths. It is circumpolar boreal, in North America reaching the tundra only at its border with the northern forest. It ranges southward in New England into Massachusetts and Connecticut. This species is so rarely collected that its true range in North America is largely conjecture.

29. Cladonia gracilis (L.) Willd.
Fl. Berol. Prodr. 363. 1787. *Lichen gracilis* L., Sp. Pl. 1152. 1753.

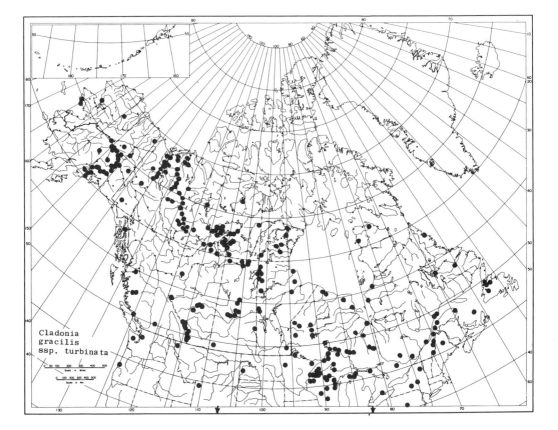

Cladonia
gracilis
ssp. turbinata

Primary squamules persistent, or persistent only at the base of the podetia, or disappearing; small to middle-sized, rarely large, 2–5 (–10) mm long, 0.8–6 mm broad; irregularly lobed or crenate or sinuate, flat or convolute, ascending; upper side glaucescent or olive-green, more greenish than in *C. verticillata* which tends to be bluish; underside white. Podetia arising from the upper side of the primary squamules, very variable, the size depending partly upon the variation collected, up to 80 mm tall; cup-bearing in most forms, subulate in some, the inner membrane of the cups closed, cups shallow and flaring quite rapidly, usually proliferate from the margins, the proliferations bearing apothecia, or cups, or occasionally subulate; esorediate with a smooth cortex which is continuous or of slightly dispersed flat areoles which are separated by smooth white bands or lines, varying in color from greenish glaucescent to olive or in sun forms to brownish green or brown,

the base becoming black but not spotted; squamulose or not.

Apothecia varying from small to large; on the margins of the cups or on marginal proliferations; dark brown. Pycnidia on the margins of the cups.

Reactions: K− or K+ yellow in the rapidly growing tips, KC−, P+ red.

Contents: fumarprotocetraric acid.

This species grows on soil, either sandy or high in humus, on rotting logs or over rocks with some soil over them. It is a nearly cosmopolitan species. In the boreal forest and temperate forests ssp. *turbinata* (Ach.) Ahti (var. *dilatata* (Hoffm.) Vain.) is common. This subspecies has short podetia, usually less than 50 mm tall with well-expanded cups and commonly with proliferations bearing apothecia on the margins of the cups. This subspecies ranges into the low arctic usually in heath tundras and willow thickets as well as in small isolated forest stands. Subspecies *gracilis* is more common in tundras and

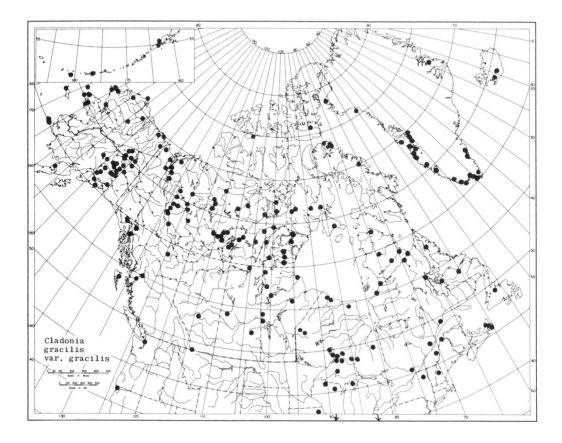

Cladonia
gracilis
var. gracilis

lichen woodlands northward. In ssp. *gracilis* (var. *chordalis* (Flörke) Schaer.) the podetium is slender with narrow or no cups and subulate podetia 30–80 mm tall. It may form quite large mats despite the few branches which are formed by the plants.

30. Cladonia granulans Vain.

Bot. Mag. Tokyo 35:65. 1921. *Cladonia blakei* Robb., Rhodora 30:138. 1931.

Primary squamules small to medium-sized, yellow above, white below, esorediate. Podetia cup-forming, the cups goblet-shaped, regular or irregular, imperforate, the margins becoming proliferate; the cortex smooth to rugose and covered with small subglobose yellowish verruculae which appear as tiny squamules, 0.2–0.3 mm long, esorediate.

Apothecia red, at the tips of proliferations.

Reactions: K−, C−, KC+ yellowish, P−, UV+.

Contents: usnic and squamatic acids and accessory bellidiflorin.

This species grows on soil. It is a member of the Beringian element which has been reported from Alaska (Krog 1968), Yellowstone Park, Wyoming (Yoshimura 1968), and the Reindeer Reserve, N.W.T. (Ahti et al. 1973). Although Evans had annotated the type material of *C. blakei* as containing barbatic acid, Yoshimura (1968) found that it contained squamatic acid instead and corrected the Yellowstone material. *C. granulans* was originally described from material from Japan and is also known from Kamchatka. It is apparently very close to *C. metacorallifera* but lacking the didymic acid present in that speices.

31. Cladonia grayi Merrl.

ex Sandst., Clad. Exs. 1847. 1929.

Primary squamules usually persistent, small to middle-sized. Podetia goblet-shaped, much like those of *C. chlorophaea;* sorediate with soredia about the same size as those of *C. conista* but smaller than those of *C. cryptochlorophaea* (Kristinsson 1971), the podetia tending to be more rugose than those of *C. chlorophaea* and with a browner tinge (Ahti 1966).

Reactions: K−, KC−, P− or P+ red.

Contents: grayanic acid with accessory fumarprotocetraric acid.

This species grows on soil and on bases of trees. It is circumpolar, boreal, and temperate, reaching the Arctic along the north edge of the tree line. Although its range was thought to be disjunct from Manitoba to Alaska (Bowler 1972), more recent data suggest that the disjunction is probably only due to lack of collections. Southerly the range extends to Florida and Texas.

32. Cladonia kanewskii Oxn.

Ukrain. Bot. Zhurn. 3:9. 1926. *Cladonia nipponica* var. *aculeata* Asah., Lich. Japan 1:123. 1950. *Cladonia nipponica* var. *sachaliensis* Asah., ibid. *Cladonia nipponica* var. *aculeata* f. *prolifera* Asah., ibid.

Primary thallus unknown. Podetia pale greenish gray to yellowish gray, 5–12 cm tall, to 3 mm thick, branching sparse, by isotomic dichotomy, the branches ascending, subulate; smooth, slightly glossy; inner surface smooth with narrow cracks which expose floccose hyphae, with inner cartilaginous strands forming a continuous uneven layer around the central canal, some strands scattered in the medulla; cups usually lacking but if present to 3 mm broad, axils closed or perforate.

Apothecia dark brown, peltate, soon convex; spores 8 × 3–4 u.

Reactions: K−, C−, KC+ yellow, P−.

Contents: only usnic acid plus abundant ''diterpene'' crystals on tips of podetia.

This species grows mainly in alpine tundras in moist sites. It is a member of the Beringian element in North America growing on montane coastal and inland tundras and on the coastal tundras in western Alaska but not as yet known from the north slope. My previous reports were under *C. nipponica* (1967). The situation regarding these two species has been clarified by Ahti (1973). Ahti and Brodo (1981) discuss this species and a related new species of the boreal forest, *Cladonia labradorica* Ahti & Brodo, which is more branched, less robust, with poorly distinct cortex, and containing usnic acid plus an unknown. It grows on the Ungava Peninsula.

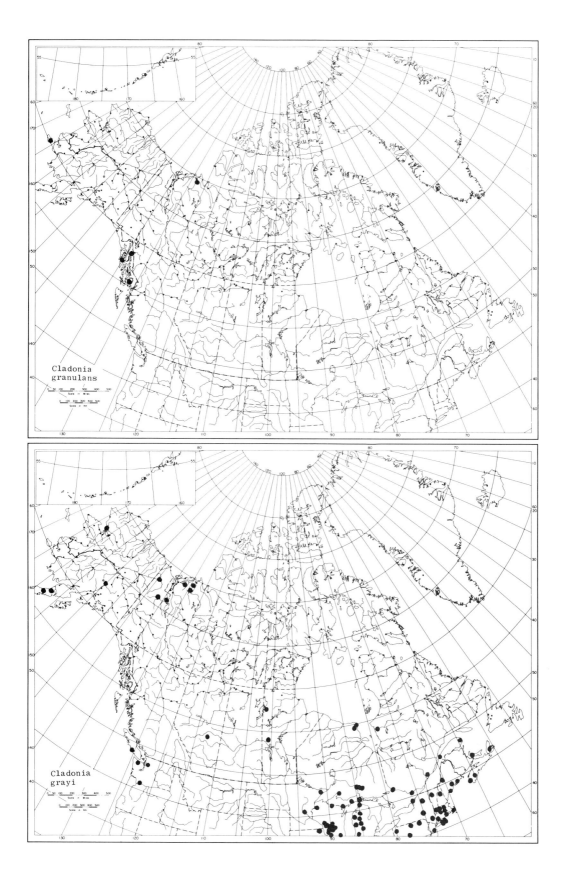

Cladonia
granulans

Cladonia
grayi

Cladonia kanewskii Oxn.

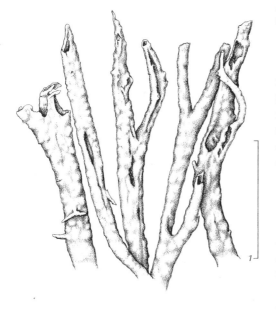

33. **Cladonia macrophylla** (Schaer.) Stenham.

Öfvers. Kungl. Vet. Akad. Forhandl. 1865(4):231. 1865. *Cladonia ventricosa β. macrophylla* Schaer., Lich. Helv. Spicil. 316. 1833. *Cladonia decorticata* var. *alpicola* Flot., Flora 8:340. 1825. *Cladonia alpicola* (Flot.) Vain., Acta Soc. F. Fl. Fenn. 10:58. 1894.

Primary squamules persistent, large or rarely middle-sized, length and breadth 3–8 mm; irregularly lobed, crenate or entire or lobate-incised, concave or involute; upper side glaucescent or rarely olive-green; underside white or darkening toward the base; esorediate. Podetia arising from the primary squamules, solitary or rarely several per squamule, 10–60 mm tall, up to 5 mm in diameter but usually less than the maximum; cylindrical, cupless, commonly tipped with apothecia; those podetia with apothecia with fewer branches than those which are sterile, the branches widely spreading, the fertile ones thickening, the sterile ones blunt or subulate, the sides more or less fissured as in *C. cariosa;* esorediate to verruculose, granulose

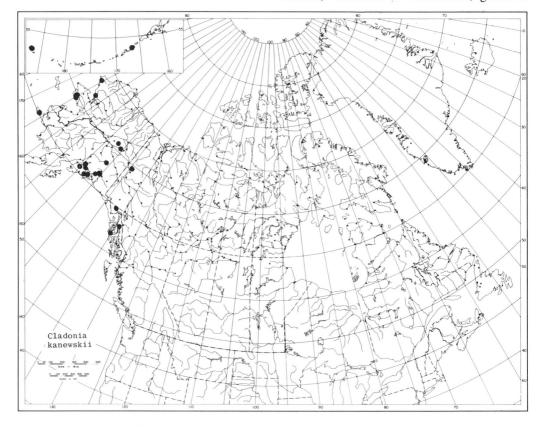

Cladonia
kanewskii

Cladonia macrophylla (Schaer.) Stenham.

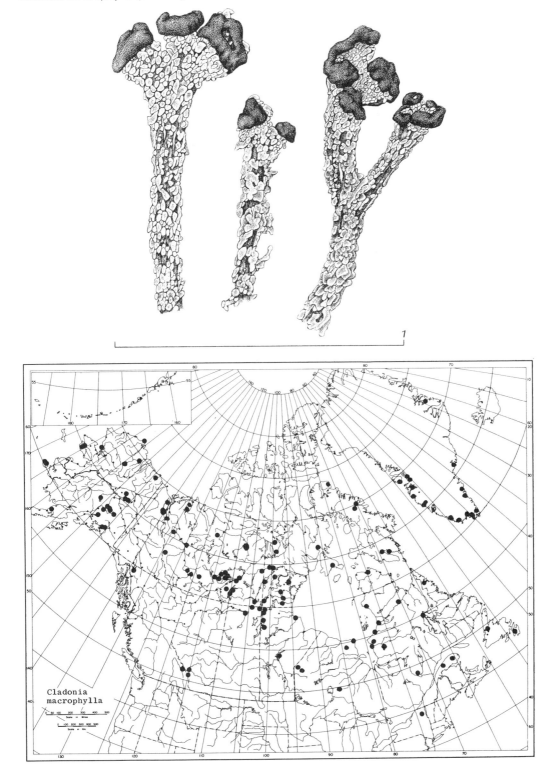

above but not sorediate; cortex areolate to verruculose with *peltate* squamules which are usually dispersed and up to 1 mm broad, the decorticate part of the podetium opaque, ashy to pale brown, the verruculae and squamules glaucescent to olive-green to brownish-green.

Apothecia brown, confluent or conglomerate or fissured-lobate, sometimes with squamules among the disks. Pycnidia on the apices and sides of the podetium.

Reactions: K−, KC−, P+ yellow.

Contents: psoromic acid plus an unknown acid.

This species grows on soils high in humus and on soil over rocks. It is circumpolar, arctic, alpine, a few disjunct stations extending southward in the boreal forest and on mountains as Mt. Marcy in the Adirondacks and in the Rocky Mountains in Alberta and British Columbia.

34. Cladonia macrophyllodes Nyl.
Flora 58:447. 1875.

Primary squamules persistent; large, 8–15 mm long and 1–8 mm broad; lobate or incised lobed, concave or involute, ascending; dispersed or caespitose; upper side pale to olive-green; underside white, the base blackening esorediate. Podetia from the upper side of the primary squamules; 7–15 mm tall, up to 2 mm in diameter; bearing cups which are up to 7 mm broad, the cups gradually or quite abruptly flaring, shallow, simple, or with short proliferations from the center, entire to becoming split on the margins; the cortex continuous or smoothly areolate, the areoles close or becoming somewhat dispersed; cortex whitish or olivaceous green, the base becoming black; esorediate.

Apothecia on the margins of the cups or on short stipes on the margins; dark brown. Pycnidia on the margins of the cups and on the primary squamules.

Reactions: K+ yellow, KC−, P+ red.

Contents: atranorin and fumarprotocetraric acid.

This species grows on mineral soil and on rocks. It is an arctic alpine circumpolar species but lacking in the high arctic.

35. Cladonia major (Hag.) Sandst.
Abhandl. Naturw. Ver. Bremen 25:223. 1922. *Lichen pyxidatus* b. *major* Hag., Tent. Hist. Lich. 113. 1782.

The description of this species is as for *C. fimbriata,* except that the podetia are 20–40 mm tall and up to 3 mm in breadth of stalk.

Reactions: K−, or K+ yellowish in the young parts, KC−, P+ red.

Contents: fumarprotocetraric acid.

This species grows on soil and rotting wood. It is known from Europe and North America. Collections of this species are so uncommon that the range is ill defined. It appears to be mainly boreal to temperate with the southern range extending to North Carolina, Wisconsin, Nebraska, and to California.

36. Cladonia maxima (Asah.) Ahti
Ann. Bot. Fenn. 15:12.1978. *Cladonia gracilis* var. *elongata* f. *maxima* Asah. Atlas Japan. Cladon. 19. 1971.

Primary squamules disappearing. Podetia tall and stout, $50-100 \times 1-5$ mm, unbranched or with few branches, cupless or with short narrow cups 2–6 (–12) mm broad, quite abruptly flaring, regular or with proliferations on the edges, the proliferations bearing pycnidia or apothecia, with few squamules; the cortex continuous to low areolate, whitish or bluish, to brownish olive, the base dying and turning yellowish brown. The apothecia at the tips of the proliferations, brown to brownish black.

Reactions: K− or K+ yellow at the tips, KC−, C−, PD+ red.

Contents: fumarprotocetraric acid and rarely small amounts of atranorin in TLC according to Ahti (1978a).

This species grows among mosses, usually in very moist habitats. It is probably circumpolar and according to Ahti comprises the bulk of the material that has been classified as *C. gracilis* var. *elongata.* Other material which has been referred to *C. gracilis* var. *elongata* belongs to *C. gracilis* var. *nigripes* according to Ahti and is to be distinguished according

Cladonia macrophyllodes Nyl.

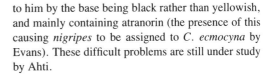

to him by the base being black rather than yellowish, and mainly containing atranorin (the presence of this causing *nigripes* to be assigned to *C. ecmocyna* by Evans). These difficult problems are still under study by Ahti.

37. Cladonia merochlorophaea Asah.
J. Jap. Bot. 16:713. 1940.

Primary squamules persistent. Podetia cup-shaped as in *C. chlorophaea* usually in looser and more brittle mats, darker in color, brownish green to chocolate brown or even black. Ahti (1966) distinguished two types of this species: type I with the entire surface rough, corticate, minutely verrucose, and dark in color, the inner surface of the cups areolate corticate or obscurely sorediate with coarse gray-brown granules, proliferations flattened and with large dark brown apothecia, the outer surface distinctly corticate; type II dirty grayish green rather than brown and covered with farinose soredia.

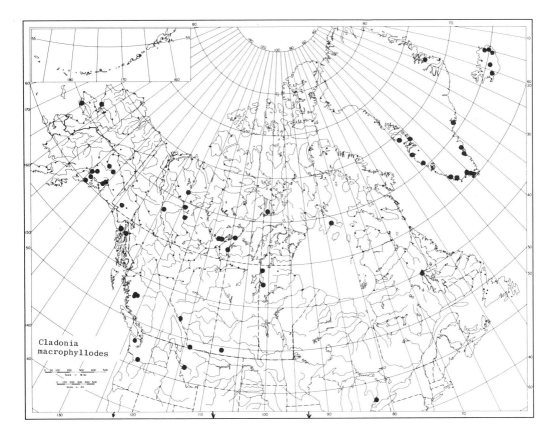

Cladonia
macrophyllodes

Cladonia major (Hag.) Sandst.

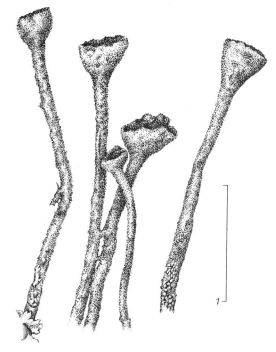

Reactions: K−, KC+ purplish-red in extracts only, P+ red or P− in type I podetia, P+ red in type II.

Contents: merochlorophaeic, 4−0-methylcryptochlorophaeic acid and accessory fumarprotocetraric acid. The "novochlorophaeic acid" strain actually contains sekikaic acid and homosekikaic acid which cocrystallize (Culberson & Kristinsson 1969).

This species grows on soil containing much humus, on tundra heaths, and on bogs. Type II is reported by Ahti (1966) to be more mesic than type I, occurring among mosses and on rotting wood. The species is circumpolar and boreal but with a few collections having been made in tundras. It seems to be a northern representative of the *C. chlorophaea* complex, ranging south only to Pennsylvania and Washington (Bowler 1972).

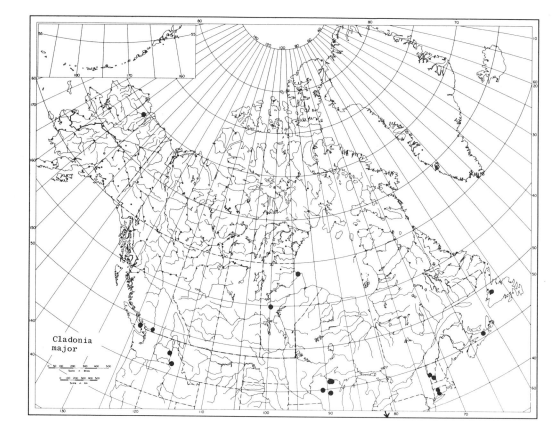

Cladonia major

Cladonia maxima (Asah.) Ahti

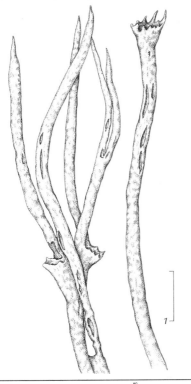

38. Cladonia metacorallifera Asah.
J. Jap. Bot. 15:612. 1939.

Primary squamules persistent; small, 1–2 mm long, 2–4 mm broad, crenate or incised, becoming involute, sparse or dense, appressed or ascending; upper side yellowish or yellowish green; underside white; esorediate. Podetia cup-forming, the cups flaring gradually from the base and goblet-shaped or abruptly from a long base and trumpet-shaped; the margins subentire, rarely proliferate, yellowish, esorediate; but most of the surface with crowded verruculae which are smaller and more protuberant than in *C. coccifera,* many showing a tendency to rise from the deeper tissues and give rise to minute, more or less appressed squamules.

Apothecia scarlet, on marginal stipes or short proliferations. Pycnidia on the margins of the cups.

Reactions: K−, KC+ yellow, P−.

Contents: Usnic, squamatic, didymic acids and bellidiflorin.

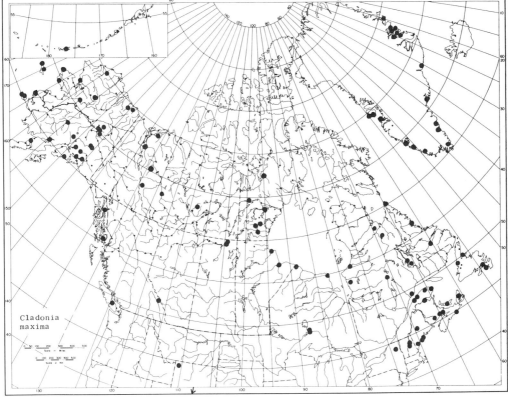

Cladonia
maxima

This species grows on soil rich in humus, often on hummocks in the tundra. The world range includes Chile and Argentina as well as Japan and North America. In the latter it behaves as a Beringian radiant with the easternmost colonies in the Great Slave Lake area and southward to southern Alberta.

39. Cladonia mitis Sandst.

Abh. Nat. Ver. Bremen 25:105. 1922. *Cladonia mitis* Sandst. Clad. Exs. no. 55. 1918 *(nom. nud.). Cladonia arbuscula* ssp. *arbuscula* chem. strain II, Ahti, Ann. Bot. Soc. Zool. Bot. "Vanamo" 32. 1:108. 1961. *Cladina mitis* (Sandst.) Hale & W. Culb., Bryologist 73:510. 1970.

Primary thallus unknown. Podetia in large mats or in tufts, intricately, loosely branched 30–70 (–100) mm tall and 1–1.5 (–4) mm in diameter; the anisotomic branching typically in whorls of 3 or 4, sometimes dichotomous, forming a sympodium which has more remote nodes than in *C. arbuscula,* each node with 1–2 short branches ramified in turn in the same unequal fashion; toward the tips the branching pattern becoming equal, the sterile branchlets ending with 2–5 points all recurved to one side, or in the case of unequal branches with some recurved and others spreading, the fertile branchlets mainly in short corymbs, the entire appearance less tufted and less well combed than in *C. arbuscula;* the axils mainly open; clear yellow or with a bluish cast varying toward glaucescent, rarely to sulfur yellow, the extreme tips scarcely browned; the surface dull, scarcely arachnoid-tomentose, smooth or verruculose with age; mainly impellucid, rarely semipellucid; normally lacking soredia.

Apothecia small, borne singly or more or less in corymbs at the tips of the branches, brown. Pycnidia at the tips of the branches; containing colorless jelly.

Reactions: K−, KC+ yellow, P−.

Contents: usnic acid constant plus accessory rangiformic acid and substances A and B and rarely psoromic acid.

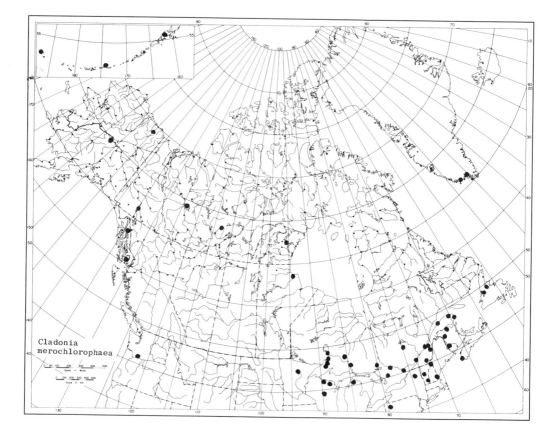

Cladonia merochlorophaea

Cladonia metacorallifera Asah.

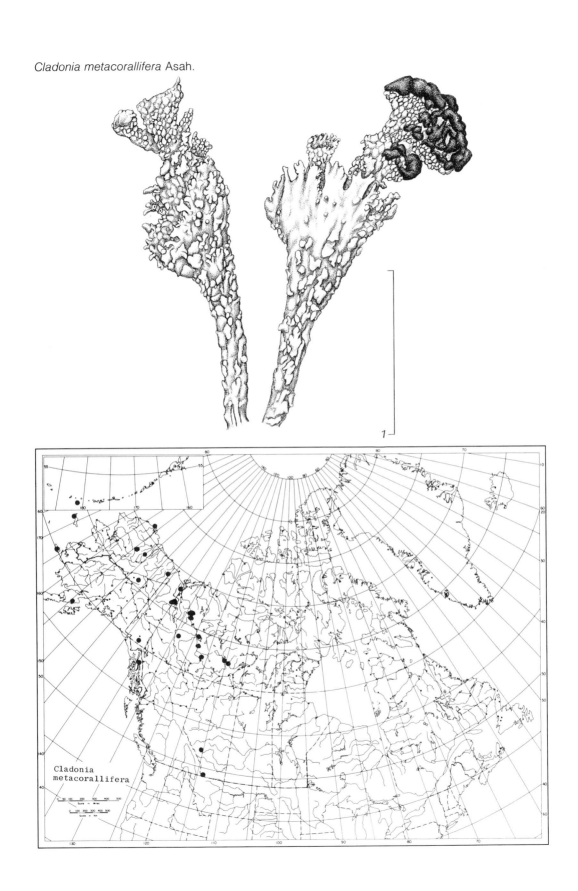

Cladonia mitis Sandst.

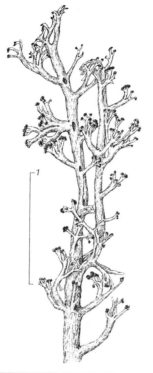

This species grows on sandy soils and in bogs as well as on soil rich in humus; it is especially common in the spruce forests of the north. It may also be found over earth on rock outcrops. This is an arctic to temperate circumpolar species, in North America ranging southward to Virginia, Wisconsin, Colorado, and Oregon.

40. Cladonia multiformis Merr.

Bryologist 12:1. 1909. *Cladonia multiformis* Merr., Bryologist 11:110. 1908 *(nom. nud.)*. *Cladonia furcata* var. *subcrispata* Nyl. in Linds., Trans. Royal Soc. Edinburgh 22:168. 1859. *Cladonia crispata* var. *subcrispata* (Nyl.) Vain. Acta Soc. F. Fl. Fenn. 4:385. 1887. *Cladonia subcrispata* (Nyl.) Vain., Acta Soc. F. Fl. Fenn. 53:52. 1922.

Primary squamules persistent or disappearing, middle-sized, 1–4 mm long, up to 0.5 mm broad, digitately lobed, ascending, flat to involute; the upper side glaucescent to olive-green or brownish; the underside white, darkening toward the base; esorediate. Podetia from the upper side of the primary squamules, up to 45 mm tall and about 1–2 mm in diameter; cup-bear-

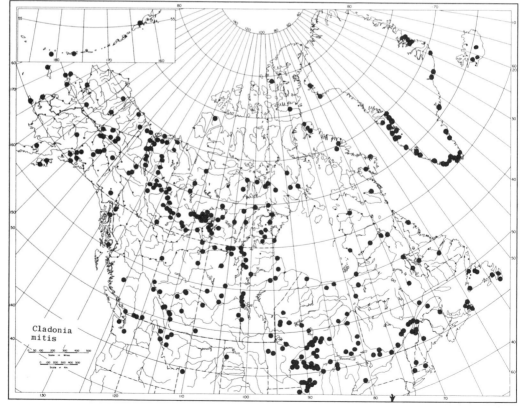

Cladonia
mitis

Cladonia multiformis Merr.

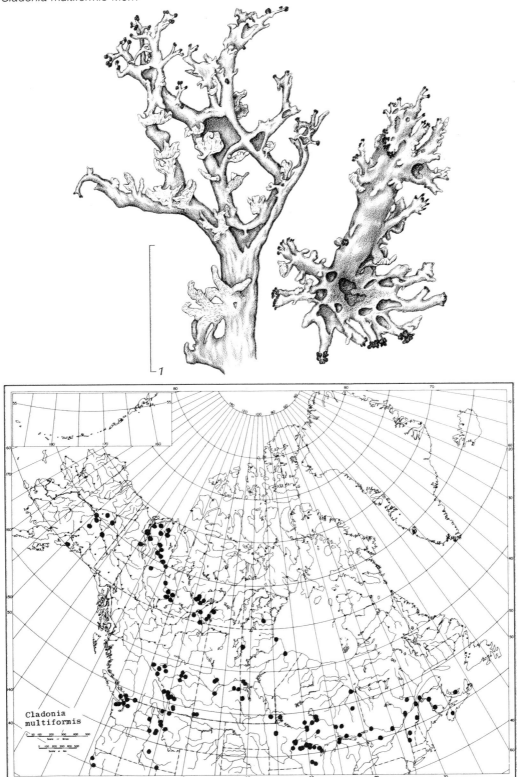

Cladonia
multiformis

ing, the cups flaring rapidly, the margins often proliferate and these sometimes developing into secondary ranks of cups; the inner sieve-like perforated membranes of the cups forming the most characteristic feature of the species, the perforations usually radially lengthened; the cortex subcontinuous to slightly contiguous areolate; esorediate; squamulose or not; glaucescent or varying toward brown or reddish brown.

Apothecia on the apices of the branches or proliferations, sometimes subcorymbosely arranged; dark brown. Pycnidia on the margins of the cups of the tips of the branchlets.

Reactions: K−, KC −, P+ red.

Contents: fumarprotocetraric acid and accessory ursolic acid.

This species grows on soil, among mosses, on rotting logs in moist habitats. It is usually considered a North American endemic but has been reported from South Africa. It is characteristically a member of the boreal forest, entering the tundra only along the ecotone between it and the forest as in the Great Slave Lake area and the Reindeer Reserve, N.W.T. Southward it ranges to Pennsylvania in the east, to Iowa in the center, and to Colorado and Washington in the west.

41. Cladonia nipponica Asah.

Lich. Japan 1:123. 1950. *Cladonia nipponica* var. *turgescens* Asah., ibid. (non *Cladonia nipponica* "Asah. em. Asah." Asah., J. Jap. Bot. 32:259. 1957).

Podetia branching divaricately, anisotomically trichotomous or dichotomous, sometimes polytomous, 4–10 cm tall 2–5 (–8) mm thick; cupless or with cups 1–4 mm broad, axils open or closed, podetia yellow-gray to yellow; outer surface smooth, inner wall with more or less loosely reticulate shallow network of strands, medulla loose; the cartilaginous strands more numerous and better developed than in *C. boryi.*

Reactions: K−, C−, KC+ yellow, P+ bright yellow at tips of podetia only.

Contents: usnic and psoromic acids.

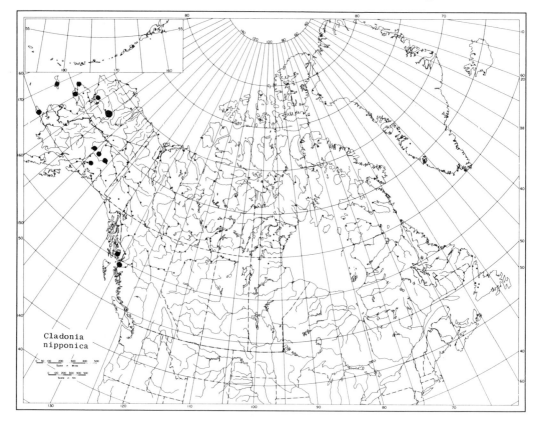

Cladonia
nipponica

Cladonia nipponica Asah.

This species is in fell fields and over boulders as well as on soils. It is a Beringian radiant, widely distributed in eastern Asia, in North America from Alaska and British Columbia. Specimens which I reported as *C. boryi* from Alaska were this species (Ahti 1973). It differs from that species in the more numerous cartilaginous strands forming a more continuous layer on the inner wall and the presence of psoromic acid.

42. Cladonia norrlinii Vain.
Acta Soc. F. Fl. Fenn. 53:86. 1922.

Primary squamules persistent; small to large, up to 10 or even 17 mm long and 2 or rarely 5 mm wide, subwedge-shaped, sinuate edged, to crenate, narrowly lobed, becoming involute-concave, ascending; upper side glaucescent; underside white; esorediate or sparsely granulose-sorediate on the margins and underside. Podetia arising from the upper side and margins of the primary squamules, 15–45 mm

tall and up to 3 mm in diameter; cylindrical and cupless, toward the apices irregularly or fastigiately branched or simple, the sterile branches blunt, those tipped with apothecia becoming dilated; the sides becoming entirely decorticate or the base verruculose areolate, granulose sorediate or the soredia subfarinose; partly or widely squamulose at the base, the squamules as narrowly lobed as the primary squamules, commonly involute, dull, not transparent; the decorticate areas chalky white, the granules and verruculae ashy or whitish glaucescent.

Apothecia small to middle-sized, up to 3 mm broad, commonly perforate or lobate, or conglomerate, bulging over the edge of the podetia; dark brown-red or reddish. Pycnidia on the apices of the podetia or on the upper side of the primary squamules.

Reactions: K+ yellow, KC−, P+ yellow.

Contents: atranorin plus psoromic acid.

This species grows on earth rich in humus, on sandy soils, and on thin soil over rocks. It has been reported to also grow on rotten wood. It is a circumpolar, boreal, and arctic species which in North America ranges southward to Connecticut, Wisconsin, Nebraska, and Colorado.

43. Cladonia phyllophora Hoffm.
Deutschl. Fl. 2:123. 1796. *Lichen phyllophorus* Ehrh., Pl. Crypt. 287. 1793 *(nom. nud.). Cladonia cristata* Hoffm., Deutschl. Fl. 2:124. 1796. *Baeomyces degenerans* Flörke, Berl. Magz. 283, 285, 290, 292. 1807. *Cladonia degenerans* (Flörke) Spreng., in L., Syst. Veg. 4:273. 1827.

Primary squamules persistent or disappearing; small to middle-sized, rarely large, 2–5 mm long, rarely up to 15 mm, the breadth up to 4 mm, irregularly broadening from the base, crenate or incised, sinuate, flat or involute, ascending; upper side glaucescent or olive-glaucescent; underside white or the base becoming black; esorediate. Podetia from the upper surface of the primary squamules; 8–80 mm tall and up to 4 mm in diameter; cup-bearing, the cups irregular, rarely regular, sometimes scarcely recognizable as cups; the interior closed, the margins commonly proliferate, the sides with a continuous cortex or becoming areolate with

Cladonia norrlinii Vain.

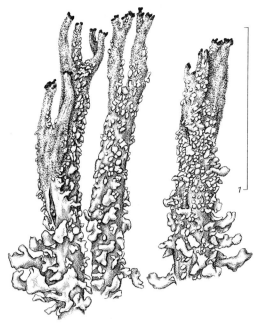

small, hardly elevated areoles, the part between the areoles decorticate and subtomentose, in the older parts the patches of cortex appearing as whitish or pale spots in a dark background; esorediate, with or without squamules, if present, the squamules small and similar to those of the primary type.

Apothecia small to middle-sized, up to 2 mm in diameter, on the tips of the proliferations from the margins of the cups, sometimes fastigiately clustered but usually not; dark brown, rarely pale brown or reddish brown. Pycnidia on the margins of the cups or the tips of the proliferations.

Reactions: K−, KC−, P+ red.
Contents: fumarprotocetraric acid.

This species grows on soil, either sandy or rich in humus. In the north it prefers boggy sites and willow thickets, on hummocks and in late snow areas. Within the forest it is more likely to be found on sandy soils. It is circumpolar and boreal as well as

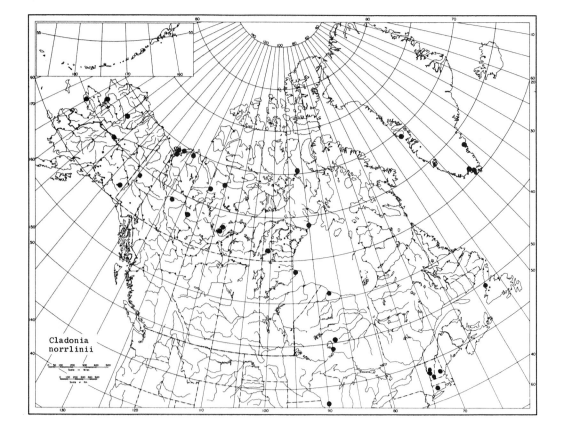

Cladonia
norrlinii

Cladonia phyllophora Ehrh. ex Hoffm

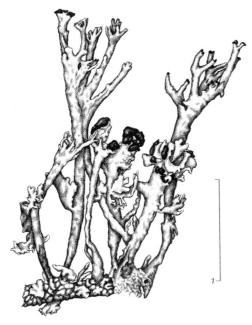

low arctic. In North America it ranges southward to West Virginia, Wisconsin, Colorado, and Oregon.

It sometimes is difficult to distinguish from *C. gracilis* but has more irregular cups and the base has whitish spots of cortex distributed over a black background instead of a uniformly blackened dying base.

44. Cladonia pleurota (Flörke) Schaer.

Enum. Lich. Eur. 186. 1850. *Capitularia pleurota* Flörke, Ges. Naturf. Freund. Mag. 2:218. 1808. *Cladonia coccifera* var. *pleurota* (Flörke) Schaer., Lich. Helvet. Spic. 25. 1823.

Primary squamules persistent or disappearing; small to large, 1–7 mm long, up to 5 mm broad, irregularly crenate-incised to lobate; upper side yellowish to olivaceous or pale glaucescent; underside pale or brownish toward the base; esorediate or with scattered granules below. Podetia varying up to 40 mm tall but usually much less; cup-bearing, the cups flaring quite soon and gradually from the base, goblet-shaped, regular and entire or dentate to proliferate from

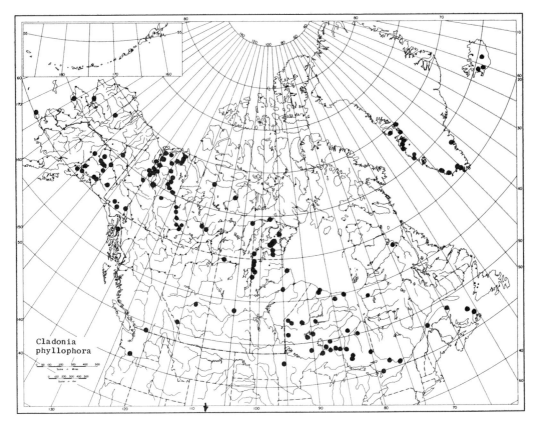

Cladonia
phyllophora

Cladonia pleurota (Flörke) Schaer.

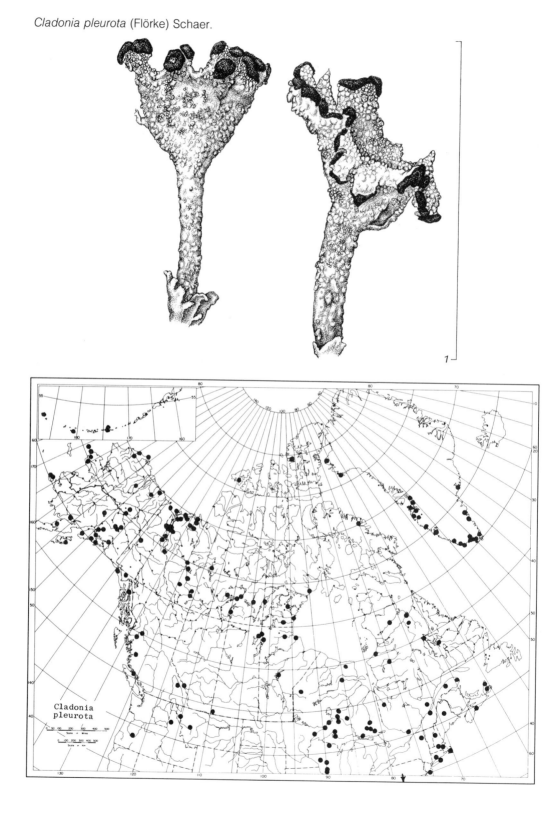

the margin, the proliferations bearing apothecia or rarely small cups, very rarely with small cups proliferate from the center; the inside of the cups and the upper part of the podetia with granular soredia, the base corticate and the cortex continuous or areolate to verruculose.

Apothecia scarlet, on the margins of the cups or on stipes from the margins of the cups, convex. Pycnidia usually on the margins of the cups, rarely on the upper side of the primary thallus.

Reactions: K−, KC+ yellow, P−.

Contents: usnic acid, zeorin and accessory bellidiflorin.

This species grows on soil, particularly with heavy clay content, and also over rocks with a thin coating of soil. It is circumpolar in the arctic and the boreal forest, the range there corresponding with that of *C. coccifera* and like it becoming rarer northward. The southern edge of the range is far southward of that of *C. coccifera,* however, extending to South Carolina, Texas, Colorado, and California.

45. Cladonia pocillum (Ach.) O. Rich.
Catal. Lich. Deux-Sèvres 8. 1878. *Baeomyces pocillum* Ach., Meth. Lich. 336. 1803.

Primary squamules thick, adnate forming little appressed rosettes around the base of the podetia. Podetia growing from the upper side of the primary squamules, usually short, less than 20 mm tall, simple or with marginal proliferations; cup-bearing, the cups goblet-shaped, the interior closed, both the interior and the outer surface covered with small to coarse peltate squamules; the podetia slaty gray to olive-green or with brownish shades.

Apothecia not common, brown, on the edges of the cups or the upper side of the primary squamules.

Reactions: K−, C−, KC−, P+ red.

Contents: fumarprotocetraric acid. Atranorin has been reported as a rare constituent in some arctic specimens by Dahl (1950).

This species grows on mineral soil containing lime and is thus strongly calciphilous; it may also grow on peats, mosses, and over rocks. Like *C. pyxidata* it is circumpolar, arctic to temperate. The ro-

settes of primary squamules are often found without podetia and are very distinctive.

46. Cladonia pseudorangiformis Asah.
J. Jap. Bot. 18:549. 1942. *Cladonia rangiformis* var. *versicolor* Elenk., Acta Hort. Petropol. 31:254. 1912.

Primary thallus unknown. Podetia cupless, branching at the base and repeatedly sympodially, divaricate, sometimes grouped and laterally deflexed, the axils perforate, lacking squamules or slightly squamulose at base, forming large colonies; gray to gray-green or with the tips brownish, the tips pointed; the cortex uniform or minutely areolate, toward the base the areolae becoming separate and showing the translucent cartilaginous layer between, the inner cartilaginous layer continuous.

Apothecia not known.

Reactions: K+ yellow, C−, P− podetia but extracts P+ yellow.

Contents: atranorin, merochlorophaeic acid, psoromic acid.

This is a species of subalpine and maritime bogs and heaths, or in old sandy lichen woodlands in middle and boreal zones (Ahti 1962). It is an east Asian and North American species, ranging from Alaska to Newfoundland south to New England mountains and around the north side of Lake Superior.

47. Cladonia pseudostellata Asah.
J. Jap. Bot. 18:620. 1942.

Primary thallus not known. Podetia in tufts or mats, 30–60 mm tall, 1–4 mm diameter, cupless; sympodially branched with the upper branches whorled, the branchlets short, dichotomous or in whorls, spinescent, the axils usually perforate; surface smooth, straw-colored to yellowish, inner cartilaginous layer continuous, 40–50 μ thick, the outer cortex 30–60 μ thick. Apothecia not known.

Reactions: K−, C−, KC+ yellowish, P−.

Contents: usnic and hypothamnolic acids.

This species grows on soils rich in humus. It appears to prefer open stands of spruce. It is an amphi-Beringian species known from Japan and Alaska.

Cladonia pocillum (Ach.) O. Rich.

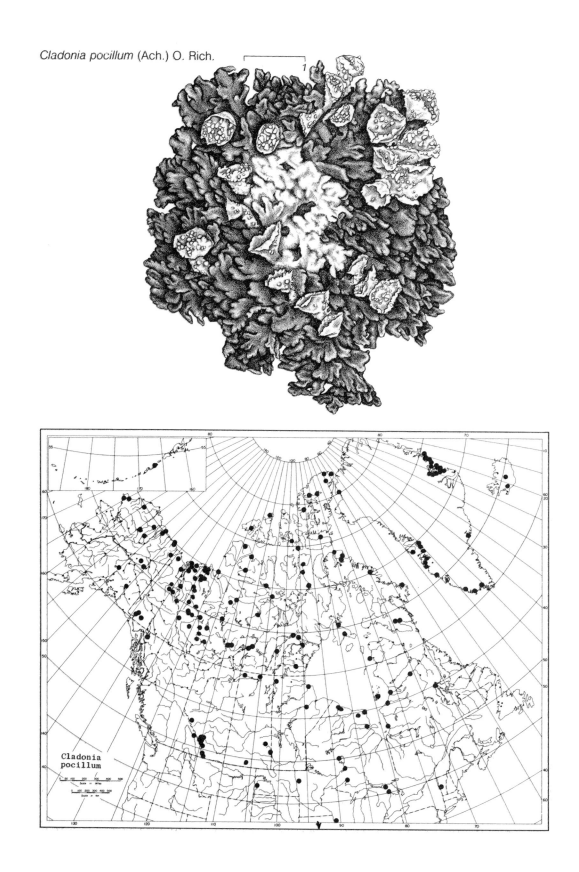

Cladonia
pocillum

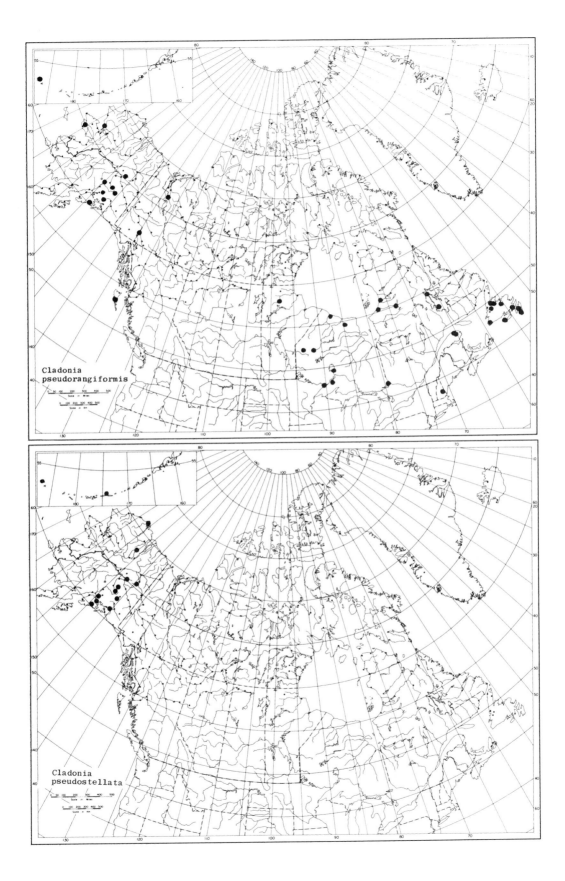

Cladonia
pseudorangiformis

Cladonia
pseudostellata

48. **Cladonia pyxidata** (L.) Hoffm

Deutschl. Fl. 2:121. 1796. *Lichen pyxidatus* L., Sp. Pl. 2:1151. 1753.

Primary squamules persistent, rarely disappearing; small to medium-sized, 2–7 (–15) mm long and up to 4 mm broad; irregularly lobed or incised, the tips rounded, the sides crenate or sinuate; quite thick; ascending or appressed; upper side glaucescent to pale olive-green or brownish; underside white, darkening at the base; esorediate. Podetia growing from the upper side of the primary squamules, 4–40 (–70) mm tall; simple or with short marginal proliferations bearing apothecia; cup-bearing, the cups flaring gradually and goblet-shaped, quite deep, the interior of the cups closed and decorticate in part with small peltate squamules covering the interior as well as the outer side; podetia slaty gray to olive-green or with brownish shades.

Apothecia uncommon; small to quite large; borne on the margins of the cups or on short stipes on the margins; brown or reddish brown, rarely pale. Pycnidia on the margins of the cups or on the upper side of the primary squamules.

Reactions: K−, KC−, P+ red.
Contents: fumarprotocetraric acid.

This species usually grows on acid mineral soils but may also be found on rotting logs, the bases of trees and over rocks. It is a cosmopolitan species, growing on all continents. In North America it occurs in all of the arctic, the boreal forest, and in temperate regions south to South Carolina, Wisconsin, Nebraska, and California.

49. **Cladonia rangiferina** (L.) Wigg.

Prim. Fl. Holsat. 90. 1780. *Lichen rangiferinus* L., Sp. Pl. 1153. 1753. *Cladina rangiferina* (L.) Harm., Bull. Soc. Sci. Nancy ser. 2, 14:387. 1896.

Primary thallus soon disappearing and seldom seen; crustose, thin, of subglobose verrucules 0.22–0.4 mm in diameter; ashy-white. Podetia in tufts or in extensive mats; usually robust, 5–10 (–15) cm tall and 1–1.6 (–3) mm in

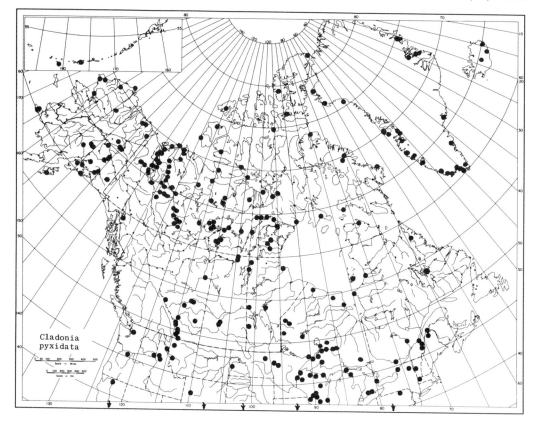

Cladonia
pyxidata

Cladonia rangiferina (L.) Wigg.

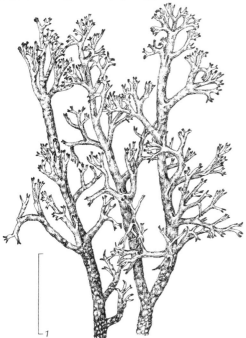

1

diameter; the branching anisotomic dichoto-mous, trichotomous, or tetrachotomous, the branches rebranched, the tips of sterile branch-lets with whorls of 2–3 branchlets around an open axil and recurved to one side, the tips of fertile branches in a corymbose arrangement; the axils open; ashy gray, more gray than the color of the other *Cladinae,* becoming whitish or pale greenish in shaded places, occasionally black-ened in the sun, the base blackening on dying; the surface markedly arachnoid-tomentose with the greenish areoles scattered and more or less verruculose, especially toward the base, the young parts more or less smooth; usually lack-ing soredia; usually not pellucid.

Apothecia small; solitary or in corymbs at the tips of the branchlets; dark brown or black. Pycnidia at the tips of the branchlets; containing colorless jelly.

Reactions: K+ yellow, KC−, P+ red.

Contents; atranorin and fumarprotoce-traric acid, lacking usnic acid.

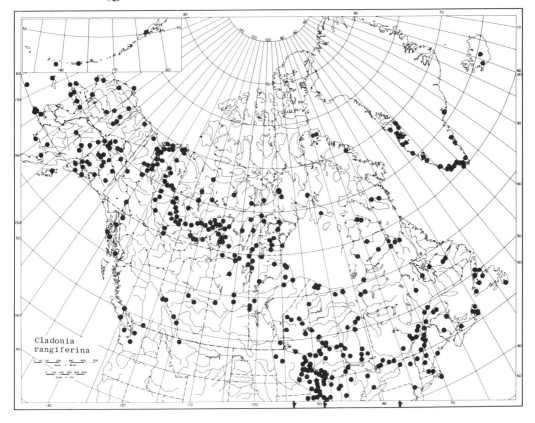

Cladonia
rangiferina

This species grows on sandy soils in woodlands, on soil rich in humus, and on thin soil over rock outcrops. It is one of the most common species in bogs and on tundras. It is circumpolar arctic to temperate, in North America lacking in the high arctic but common from mid-arctic southward to Georgia, Albama, Arkansas, Montana, and Oregon. It is especially abundant in the boreal forest.

50. Cladonia scabriuscula (Del. in Duby) Nyl.

Flora 58:447. 1875. *Cenomyce scabriuscula* Del. in Duby, Bot. Gall. 2d ed. 623. 1830.

Primary squamules disappearing; middle-sized, 2–5 mm long and as broad; irregularly lobed, crenate, ascending; the upper side glaucescent; the underside white; sorediate or not. Podetia 30–110 mm tall, up to 2–5 mm in diameter; mainly dichotomously branching, to some extent sympodial, the branches cylindrical, dilating a little at the axils; the axils open or closed; the apices usually subulate; partly or entirely isidiose-sorediate or sorediate, the soredia accompanied by minute spreading and appressed squamules, the apices granulose-sorediate or verruculose-granulose; squamulose or not; the cortex chinky to areolate becoming decorticate, the interspaces dull, white; whitish glaucescent to ashy olive or ashy bluish and variegated.

Apothecia small, dark brown; sometimes on corymbose branchlets, at the tips of the podetia. Pycnidia on the tips of the branches.

Reactions: K−, KC−, P+ red.

Contents: fumarprotocetraric acid and accessory ursolic acid

This species grows on damp soil in woods, sometimes on thin soil over rocks. It has been reported from all continents except Africa; in North America it appears to be mainly boreal but with a few outliers in the tundra in Greenland, Ungava Peninsula, northern Ontario, and Alaska. To the south it is in the New England states to Pennsylvania, to Wisconsin, Michigan, and Minnesota and in the west to

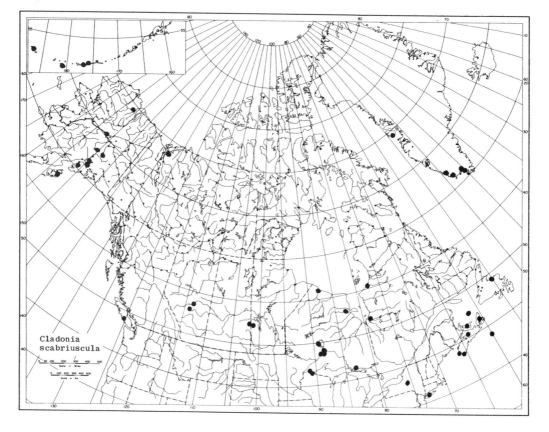

Cladonia
scabriuscula

Cladonia squamosa (Scop.) Hoffm.

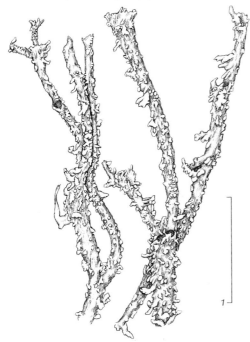

1

Washington and California. In the east it is easily confused with the boreal species *C. farinacea* Evans which lacks the tiny squamules that accompany the soredia in the upper part of the podetia. *C. farinacea* does not reach the Arctic.

51. Cladonia squamosa (Scop.) Hoffm.
Deutschl. Fl. 2:125. 1796. *Lichen squamosus* Scop., Fl. Carn. 2d ed. 368. 1772.

Primary squamules persistent or disappearing; middle-sized or rarely large, 2–7 (–10) mm long and 1 mm broad; irregularly subpinnate or subdigitate laciniate, crenate, often more or less wedge-shaped, ascending, flat or involute, scattered or tufted, rarely caespitose or forming a compact crust; the upper side glaucescent or whitish or olive-green to brownish; underside white, esorediate, or the underside partly granulose. Podetia growing from the upper side of the primary squamules; very variable; the bases persistent or dying off and the growth continuing from the apices; usually 10–

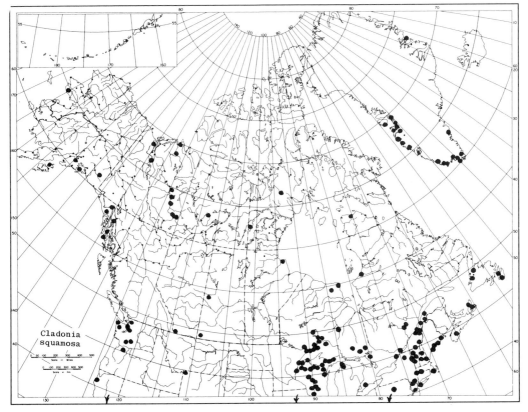

Cladonia
squamosa

90 mm tall and the diameter up to 2.5 mm; cups common, rather abruptly flaring, small to middle-sized, usually widely perforated, the margins becoming repeatedly proliferated; the cupless podetia branching radiately or irregularly with sterile proliferations which may be blunt or subulate, or the branches may be tipped with apothecia, the proliferations simple to clustered and grouped in close tufts or clusters, sometimes scattered; strongly tending to become decorticate but not sorediate, the cortex thus varying all the way from almost continuous, through scattered corticate areas, to entirely decorticate; the decorticate parts retaining the inner part of the medullary layer and showing an arachnoid appearance unlike the smooth appearance of the decorticate parts of the very similar *C. crispata;* the corticate parts smooth or rugose and colored from whitish through usually bluish or glaucescent greenish to olive-green or brown, or even blackened in the sun; the decorticate parts white; the sides of the podetia commonly quite squamulose furfuraceous with fragile laciniate squamules.

Apothecia small; on the margins of the cups or the tips of the branchlets; pale to yellowish or dark brown. Pycnidia on the margins of cups and on the apices or the branchlets.

Reactions: K−, KC−, P−.

Contents: squamatic acid.

This species grows on earth, over rocks with small accumulations of soil, on rotting logs, occasionally on sandy soils. It is circumpolar, arctic, and temperate. It is lacking in the high arctic but ranges southward to Florida and California. The morphology is extremely variable; the finely divided squamules, the decorticate podetia with squamules rather than soredia, and the squamatic acid (UV+) all assist in identifying it.

52. Cladonia stellaris (Opiz) Pouz. & Vezda

Preslia 43:193. 1971. *Cenomyce stellaris* Opiz, Vollständ. umriss statis. Topogr. Königreich. Böhm. 493. 1823. *Lichen rangiferinus alpestris* L., Sp. Pl. 1153. 1753. *Lichen rangiferinus* ssp. *alpestris* (L.) Ehrh., Hannov. Magaz. 239. 1790. *Cladonia alpestris* (L.) Rabenh., Clad. Europ. 11. 1860. *Cladina alpestris* (L.) Harm., Bull. Soc. Sci. Nancy 14:389. 1896.

Primary thallus crustose, soon disappearing and rarely seen, of dispersed or glomerate granules or verrucules, 0.16–0.28 mm in diameter, pale yellow. Podetia 5–10 (−18) cm tall, 1–1.5 (−2.5) mm in diameter; forming compact rounded, subglobose heads growing singly or in groups or in mats; the branching very dense, isotomic, rarely dichotomous, mainly polychotomous, (3−)4 (−6) branches in a whorl, the branches mainly equal in size but sometimes forming sympodia, spreading widely and showing little tendency to curve in one direction, the axils almost always perforate, the ultimate branchlets in small whorls of (3−) 4–6 members around a perforate axil; pale yellowish to whitish, darker gray in deeper woods, the base blackening; the surface dull, arachnoid to subtomentose with vaguely defined dispersed algal areolae, the older parts becoming translucent as the outer layers disperse.

Apothecia small, on the tips of subcorymbosely grouped branchlets at the tips of the podetia, brown or dark brown. Pycnidia at the tips of branchlets and containing a red jelly.

Reactions: K−, KC+ yellow, P−.

Contents: usnic acid constant; accessory perlatolic acid, pseudonorrangiformic acid (sub. C), and substances A and B.

This species grows on rocks and soil and occasionally old wood and stumps, on tundras and in open woods. It is circumpolar. In North America it is common across Canada and in Alaska, ranging southward in the east to West Virginia, around the Great Lakes, and in the west to Wyoming and California.

53. Cladonia stricta (Nyl.) Nyl.

Flora 52:294. 1869. *Cladonia degenerans* var. *stricta* Nyl. in Middendorff, Reise Sibir. 4:4. 1867. *Cladonia lepidota* var. *stricta* (Nyl.) DuRietz, Bot. Not. 1924:67. 1924. *Cladonia lepidota* auct. non (Ach.) Nyl. *Cladonia gracilescens* auct., non (Flörke) Vain.

Primary squamules persistent or disappearing; small to large, reported up to 20 mm long but American material much smaller in size, rounded to wedge-shaped, crenated or entire, flat; the upper side whitish glaucescent to glaucescent or olive-green; underside white, blackening at the base; esorediate. Podetia arising from the upper side of the primary squamules, 60–

Cladonia stellaris (Opiz) Pouzar & Vezda

1

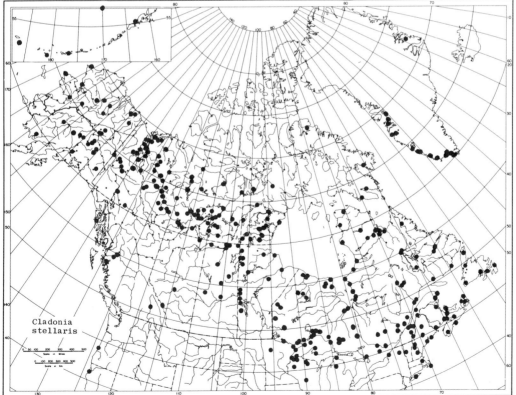

Cladonia
stellaris

Cladonia stricta (Nyl.) Nyl.

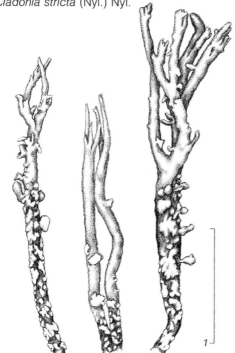

100 (–150) mm tall, up to 1.5 mm broad, rarely 2 mm; with narrow cups, flaring gradually or abruptly, at the tips 2–5 mm broad, with tiers from the centers or from the margins of the cups; inner surface of cups closed or becoming cribrose; cups shallow, the margins becoming split or denticulate; cortex continuous or becoming areolate-chinky, the areoles slightly raised, more or less dispersed, the narrow decorticate part more or less distinctly subtomentose, whitish glaucescent to pale ashy brown, with the base blackening but with white and black corticate spots at the base as in *C. phyllophora;* esorediate, squamulose, and with grouping of squamules toward the edges of the cups.

Apothecia on the margins of the cups or on short stipes, brown or reddish brown. Pycnidia on the margins of the cups or denticulations.

Reactions: K+ yellow, KC–, P+ red.

Contents: atranorin and fumarprotocetraric acid.

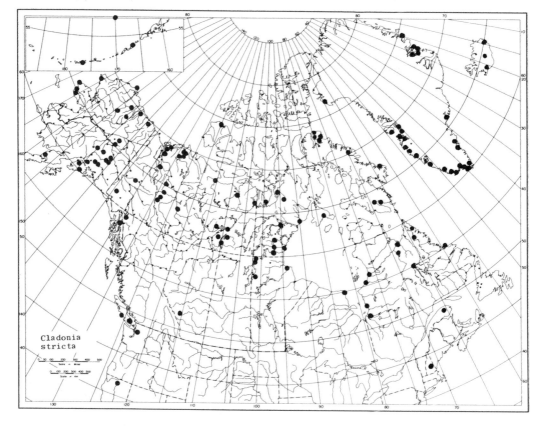

Cladonia
stricta

Cladonia subcervicornis (Vain.) Kernst.

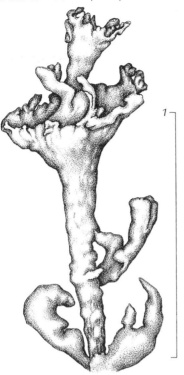

1

This species grows on soils rich in humus, in moist places between boulders, and is especially characteristic of late snowbank areas on the tundras. It is a circumpolar, arctic, alpine species which in North America ranges southward in the eastern mountains to New Hampshire and in the west to northern California.

54. **Cladonia subcervicornis** (Vain.) DuRietz

Bot. Not. 1922. 217. *Cladonia verticillata* var. *subcervicornis* Vain. Acta Soc. F. Fl. Fenn. 10:197. 1894.

Primary squamules persistent; large, 2–10 mm long, 2–4 mm broad, laciniate or shorter and rounded; upper side bluish or leaden green; underside whitish or with a bluish tinge, the base blackened. Podetia from the upper side of the primary squamules, 6–10 mm tall; cup-bearing, the cups flaring rapidly, their margins splitting into numerous proliferations, simple or with a few tiers from the centers of the closed cups, esorediate, lacking squamules, bluish green; the

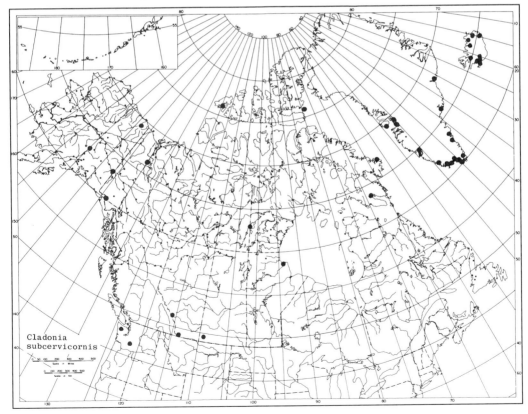

Cladonia
subcervicornis

Cladonia subfurcata (Nyl.) Arn.

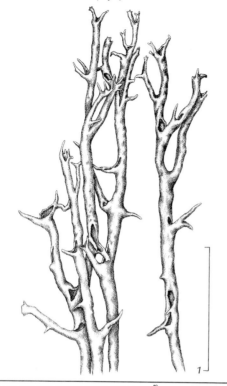

1

cortex continuous or contiguous areolate.

Apothecia from the margins of the cups, minute, dark brown. Pycnidia on the margins of the cups or of the primary squamules.

Reactions: K+ yellow, KC−, P+ red.

Contents: atranorin and fumarprotocetraric acid.

This species grows on soils containing humus and on granitic and gneissic cliffs. It is arctic-alpine and in Europe and North America. In the latter it is arctic and in alpine areas in the west.

55. Cladonia subfurcata (Nyl.) Arn.

in Rehm, Clad. exs. 263. 1885. *Cladonia degenerans* *Cl.* [= ssp.] *trachyna* f. *subfurcata* Nyl. in Norrl., Notis. Sällsk. Fauna et Flora. Fenn. Forhandl. 13:320. 1873. *Cladonia delessertii* (Del. in Nyl.) Vain., Acta. Soc. F. Fl. Fenn. 4:397. 1887. *Cenomyce delessertii* Del. in Nyl., Syn. Lich. 1:208. 1860 (lacking diagnosis).

Primary squamules disappearing; middle-sized, 1–5 mm long; irregularly crenate-

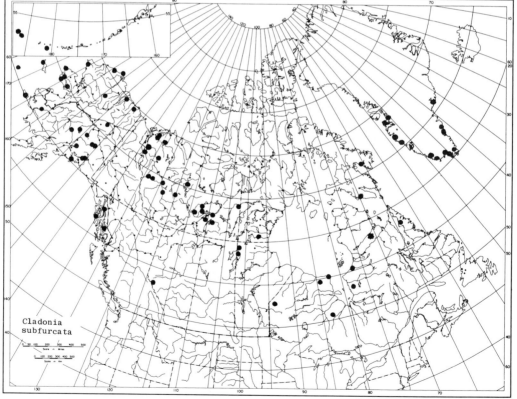

Cladonia
subfurcata

lobate or laciniate, ascending, flat, tufted, esorediate; upper side glaucescent to olive or pale glaucescent; underside white. Podetia growing from the upper side of the primary squamules; the base dying and growth continuing from the apices, 15–110 mm tall and 1–1.5 (–2.5) mm in diameter; subcylindrical and somewhat thickened at the axils; cupless; branching in sympodia, dichotomously or trichotomously, the axils perforated or not, the branchlets divaricate or ascending, usually short; the apices obtuse or shortly attenuate, erect, branching widely or subradiately spinulose around a perforate axil; forming tufts or small mats, esorediate; the cortex partly subcontinuous, partly contiguous-areolate, smooth or low rugose; podetia lacking squamules or with a few at the base; dull or shining, not pellucid; pale or olive-glaucescent or reddish brown to brown, also variegated at times, the dying parts usually spotted, the decorticate portion black and the spots of cortex pale to olive green.

Apothecia small, up to 0.7 mm in diameter, at the apices of the branchlets. Pycnidia at the apices of the branchlets; containing reddish material which turns violet with KOH.

Reactions: K–, KC–, P–.

Contents: squamatic acid.

This species grows on soil rich in humus. It is circumpolar, arctic, and boreal, lacking in the high arctic and not yet found in North America south of Canada, Alaska, and Greenland, although it is reported from South America.

56. Cladonia subulata (L.) Wigg.

Prim. Fl. Holsat. 90. 1780. *Lichen subulatus* L., Sp. Pl. 1153. 1753. *Cladonia cornutoradiata* (Coem.) Sandst., Abhandl. Naturw. Ver. Bremen 21:373. 1912. *C. fimbriata* f. *cornuto-radiata* Coem., Bull. Acad. Roy. Belgique II. 19:40. 1865.

Primary squamules persistent or often disappearing, small; upper side whitish glaucescent to blackening; underside white. Podetia from the upper side of the primary squamules, tall and slender, 30–100 mm tall and up to 3.5 mm thick, cylindrical, cupless or with irregular cups which are formed by circles of long proliferations; with a slight amount of cortex at the base, or more usually totally decorticate and covered

with very fine farinose soredia; white to ashy or pale glaucescent or with brownish variegation in the ashy coloring.

Apothecia rare; sessile on the margins of the cups or else on stipes from the margins or on the tips of the podetia; dark brown, brown, or reddish-brown. Pycnidia on the margins of the cups or more often the tips of the proliferations or podetia.

Reactions: K–, KC–, P+ red.

Contents: fumarprotocetraric acid.

This species is found on sandy soil rich in humus, on peats, and on rotting logs. It is circumpolar and in both northern and southern hemispheres. It is a boreal species found in the boreal forest and reaches into the tundra only along the northern tree line.

57. Cladonia sulphurina (Michx.) Fr.

Lich. Eur. Reform. 237. 1831. *Scyphophorus sulphurinus* Michx., Flora Bor.- Amer. 2:238. 1803. *Baeomyces deformis* γ *B. gonechus* Ach., Meth. Lich. 335. 1803. *Cladonia gonecha* (Ach.) Asah., J. Jap. Bot. 15:608. 1940.

Primary squamules larger than in *C. deformis,* 5–10 mm long and 2.5 to 5 mm wide, incised or lobate; upper side yellowish glaucescent; underside white to pale brown or dark; sometimes sorediate. Podetia about the same length as in *C. deformis,* about 25–85 mm tall, but stouter and the cups more irregular, the marginal proliferations more prominent and the sides more often fissured and split; the surface covered, except at the base, with farinose yellowish soredia; the basal corticate area as in *C. deformis.*

Apothecia and pycnidia similar to those of *C. deformis.*

Reactions: K–, KC+ yellowish, P–.

Contents: usnic acid, squamatic acid, and accessory bellidiflorin.

This species has much the same requirements as *C. deformis,* a fairly high humus content. Yet the two species do not seem to occur together. There should be studies made of the autecology of this pair of species. *C. sulphurina* has a circumpolar range but slightly less far north into the tundras than that of *C. deformis.* It has been well known under the name *C.*

Cladonia subulata (L.) Wigg.

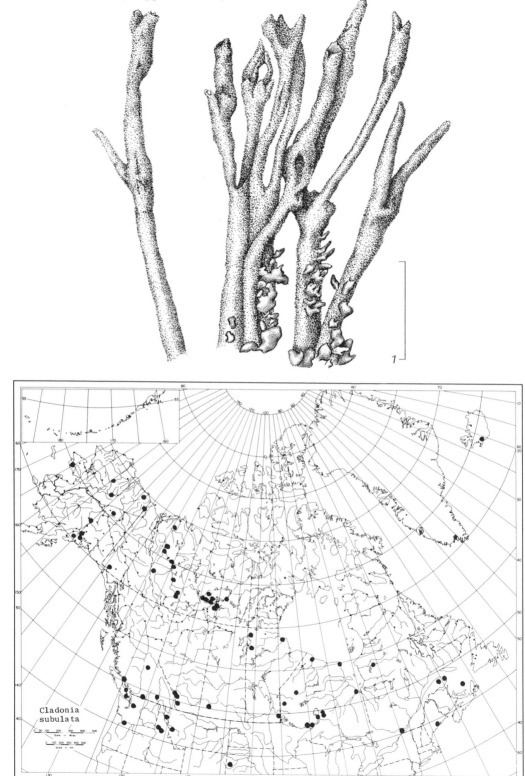

Cladonia sulphurina (Michx.) Fr.

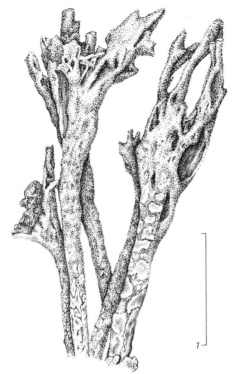

1

gonecha and the necessary shift caused by the prior name, *C. sulphurina*, discussed by Ahti (1978a).

58. Cladonia symphycarpa (Ach.) Fr.

Lich. Suec. Exs. 232. 1828. *Lichen symphycarpus* Ach., Lichenogr. Suec. Prodr. 198. 1798.

Primary squamules large and abundant, larger than in *C. cariosa,* usually more or less appressed at the periphery and erect toward the center of the thallus. Podetia rarer than in *C. cariosa* but resembling them, cupless and terminating with apothecia, branched toward the top and somewhat broadening upwards; cortex soon verruculose areolate and fissured, bluish green to olive green.

Apothecia dark brown, much larger than the tips of the podetia.

Reactions: K+ yellow turning red, KC−, P+ orange-red.

Contents: atranorin and norstictic acid.

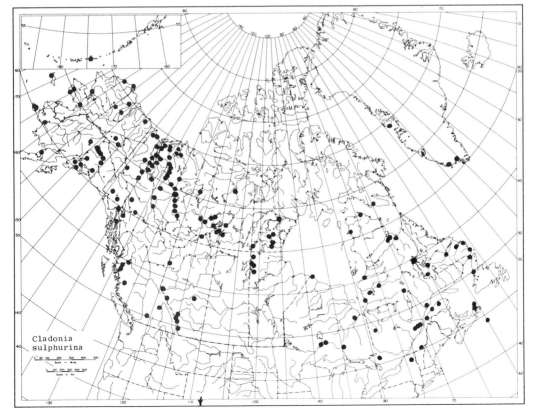

Cladonia
sulphurina

This species grows on mineral soil, usually calciferous. It appears to be circumpolar, arctic to temperate, but the North American collections are too scant to give more than an approximation of the range. In the Arctic it appears to range just north of the tree line.

The resemblance of this species to *C. cariosa* is very close. The presence of the norstictic acid in addition to the atranorin is the strongest character, although the larger, thicker primary squamules also help.

59. Cladonia thomsonii Ahti.
Bryologist 81:334. 1978

Primary thallus unknown. Podetia growing in dense to loose cushions, dying at the base, usually 4–6 cm tall, the branches ashy white, 1–3 mm thick, with irregular anisotomic polytomy with tetrachotomy dominant but dichotomy frequent; cupless or with indistinct cups at the axils, the axils usually perforate; the podetial wall also may become split with age, podetial squamulus rare, scarcely lobate, 1–3 mm long; podetial surface areolate corticate with rather wide white, arachnoid medullary interspaces; the base sometimes with tuberculate outgrowths, becoming yellowish gray to black.

The podetial wall 220–300 μ thick with 25–40 μ cortex, the inner cartilaginous layer 50–75 μ, the inner surface slightly longitudinally furrowed, somewhat glossy and with scattered, minute papillae.

Apothecia brown, the margins persistent, spores not seen. Pycnidia at tips of podetia, containing hyaline jelly, the conidia 6×1 μ.

Reactions: K+ yellow, C−, KC−, P−.

Contents: atranorin usually with one or two of three unknowns in TLC tests, differing from *C. wainii* Sav. in that the latter Siberian species contains the atranorin plus two different unknowns.

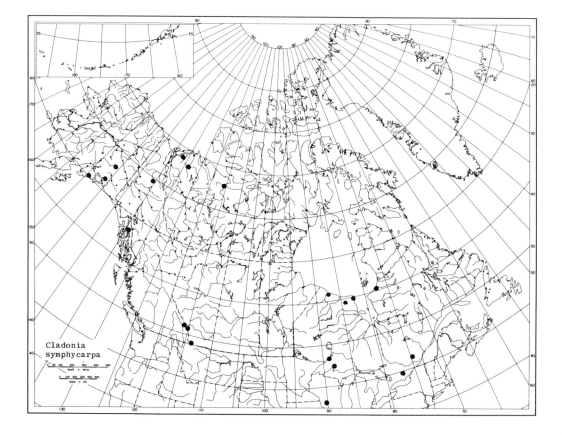

Cladonia
symphycarpa

Cladonia thomsonii Ahti

1⌐

This species grows on gravelly tundras and well-drained alpine tundras. It is so far known only from Alaska and the Northwest Territories near the mouth of the Mackenzie River. It is probably closely related to the poorly known *C. wainii* Sav., which is reported only from the type locality in Kamchatka. It differs in being whitish rather than grayish, with narrowly isotomic instead of widely anisotomic branching, and with one dark brown spot at Rf 0.43 instead of two purple-brown spots at Rf 0.5 and 0.47 in n-hexane-ether-formic acid, 5:4:1. Yoshimura (1968) suggested that it might be a new species and Krog (1968) and Ahti (1973) agreed and so described it in 1978b.

60. **Cladonia turgida** Ehrh. ex Hoffm.
Deutschl. Fl. 2:124. 1796.

Primary squamules persistent; large, 5–25 mm long, 1–7 mm broad; coarse, rugose, with broad and rounded divisions; upper side glaucescent; underside cream-colored. Podetia with more or less complex branching, some-

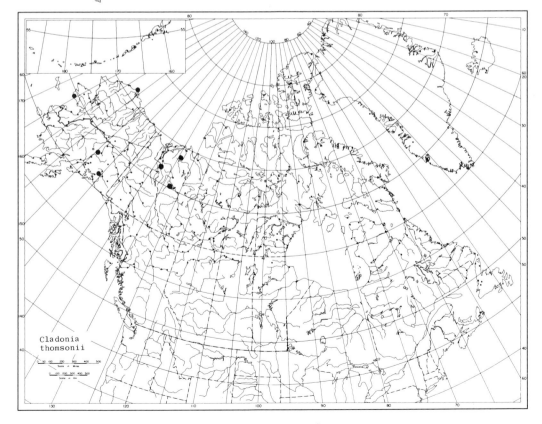

Cladonia
thomsonii

Cladonia turgida Ehrh. ex Hoffm.

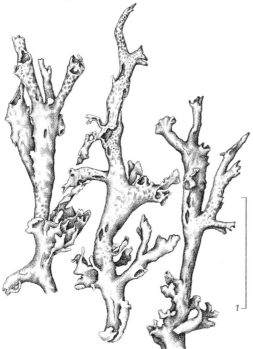

1⌐

times dying at the base, the branching dichotomous or polytomous, the axils closed or open
and often dilated, with short terminal branchlets
in pairs or small groups, the branchlets usually
subulate and sterile, sometimes with apothecia
at the tips; small cups occasionally present, the
inner membranes variously lacerate or perforate, the margins sparingly proliferate with subulate or small cup-bearing apices, esorediate; the
cortex smooth, ashy glaucescent, either continuous or composed of contiguous or dispersed
areoles separated by whitish bands or lines;
podetial squamules sometimes present.

Apothecia brown, at the tips of the podetia. Pycnidia on the tips of the podetia or the
margins of the cups.

Reactions: K+ yellow, C−, KC−, P+
red.

Contents: atranorin and fumarprotocetraric acid.

This species grows on mineral soils as well as
soils rich in humus. It is a circumpolar, arctic to north

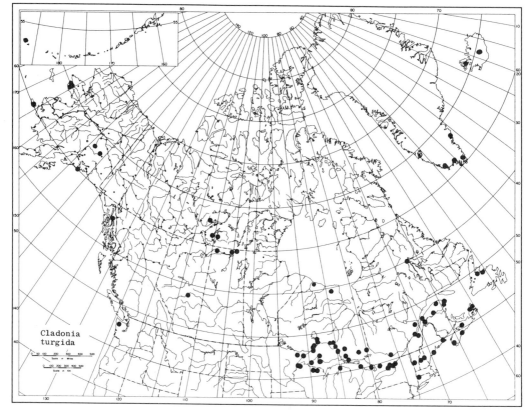

Cladonia
turgida

temperate species mainly in the boreal forest in North America.

Cladonia uncialis (L.) Wigg.

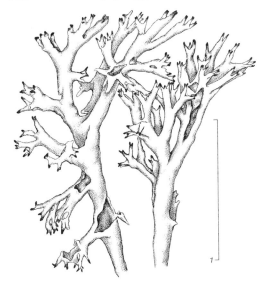

61. **Cladonia uncialis** (L.) Wigg

Prim. Fl. Holsat. 90. 1780. *Lichen uncialis* L., Sp. Pl. 1153. 1753.

Primary squamules rarely seen, usually disappearing; small, up to 1 mm long, broad crenate or incised crenate or rarely incised, ascending, flat, sparse or tufted; the upper side yellowish glaucescent; the underside white. Podetia appearing to arise from above the remains of dead podetia, or from fragments of other podetia; the base dying and the growth continuing from the apices; 20–80 (–110) mm tall and 1–1.5 (–4) mm in diameter; subcylindrical; cupless; dilated at the axils; dichotomously or sympodially or verticillately branched, repeatedly branched, the whole forming dense mats or tufts; the ultimate branchlets spine-like and in small whorls around open perforations; most of the

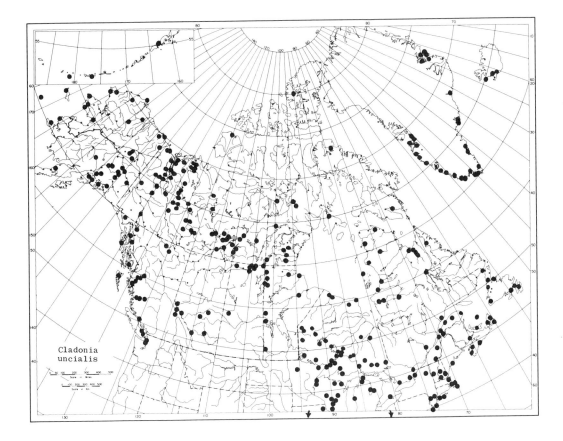

Cladonia
uncialis

Cladonia verticillata (Hoffm.) Schaer.

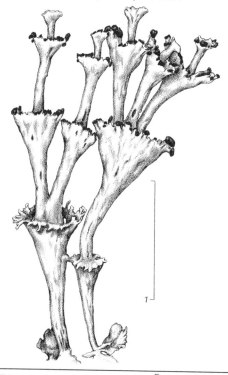

axils usually perforate; the inner cartilaginuous layer continuous; the inner surface smooth; the outer surface smooth and shiny, sometimes with the greenish areolate projecting slightly; light green or deep or pale yellowish; the podetia very brittle.

Apothecia small, up to 0.8 mm in diameter, brown; at the apices of the branches in usually a radiate or subcymose arrangement of the branchlets. Pycnidia at the tips of the branchlets.

Reactions: K−, KC+ yellowish, P−, UV+ in those specimens with squamatic acid.

Contents: usnic acid plus accessory squamatic acid, the latter in a high percentage of the more southern specimens and in a lower percentage of the northern specimens. Substance A is a rare accessory component in the species.

This species grows on sandy soil in the open and among mosses in bogs and tundras. It is circumpolar, arctic to temperate, in North America growing

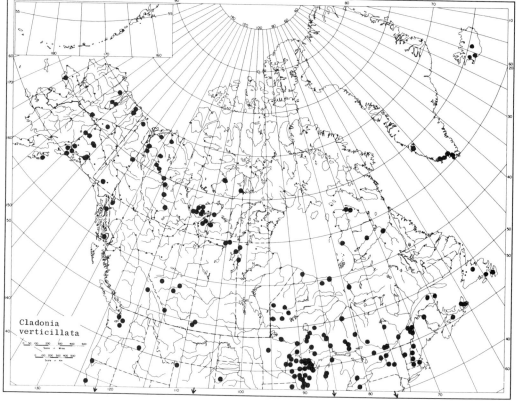

Cladonia
verticillata

from Alaska to Greenland and south to Georgia, Arkansas, and Washington.

62. **Cladonia verticillata** (Hoffm.) Schaer.

Lich. Helvet. Spic. 31. 1823. *Cladonia pyxidata C. verticillata* Hoffm., Deutschl. Fl. 2:122. 1796.

Primary squamules persistent or disappearing; varying from small to large, depending on the form and the substratum, up to 8 mm long and 4 mm broad; irregularly wedge-shaped or lobed, the lobes crenate or slightly incised, flat or convolute, often ascending, rarely caespitose; the upper side olive-green or reddish or brownish glaucescent or slaty green; underside white, or black toward the base; esorediate. Podetia from the upper side or the margins of the primary squamules, up to 50 mm tall and 3 mm in diameter, flaring quickly at the tops into short, broad cups, up to 9 mm across, shallow and with small, pointed or cup-bearing proliferations growing from the centers of the closed cups; the margins entire or with apothecia or pycnidia; sometimes with several tiers of cups, each arising from the center of the previous tier; the cortex continuous or chinky areolate, the areoles smooth, subcontiguous, the narrow interspaces white, dull, whitish, green to olivaceous, or ashy or bluish green, or even quite brown; esorediate; with or without squamules.

Apothecia sessile or on short stipes; small to middle-sized, usually less than 3 mm broad; brown or reddish brown; rounded; broader than the stipes which support them. Pycnidia in the centers of the cups or on the margins.

Reactions: K−, KC−, P+ red.
Contents: fumarprotocetraric acid.

This species grows on mineral soils, on thin soil over rock outcrops, rarely on rotting wood. It is cosmopolitan in lower latitudes. In the Arctic it is lacking in the high arctic but is widely distributed south of there from Greenland to Alaska to the southern states except for Florida.

COCCOCARPIA Pers.

In Gaudichaud, C., Voyage autour du monde. . . . . . 206–207. 1826.

Thallus foliose, heteromerous, dorsiventral, the lobes imbricate or side by side, broadly wedge-shaped to fan-shaped, the margins usually turned under and sligthly thickened; upper side glabrous, smooth and shining or with minute wrinkles more or less concentric with the thallus edge, or scabrid, lead gray to bluish or black; upper cortex paraplectenchymatous, the cells arranged in rows more or less parallel and radiating outwards in the thallus to show faint striae under high magnification; lower surface pale tan to black, often dense with long rhizines which may be pale or dark blue-black or black with white tips. the rhizines often projecting beyond the thallus margins.

Apothecia adnate, biatorine, disk red-brown to becoming black; spores 8, hyaline, ellipsoid (or globose in some tropical species), simple but in our species with two oil droplets which give the appearances of two cells. Pycnidia in warts on fertile specimens; conidia cylindrical, 3–6 × 1–1.5 μ.

Alga: *Scytonema*.
Type species: *Coccocarpia pellita* (Ach.) Müll. Arg.

REFERENCES
Arvidsson, L. & D. J. Galloway. 1979. The lichen genus *Coccocarpia* in New Zealand. Bot. Not. 132:239–246.
Swinscow, T. D. V. & H. Krog. 1976. The genus *Coccocarpia* in East Africa. Norw. J. Bot. 23:251–259.

Only one species of this primarily tropical genus reaches the tundra.

1. **Coccocarpia erythroxyli** (Spreng.) Swinsc. & Krog

Norw. J. Bot. 23:256. 1976. *Lecidea erythroxyli* Spreng., K. Vetens. Acad. Nya Handl. 1:47. 1820.

Thallus foliose, the lobes abutting to somewhat imbricate, broadly rounded, to 1 cm wide, ridged concentrically to lobes; dark slaty blue to blue-black; lacking isidia or soredia but often with small squamules or lobules, particularly centrally on the thallus; lower side with blue black or white tipped rhizinae.

Apothecia waxy or brownish red to black; margin sinuate; spores 8–15×4–5 μ (Swinscow & Krog 1976).

In the tundra this species grows on humus among mosses apparently in rather sheltered spots.

The main distribution of this species is apparently pantropical to subtropical, in North America hitherto being known from the southeastern states. The two disjunct collections known from Alaska compare with the distributions of *Umbilicaria caroliniana, Sticta weigelii,* and *Baeomyces roseus.* They are from Anaktuvuk Pass, Nash (Moser, Nash, Thomson, Bryologist (1979) sub *C. pellita*) and Noluck Lake, Krog (pers. comm. Arvidsson to Thomson). Are these relics of an ancient Arcto-Tertiary flora or do they represent long-distance dispersal?

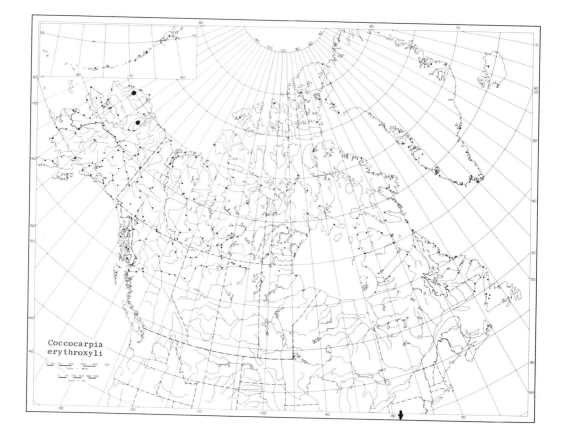

Coccocarpia
erythroxyli

COLLEMA G. H. Web.

In Wigg., Primit. Flor. Holsat. 89. 1790. *Synechoblastus* Trev., Caratt. tre nuov. Gener. Collem. 2–3. 1853. *Collemodes* Fink, Mycologia 10:236. 1918.

Foliose, rarely subfruticose or crustose lichens, gelatinous when wet, lobed and often also lobulate, olive-green or blackish; underside attached by hapters, homiomerous (the algae distributed throughout the thallus), the hyphal context lax; cortex lacking except in apothecial tissues.

Apothecia round, sessile, flat, concave or convex. lecanorine or with proper margin; disk reddish or darker, the thalline exciple lacking cortex or with pseudocortex, proper exciple with the cells more or less parallel to the surfaces (euthyplectenchymatous) or distinctly irregularly cellular (paraplectenchymatous); paraphyses simple or branched, septate, the apices thickened; asci subcylindrical to clavate, the apical membrane thick; spores 4 or 8, variable in shape, 1 to several septate to usually muriform, hyaline or pale colored. Pycnidia immersed, globose, fulcra endobasidial, conidia scarce, oblong, usually straight, the ends usually slightly swollen, some species with internal conidia (*C. bachmanianum* in the Arctic).

Algae: *Nostoc.*

Type species: *Collema* nigrescens (Huds.) DC.

REFERENCES

Degelius, G. 1954. The lichen genus *Collema* in Europe. Symbol. Bot. Upsal. 13(2):1–499.
Degelius, T. 1974. The lichen genus *Collema* with special reference to the extra-European species. Symbol. Bot. Upsal. 20(2):1–215.

Key to the Arctic Species of Collema
(based on Degelius 1974)

1. Thallus with isidia.
    2. Isidia squamiform.
        3. End lobules large and broad, 0.5–3.0 cm broad . . . . 6. *C. flaccidum*
        3. End lobules small, less than 0.5 cm broad . . . . 4. *C. crispum*
    2. Isidia not squamiform.
        4. Isidia terete or coralloid.
            5. Isidia coarse, coralloid . . . . 8. *C. glebulentum*
            5. Isidia slender, simple or branched, on ridges and pustules of the thallus . . 7. *C. furfuraceum*
        4. Isidia globular.
            6. Thallus with lobules swollen and plicate.
                7. Spores submuriform, pale yellowish brown, 13 μ broad, pycnidia absent . . . .
                . . . . 1. *C. bachmanianum*
                7. Spores 4-celled, hyaline, 6.5–10.5 μ broad, pycnidia present . . . . 12. *C. tenax*
            6. Thallus with small more or less concave lobules.
                8. Spores 4-celled, linear oblong . . . . 14. *C. undulatum var. granulosum*
                8. Spores muriform, more or less ellipsoid . . . . 13. *C. tuniforme*
1. Thallus lacking isidia.
    9. Spores 4-celled, not muriform.
        10. Lobules, at least the tips, more or less swollen and plicate, raised on edge, on calcareous rocks . . . . 11. *C. polycarpon*
        10. Lobules not swollen or plicate.
            11. Exciple subparaplectenchymatous, spores broad, 13–15 μ . . . . 4. *C. crispum*
            11. Exciple euparaplectechymatous, spores narrow, less than 9 μ broad.
                12. Lobules narrow and linear to 1.5 mm broad, upper side convex; spores 4.5–6.5 μ broad . . . . 10. *C. multipartitum*
                12. Lobules nonlinear 2–4 μ broad, upper side convex; spores 6.5–9 μ broad . . . . . .
                . . . . 14. *C. undulatum*

9. Spores muriform.
   13. Lobules swollen and plicate.
      14. Thallus crustose, spores 4 per ascus .... 9. *C. limosum*
      14. Thallus foliose, spores 8 per ascus.
         15. Margin of apothecium coarsely crenulate, spores broad, 13 μ, pale yellowish brown .... 1. *C. bachmanianum*
         15. Margin of apothecium even, spores narrow, 8.5–10.5 μ, hyaline ...... 12. *C. tenax*
   13. Lobules not swollen and plicate.
      16. Thallus crustose, subumbilicate, less than 5 mm broad .... 2. *C. callopismum*
      16. Thallus foliose or subfruticulose, larger.
         17. Thallus subfruticulose, spores 4 per ascus, on soil .... 3. *C. ceraniscum*
         17. Thallus foliose, spores 8 per ascus, on rocks or soil.
            18. Exciple euthyplectenchymatous or subparaplectenchymatous spores 4-celled, 13–15 μ, lobes concave................................. 4. *C. crispum*
            18. Exciple paraplectenchymatous, spores submuriform, 8–13 μ, lobes narrow and canaliculate .... 5. *C. cristatum*

1. Collema bachmanianum (Fink.) Degel.

Symbol. Bot. Upsal. 13:189. 1954. *Collemodes bachmanianum* Fink, Mycologia 10:237. 1918.

Thallus foliose, lobate, the lobes to 6 mm broad, flattened or slightly concave, deeply or shallowly lobulate with more or less swollen or knotty margins; upper surface smooth or with few coarse plicae and often with erect dense divided lobules, dark olive-green to blackish or occasionally yellowish or bluish; lower surface paler and usually with dense hapters.

Apothecia usually numerous, sessile with constricted base, to 3 mm broad; margin thin to thick, coarsely crenulate; disk flat to concave, light or dark red, smooth; proper exciple euthyplectenchymatous, hymenium 100–135 μ, I + blue; paraphyses simple or branched, the apices thickened; asci narrowly clavate; spores 8 or sometimes 4 or 6, ellipsoid with acute or obtuse ends, straight, submuriform with 3 transverse and 1 longitudinal septa, pale yellowish brown, 20–36 × 8.5–15 μ. Pycnidia lacking and with internal conidia instead.

This species grows on more or less calciferous soil or sand. It is circumpolar, arctic to temperate, in North America occasionally collected in the Arctic as well as southward where it becomes more common. The Arctic material is of var. *millegranum* Degel. with more or less globular isidia on the thallus and apothecial margins.

2. Collema callopismum Mass.

Miscell. Lichenol. 23. 1856.

Thallus very small, to 5 mm broad, subumbilicate, lobes few, of small lobules flat or more or less terete, smooth or sometimes with isidia-like granules, becoming pulvinate, dark olive-green or blackish thallus partly paraplectenchymatous.

Apothecia sparse, partly immersed to sessile with constricted base, margin thin, entire to papillose or lobate, disk flat or concave, red, red-brown, or blackish; proper exciple paraplectenchymatous or nearly so; hymenium 90–150 μ; paraphyses branched or not, with thickened apices; asci narrowly clavate; spores 8 per ascus, oval to ellipsoid, occasionally nearly globose, with rounded or acute ends, more or less constricted at the septa, 4 celled to submuriform with 3(–4) transverse septa and 1 longitudinal septum, or rarely muriform with 2 longitudinal septa, hyaline, 17–26 × 8.5–13 μ.

This species grows on calciferous rocks. It is a very rare species, apparently circumpolar but with very disjunct range. The only known North American and Arctic specimens are from Greenland, Diskofiord, and from Yoho National Park, British Columbia. The latter specimen is of var. *rhyparodes* (Nyl.) Degel. which is distinguished by having larger spores (20–45 × 9–15 μ) and more transverse septa (3–6) than the typical var.

Collema bachmanianum (Fink) Degel.

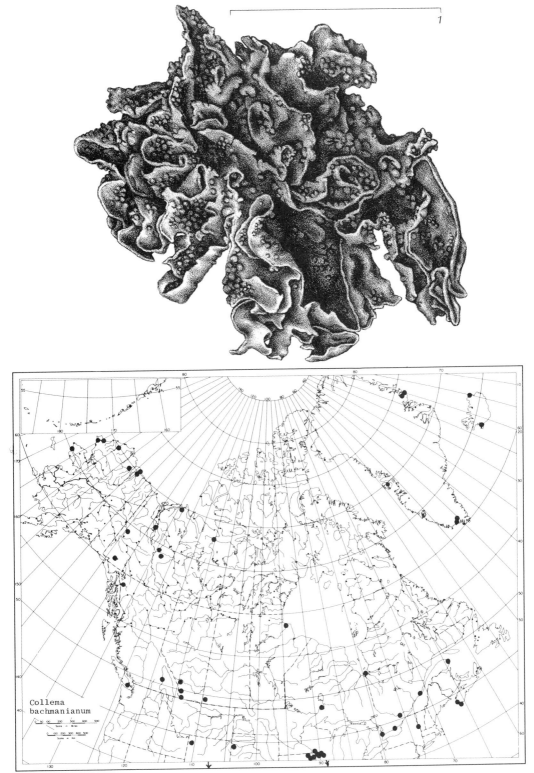

Collema callopismum Mass.

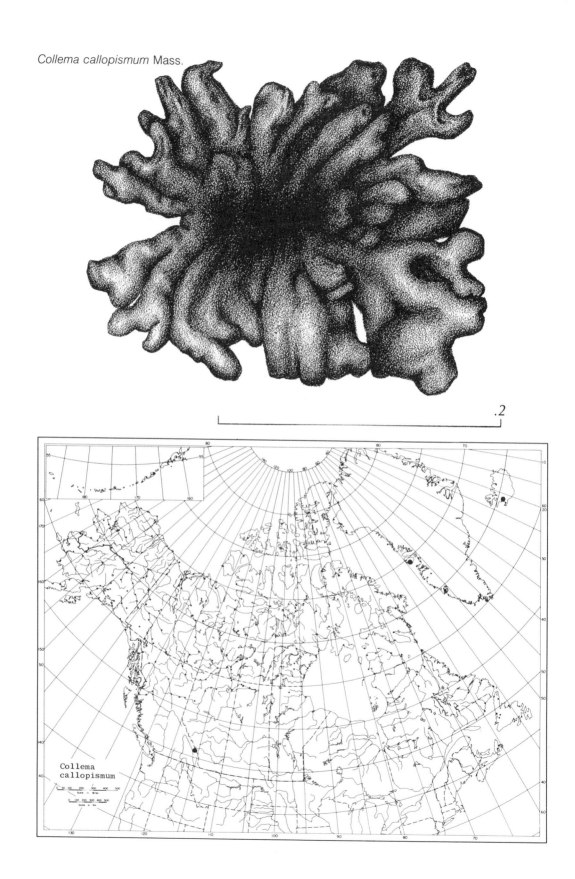

.2

Collema
callopismum

3. **Collema ceraniscum** Nyl.

Flora 48:353. 1865. *Collema arcticum* Lynge, Norske Vid.-Akad Oslo, Result. norske statsunderst. Spitsbergeneksp. 1(9):45. 1926.

Thallus foliose to usually subfruticulose, pulvinate to 1 cm thick, the lobes ascendent or erect, the lower ones flattened, to 2 mm broad, smooth or knotty; the end branches of the lobules teretiform, papilliform or knotty, forming a thick cushion, brown, blackish brown or blackish, paler within the cushion.

Apothecia numerous, sessile with constricted base, near apices of erect lobules, subglobose or flattened, to 0.8 mm broad; margins thick or thin, entire or papillose or lobulate; disk strongly concave, red or red-brown to blackish, smooth, epruinose; proper exciple nearly or entirely paraplectenchymatous; hymenium 150–190 μ; paraphyses simple or branched the tips clavate to subglobose and often with the upper cells constricted; asci narrowly clavate; spores (2–) 4 per ascus, broadly oblong to oval or almost cubic, muriform with 3–4 transverse and 1–3 longitudinal septa and the septa not constricted, hyaline, 20–40 × 13–22 or in subglobose ones 17–28 μ.

This species grows on humus soil, on decaying vegetation and on boulders, often in lichen heaths. It is circumpolar arctic, ranging southward in North America in the Rocky Mountains to Alberta, Miette Hot Springs.

4. **Collema crispum** (Huds.) Wigg.

Primit. Flor. Holsat. 89. 1790. *Lichen crispus* Huds., Flora Anglia ed. 1:447. 1762.

Thallus foliose, small to medium sized, deeply and broadly lobate, the lobes rounded, concave; upper surface smooth or with isidia, the isidia superficial or marginal, squamiform, bright green to usually dark green or blackish; lower side of same color or paler bluish to gray-blue, more or less covered with hapters.

Apothecia often lacking, appressed with the base constricted or not, 1–2 mm broad; margin thin, granulose to lobulate, sometimes with white rhizinae on the lower side; disk flat, light to dark red-brown, smooth, epruinose;

proper exciple subparaplectenchymatous, to euthyplectenchymatous, hymenium 130–170 μ, I+ blue; paraphyses simple or branched, with globose or clavate tips; asci narrowly clavate; spores 8, straight with round ends, 4 celled or submuriform, the cells constricted at the septa, hyaline, 17–47 × 8.5–18 μ.

This species grows on soil and rocks containing more or less calcium, especially in disturbed areas. It appears to be circumpolar temperate but occurs in the Arctic in Iceland according to Degelius (1954). In North America it has not been collected north of the southern edge of Canada except in the Arctic Red River valley.

5. **Collema cristatum** (L.) G. H. Web. in Wigg.

Primit. Florae Holsat. 89. 1790. *Lichen cristatus* L., Sp. Pl. 1143. 1753.

Thallus foliose, deeply lobate, radiating, irregularly branched lobules canaliculate, entire or incised, the margin not swollen, dark olive-green or brownish to blackish, lower surface with scattered tufts of hapters.

Apothecia usually marginal, sessile and finally constricted at the base, the margin thick, entire or markedly crenate or verrucose, paraplectenchymatous, disk flat to convex or concave, red, red-brown or blackish; proper exciple thick and paraplectenchymatous; hymenium 90–130 μ; paraphyses simple or branched, thickened at the apices; asci narrowly clavate; spores 8 (sometimes 4 or 6), ellipsoid, straight, the ends acute, submuriform with 3–4 transverse and 1–2 longitudinal septa, not or slightly constricted at the septa, hyaline, 18–40 × 8–13 μ.

This is a species of calcareous rocks, but may occasionally grow on soil, rarely on old wood or bases of plants. It is circumpolar, arctic to temperate, but appears to be rare in North America where it has been collected in Manitoba, Alberta, Utah, Colorado, Texas, and also in Mexico. It occurs in northeast Greenland and has been reported from an unspecified locality in Iceland. It is much more common in Europe as an arctic-alpine species which ranges into the Mediterranean region.

Collema ceraniscum Nyl.

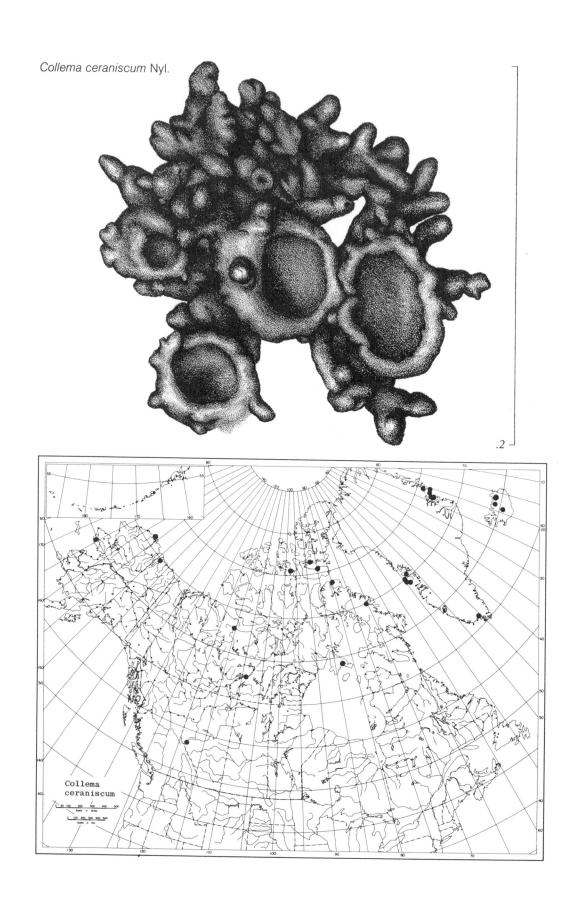

.2

Collema
ceraniscum

Collema crispum (Huds.) Wigg.

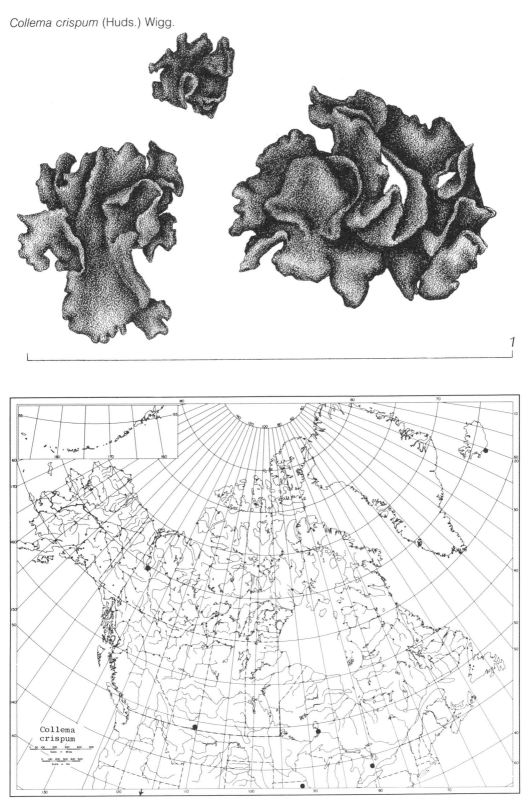

1

Collema
crispum

Collema cristatum (L.) G.H. Web. in Wigg.

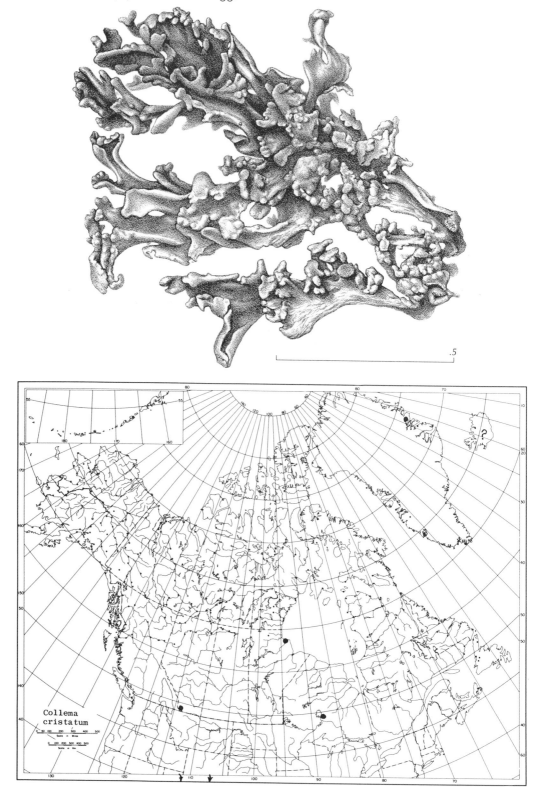

.5

Collema
cristatum

6. Collema flaccidum (Ach.) Ach.

Lich. Univ. 647. 1810. *Lichen flaccidus* Ach., K. Vet. Acad. Nya Handl. 1795:14. 1795.

Thallus foliose, thin, more or less rounded, loosely attached, broadly lobate, the margins sometimes bent down; upper surface smooth to slightly pustulate, olive-green to blackish; isidiate with flattened horizontal or vertical squamiform isidia; underside paler, grayish or bluish.

Apothecia rare, on upper surface of thallus, with constricted base, nonstipitate, usually to about 1.5 mm broad, disk flat, pale to dark red, smooth, sometimes white pruinose; margin thin, entire; exciple with pseudocortex, proper exciple subparaplectenchymatous; hymenium 90–130 μ, I+ blue, upper part yellowish brown, paraphyses simple the apices thickened, asci narrowly clavate; spores 8, with acute ends, narrowly ellipsoid to fusiform, 4, 5, or 6 celled, hyaline, 20–40 × 6–8.5 μ.

This species grows on rocks and on trees. It is circumpolar boreal to temperate but in Europe reaches the Arctic. In the American Arctic it has been collected only at the Ogotoruk Creek site in Alaska by Krog. It also occurs in Iceland but not yet in Greenland according to Degelius (1954).

7. Collema furfuraceum (Arn.) DuRietz

Ark. Bot. 22A(13):3. 1929. *Synechoblastus nigrescens* a. *furfuraceum* Arn., Flora 64:115. 1881.

Thallus foliose, adnate, deeply and broadly lobate, the lobes to 1 cm broad; strongly ridged, the ridges radiate, sometimes pustulate; upper side with numerous terete isidia, dark olive-green to blackish; lower side paler, typically with pseudocortex on both sides and in old parts paraplectenchymatous throughout.

Apothecia rare, to 1.5 mm broad, adnate; margins isidiate; proper exciple euthyplectenchymatous to subparaplectenchymatous; spores 8, straight or curved or flexuous, acicular, 5–6 celled, hyaline, 40–80 × 3–6.5 μ.

This species grows on barks of *Populus, Picea,* or *Quercus,* less often on rocks. It is circumpolar temperate. While it has not yet been collected in the American Arctic it occurs in alpine parts of Alaska that suggest that it should eventually be found in the Arctic. Hence it is included here. In Europe it occurs on the Arctic coast.

8. Collema glebulentum (Cromb.) Degel.

In Magn., Ark. Bot. ser. 2, 2:88. 1952. *Leptogium glebulentum* Nyl. ex Cromb., J. Bot. 20:272. 1882.

Thallus large, foliose, deeply lobate, dark olive-green to blackish; upper side smooth, with numerous superficial and marginal coralloid isidia forming a thick layer on the thallus; lower surface paler and with broad areas of hapters; parts of the thallus may be paraplectenchymatous throughout, a pseudocortex of several layers of cells sometimes at the lower surface.

Apothecia not known.

This species grows on siliceous or calcareous rocks, sometimes over mosses on the rocks. It is circumpolar arctic-alpine and in North America ranges south to the north shore of Lake Superior. It has yet to be collected in the mountains, either eastern or western.

9. Collema limosum (Ach.) Ach.

Lich. Univ. 629. 1810. *Lichen limosus* Ach., Lichenogr. Suec. Prodr. 126. 1798.

Thallus crustose to subfoliose, almost like a thin membrane on the soil, the margins swollen, entire to crenate or sublobate around the isolated apothecia, starting as granules which coalesce to form more continuous thallus, dark olive-green, brownish or bluish, smooth to rugose verrucose.

Apothecia numerous, innate to appressed, sometimes sessile with constricted base, to 4 mm broad; disk flat to convex, sometimes concave, light or dark red to red-brown, smooth, epruinose; margin thick, entire to verrucose or lobulate; proper exciple euthyplectenchymatous; hymenium 90–120 μ I+ blue; paraphyses simple or branched, with clavate to subglobose apices; asci narrowly clavate; spores (2–) 4 per

Collema flaccidum (Ach.) Ach.

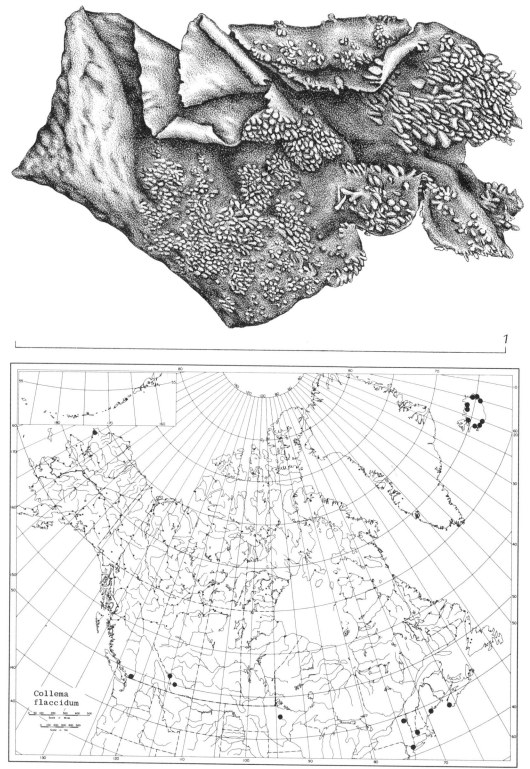

1

Collema
flaccidum

ascus, oblong to ovoid with rounded or acute ends, the cells constricted at the septa, muriform with 3–5 transverse and 1–2 longitudinal septa, hyaline, $20–40 \times 8.5–17 \ \mu$.

This species grows on soils, preferably clays and more or less calciferous. It is circumpolar, boreal and temperate, reaching the American Arctic only near the mouth of the Mackenzie River. Its North American range extends south to South Carolina and to California according to Degelius (1974) but this is a rarely collected species. Other stations reported by Degelius were in Illinois, Iowa, and New York.

10. **Collema multipartitum** Sm.
In Smith & Sowerby, Engl. Bot. 36. 1814.

Thallus foliose, thick, deeply multifid lobed, the lobes discrete, more or less linear, quite convex, the margin more or less entire, not thickened, commonly minutely striate, dark olive or olive-brown to black; lower surface with scattered hapters. Lobes 200–350 μ thick.

Apothecia sparse to numerous, 1–2 mm broad, sessile to slightly stipitate, the margin entire to lobulate, usually persistent, slightly striate, disk flat to slightly convex, dark red to brownish or blackish; the proper exciple parapletenchymatous of more or less isodiametric cells, subhymenium of about the same thickness (110–150 μ); hymenium 90–100 μ I+ blue; paraphyses simple or irregularly branched, 2–4.5 μ, the apices thickened or not, asci clavate, $60–80 \times 17–22 \ \mu$; spores 6, linear oblong, straight or curved, 3-septate, hyaline, $20–60 \times 4.5–6.4 \ \mu$.

This species grows on calcareous rocks. It is distributed in Europe, North Africa, and in western North America, in boreal and temperate areas. It reaches the Arctic at Anaktuvuk Pass and in the valley of the Arctic Red River.

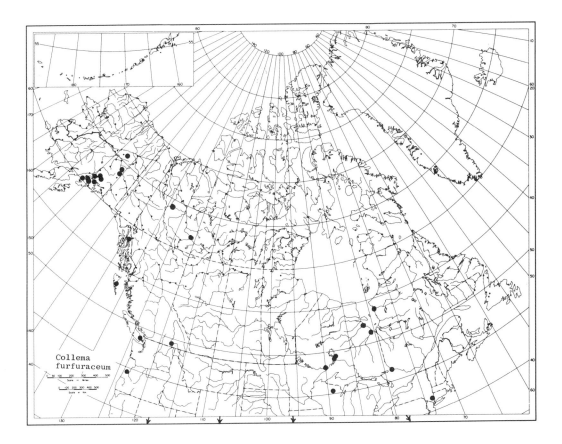

Collema furfuraceum

Collema glebulentum (Nyl. ex Crombie) Degel.

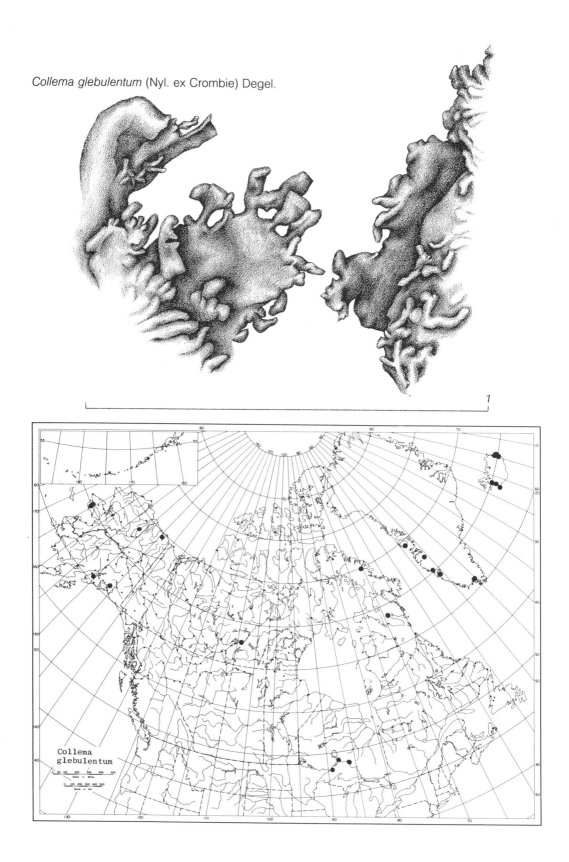

1

Collema
glebulentum

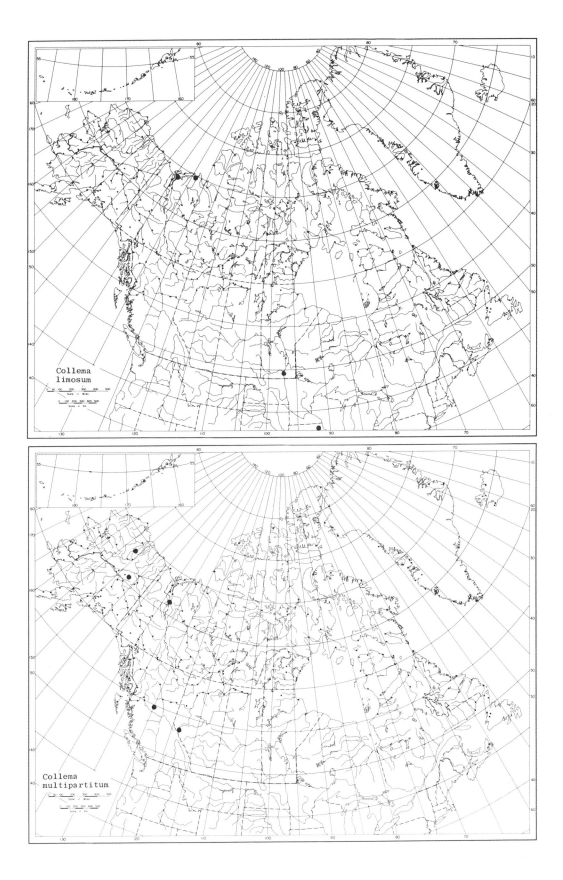

Collema
limosum

Collema
multipartitum

11. Collema polycarpon Hoffm.
Deutschl. Flora 102. 1796.

Thallus foliose, deeply lobate, the lobes to 2.5 mm broad, flattened with the edges raised, the tips swollen and plicate with lamellate lobules with erect branches making a cushion-like thallus, dark olive-green to blackish; underside paler and with rounded hapters.

Apothecia numerous, covering the thallus except for the lobe tips, sessile on tips of short branches, nearly stipitate, to 1.5 mm broad; disk red, red-brown, or black, glossy, epruinose; margin thin, smooth; proper exciple subparaplectenchymatous, hymenium 65–106 μ, I+ blue; paraphyses simple or branched, the tips thickened; asci subcylindrical; spores 8, fusiform, straight, with acute ends, 4 celled (with some 2–3 celled) constricted or not at septa, hyaline, 13–30 × 5–8.5 μ.

This species grows on calcareous rocks. It is circumpolar arctic to temperate and also occurs in the southern hemisphere. In North America it has not yet been collected in the high arctic but occurs from Alaska and northern Quebec–Labrador south to Mississippi and Arizona.

12. Collema tenax (Sw.) Ach. em. Degel.
Lich. Univ. 635. 1810. *Lichen tenax* Sw., Nova Acta Reg. Soc. Sci. Upsal. 4:249. 1794.

Thallus foliose, very variable, usually rather thick, lobes radiating or irregular, contiguous or separate, crenate to lobulate, flattened or concave, usually with a few coarse plicae toward the ends; upper side smooth or sometimes rugulose when dry, sometimes with accessory lobules, dark olive-green to blackish.

Apothecia sessile with constricted base, to 3 mm broad; margin thin to thick, usually smooth but may be a little crenulate to granulose or fine-rugulose; proper exciple euthyplectenchymatous; hymenium 65–100 μ, I+ blue; paraphyses simple or branched, the tips globose to clavate; asci narrowly clavate; spores 8, fusiform or ellipsoid submuriform, may be 3–4 celled or with 3 (–4) transverse and 1 longitudinal septa, usually constricted at the septa, 17–30 × 8.5–13 μ.

This species grows on soil, usually calciferous, including clays to sands. It is circumpolar arctic to temperate. Northern material tends to be var. *expansum* Degel. with broad glossy lobes bearing large lobules or var. *corallinum* (Mass.) Degel. with narrow lobes to 2 mm broad, non-isidiate, the lobes incised, erect and forming clusters.

13. Collema tuniforme (Ach.) Ach. em. Degel.
Lich. Univ. 649. 1810. *Lichen tunaeformis* Ach., K. Vet. Acad. Nya Handl. 17. 1795.

Thallus foliose, deeply and broadly lobate, with sparse lobules 2–5 mm broad, the lobules broadly canaliculate along their length except for the tips, the margins coarsely undulate, entire to slightly crenate; upper surface with globular to clavate isidia, dark olive-green to blackish; lower surface paler, to grayish or bluish, with isidia on the erect lobules, with hapters or sometimes even downy.

Apothecia sessile with more or less constricted base, to 2 mm broad; disk flat or concave or convex, red-brown or dark red, smooth, sometimes glossy; margin thin to thick, entire smooth or isidiate; proper exciple paraplectencnymatous; hymenium 85–130 μ I+ blue; paraphyses simple or branched, more or less thickened at tips; asci narrowly clavate; spores 8, broadly oval with rounded or occasionally acute tips, submuriform to muriform with 3 transverse and 1–3 longitudinal septa, not constricted at the septa, hyaline, 15–28 × 6.5–15 μ.

This species grows on calcareous rocks or rocks with calcareous dust, preferably in places with periodically wet rocks, sometimes even inundated. It is circumpolar arctic-alpine but with broad requirements so that it occurs south to Maryland, Illinois, Utah, and Colorado.

14. Collema undulatum Laur. ex Flot.
Linnaea 23:161. 1850.

Thallus foliose, lobate, the lobes elongate with small lobules 2–4 mm broad, concave, the margin coarsely undulate, entire not swollen; upper surface smooth not glossy, lack-

Collema polycarpon Hoffm.

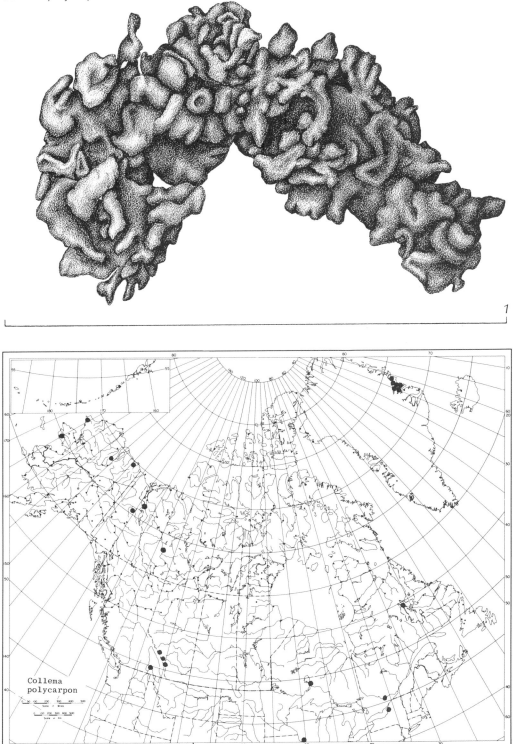

1

Collema tenax (Sw.) Ach.

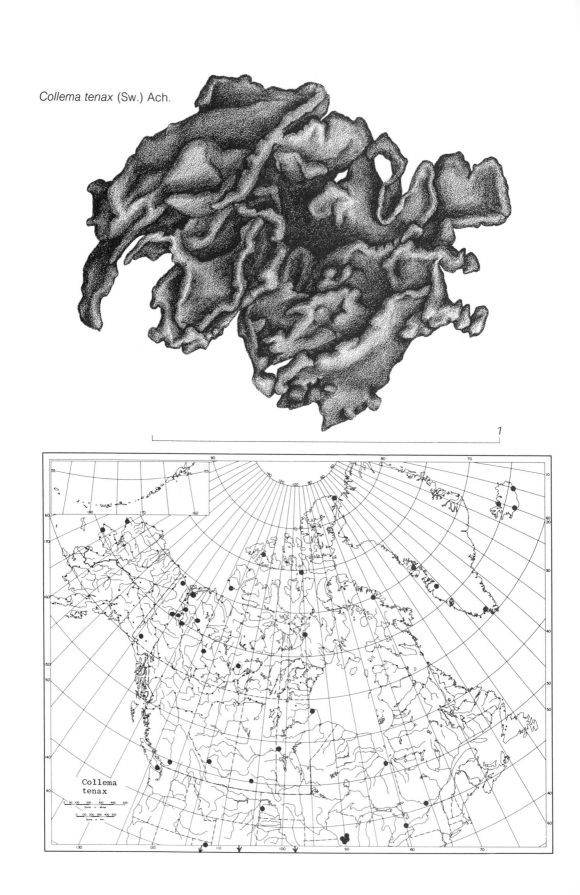

1

Collema
tenax

Collema tuniforme (Ach.) Ach.

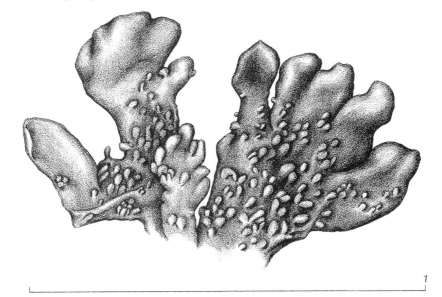

1

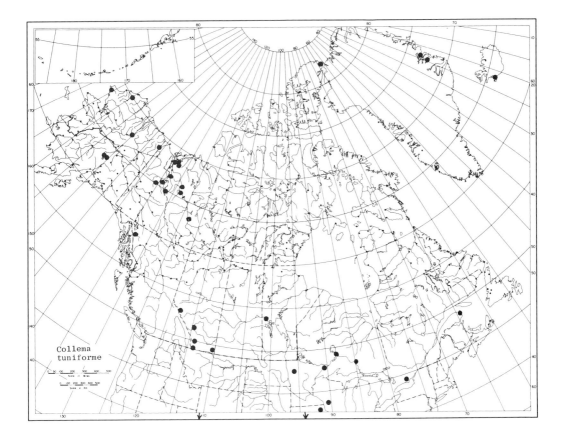

Collema
tuniforme

Collema undulatum Laur. ex Flot.

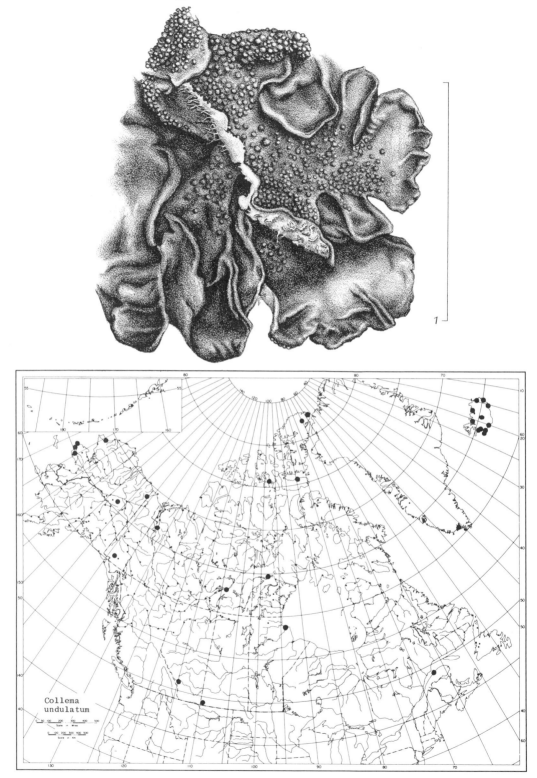

1

ing isidia, dark olive-green to blackish; lower surface paler with groups of hapters.

Apothecia numerous, superficial or marginal, sessile, to 1.5 mm broad; disk flat to convex dark red to red-brown, smooth; margin thick, entire or crenulate; proper exciple thick and paraplectenchymatous; hymenium 85–105 μ, I+ blue; paraphyses simple or branched clavate at apices; asci clavate; spores 8, linear oblong with rounded ends, not or slightly constricted at septa, 3 septate, hyaline, 17–30 × 6.5–9 μ.

This species is calciferous and grows on soil or rocks. It is primarily European and North American but has been reported from Japan. It is arctic-alpine and in North America occurs south to the Gaspé Peninsula and to Alberta and Saskatchewan.

There are two varieties in this species, var. *granulosum* Degel. has the thallus with globular isidia on the surface and is the more arctic variant. Var. *undulatum,* the typical material lacks the isidia and is more southerly.

CORISCIUM Vain.

Acta Soc. Fauna Flora Fenn. 7:188. 1890.

Thallus foliose, squamulose, of small cochleate lobes; the upper cortex paraplectenchymatous, the algal layer thick, glomerulate, the medulla loose; the lower cortex discontinuous, lacking rhizinae, attached by a weft of hyphae. Fruiting bodies uncertain, suggested as the basidiomycete *Omphalina* by Gams (1962) and by Heikkila and Kallio (1966).

Algae: *Polycoccus* (Keissler 1938).

Type species: *Coriscium viride* (Ach.) Vain.

REFERENCES

Gams, H. 1962. Die Halbflechten *Botrydina* und *Coriscium* als Basidiolichenen. Österr. Bot. Z. 109:376–380.

Heikkila, H. & P. Kallio. 1966. On the problem of subarctic Basidiolichens. I. Ann. Univ. Turku A. II. 36:9–35.

Poelt, J. & F. Oberwinkler. 1964. Zur Kenntnis der flechtenbildenden Blätterpilze der Gattung *Omphalina*. Österr. Bot. Zeit. 111:393–401.

1. Coriscium viride (Ach.) Vain.

Acta Soc. Fauna Flora Fenn. 7:188. 1890. *Endocarpon viride* Ach., Lich. Univ. 300. 1810.

Thallus small, squamulose, of rounded cochleate lobes, circular or becoming slightly lobate with rounded lobes, to 5 mm broad, usually less, the margins inrolled and white, the surface blue-green, slightly pruinose; rhizinae lacking but the underside attached to the substratum by a loose network of hyphae. Fruiting bodies uncertain.

Upper cortex paraplectenchymatous, several cells thick, uneven in thickness, algae distributed through the medulla in glomerules; the lower cortex plectenchymatous to paraplectenchymatous, thin, the hyphae anastomosing in part into the hyphae penetrating the ground, at the edges the lower cortex is more continuous.

This species seems to always grow on humus, mossy or woody in origin, in partially shaded sites, as the sides of lemming burrows, sides of hummocks, edges of solifluction lobes, etc.

This is a circumpolar species, known from Scotland, Fennoscandia, central Europe, Siberia, southeast Greenland. In North America it is known from Baffin Island, Labrador, Ungava Peninsula, Northwest Territories, Alaska, and south in the eastern mountains to Mt. Washington and Mt. Adams in the White Mountains and Mt. MacIntyre in the Adirondacks.

This species would easily be confused in gross appearance with *Normandina pulchella,* a temperate zone lichen, but that species has no upper cortex, is homiomerous, and its alga is *Pleurococcus.*

In European treatments this is generally placed as a lichenized thallus of the gilled basidiomycete *Omphalina luteolilacina* (Favre) Henderson.

Coriscium viride (Ach.) Vain.

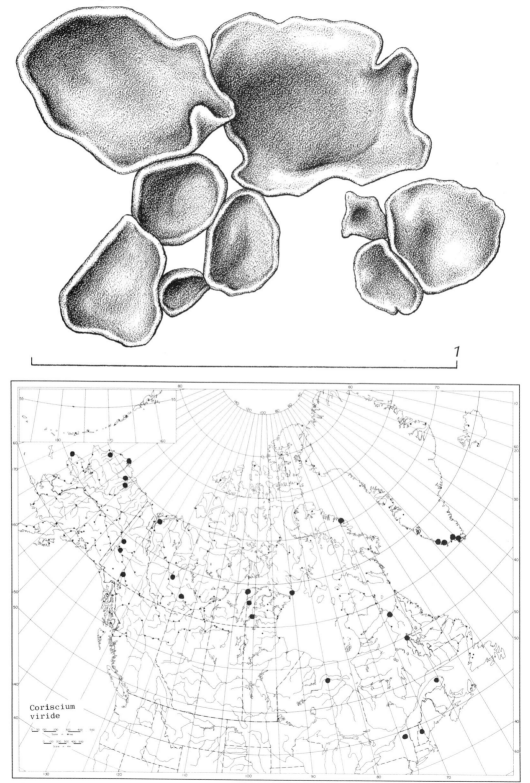

1

Coriscium
viride

CORNICULARIA Ach.

Kgl. Vetensk.-Akad. Nya Handl. 15:259. 1794. *Cetraria* sect. *Cornicularia* (Ach.) Fr., Syst. Orb. Veget. 1:239. 1825. *Coelocaulon* Link, Grundrisz der Kraüterkunde 3:165. 1833. *Evernia* sect. *Cornicularia* (Ach.) Tuck., An enumeration N. Amer. Lich. 47. 1845. *Alectoria* sect. *Cornicularia* (Ach.) Flagey, Mem. Soc. Emulat. Doubs 354. 1882. *Pseudocornicularia* Gyeln., Acta Fauna Flora Univ. 1:7. 1933.

Fruticose erect lichens with irregularly branching terete to somewhat angled podetia; cortex with irregularly or radially, not longitudinally oriented hyphae. Apothecia terminal or subterminal, thalloid margins with teeth or short branchlets or crenulate; spores 8, hyaline, simple, small, ellipsoid.

Algae: *Trebouxia.*

Type species: *Cornicularia divergens* (Ach.) Ach.

REFERENCES
DuRietz, G. E. 1926. Vorarbeiten zu einer "Synopsis Lichenum." 1. Die Gattungen *Alectoria, Oropogon* and *Cornicularia*. Ark. Bot. 20A(11):1–43.
Hakulinen R. 1965. Über die Verbreitung und das vorkommen einiger Nördlichen Erd- und Steinflechten in Ostfennoskandien. Aquilo 3:22–66.
Keissler, K. 1960. Usneaceae in Rabenh. Kryptogamen-Flora 9:5:4:1–755.
Räsanen, V. 1952. Studies on the species of the lichen genera *Cornicularia, Cetraria* and *Nephromopsis*. Kuopio Luonn. Ystav. Yhdist. Julks. Ser. B, 2:1–53.
Weber, W. A. & S. Shushan. 1955. The Lichen flora of Colorado: *Cetraria, Cornicularia, Dactylina,* and *Thamnolia*. Univ. Colo. Studies 3:115–134.

1. Thallus black, attached to rocks, apothecia common and with a spinulose corona on the margin . . . . 4. *C. normoerica*
1. Thallus brown, loosely growing on soil or over rocks, apothecia rare.
   2. Medulla C+ red, containing olivetoric acid; branches elongate and tapering, with raised white pseudocyphellae . . . . 2. *C. divergens*
   2. Medulla C−, containing protolichesterinic acid; branches short and spinescent.
      3. Thallus with few usually flattened branchlets in addition to the main branches; pseudocyphellae deeply concave. . . . . 1. *C. aculeata*
      3. Thallus with many small terete side branchlets; pseudocyphellae flattened . . . . . 3. *C. muricata*

Cornicularia aculeata (Schreb.) Ach.

Meth. Lich. 302. 1803. *Lichen aculeatus* Schreb., Flora Lips. 125. 1771. *Cetraria aculeata* (Schreb.) Fr., Syst. Orb. Veg. 239. 1825. *Lichen islandicus* γ *tenuissimus* L., Sp. Pl. 1145. 1753. *Cetraria tenuissima* (L.) Vain., Ark. Bot. 8(4):25. 1909. *Cornicularia tenuissima* (L.)Zahlbr., Annal. Naturh. Museum Wien 42:64. 1928.

Thallus fruticose, erect, very stiff and brittle; broadly dichotomously branched, usually only a few cm tall, 1–2 cm in poor condition, the branches angular, scantily pseudocyphellate, the pseudocyphellae usually concave, 1 mm thick on main axes, the smaller branches becoming pointed, often flattened, lacking soredia or isidia; medulla very thinly cobwebby to hollow, white.

Apothecia rare, not seen, described (Keissler 1960) as terminal, to 3 mm broad, the margin dentate, or setose; disk flat to convex; hymenium 60 μ; epithecium brownish; asci saccate; spores 6–8, hyaline, simple, ellipsoid, 6–8 × 4 μ.

Reactions: K−, C−, KC−, P−.
Contents: protolichesterinic acid.

This species grows on sandy soil or among mosses at the edge of frost boils, sometimes over rocks with thin soil, occasionally at the base of shrubs. It is circumpolar arctic and alpine, in North America ranging southward to Mt. LeConte, Tennessee, in the east and into California in the west. It is also known from South America.

2. Cornicularia divergens Ach.

Meth. Lich. 303. 1803. *Alectoria divergens* (Ach.) Nyl., Mem. Soc. Imp. Sci. Nat. Cherbourg 3:171. 1855.

Thallus fruticose, erect, branching, fragile, shining, dark-red-brown to blackish

Cornicularia aculeata (Schreb.) Ach.

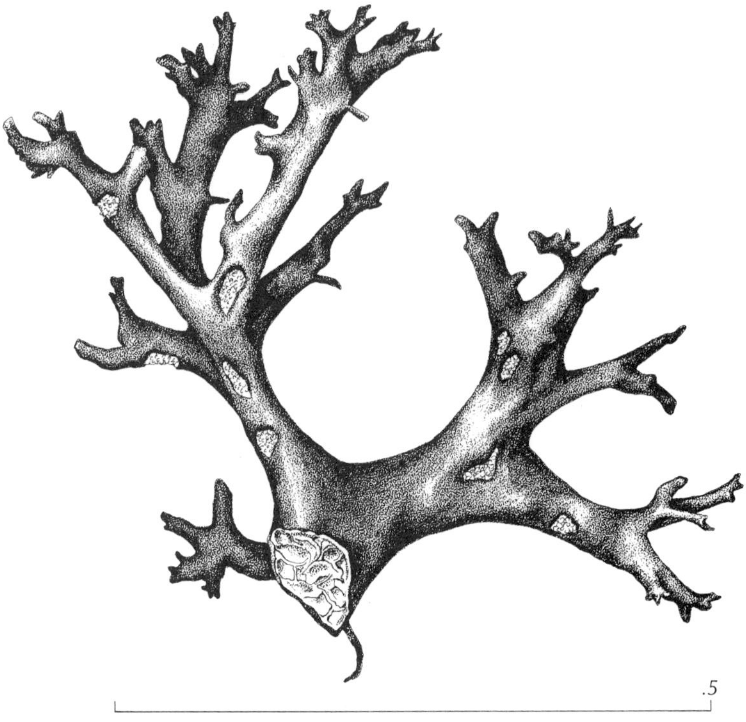

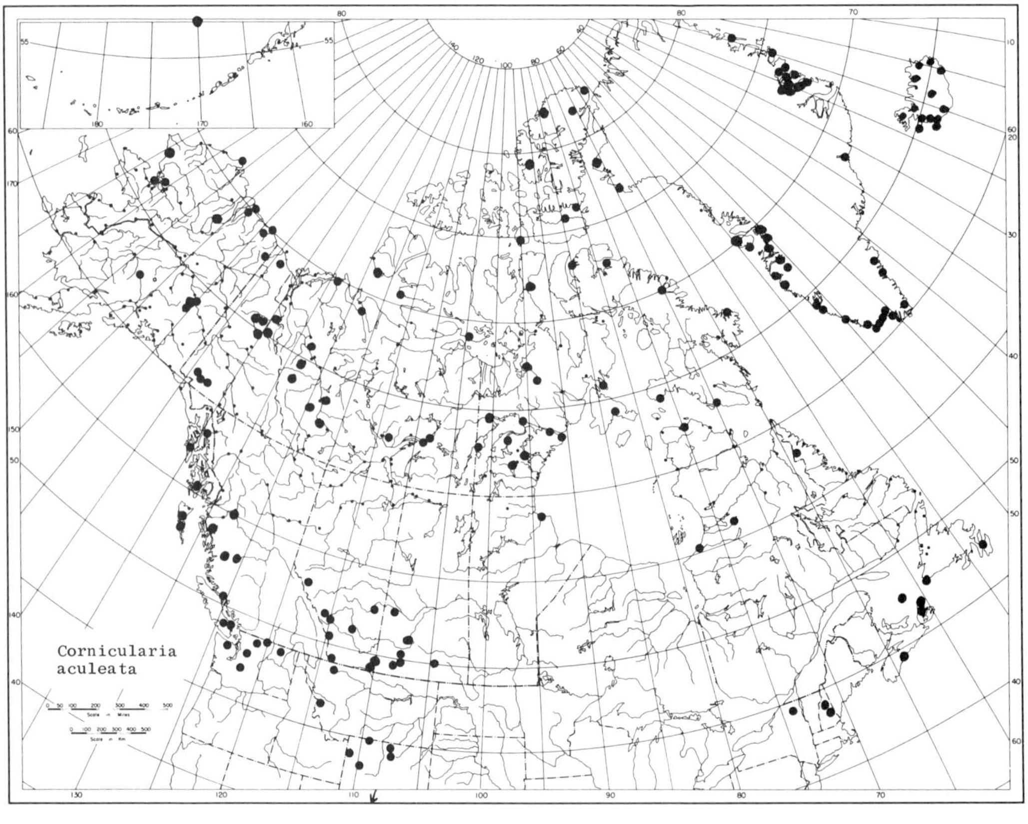

Cornicularia divergens Ach.

brown, 5–8 cm tall, the main axes to 2 mm, the branches dichotomous, with raised white pseudocyphellae; interior with fairly dense cottony hyphae.

Apothecia rare; lateral and making the tips geniculate, adnate with margin concolorous with the thallus, crenulate to toothed; disk to 10 mm broad, chestnut brown, smooth, shining; epithecium brownish, hymenium hyaline, 45–50 μ; asci saccate; paraphyses slender, the tips not much thickened; spores 8, simple, ellipsoid, hyaline, 7–8 × 3.5–4 μ.

Reactions: K–, C+ red, KC+ red, P–.
Contents: olivetoric acid.

This species grows on soil, on humus, forming large mats in heath tundras, occasionally on the lower branches of trees at the tundra border. It is usually much intermixed with *Bryoria nitidula*. It is circumpolar arctic alpine and is also in South America. Its North American range is south to Newfoundland in the east, to British Columbia in the west.

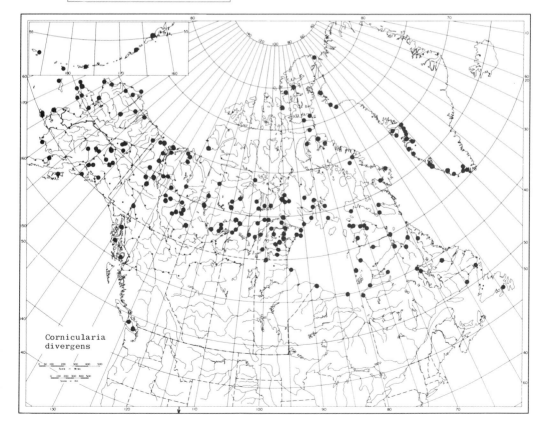

Cornicularia
divergens

3. Cornicularia muricata (Ach.) Ach.

Method. Lich. 302. 1803. *Lichen muricatus* Ach., Lich. Suec.
Prodr. 214. 1798. *Cornicularia aculeata* var. *muricata* (Ach.)
Ach., Lich. Univ. 612. 1810. *Cetraria aculeata* var. *muricata*
(Ach.) Schaer., Enum. Crit. Lich. Europ. 17. 1850.

Thallus fruiticose, erect, richly branched,
1–2 cm tall, fragile, dichotomously branched,
the branches with many short thorn-like lateral
branches; with flat pseudocyphellae.

Apothecia not seen but reportedly as in
C. aculeata.

Reactions: K−, C−, KC−, P−.

Contents: protolichesternic acid and rangiformic acid (Krog, 1968).

This species grows on sandy soil and among
other lichens in heath tundras. It appears to be high
arctic and alpine with a range in North America
northerly as compared with that of *C. aculeata,* coming south only in the west in British Columbia and to
Newfoundland in the east. Much more information is
needed on this species.

4. Cornicularia normoerica (Gunn.) DuRietz

Ark. Bot. 20A(11):39. 1926. *Lichen normoericus* Gunn., Fl.
Norveg. 2:123. 1766. *Lichen tristis* Web., Spicil. Flor. Goettin-
gens 209. 1778. *Cornicularia tristis* (Web.) Hoffm., Descript.
Adumb. Plant. 2:36. 1794. *Cetraria normoerica* (Gunn.) Lynge,
Vidensk. Skr. I. Mat.-naturv. Klasse 7:185. 1921.

Thallus short, fruticose, erect forming
tufts on rock surfaces, dull or shining black or
chestnut brown, dichotomously branched sparingly, 1–2 cm tall, usually to 3 mm broad lobes,
somewhat flattened, knotty, lacking isidia, soredia, or pseudocyphellae.

Apothecia usually abundant, terminal, to
3 mm broad, the margins verrucose to with a
thorny corona; disk dark brown to black, somewhat shiny, flat to convex; epithecium dark
brown; hymenium hyaline; hypothecium grayish; asci clavate; spores 8, ellipsoid, hyaline,
5–9.5 × 4–6 μ.

Reactions: K−, C−, KC−, P−.

Contents: no substances.

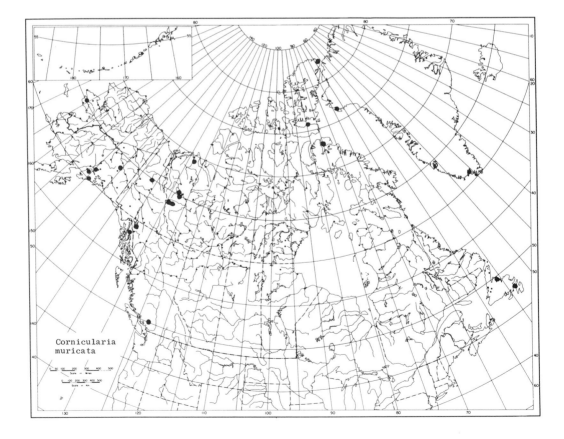

Cornicularia
muricata

Cornicularia normoerica (Gunn.) DR.

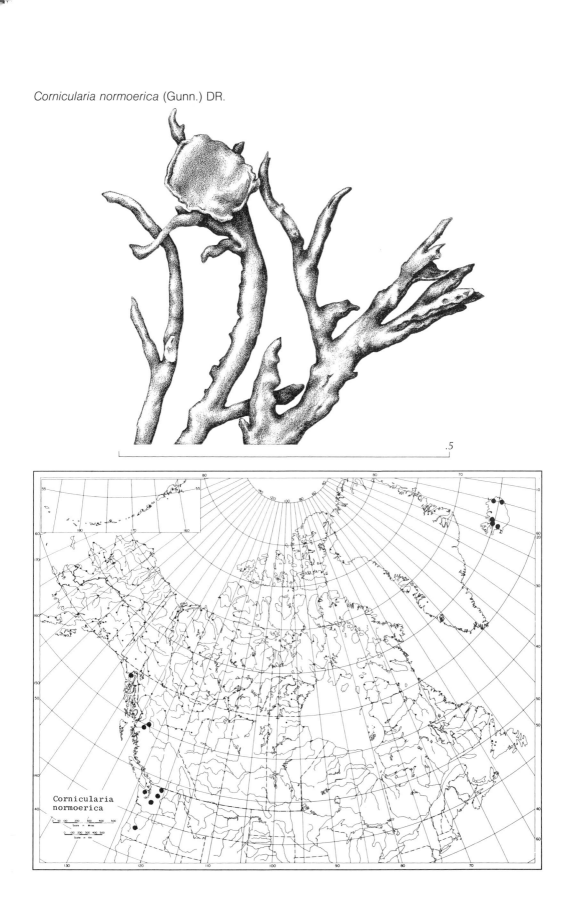

.5

Cornicularia
normoerica

This species grows on cliffs and boulders. It ranges in Europe as an arctic-alpine, and in western North America from Alaska to Oregon. It reaches the Arctic in Iceland. A report by Macoun from Digges Island, in the Canadian Eastern Arctic is incorrect; the specimen is *Pseudephebe pubescens*.

DACTYLINA Nyl.

Synopsis Lichenum 1:286. 1858.

Thallus of erect, fruticose podetia, sparingly dichotomously branched, lacking rhizinae, corticate, the cortex of columnar, radiate, dense hyphae, prosoplectenchymatous; medulla of pachydermatous hyphae close under the cortex, the interior becoming hollow or loosely arachnoid; algae dispersed or in small glomerules in the outer medulla.

Apothecia on lateral branches, terminal, lecanorine; hypothecium hyaline to yellowish, paraphyses unbranched, capitate, asci clavate, 8 spored, spores simple, spherical to slightly ellipsoid, hyaline, thick walled, pycnidia immersed, globose, sterigmata exobasidial, conidia straight or slightly curved.

Algae: *Trebouxia*.

Type species: *Dactylina arctica* (Richards.) Nyl.

REFERENCES
Krog, H. 1968. *Dactylina madreporiformis* in Alaska. Bryologist 73:628–630.
Lynge, B. 1933. On *Dufourea* and *Dactylina*, three Arctic lichens. Skrift. Svalbard Ishavet (Oslo) 59:1–62.
Thomson, J. W. & C. D. Bird. 1978. The lichen genus *Dactylina* in North America. Can. J. Bot. 56:1602–1624.

1. Plants well branched, usually less than 2 cm. tall, yellow or brownish yellow or partly violet pruinose, hollow or with cobwebby hyphae filling the center.
   2. Plants sparsely dichotomously branched, branches with few short lateral branches, filled with cobwebby hyphae, pycnidia rare; medulla P−, C−; containing usnic and protolichesterinic acids. . . . . 3. *D. madreporiformis*
   2. Plants dichotomously or sympodially branched, yellowish or brownish, usually with light violet pruina towards the tips, the branches commonly muricate knobbed with short lateral branchlets, hollow at least in part; pycnidia common; medulla PD+ red or PD−, acetone extract UV+ yellowish or UV−; usnic acid present, physodalic and physodic acid present or absent, gyrophoric acid absent . . . . 4. *D. ramulosa*
1. Plants little branched or unbranched, turgid, hollow, usually over 2 cm tall, yellow or brownish yellow; gyrophoric acid present.
   3. Medulla P+ orange-red, physodalic acid present, cortex C+ pink, medulla C−, the gyrophoric acid limited to the inner cortex above the algal layer . . . . 2. *D. beringica*
   3. Medulla P−, physodalic acid absent, cortex and medulla both C+ red, the gyrophoric acid distributed through the inner cortex and the entire medulla . . . . 1. *D. arctica*

1. Dactylina arctica (Richards.) Nyl.

Synopsis Lichenum 1:286. 1858. *Dufourea arctica* Richards. in J. Franklin, Narrative of a journey to the shores of the Polar Sea. 775. 1823.

Podetia erect, finger-like, unbranched or little branched, when decumbent regenerating along the length of the podetium and thus forming tufts, varying from very small to 7 cm tall and 14 mm broad, pale yellow-green, brownish in exposed situations; the base dying and becoming dark brown, inflated, hollow, more or less shiny, usually smooth, sometimes slightly foveolate, lacking soredia or isidia, the apices rounded or obtuse-pointed.

Outer cortex firm, 25–40 μ thick, of radiating-oriented hyphae which are prosoplectenchymatous, medullary hyphae pachydermatous, 5–7.5 μ thick, lax, the entire center hollow, the algae dispersed in the loose medulla, 10–15 μ in diameter.

Dactylina arctica (Hook.) Nyl.

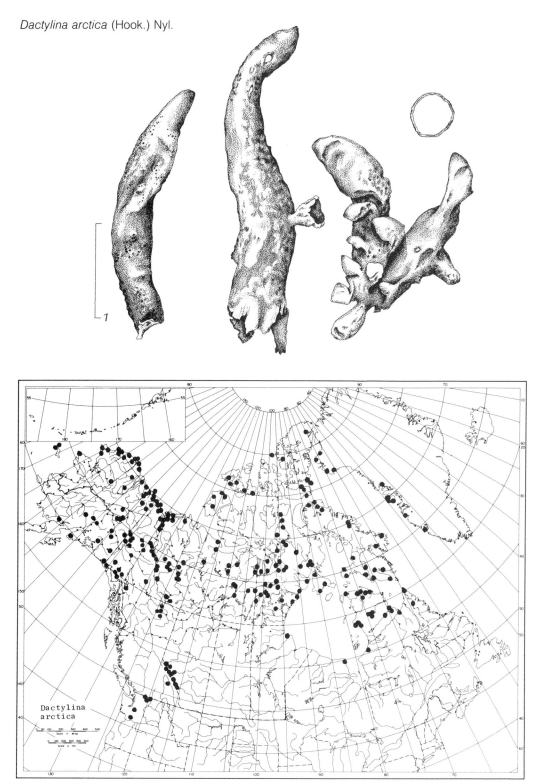

Apothecia rare, on the tips of lateral branches, to 5 mm broad, the disk chestnut-brown, shining, epruinose, lecanorine, the margin colored as the thallus, slightly crenulate. Subhymenium and hypothecium thin, 25 μ thick, grading into a very large celled paraplectenchymatous layer below, this becoming arachnoid next to the hollow interior. Hymenium ca 100 μ thick, paraphyses conglutinate, the tips light brown, capitate, 4–5 μ at the tip, asci clavate, 10–18 × 25–65 μ, spores biseriate, 8, spherical, 4–6 μ, the wall thick and the lumen irregular. Pycnidia rare, immersed, globose, the ostiole dark bordered, conidia straight, bacilliform, 5–7 × 1 μ.

Reactions: K−, C+ (rarely C−), KC+ yellow, P+ red-orange or P−.

Contents: gyrophoric and usnic acids, accessory physodalic and barbatic acids.

Although this species may grow scattered as smaller individuals in dry tundras it is more luxuriant when a more constant supply of water as late snow-banks is available. Here its preferred habitat is just above the snowbank and among arctic heather *(Cassiope)*. Its fruiting is confined to such habitats. This is a circumpolar arctic-alpine species found south to northern Labrador, northern Quebec, and Ontario and in the western mountains to Alberta, British Columbia, and Washington.

2. **Dactylina beringica** Bird & Thoms.
Can. J. Bot. 56:1612. 1978.

Podetia similar to those of *D. arctica*, finger-like, unbranched or little branched, when decumbent regenerating along the length of the podetium, varying in size; pale yellowish green to brown, the base dying and becoming dark brown, inflated and hollow, more or less shiny, smooth, the tips blunt or acute.

Apothecia not seen.

Reactions: cortex K−, C+ pink, P−; medulla K−, C−, KC+ pink, P+ red-orange.

Contents: usnic acid and gyrophoric acid

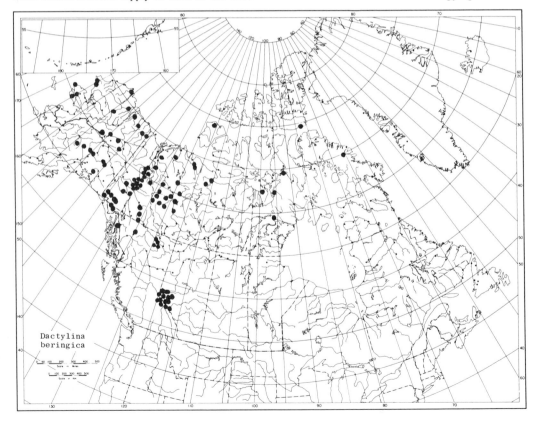

Dactylina
beringica

in the cortex only, physodalic and physodic acids in the medulla.

The ecology of this species is similar to that of *D. arctica* and indeed both may occur together in collections made where they are sympatric. The range of this species is amphi-Beringian, extending from the USSR across the Bering Straits into North America where it ranges east to Baffin and Devon islands in the Northwest Territories and south into Alberta and British Columbia in the Rocky Mountains.

3. Dactylina madreporiformis (Ach.) Tuck.

Proc. Amer. Acad. Arts Sci. 5:398. 1862. *Dufourea madreporiformis* Ach. Lich. Univ. 525. 1810. *Non Lichen madreporiformis* Wulf. in Jacquin Collectanea 3:105. 1791. = *Cladonia papillaria.*

Podetia fruticose, to 3.5 cm tall, 2 mm in diameter, soft rather than fragile, inflated, cylindrical, foveolate; branching sparsely dichotomously from near the base to form tufts, tips rounded, the branches becoming erect except in f. *irregularis* (Vain.) Lynge in which the podetia become flattened, dorsiventral and decumbent; yellow to straw-colored or greenish yellow, often browned in the sun, shining, lacking soredia, isidia, or pruina.

Cortex of radiate prosoplectenchymatous hyphae, 25–50 μ thick, medulla more or less solid, becoming arachnoid to the center but not hollow, the hyphae 7 μ in diameter, algae scattered in the outer medulla, 13–15 μ in diameter.

Apothecia rare, seen in Lich. Colo. Exs. 210, at the tips of branches, disk pale chestnut, shining, 2–4 mm broad, margin thick, lecanorine, hypothecium hyaline, 35 μ thick, upper part of the hymenium pale brownish, the lower part hyaline, 35–50 μ thick, paraphyses conglutinate, very thick walled with slender lumen, septate, tips capitate, 6 μ thick, asci pyriform, 12–14 × 25 μ, spores biseriate, 8, simple, more or less spherical, thick walled, 10 μ in diameter. Pycnidia distributed over the thallus, immersed, spherical, conidia slightly curved, 13–20 × 0.5 μ.

Reactions: K−, C−, KC+ yellow, P−.

Contents: usnic and protolichesterinic acids.

This species occupies a variety of habitats within the tundras, occurring as smaller individuals in very dry heath tundras and becoming more abundant in moister open tundras but seldom in the very wet tussock types of tundra. It usually appears more robust near snowbank accumulations in sheltered habitats, probably requiring some winter snow protection. Like *D. ramulosa* it is calciphilous.

This circumpolar species ranges at higher altitudes south into the Alps in Europe and into Colorado, Utah, and New Mexico in North America. It seems to be rare in Alaska and the Yukon, more common southward through Alberta and in the Rocky Mountains. It is absent in Greenland.

The forma *irregularis* is known from the central Rocky Mountains and from Asia.

4. Dactylina ramulosa (Hook.) Tuck.

Proc. Amer. Acad. Arts & Sci. 5:397. 1862. *Dufourea ramulosa* Hook. in Botanical Appendix to W. E. Parry. Journal of a second voyage for the discovery of a Northwest Passage from the Atlantic to the Pacific; performed in the years 1821–22–23. p. 414. 1825.

Podetia fragile, erect, to 2 cm tall, the base dying and growing from the apices; subterete or flattish, more or less sympodial with few branches but many short, divergent, muricate black-tipped side branches, usually forming tufts when undisturbed; yellowish straw-colored, to brown in strong illumination, usually with a whitish or violet pruina; lacking soredia or isidia, only slightly if at all foveolate, the interior hollow with very loose arachnoid hyphae scattered in the cavity.

Cortex firm, 25–50 μ thick, of radiatingly oriented prosoplectenchymatous hyphae, the cells very thick walled and not clearly distinguishable, yellowish or brownish, sharply differentiated from the loosely arachnoid medullary hyphae which are pachydermatous, 7 μ diameter; the algae scattered or in small glomerules in the outer medulla, 7.5–16 μ.

Apothecia on the tips of side branches, 1–3 mm broad (Lynge reported to 6 mm), the disk flat, smooth, shining, chestnut colored; the rather thick lecanorine margin coarsely crenulate or folded, the interior hollow with loosely arachnoid filaments as in the podetia; the cortex as in the podetia, hymenium thin, 50 μ, the

Dactylina madreporiformis (Ach.) Tuck.

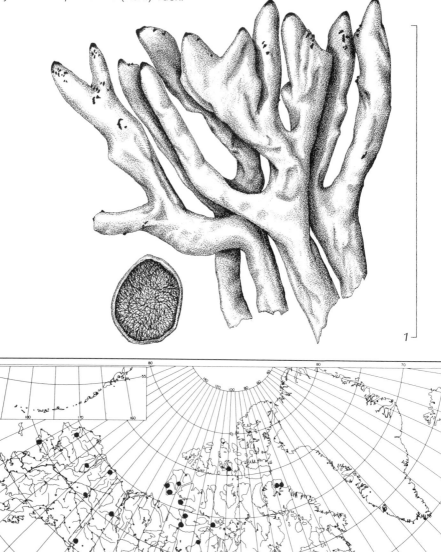

Dactylina
madreporiformis

Dactylina ramulosa (Hook.) Tuck.

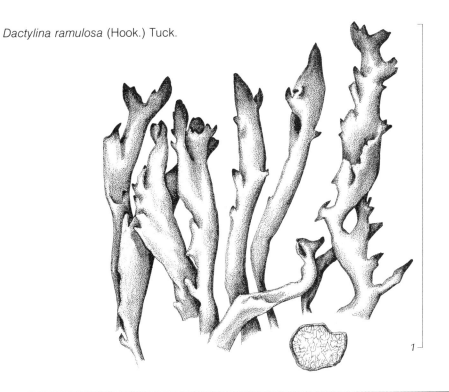

1

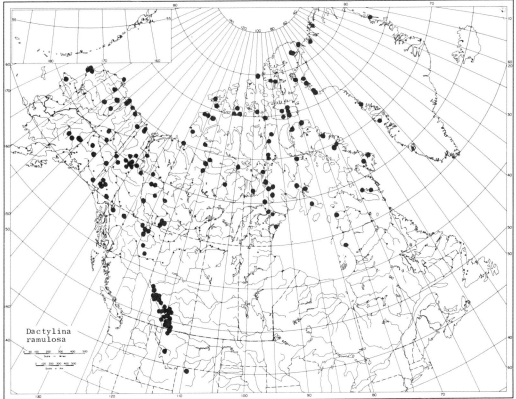

Dactylina
ramulosa

subhymenium a hyaline layer, 25 μ thick under the hymenium of short cells more or less oriented flatwise to the hymenium; the hypothecium brownish or yellowish, 25 μ thick, dense, a layer of dispersed algae in the arachnoid medulla under the hypothecium; asci pyriform with a very thick gelatinuous tip, 10×30 μ; spores biseriate, spherical, 5–6 μ diameter with thick wall; paraphyses unbranched except at base, septate, clavate-capitate, the tips brown, 4 μ thick, with thick walls and very slender lumina. Pycnidia rare, globose, conidia cylindrical, straight, $5-6 \times 1$ μ.

Reactions: cortex K−, C−, KC− or + yellow to pink; medulla K−, or KC+ red and P+ red when physodalic acid is present.

Contents: usnic acid in the cortex and usually physodalic and physodic acids in the medulla (see Thomson & Bird 1978 for details).

D. ramulosa is a calciphilous lichen which grows in a wide variety of tundra habitats, reflecting this also in the variability of its morphology. In very dry *Empetrum-Arctostaphylos* heaths it grows short and fragile, the branches becoming scattered rather than in dense tufts. In the more moist *Carex* and *Eriophorum* tussock tundras it grows on the tops of the tussocks among the sedges and tends to be larger, more luxuriant, and tuft-forming. The formation of apotheica, like that in *D. arctica,* is more frequent adjacent to late snow patches, above the patches where the snow melts away. This is a circumpolar arctic-alpine lichen which ranges southward in North America only to Alberta in the west and to northern Quebec in the east.

DERMATOCARPON Th. Fr.

Genera Heterolich. 105. 1861.

Thallus squamulose to umbilicate, with upper cortex and usually a well-developed lower cortex, heteromerous. Perithecia immersed in the thallus with only the ostiole showing or a small portion of the tip; with perithecial wall hyaline or the upper part black and dimidiate, or entirely black; periphyses abundant, slender; lacking algae in the hymenium; the paraphyses soon gelatinizing or rarely branching and interwoven; asci cylindrico-clavate to saccate; spores 8 (rarely 16), simple, hyaline, ellipsoid.

Algae: *Pleurococcus.*

Type species: *Dermatocarpon miniatum* (L.) Mann.

REFERENCES
Degelius, G. 1934. Über *Dermatocarpon rivulorum* (Arn.) DT. & Sarnth. und *D. arnoldianum* Degel. n. sp. Nyt Mag. Naturvid. 75:151–161.
Zschacke, H. 1934. Dermatocarpaceae in Rabenh. Kryptogamen-Flora 9: 1: 1:590–656.

1. Thallus well developed, foliose to umbilicate.
   2. Underside of well-developed specimens with plicate veins, usually single lobed, not pruinose, green when wet.
      3. Spores elliptical or elongate elliptical.
         4. Thallus thin (wet: 0.1–0.3 mm thick); cortical cells of upper side ca. 6.5 μ broad; spores (13–) $15-21.5 \times 6.5-8.5$ μ .... 10. *D. rivulorum*
         4. Thallus thick (wet: 0.4–0.7 mm thick); cortical cells of upper side ca. 8.5 μ broad; spores $13-15 \times 4.5-6.5$ μ .... 1 *D. arnoldianum*
      3. Spores spherical or subspherical, $10-17 \times 8-10$ μ, thallus weakly pruinose .... 8. *D. lyngei*
   2. Thallus without plicate veins.
      5. Thallus not pruinose, green when wet, always polyphyllous, cortical cells of upper side ca. 6.5 μ broad; spores $10-16$ $(-19) \times 5-8$ μ................... 4. *D. fluviatile*
      5. Thallus blue-gray or gray pruinose, not changing color when wet.
         6. Thallus margins not downrolled, thallus mono- to polyphyllous, cells of upper cortex ca. 8.5 μ, spores elongate elliptical, $8-14 \times 5-6$ μ .... 9. *D. miniatum*

6. Thallus margins downrolled, thallus always polyphyllous; cells of upper cortex ca. 6.5 μ; spores almost spherical, 9–15×7–9 μ . . . . 6. *D. intestiniforme*

1. Thallus not foliose, but squamulose to crustose.

     7. Thallus small foliose, lobes less than 1.5 mm broad; spores spherical or sub-spherical, 10–17×8–10 μ; thallus weakly pruinose, the underside brown, rough, and veined . . . . 8. *D. lyngei*

     7. Thallus squamulose.

         8. Spores large, 16–23×6–7 μ; thallus gray-white to gray-brown, more or less pruinose, margins lobate, underside black, perithecium brown to black . . . . 2. *D. cinereum*

         8. Spores smaller, 10–17×6–8 μ, perithecium pale below, dark only near the mouth.

             9. Thallus brown or red-brown, often shining.

                 10. Thallus red-brown, lobes concave to flat, 3–5 mm broad, margins ascending, cortical cells 4–8 μ, spores 11–14×6–8 μ . . . . 7. *D. lachneum*

                 10. Thallus brown, lobes concave to flat, adnate or just dark edge ascending, 2–7 mm broad, cortical cells 8–10 μ; spores 10–16×5–7 μ . . . . 5. *D. hepaticum*

             9. Thallus brown-gray or pale, thickly pruinose, lobes concave to flat, adnate or ends ascending, spores 12–17×6–7 μ . . . . 3. *D. daedaleum*

1. Dermatocarpon arnoldianum Degel.

Nyt. Mag. Naturvid. 75:157. 1934.

Thallus foliose, monopolyllus, umbilicate, rigid, irregularly lobed and incised, thick; upper side smooth, epruinose, brownish ashy dry, green when wet; underside rugose to veined, pale brownish, lacking pruina or rhizinae; upper cortex paraplectenchymatous, 20–35 μ, the cells 8.5 μ.

Perithecia immersed, globose, pale, the gelatin I+ blue; periphyses numerous, asci subcylindrical; spores 8, hyaline, oblong, 13–15×4.5–6.5 μ.

This species grows on rocks overflowed by snow meltwater. It is arctic and alpine in Europe in Scandinavia and the Alps. The sole reported occurrence in the American Arctic in the Ukinyik Creek drainage, Alaska, collected by Viereck and Bucknell (4419), has been verified by Degelius.

2. Dermatocarpon cinereum (Pers.) Th. Fr.

Nova Acta Reg. Soc. Scient. Upsal. Reihe 3, 3:356. 1861. *Endocarpon cinereum* Pers. Neue Annal. Bot. I. Stück. 28. 1794.

Thallus adnate squamulose, chinky between the squamules in the center, gray-brown, ashy brown, ashy gray with whitish gray pruina when young, this disappearing; the squamules 0.2–1.5 mm long, 0.2–0.4 mm broad, the margins entire, over a black hypothallus; upper cortex 30–50 μ of paraplectenchymatous cells with 3–8 μ diameter.

Perithecia immersed in the thallus, the summit forming 0.2 mm-broad warts; perithecium brown to black; periphyses numerous; paraphyses gelatinizing, the gelatin I+ pale violet, at the base of the perithecium blue; asci cylindrico-clavate; spores 8, biseriate, simple hyaline, ellipsoid, 16–23×6–7 μ.

This species grows on soil or humus, especially in calcareous habitats. It is circumpolar, arctic-alpine, and boreal, ranging into New England, the North Central states, and California.

3. Dermatocarpon daedaleum (Kremph.) Th. Fr.

Nova Acta Reg. Soc. Scient. Upsal. Reihe 3, 3:355. 1861. *Endocarpon daedaleum* Kremph. Flora 38:66. 1855.

Thallus slightly foliose, the margins adnate or slightly raised, with lobate early stages with 0.7–1.5 mm-broad lobes, becoming a broad crust; upper side gray-brown, thickly pruinose

Dermatocarpon arnoldianum Degel. underside.

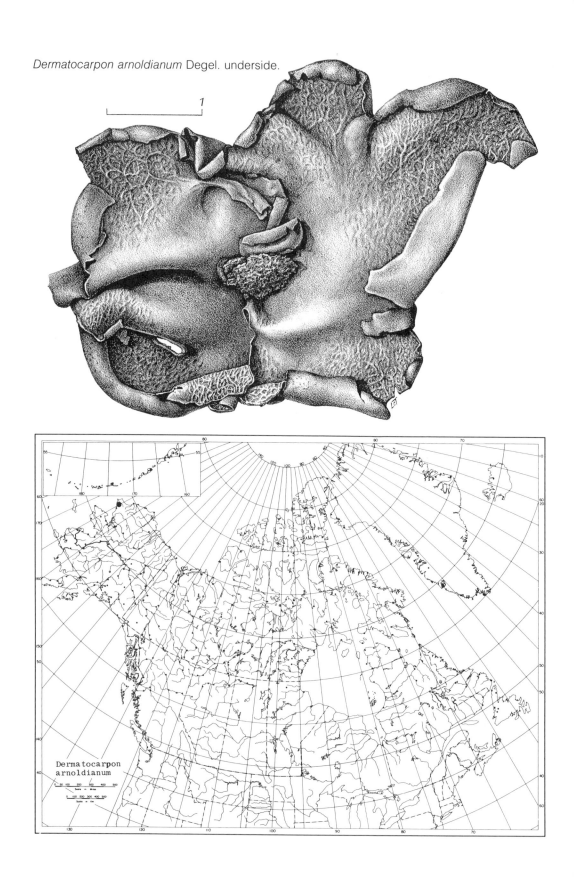

Dermatocarpon
arnoldianum

Dermatocarpon cinereum (Pers.) Th. Fr.

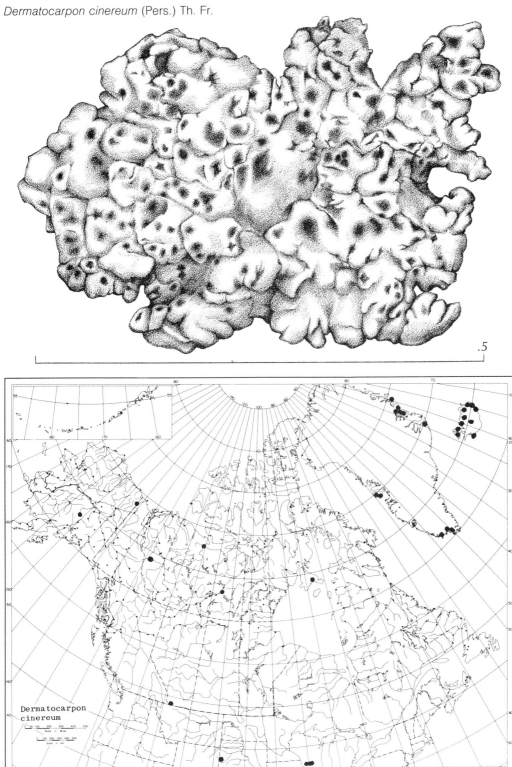

.5

Dermatocarpon
cinereum

at first, becoming bare, dull; underside pale or whitish and over a black hypothallus; upper cortex paraplectenchymatous, 60 μ, the cells 3–5 μ in diameter; lower cortex undeveloped.

Perithecia abundant, immersed, becoming slightly projecting in black warts, with brown-red mouth; periphyses short, 10 μ; hymenial gelatin I+ dirty reddish; asci clavate or clavate-saccate; spores 8, hyaline, biseriate, simple, ellipsoid or elongate egg-shaped with blunt ends, 12–17 × 6–7 μ.

This species grows over mosses and on soil rich in humus, especially over calcareous soils. It appears to be an amphi-Atlantic species in Scandinavia and the Alps in Europe and in Iceland and Greenland. Thus far it has not been found in North America.

4. Dermatocarpon fluviatile (G. H. Web.) Th. Fr.

Nova Acta Reg. Soc. Scient. Upsal., 3:354. 1861. *Lichen fluviatilis* Web., Spicil. Flor. Göttingens 265. 1778. *Lichen aquaticus* Weis., Plant. Cryptogam. Flor. Göttingens 77. 1770. *non Lichen aquaticum* L., Sp. Pl. 1148. 1753. *Dermatocarpon aquaticum* (Weis.) Zahlbr. Annal. Naturhist. Hofmuseum Wien 16:81. 1901.

Thallus foliose, polyphyllous, cartilaginous, light to dark brown when dry, green when wet, the lobes imbricate, rounded undulate; upper side smooth, epruinose, underside smooth, of same color as upper or slightly paler; upper cortex 25 μ paraplectenchymatous, the cells 4–6.5 μ.

Perithecia immersed or in slightly elevated warts; periphyses elongate; gelatin I+ blue; asci clavate; spores ellipsoid or elongate with rounded ends, 10–16 (–19) × 5–8 μ.

This species grows next to the water, at times immersed on stream and lake edges on rocks. It is circumpolar boreal and arctic to temperate, rarely reaching the arctic. It has been reported from Spitzbergen by Lynge (1924).

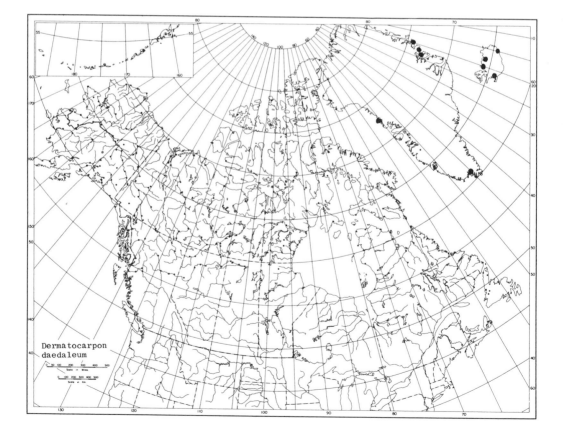

Dermatocarpon daedaleum

Dermatocarpon fluviatile (G.H. Web.) Th. Fr.

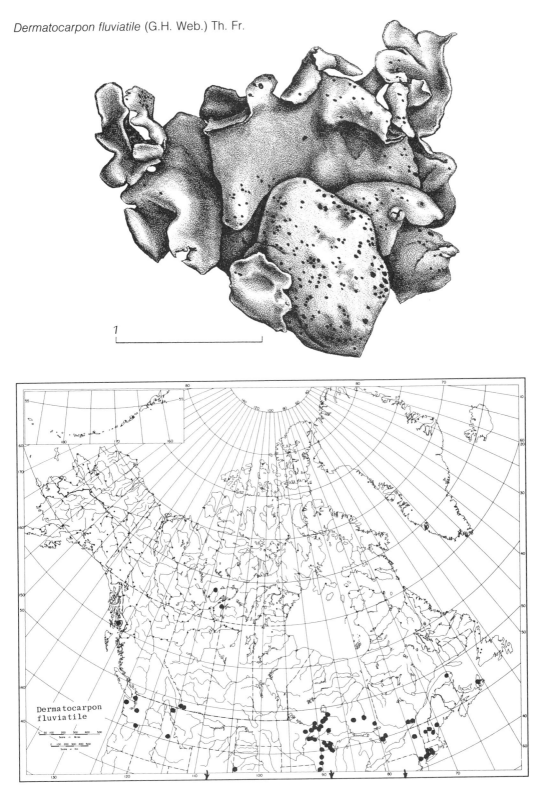

1

5. Dermatocarpon hepaticum (Ach.) Th. Fr.

Nova Acta Reg. Soc. Scient. Upsal., Reihe 3, 3:355. 1861. *Endocarpon hepaticum* Ach., Kgl. Vetensk.-Akad. Nya Handl. 156. 1809.

Thallus adnate squamulose, the squamules forming an almost continuous crust with chinks between, usually 1–3 mm broad squamules, flat or slightly convex, only the very edge sometimes ascending, rounded, or angular, light or dark reddish brown, smooth, epruinose; underside blackish, smooth, attached by hapters; upper cortex 20 μ, paraplectenchymatous, the cells 8–10 μ; lower cortex 80–100 μ.

Perithecia immersed, only the dark ostiole visible, the perithecium pale; asci cylindrical; periphyses abundant; hymenial gelatin I−; spores 8, hyaline, simple, ellipsoid 11–16 × 5–7 μ.

This species grows on calcareous soils. It is circumpolar, arctic, boreal, and temperate. In North America it ranges through much of the continent on appropriate substrata.

6. Dermatocarpon intestiniforme (Körb.) Hasse

Bryologist 15:46. 1912. *Endocarpon intestiniforme* Körb., Parerg. Lich. 42. 1859. *Lichen polyphyllus* Wulf., Schrift. Gesellsch. Naturforsch. Freunde Berlin 8:142. 1787. non *Lichen polyphyllus* L., Sp. Pl. 1150. 1753. *Endocarpon polyphyllum* (Wulf.) Arn., Die Lich. Fränk. Jura, Sonderdr. 235. 1885. *Dermatocarpon polyphyllum* (Wulf.) DT. & Sarnth., Die Flecht. Tirol. 504. 1902.

Thallus foliose, umbilicate but very polyphyllus, rigid, the lobes downrolled giving an intestinal appearance to the thallus; upper side smooth, brown, thickly blue-gray pruinose; underside smooth, light to dark brown; cortex with paraplectenchymatous cells 6.5 μ diam.

Perithecia immersed, with black mouth showing, hyaline, spores 8, hyaline, nearly spherical, 9–15 × 7–9 μ.

This species grows on calcareous rocks especially in dry habitats. It is circumpolar arctic-alpine and rather rarely collected in North America south to Montana and Washington.

7. Dermatocarpon lachneum (Ach.) A. L. Smith

Monogr. British Lich. 2:293. 1926. *Lichen lachneus* Ach., Lich. Suec. Prodr. 140. 1798. *Endocarpon lachneum* (Ach.) Ach., Meth. Lich. 127. 1802. *Endocarpon rufescens* Ach., Lich. Univ. 304. 1810. *Dermatocarpon rufescens* (Ach.) Th. Fr., Nova Acta Reg. Soc. Scient. Upsal. Reihe 3, 3:354. 1861.

Thallus squamulose with the squamules becoming free at the edges, overlapping, shining red-brown, 7–9 mm long, 3–5 mm broad, corticate above and below; the cortex 50 μ, paraplectenchymatous, the cells 4–8 μ in diameter, the lower cortex 70–100 μ; over a black hypothallus.

Perithecia immersed in the thallus, only the mouth showing and dark, the perithecium pale; periphyses numerous; hymenial gelatin I+ red-brown; asci cylindrical; spores 8, uniseriate or biseriate, simple, hyaline, ellipsoid, 11–14 × 6–8 μ.

This species grows on calcareous soils and on humus, sometimes on vertical cliff faces and crevices in rocks. It is circumpolar arctic and alpine, in North America ranging south to New England, Arizona, and California.

8. Dermatocarpon lyngei Servit

Beih. Bot. Centralb. 55 (B) 266. 1936. *Dermatocarpon sphaerosporum* Lynge, Skrift. Svalb. Ishavet 41:17. 1932. non *Dermatocarpon sphaerosporum* (Bouly de Lesd.) Zahlbr., Cat. Lich. Univ. 8:72. 1931.

Thallus foliose, rigid, umbilicate, deeply incised margins appearing polyphyllous with lobules; the marginal lobes undulate crisped overlapping; upper side smooth, epruinose, pale brown with the margins brownish-black, green when wet; underside with raised network of veins, brown, lacking rhizinae; upper cortex 20–25 μ, paraplectenchymatous; lower cortex 40–50 μ.

Perithecia numerous, 200–250 μ diameter, the ostiole raised, brown, the gelatin I+ reddish; asci saccate; spores 8, hyaline, globose and 8–10 μ diameter or subglobose 11 × 8–9 μ.

This species grows on rocks. It is currently known only from Greenland and Iceland.

Dermatocarpon hepaticum (Ach.) Th. Fr.

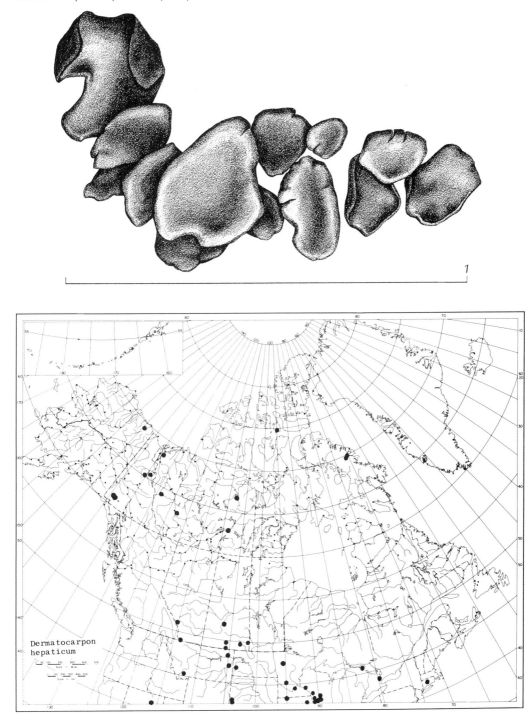

1

Dermatocarpon
hepaticum

Dermatocarpon intestiniforme (Körb.) Hasse

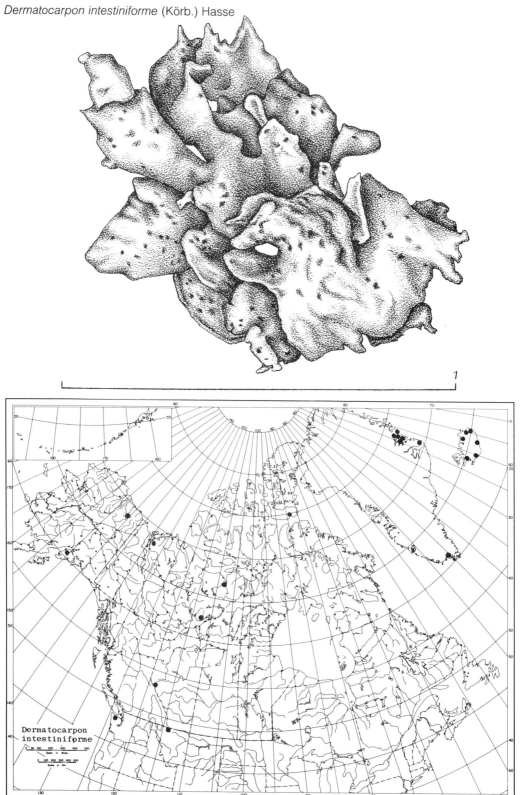

1

Dermatocarpon lachneum (Ach.) A. L. Smith

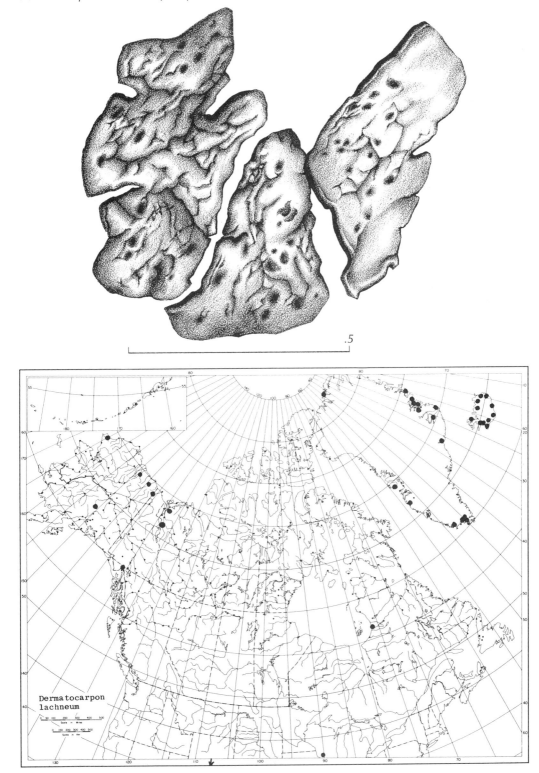

.5

Dermatocarpon
lachneum

9. **Dermatocarpon miniatum** (L.) Mann

Lich. in Bohem. observ. Dispos. 66: 1825. *Lichen miniatus* L., Sp. Pl. 1149. 1753., *Endocarpon miniatum* (L.) Gert., Meyer & Schreb., Ökon-techn. Flora Wetterau 3:230. 1801.

Thallus foliose, umbilicate, monophyllus or much divided in some forms with many lobules close together, the lobes 1–7 cm broad, flat or slightly rippled, reddish brown or gray-brown, gray pruinose; 0.3–0.5 mm thick; underside smooth to warty, not veined, light to dark brown; the upper cortex and lower cortex about the same thickness of paraplectenchymatous cells with diameter of 4–8.5 μ.

Perithecia immersed, pale with only the tips brownish, 200 μ diameter, the gelatin I+ blue; spores 8, hyaline, elongate elliptical 8–14 × 5–6 μ.

This species grows on calcareous rocks often in very dry habitats with much sunlight. It is circumpolar, arctic to temperate, ranging over most of North America where there are appropriate rock substrata.

10. **Dermatocarpon rivulorum** (Arn.) Dalla Torre & Sarnth.

Die Flecht. Tirol 504. 1902. *Endocarpon rivulorum* Arn., Verhandl. zool.-bot. Gesell. Wien 24:249. 1874.

Thallus foliose, monophyllous, rigid, umbilicate; upper side smooth, epruinose, brownish-gray, green when wet; underside with plicate veins, brownish-black or brown, epruinose, lacking rhizinae; upper cortex thin, 13–21 μ, the paraplectenchymatous cells 6.5 μ.

Perithecia immersed, globose, pale, the gelatin I+ blue; periphyses numerous; asci subcylindrical; spores 8, hyaline, oblong, (13–) 15–21.5 × 6.5–8.5 μ.

This species grows on rocks in drainage areas where snow melt provides moisture for some time in spring, or in occasionally flooded areas beside brooks. It is a rare arctic-alpine species found in Europe in Scandinavia and the Alps, in North America in Alaska (in the White Mts., and Lake Peters), Colorado, and in Greenland.

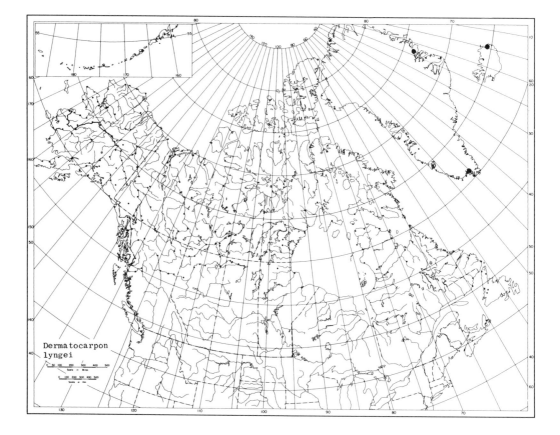

Dermatocarpon lyngei

Dermatocarpon miniatum (L.) Mann.

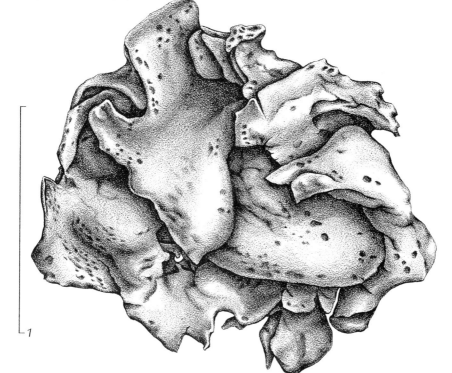

1

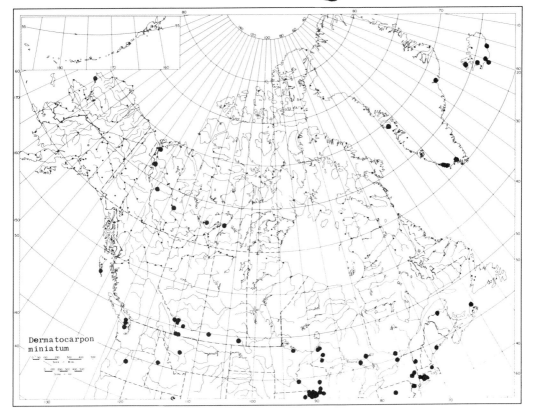

Dermatocarpon
miniatum

Dermatocarpon rivulorum (Arn.) Dalla Torre & Sarnth. upper side.

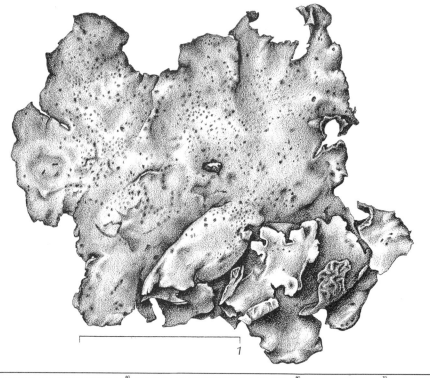

1

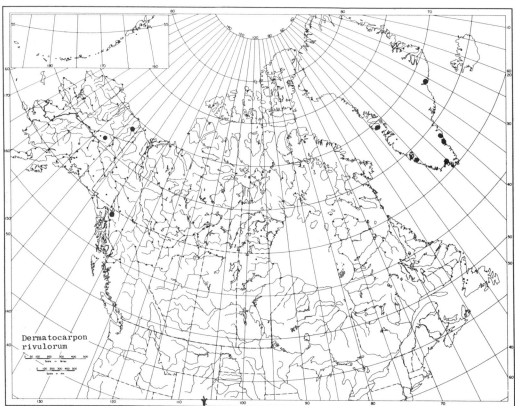

Dermatocarpon
rivulorum

Dermatocarpon rivulorum (Arn.) Dalla Torre & Sarnth. underside.

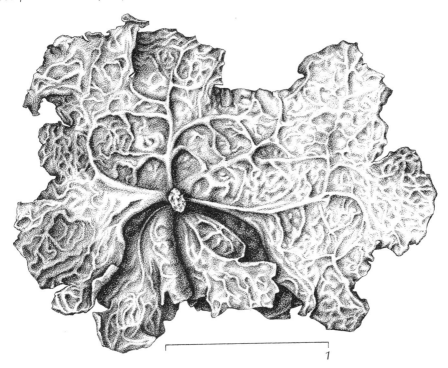

1

ENDOCARPON Hedw.

Descript. et Adumbr. Muscor. Frond. 2:56. 1789.

Thallus squamulose to subfruticose, adnate to ascending; upper cortex paraplectenchymatous; lower cortex present or not; attached by rhizinae.

Perithecia immersed in the thallus; perithecial wall dark; periphyses present; hymenial algae present; paraphyses soon gelatinizing, the hymenial gelatin I+ blue; asci saccate; spores 2, becoming brown, muriform with many cells, ellipsoid.

Alga: *Stichococcus diplosphaera* (Bialosuknia) Chodat in both thallus and hymenium (Ahmadjian & Heikkila 1970).

Type species: *Endocarpon pusillum* Hedw.

1. Thallus crustose, squamulose areolate, tightly adnate .... *(E. pusillum)*
1. Thallus tortuose nodulate, more or less stipitate. .... 1. *E. tortuosum*

1. Endocarpon tortuosum Herre

Bot. Gaz. 51:288. 1911.

Thallus of small tortuose squamules to 7 mm tall, forming a rough thick crust or scattered, the squamules more or less dichoto-mously dissected, dull dark brown, more greening when wet; lower cortex of vertically oriented dense hyphae.

Perithecia immersed in the lobes, the ostiole only slightly visible; the wall pale brown;

Endocarpon tortuosum Herre

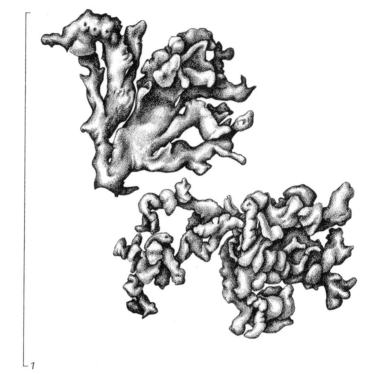

1

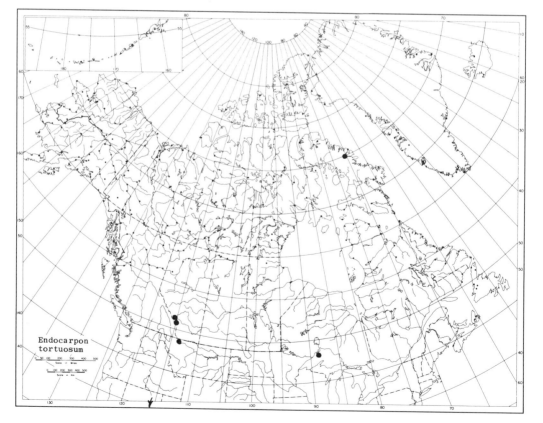

Endocarpon
tortuosum

asci subcylindrical to saccate; spores 2, becoming brown, muriform with numerous cells, 42–66 × 18–24 μ.

Reactions: thallus K–, C–, P–; hymenium I+ blue, asci I+ coppery red.

Contents: not known.

This species grows on rocks on trickle surfaces.

It is a North American species recorded from Nevada, Montana, and Baffin Island. It also occurs in Alberta at altitudes of 7950 ft and 5900 ft according to specimens collected by C. D. Bird and collaborators. Recently, on the 1972 foray of the American Lichenological and Bryological Society foray to Isle Royale, Michigan, Dr. Clifford Wetmore collected it on Davidson Island. The distribution is so scantily known that it is difficult to assess. It seems suggestive of those species which have a western range and disjunct occurrences either in the eastern Arctic or in the Lake Superior region. Yet this species appears to combine both range types.

EPHEBE Fr.

Syst. Orbis Vegetabilis 256. 1825. *Ephebeia* Nyl., Flora 58:6. 1875. *Spilonematopsis* E. Dahl, Medd. om Grønl. 150:34. 1950.

Thallus fruticose, filamentous, greenish brown or black, lacking rhizinae, attached by a holdfast; hyphae various, with angular, rounded, or elongate cells in parallel, right-angled or net-like array.

Apothecia pycnoascocarps with proper margin, mouth green or brown; hymenium gelatinous, the upper part brown; hypothecium dense; asci cylindrical or obclavate, thin walled, greening in age; paraphyses septate, dissolving or with vacuolate cells, tips thickened; spores 8 or 16, 1-celled, hyaline, round, oval, or bone-shaped. Conidiophores thin, elongate celled, the conidia terminal. Hymenial gelatin I+ blue.

Algae: *Stigonema*.

Type Species: *Ephebe lanata* (L.) Vain.

REFERENCE
Henssen, A. 1963. Eine revision der Flechtenfamilien Lichenaceae und Ephebaceae. Symbol. Bot. Upsal. 18:1–123.

Key to the American Arctic species (based on Henssen)
1. Sterile specimens.
   2. Thallus 1–2 mm tall, filaments 15–20(–40) μ thick........................ 3. *E. multispora*
   2. Thallus larger than 2 mm, filaments at least partially over 40 μ thick.
      3. Hyphae thin and with longitudinally extending cells, parallel and forming strands in the old parts... 2. *E. lanata*
      3. Hyphae when young in an irregular net with angled cell walls, when older forming an irregular cell mass... 1. *E. hispidula*
1. Fertile specimens.
   4. Asci with 16 spores.
      5. Thallus 1–2 mm tall, pycnoascocarps terminal .... 3. *E. multispora*
      5. Thallus over 2 mm tall; pycnoascocarps lateral .... 1. *E. hispidula*
   4. Asci with 8 spores... 2. *E. lanata*

1. Ephebe hispidula (Ach.) Henss.

Sym. Bot. Upsal. 18:49. 1963. *Cornicularia hispidula* Ach., Lich. Univ. 617. 1810. *Ephebeia hispidula* (Ach.) Nyl. Flora 60:231. 1877. *Ephebe spinulosa* Th. Fr., Bot. Not. 1866:59. 1866.

Thallus filamentous, blackish, 5–10 (–30) mm tall, filaments branching at the base and higher with thick axes, roughly beset with smaller branchlets, filaments 70–100 μ thick at the base; hyphae in the young parts forming an irregular net with angular cells, later becoming an irregular mixture of shorter and longer cells. Pycnoascocarps lateral, seldom terminal, to 0.25 mm broad, with a thalloid margin and a black-green proper margin; hymenium 125–155 μ;

Ephebe hispidula (Ach.) Henss.

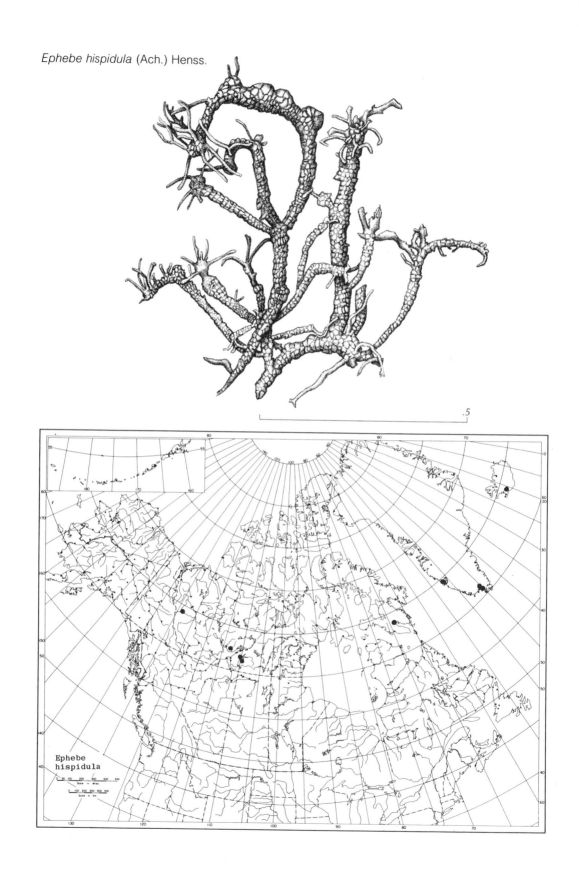

.5

Ephebe
hispidula

Ephebe lanata (L.) Vain.

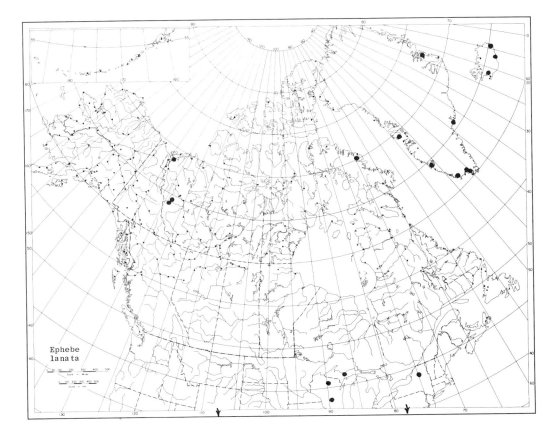

paraphyses septate with thickened tips and joints; hypothecium 20–55 μ; asci cylindrical, tips rounded, 65–100 $\times$ 7–11 μ; spores to 16, 1-celled, hyaline, broadly oval, 7–9.5 $\times$ 4–5 μ.

This species grows primarily on rocks on the banks of rivers and lakes. It is reported from northern Europe and Greenland. In North America George Scotter has found it along the shores of several lakes to the east of Great Slave Lake.

The difference in the internal anatomy, the irregular large cells instead of the lengthwise oriented cells of *E. lanata*, are of help in distinguishing it from that species, as well as the number of spores in the ascus and the more abundant lateral short branchlets.

2. **Ephebe lanata** (L.) Vain.

Medd. Soc. Fauna Flora Fenn. 14:20. 1888. *Lichen lanatus* L., Sp. Pl. 1155. 1753. *Conferva atrovirens* Dillw., British Confervae 60. 1809. *Usnea intricata* Hoffm., Deutschl. Flora 2:136. 1796. *Lichen intricatus* Ehrh. ex Schrad., J. Bot. 1:73. 1799. *Ephebe lapponica* Nyl., Flora 58:7. 1875. *Ephebeia cantabrica* Nyl., Flora 58:6. 1875. *Ephebe intricata* (Hoffm.) Lamy,

Ephebe
lanata

Bull. Soc. Bot. France 25:338. 1878. *Ephebeia martendalei* Cromb. ex Nyl., Flora 66:104. 1883.

Thallus filamentous, pendent, in masses, seldom in rosettes, to 2–3 cm tall, blackish, dark green, or shining brown; the filaments branching irregularly, loosely interwoven, 70–140 μ thick at the base to 200 μ, smooth, seldom rough with small branchlets; hyphae parallel and longitudinally oriented, in thicker axes forming strands, the edges forming a partial pseudoparenchyma.

Pycnoascocarps lateral, with a thalloid margin and later in additional proper margin; hymenium 160–170 μ, hyaline; hypothecium 50–60 μ; asci cylindrical, the tips rounded; spores 8, 1-celled, hyaline, oval, obliquely uniseriate, $11–20 \times 3.5–7 \mu$.

This species grows on rock surfaces and cliff faces, preferably on surfaces where water trickles frequently. It is circumpolar in the northern hemisphere, extending northward into Greenland, Iceland, and the Northwest Territories.

3. Ephebe multispora (E. Dahl) Henss.

Sym. Bot. Upsal. 18:46. 1963. *Spilonematopsis multispora* E. Dahl, Medd. om Grønl. 150:34. 1950.

Thallus filamentous, brown, like a small *Stigonema* cushion, 1–2 mm thick, the lobes massed into larger groups; filaments erect, divided at the base into several equal axes about 1 mm long and 15–20 (–40) μ thick with few side branches; hyphae thin or with angular cells, poorly seen in the algal gelatin and branching at right angles.

Pycnoascocarps to 1 mm broad, entirely terminal, with thalloid margin only, with filaments originating as prickles from the margin; hymenium 70–100 μ thick; hypothecium 40 μ; asci cylindrical, rounded or apically narrowed, $30–60 \times 6–10 \mu$; paraphyses septate, 1.5 μ thick; spores 16, 1-celled, hyaline, almost spherical, $4–6 \times 3–5 \mu$.

This species grows on moist cliffs. It has been collected only in the Julianehaab District of southwest Greenland by E. Dahl.

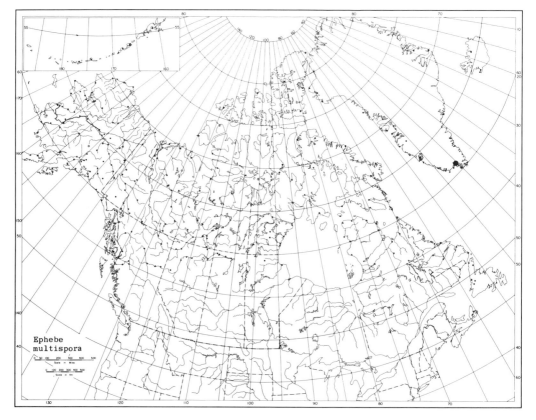

Ephebe
multispora

EVERNIA Ach.

Lich. Univ. 84. 1810.

Thallus fruticose, erect or hanging, more or less dorsiventral or radiate, corticate, the hyphae of the cortex oriented perpendicular to the surface; medulla filled with arachnoid hyphae without denser strands.

Apothecia lateral or terminal, adnate to sessile, with lecanorine margin concolorous with the thallus; hypothecium hyaline; hymenium hyaline; paraphyses unbranched, thick; asci clavate; spores 8, single-celled, small, ellipsoid, thin-walled. Pycnidia over the surface, innate, the mouth dark, fulcra endobasidial, conidia needle-shaped, straight.

Algae: *Trebouxia.*

Type species: *Evernia prunastri* (L.) Ach.

REFERENCES
Bird, C. D. 1974. Studies on the lichen genus *Evernia* in North America. Can. J. Bot. 52:2427–2434.
Bowler, P. W. 1977. *Ramalina thrausta* in North America. Bryologist 80:529–532.
Culberson, C. 1963. The lichen substances of the genus *Evernia*. Phytochemistry 2:335–340.
Howe, R. H. 1911. The genus *Evernia* as represented in North and Middle America. Bot. Gaz. 51:431–441.

1. Branches hard, brittle; medulla dense; cortex firm and unbroken; growing on soil; apothecia lacking, esorediate . . . . 3. *E. perfragilis*
1. Branches soft, not brittle; medulla loose, cortex cracking; growing on trees, soil, or rarely rocks.
  2. With abundant sorediate isidia; medulla very lax; usually growing on trees, rarely on rocks . .
  . . . . 2. *E. mesomorpha*
  2. Esorediate, medulla fairly dense; growing on soil and gravel. Alpine variant of . . 1. *E. divaricata*

1. Evernia divaricata (L.) Ach.

Lich. Univ. 441. 1810. *Lichen divaricatus* L., Syst. Veg. edit. 12. 713. 1768. *Letharia divaricata* (L.) Hue, Nouv. Archiv. Mus. ser. 4. 1:59. 1890.

Thallus pendent from trees in usual form, prostrate on ground and to 5 cm long in the alpine variant found in the tundras; branching and with the branches terete to angular, soft and flaccid with common breaks in the cortex; medulla lax but more dense than in the typical form; lacking isidia or soredia.

Apothecia not yet known in the alpine variant.

Reactions: K −, KC+ yellow, P−, UV+.

Contents: usnic and divaricatic acids.

The typical form of the species is a long pendent epiphyte on conifers in the subalpine to timberline with a disjunct range in Europe, Asia, and western North America. This alpine variant was described by Bird (1974) from two areas in the Rocky Mountains in western North America, one area in Alberta and Montana, one, southern, in Colorado, New Mexico, Utah, and Arizona. Bowler (1977) added a very disjunct record, checked by I. M. Brodo, of material collected together with *Ramalina thrausta* by Gillett on gravel at Eastern Creek 6–8 km east of Fort Churchill, Manitoba.

2. Evernia mesomorpha Nyl.

Lich. Scand. 74. 1871. *Evernia prunastri* var. *thamnodes* Flot., Die merkw. und selten. Flecht. Hirschberg und Warmbrunn 5. 1839. *Evernia thamnodes* Arn., Verhandl. zool.-bot. Gesellsch. Wien 23:110. 1873.

Thallus fruticose, much branched, erect, projecting from the substratum, partially hanging, to 10 cm long but shorter in the arctic; the branches irregular or dichotomous, angularly rounded to partly flattened but not dorsiventral, to 1 mm thick, yellow-green to yellowish brown, the tips darker, with an abundance of yellowish to gray coarse soredia in small spots over the thallus; medulla white, loose.

Apothecia not seen in arctic material; described by Keissler (1960) as adnate, lateral, with thin thalloid margin; disk to 6 mm broad, slightly concave, chestnut-brown; epithecium pale brown; hypothecium hyaline; spores 8,

Evernia divaricata (L.) Ach. alpine variant.

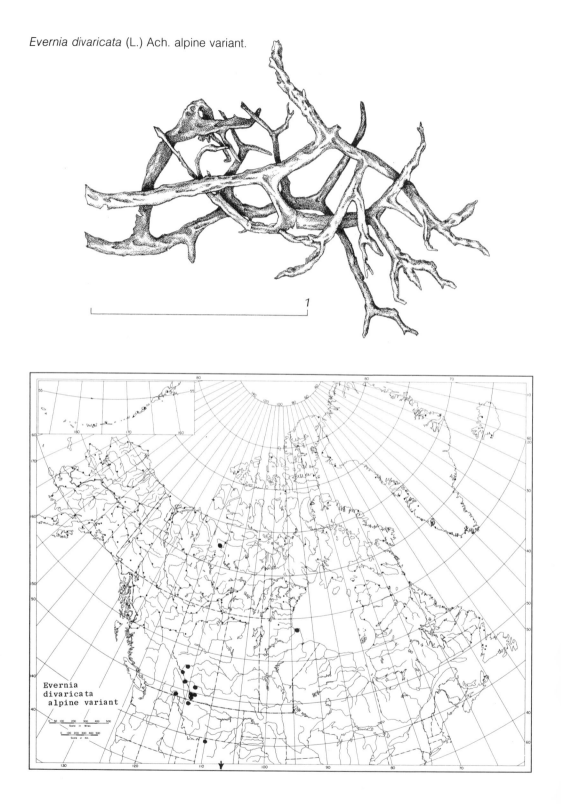

Evernia
divaricata
alpine variant

Evernia mesomorpha Nyl.

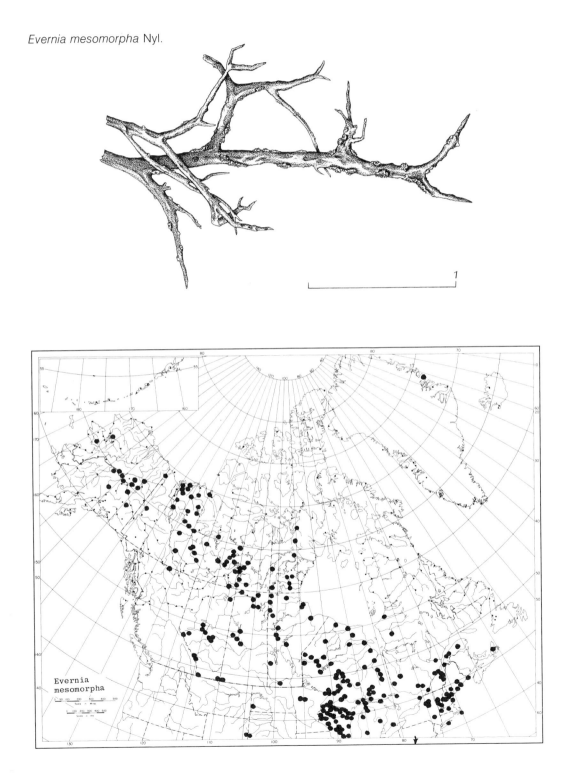

Evernia perfragilis Llano

hyaline, spherical, simple, thick-walled, 3–4 μ in diameter.

Reactions: K−, C−, KC−, P−.

Contents: usnic and divaricatic acids.

This species grows on the bark and twigs of trees and shrubs, especially *Picea, Latrix,* and *Betula*. It is rarely on rocks north of the tree line. It is a circumpolar boreal and temperate species which ranges into the Arctic along the tree line and further into the tundra on rocks or soil. The disjunct occurrence of this species on the northeast coast of Greenland at Myggbukta is astonishing in view of the lack elsewhere in Greenland and in the Canadian eastern Arctic. An unmistakable photograph accompanied the Lynge (1932) report.

3. Evernia perfragilis Llano

J. Wash. Acad. Sci. 41:199. 1951. *Alectoria arctica* Elenk. & Sav., Acta Horti Petrop. 32:73. 1912. *Evernia arctica* (Elenk. & Sav.) Lynge, Norweg. Nov. Zemlya Exp. 43:209. 1928. Non *Evernia arctica* (Hook.) Tuck. = *Dactylina arctica* (Hook.) Tuck.

Thallus fruticose, erect, in tufts, 2–5 cm tall, brittle and fragile, breaking into pieces when

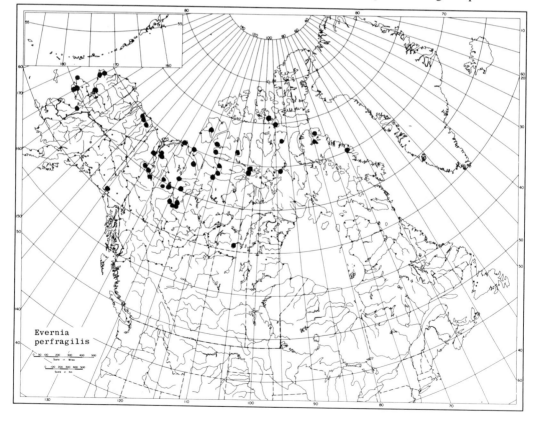

Evernia
perfragilis

dry; the main branches to 1.5 mm thick, angularly compressed, dichotomously branching, smooth or roughened to rugose, tips attenuate, lacking soredia or isidia, pale yellowish; cortex chondroid, to 55 μ thick, the cells more or less longitudinally arranged, paraplectechymatous; medulla white, dense.

Apothecia and pycnidia not known.

Reactions: K−, C−, KC−, P−.
Contents: usnic and divaricatic acids.

This species grows on calcareous soils and gravels. It is amphi-Beringian, not being found in Europe but ranging very widely in the arctic from Novaya Zemlya across the high arctic in Siberia and across northern North America to the east side of Baffin Island.

FISTULARIELLA Bowler & Rundel

Mycotaxon 6:195. 1977.

Fruticose, cylindrical to slightly compressed, inflated and hollow with oval perforations through the cortex into the lumen; cortex of rigid chondroid, longitudinally oriented prosoplectenchymatous hyphae with thin outer layer of branched cells; algae more or less in clumps, the sparse medulla lax and adherent to the interior of the cortex; lacking pseudocyphellae, the cortex smooth and shiny.

Apothecia terminal or subterminal, reflexing the branch or bearing a spur; margin thalloid; disk concolorous with the thallus; spores 8, bilocular, curved or straight.

Algae: *Trebouxia*.

Type species: *Fistulariella inflata* (Hook. f. & Tayl.) Bowler & Rundel.

REFERENCES
Bowler, P. A. & P. W. Rundel. 1975. The *Ramalina intermedia* complex in North America. Bryologist 77:617–623.
—— 1977. Synopsis of a new lichen genus *Fistulariella* Bowler & Rundel (Ramalinaceae). Mycotaxon 6:195–202.
Krog, H. & P. W. James. 1977. The genus *Ramalina* in Fennoscandia and the British Isles. Norw. J. Bot. 24:15–43.
Howe, R. H., Jr. 1913–1914. North American species of the genus *Ramalina*. Bryologist 16:65–74, 81–89; 17:1–7, 17–27, 33–40, 49–52, 65–69, 81–87.

1. Thallus sorediate.
   2. Soralia cupuliform (helmet-shaped) . . . . (*Ramalina obtusata*)
   2. Soralia either multifid-dendroid or maculiform.
      3. Corticolous; with finely branched multifid-dendroid tips; containing either divaricatic acid or sekikaic acid . . . . 2. *F. roesleri*
      3. Saxicolous; not multifid-dendroid tipped, the soralia maculiform; containing sekikaic acid . . . . 3. *F. scoparia*
1. Thallus esorediate; saxicolous or terrestrial; containing divaricatic acid . . . . 1. *F. almquistii*

1. Fistulariella almquistii (Vain.) Bowler & Rundel

Mycotaxon 6:198. 1977. *Ramalina almquistii* Vain., Ark. f. Bot. 8(4):17. 1909.

Thallus tufted fruticose 6–24 (−65) mm tall, terete or partly flattened, inflated, esorediate, with few fenestrations, mainly toward the base, pale to dark yellow, the surface smooth, shining; the medulla white, lax in part, hollow in part.

Apothecia to 4 mm broad, subterminal or lateral; margin concolorous with the thallus; reverse smooth; disk pale brownish, bare; spores oblong with rounded apices, straight, 2-celled, 11–13 × 4–5 μ. Pycnidia pale.

Reactions: K−, KC−, C−, P−.
Contents: usnic and divaricatic acids.

This species grows on soil or rocks. It is amphi-Beringian, ranging in Siberia and from Alaska to King William Peninsula, N.W.T. Bowler and Rundel (1977) suggest that this is a composite species composed of one element which is a coarse plant on

Fistulariella almquisitii (Vain.) Bowler & Rundel

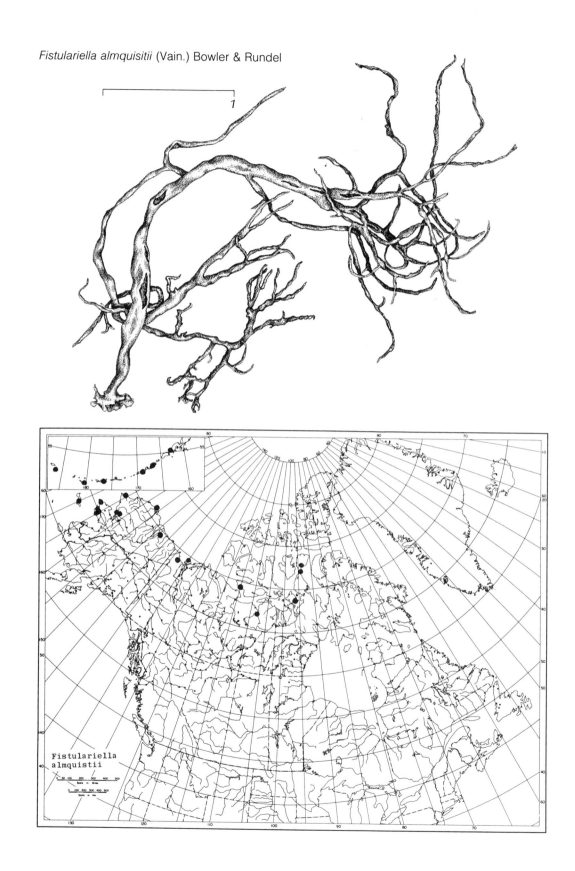

Fistulariella roesleri (Hochst. ex Schaer.) Bowler & Rundel

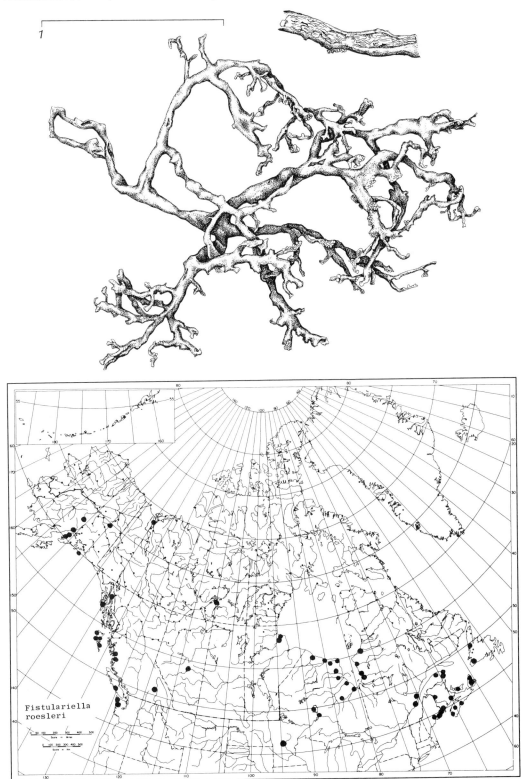

1

Fistulariella scoparia (Vain.) Bowler & Rundel

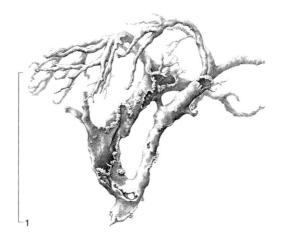

1

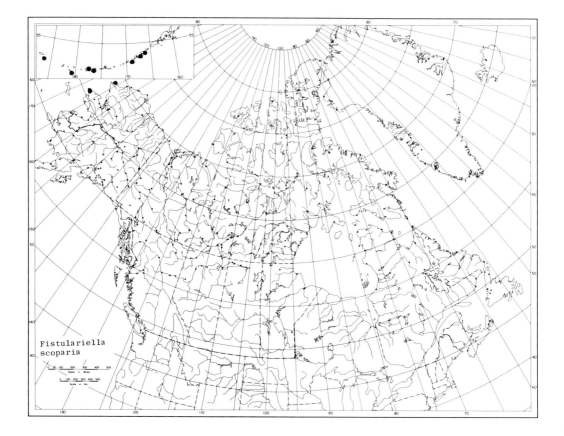

Fistulariella
scoparia

rocks with lateral apothecia, pale pycnidia, and a glossy opaque cortex. The other element is inland, on soil, is much smaller, lacks apothecia and pycnidia, has a translucent cortex and resembles a *Cladonia*. Vainio reported that this species might have soredia at the tips. If this occurs this may easily be distinguished from the sorediate *R. scoparia* by the chemistry, divaricatic acid instead of sekikaic acid aggregate.

2. Fistulariella roesleri (Hochst. ex Schaer.) Bowler & Rundel

Mycotaxon 6:199. 1977. *Ramalina fraxinea ϵ roesleri* Hochst. ex Schaer., Enum. Critic. Lich. Europ. 9. 1850. *Ramalina minuscula* var. *pollinariella* Nyl., Bull. Soc. Linn. Normandie Ser. 2, 4:165. 1870. *Ramalina pollinariella* (Nyl.) Nyl. Bull. Soc. Linn. Normandie Ser. 4, 1:202. 1887. *Ramalina roesleri* (Hochst. ex Schaer.) Hue, Rev. Bot. 6:151. 1887.

Thallus fruticose, loosely tufted with finely dissected apices, 1–3 (–6) cm tall, erect to slightly pendent, attached by a holdfast; branches 1–2 mm broad, secondary branches often at right angles to the primary, the tips terete, often apically hooked or ending in granular soredia; surface smooth, shining, more or less abundantly fenestrate, pale green; medulla at least partly hollow, more or less arachnoid; soredia in punctiform terminal or subterminal soralia, granular.

Apothecia very rare, small, subterminal; spores straight 11–16 × 5–6.5 μ.

Reactions: K−, KC+, P−.

Contents: sekikaic and homosekikaic and usnic acids in one strain, another contains divaricatic acid.

This species usually grows on twigs, especially of conifers, rarely on cliffs. It is circumpolar, boreal, occurring along the forest border.

3. Fistulariella scoparia (Vain.) Bowler & Rundel

Mycotaxon 6:199. 1977. *Ramalina scoparia* Vain., Ark. f. Bot. 8:19. 1909.

Thallus branched, tufted, short, 15–25 mm tall, flattened; dichotomously branched, the branches to 1.5 mm broad, the apices acute, the cortex smooth, shining, with no pseudocyphellae, maculiform soralia toward the ends of the lobes; the cortex with abundant yellow granules, 20–30 μ thick; medulla partly hollow, partly very lax.

Apothecia unknown, black pycnidia common.

Reactions: K −, KC+, P−.

Contents: Usnic acid, sekikaic acid aggregates (Bowler & Rundell 1977), and possibly also ramalinolic acid (Krog 1968).

This species grows on rocks near the sea. It is amphi-Beringian occurring in Siberia and Alaska.

HYPOGYMNIA (Nyl.) Nyl.

Lich. Environs. Paris 39. 1896.

Thallus foliose with elongate lobes, gray or brown, corticate above and below with the cortex prosoplectenchymatous (with tiny lumina), lacking rhizinae, hollow or solid, often with perforations into the interior.

Apothecia laminal, sessile or short pedicillate, margin thalloid, concolorous with the thallus; disk brown; hypothecium hyaline; epithecium brownish; paraphyses unbranched; spores 8 hyaline, ellipsoid, simple. Pycinidia laminal, the mouth black; conidia cylindrical with the center narrowed, straight. Thallus usually containing atranorin.

Algae: *Trebouxia*.

Type species: *Hypogymnia physodes* (L.) Nyl.

REFERENCES

Bird, C. D. & A. H. Marsh. 1973. Phytogeography and ecology of the lichen family Parmeliaceae in southwestern Alberta. Can. J. Bot. 51:262–288.

Hillmann, J. 1935. Parmeliaceae in Rabenhorst's Kryptogamenflora. 9(3) part 2:1–309.

Krog, H. 1968. The macrolichens of Alaska. Norsk. Polarinst. Skrift. 144:1–180.
—— 1974. Taxonomic studies in the *Hypogymnia intestiniformis* complex. Lichenologist 6:135–140.
Nuno, M. 1964. Chemism of *Parmelia* subgenus *Hypogymnia* Nyl. J. Jap. Bot. 39(4):97–103.
Ohlsson, K. E. 1073. New and interesting macrolichens of British Columbia. Bryologist 76:366–387.

1. Thallus hollow.
    2. Sorediate or isidiate.
        3. Thallus diffusely sorediate over the upper surface, the soredia isidioid, the lobes mottled brown and gray with black margins, K+ yellow, P−, containing atranorin and physodic acid . . . . 1. *H. austerodes*
        3. Thallus with the soralia at the tips of the lobes.
            4. Soralia labriform, thallus gray to blue-gray.
                5. Medulla P+ red, containing atranorin, physodic and physodalic acids . . . 4. *H. physodes*
                5. Medulla P−, containing atranorin and physodic acid. . . . . 6. *H. vittata*
            4. Soralia capitate; thallus blue-gray to brown, P−, containing atranorin and physodic acid. . . . . 2. *H. bitteri*
    2. Lacking soredia or isidia but one species with papilliform lobes.
        6. With small papilliform lobes toward the center of thallus; lobes brown or blackish, P−, K−, containing only physodic acid, lacking atranorin . . . . 5. *H. subobscura*
        6. Lacking such small papilliform lobes; lobes gray or blue-gray.
            7. P−, K+, containing atranorin and physodic acid. . . . . 6. *H. vittata*
            7. P+, K+, containing atranorin, physodic and physodalic acids. . . . . 4. *H. physodes*
1. Thallus with solid medulla; containing atranorin and physodic acid, P− . . . . 3. *H. oroarctica*

1. Hypogymnia austerodes Nyl.
Flora 64:537. 1881

Thallus forming rosettes, the lobes close together; upper side gray or more commonly brown, shining, the margins black as is the underside; hollow, convex above, with frequent cracks across the lobes, with isidioid warts breaking into soredia on the upper side, the lobes sometimes perforate at the ends, lacking rhizinae.

Apothecia rare, adnate, to 6 mm broad, the margins crenulate to sorediate; epithecium pale brownish; hypothecium hyaline; hymenium hyaline; asci clavate; spores 8, ellipsoid, hyaline, 6.5–7.5 × 4.5–5 μ.

Reactions: medulla P−, K−, KC+ red, C−, I−.

Contents: atranorin, physodic acid, conphysodic acid, and unknowns.

This species grows over mosses, on the bases of woody plants, on old wood, rarely on soil. It is a circumpolar arctic-alpine species, in North America ranging south to Montana, New Mexico, and into Mexico at high altitudes.

2. Hypogymnia bitteri (Lynge) Ahti
Ann. Bot. Fenn. 1:20. 1964. *Parmelia bitteri* Lynge, Skr. Vidensk.-Selsk. Christiana, Math.-Naturvidensk. Kl. 1921 (7):138. 1921.

Thallus rosette-like of deeply cut irregular lobes; upper side gray-green to brownish, shining at the tips, black spotted or in segments separated by black lines; underside black, without rhizinae; with lateral branches which rise up and are tipped with blue-gray or whitish gray capitate soralia.

Apothecia rare, not seen in American material.

Reactions: medulla P−, K−, KC+ red, C−, I−.

Contents: atranorin, physodic acid, and an unknown substance (Ohlsson 1973)

Hypogymnia austerodes Nyl.

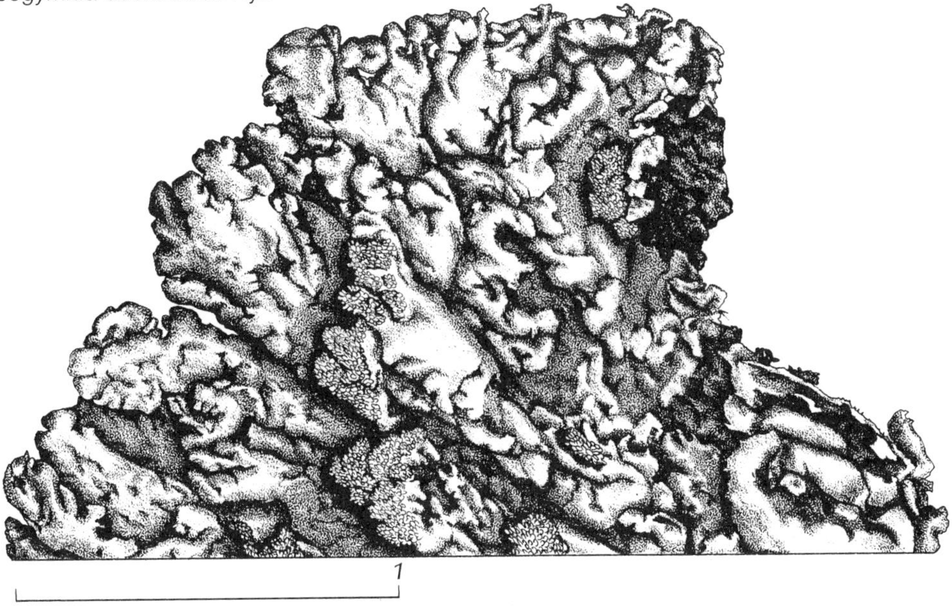

1

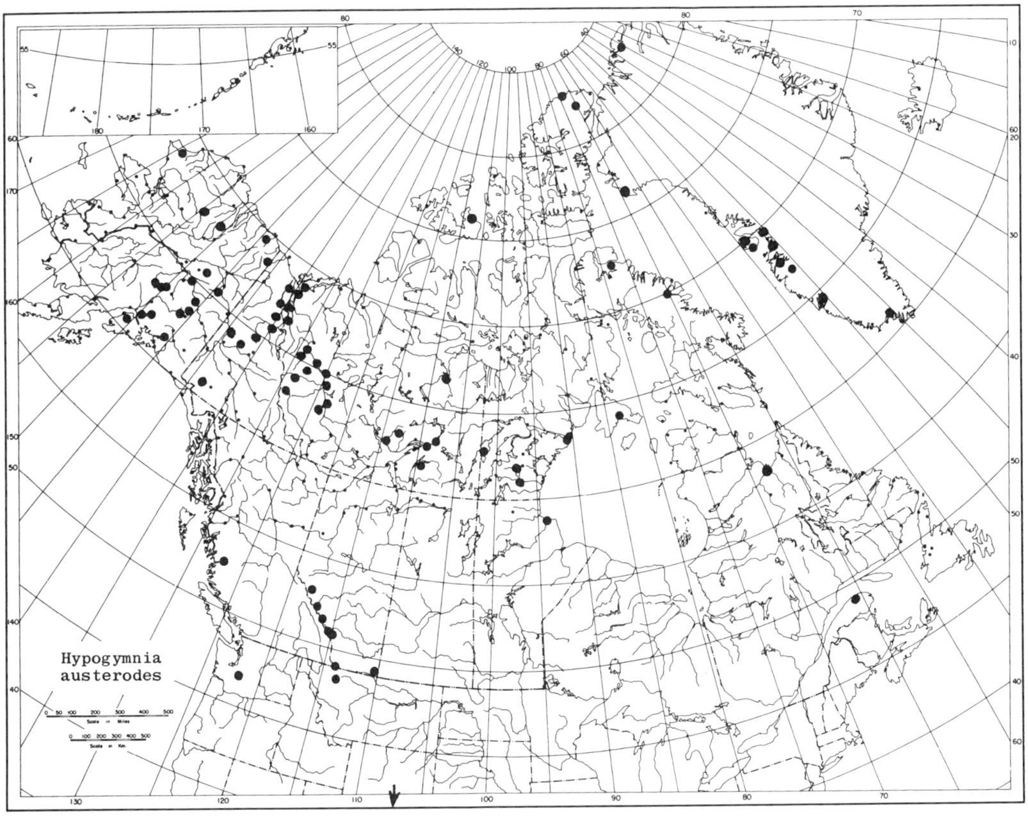

Hypogymnia
austerodes

This species grows on the bark of trees, on mosses, occasionally on rocks and soil. It is circumpolar, arctic-alpine and boreal, with its North American range south to Quebec in the east, to New Mexico and Mexico in the west.

Hypogymnia bitteri (Lynge) Ahti

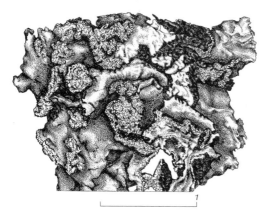

3. Hypogymnia oroarctica Krog
Lichenologist 6:136. 1974.

Thallus foliose, irregularly spreading and loosely attached; lobes solid, nodulose, 0.5–1 mm broad, terete or flattened when on stone; upper side pale gray, ashy gray, brown or black, maculate, distinctly areolate with discontinuous patches of cortex interspersed with black cracks and lines; underside black, brown toward the tips; lacking soredia or isidia.

Apothecia rare, to 5 mm broad; disk dark brown, thalloid margin thin, entire; spores subglobose to broadly ellipsoid, 10–12 × 8 μ.

Reactions: cortex K+ yellow; medulla K−, KC+ red, C−, P+ orange in layer close to cortex in apices only or P−, I−.

Contents: physodic acid, atranorin, with accessory protocetraric acid.

This species grows on rocks and over mosses in places with winter snow cover. It is arctic and northern boreal, circumpolar, in North America

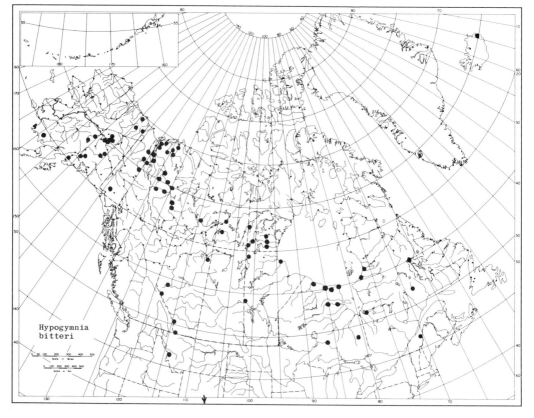

Hypogymnia
bitteri

Hypogymnia oroarctica Krog

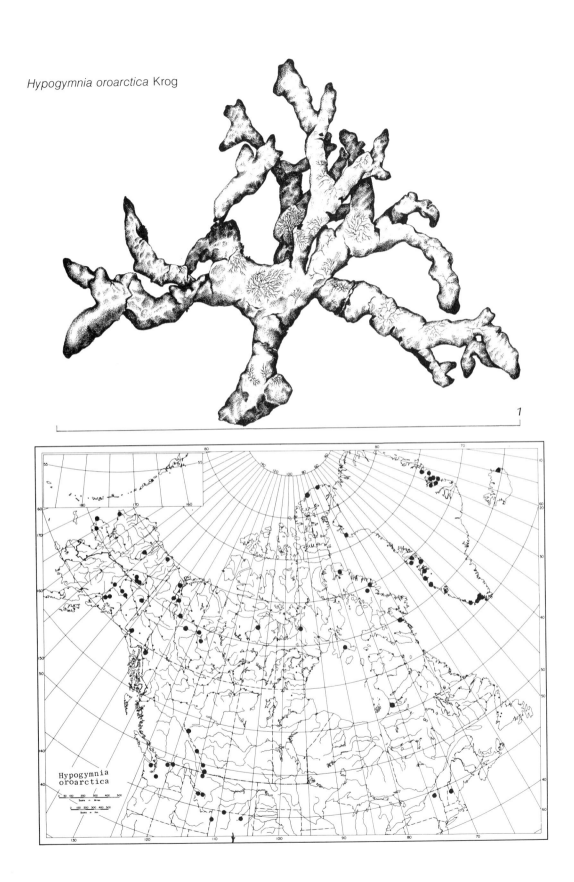

1

Hypogymnia
oroarctica

ranging south to New Mexico in the west, and the Adirondack and White mountains in the east.

4. Hypogymnia physodes (L.) Nyl.

Lich. Envir. Paris 39. 1881. *Lichen physodes* L., Sp. Pl. 1144. 1753.

Thallus forming rosettes, pinnately or irregularly branched, sometimes with smaller lobes irregularly placed, hollow, the tips usually raised, sometimes perforated, gray to blue-gray with the margins more or less black; the underside brown and shining at the tips, mainly black, strongly wrinkled, lacking rhizinae; the tips of the lobes with terminal labriform soralia, the soredia whitish gray-green; young thalli may lack the soredia.

Apothecia rare, laminal, sessile or becoming pedicillate, cup-shaped; the margin thalloid, inflexed, concolorous with the thallus; disk yellow brown to red brown, shiny; epithecium yellowish-brown; hypothecium hyaline; hymenium hyaline; spores 8, hyaline, ellipsoid, $6-9 \times 4-5 \mu$.

Reactions: medulla: P+ red, K−, KC+ red, C−, I−.

Contents: atranorin, physodic acid, physodalic acid, conphysodic acid, protocetraric acid plus unknown (Ohlsson 1973).

This species grows over mosses, on barks, on twigs of both conifers and deciduous woody plants, occasionally on soil or rocks. It is circumpolar arctic and temperate ranging south over most of the continent but rare in the Great Plains area.

5. Hypogymnia subobscura (Vain.) Poelt

Mitt. Bot. Staatssaml. Munchen 4:298. 1962. *Parmelia subobscura* Vain., Ark. Bot. 8(4):33. 1909.

Thallus more or less forming rosettes, of pinnate or dichotomously divided lobes which are elongate, hollow, with the tips more or less raised; upper side gray to more commonly brownish, the tips shining; underside black with the black rising up and showing along the sides, lacking rhizinae, without soredia or isidia but toward the center of the thallus with small pa-

pilliform lobes irregularly produced on the edges or the upper surface.

Apothecia not seen in North American material.

Reactions: P−, K−, KC+ red, C−, I−.

Contents: Physodic acid. Most authors attribute a lack of atranorin to this species. However Ohlsson (1973) stated that all 13 North American specimens that he studied with TLC contained atranorin but that 60% lacked chloratranorin which was present in all other Hypogymnia species. This species apparently requires further study.

This species grows on soil, over mosses, and at the bases of woody plants. It is circumpolar arctic and alpine. In North America it ranges south to Colorado and to northern Quebec.

6. Hypogymnia vittata (Ach.) Gas.

Act. Soc. Linn. Bordeaux 53:66. 1898. *Parmelia physodes* var. *vittata* Ach., Meth. Lich. 251. 1803.

Thallus forming small rosettes or imbricated lobes which spread irregularly and loosely over the substratum, the lobes slender, discrete, with or without small branches on the sides; upper side gray-green or brownish green, shiny, flat or convex, often edged with a black border which is a continuation of the black underside, lacking rhizinae, some of the lobes with terminal labriform soralia and more or less curved upwards; underside brown at the tips, black and reticulately wrinkled centrally, often perforated at the tips.

Apothecia not seen.

Reactions: medulla P−, K−, KC+ red, C−, I−.

Contents: atranorin, physodic acid, conphysodic acid, plus unknowns. Differing from *H. physodes* in the lack of physodalic acid.

Growing on mosses and on bark of woody plants, also on twigs, this species is alpine and temperate, probably circumpolar but less common than *H. physodes*. In North America it ranges south to Oregon in the west and into the Appalachian Mountains in the east.

Hypogymnia physodes (L.) Nyl.

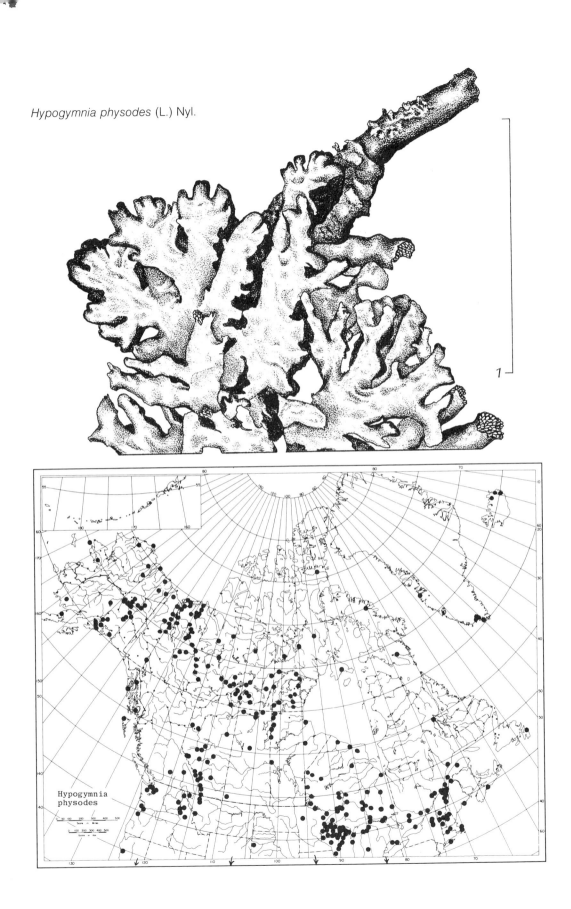

1

Hypogymnia
physodes

Hypogymnia subobscura (Vain.) Poelt

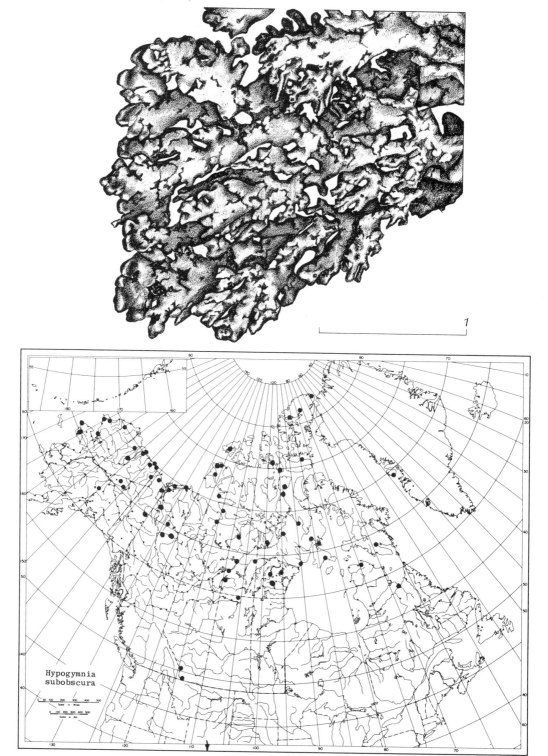

Hypogymnia vittata (Ach.) Gas.

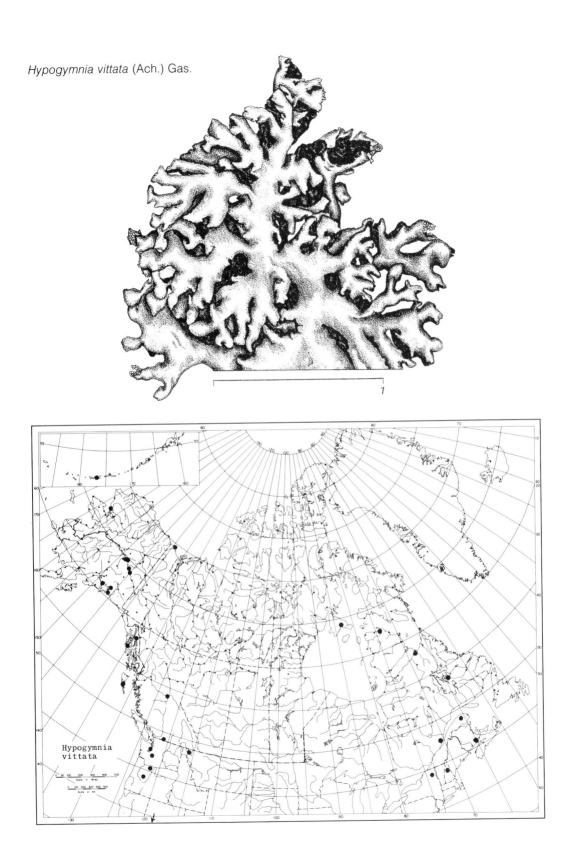

1

Hypogymnia
vittata

LASALLIA (Mérat) em. Llano

Monograph of the Umbilicariaceae 27. 1950. *Lasallia* Mérat, Nouv. Fl. Paris 2 ed., 1:202. 1821. *Umbilicaria* sect. *Lasallia* (Mérat) Schol., Nyt. Mag. Naturvid. 75:22. 1936.

Thallus foliose, umbilicate, pustulate with pustules which are raised on upper side, hollow below, smooth above to areolate, isidiate in one species (not arctic); underside smooth to areolate or papillate-verrucose, brown or black, lacking rhizines.

Apothecia common or rare, lecideine, sessile or pedicellate, with a proper margin which may become radiate; the disk flat, black, entire, smooth or rough; spores 1, rarely 2, per ascus, hyaline to brownish or dark brown, muriform with the inner cell walls thinner than the outer spore walls. A dense granular layer of fine crystals occurs between the lower part of the algal zone and upper part of the medulla.

Algae: *Trebouxia*.

Type Species: *Lasallia pustulata* (L.) Merat.

REFERENCES

Culberson, Chicita & W. L. 1958. Age and chemical constituents of individuals of the lichen *Lasallia papulosa*. Lloydia 21:189–192.

Llano, G. A. 1950. A monograph of the lichen family Umbilicariaceae in the Western Hemisphere. Navexos P-831.

1. Underside smooth to tiny-papillose, light brown to darker but not black . . . . 1. *L. papulosa*
1. Underside black-charred, strongly verrucose-areolate . . . . 2. *L. pensylvanica*

1. Lasallia papulosa (Ach.) Llano

Monograph of Umbilicariaceae 32. 1950. *Gyrophora papulosa* Ach., Lichen. Univ. 226. 1810. *Lecidea gyrophoroides* Spreng., Mantissa prim. Flor. Halens. 56. 1807. *Umbilicaria pustulata* var. *papulosa* (Ach.) Tuck., Proc. Amer. Acad. Sci. 1:262. 1848. *Umbilicaria papulosa* (Ach.) Nyl., Mem. Soc. Sci. Nat. Cherbourg. 5:107. *Umbilicaria dictyiza* Nyl., Syn. Lich. 2:5. 1863.

The thallus monophyllus, thin and fragile; margins irregular or scalloped, with blister-like pustules more or less radiating from the center, raised above, hollow below, occasionally with lobules developing over the thallus; upper side dull, brown-gray or brown, sometimes individuals quite red-brown with pigments, more or less pruinose; underside light brown to darker but not black, smooth to papillose, occasionally pruinose.

Thallus 165–250 μ thick, upper cortex ca. 20 μ, algal layer continuous; medulla loose and with crystals in the upper part, lower cortex 50 μ, scleroplectencymatous.

Apothecia to 2.5 mm broad, frequently agglomerated, sessile to subpedicellate; the disk flat, brown-black or brown-red; margin proper, smooth to crenate; hypothecium dark; hymenium 115–150 μ; paraphyses septate, 2.4 μ, the tips capitate to 4.8 μ; spores single, muriform, hyaline to becoming brown, large, 50–99 × 28–42 μ.

Reactions: K−, C+ red, KC+ red, P−.

Contents: Gyrophoric acid, papulosin, and anthraquinone pigments (Culberson 1970).

This is an especially common species in the eastern United States on sandstones, granites, gneisses, etc. In North America it ranges from northern Quebec to the Yukon, south to Florida and Alabama in the east, and into Mexico in the west. It is also reported from Africa. My reports from Coppermine and Kaminuriak Lake in Northwest Territories are incorrect and should be referred to *L. pensylvanica*.

2. Lasallia pensylvanica (Hoffm.) Llano

Monograph of Umbilicariaceae 42. 1950. *Umbilicaria pensylvanica* Hoffm., Descr. Adumb. Pl. Lich. 3:5. 1801.

Thallus to 25 cm broad, monophyllus, leathery to rigid; the margins undulate to laciniate or torn, sometimes perforated, pustulate with hollows under the pustules; upper side dull or shiny, dark brown to blackish brown or dark olive-brown, marginally more or less pruinose; underside uniformly black with coarse verrucose papillae which are sometimes paler periph-

Lasallia papulosa (Ach.) Llano

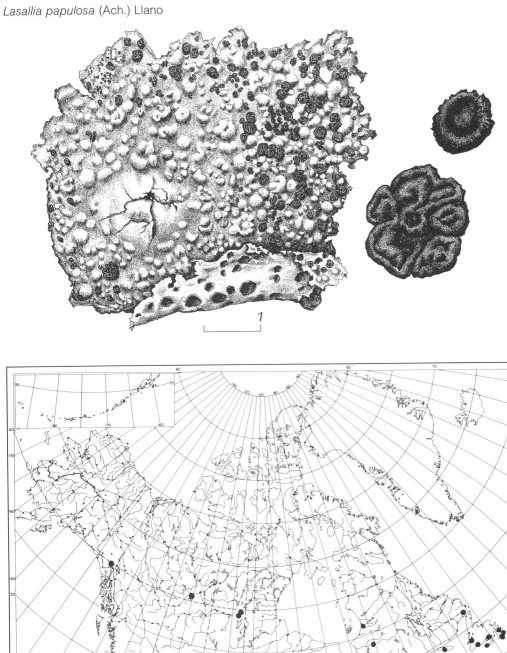

1

Lasallia pensylvanica (Hoffm.) Llano

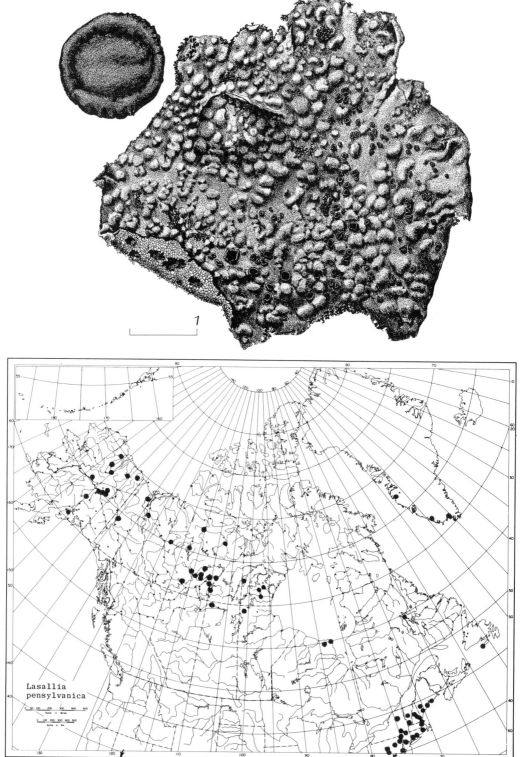

1

Lasallia
pensylvanica

erally, shaggy around the umbilicus, non-rhizinose.

Thallus 250–300 μ thick, cortex brown, 6–14 μ thick, paraplectenchymatous, occasionally palisade-like; algal layer continuous; lower cortex about 120 μ thick, scleroplectenchymatous, black.

Apothecia to 2 mm broad, sessile to subpedicellate, exciple proper, continuous; disk flat to slightly concave, black; hypothecium brown; hymenium 100 μ, with blackish epithecium; paraphyses simple to slightly branched, septate, capitate, 1.4–3.4 μ; spores single, muriform, pale to dark brown, 36–60 × 14–29 μ.

Reactions: K−, C+ red, KC+ red, P−.
Contents: Gyrophoric acid.

Growing on sandstones, granites, and other acid rocks, this is an Asiatic–North American temperate and boreal species reaching into the Arctic. Less common in North America than *L. papulosa* and reaching more northerly.

LECIOPHYSMA Th. Fr.

Bot. Not. 1865:102. 1865.

Thallus short, granulose to subfruticose, the lobes globose to cylindrical; lacking cortex, the algae distributed in the medulla; the interior hyphae reticulately arranged.

Apothecia lecideine, lacking a thalline margin, the proper margin distinct or not, with radially arranged hyphae; hymenium I+ blue; paraphyses sparingly branched, the tips dark and thickened; asci cylindrical; spores 8, hyaline, 1-celled, spherical or ellipsoid. Pycnidia in the thallus, conidia short, cylindrical.

Alga: *Nostoc*.

Type species: *Leciophysma finmarkicum* Th. Fr.

REFERENCE
Henssen, Aino. 1965. A review of the genera of the Collemataceae with simple spores (excluding *Physma*). Lichenologist 3:29–41.

1. Leciophysma finmarkicum Th. Fr.

Bot. Not 1865:102.

Thallus caespitose in small rosettes to 15 mm broad, the lobes erect, branching, to 0.8 mm tall, attached by tufts of rhizinae, verruculose, blackish brown, lacking cortex, the medulla with a loose network of hyphae forming broad lacunae, the *Nostoc* in chains and distributed through the thallus.

Apothecia black, rarely brown, to 1.0 mm broad, the proper margin indistinct; subhymenium hyaline or brownish, 110–170 μ; hymenium 120–190 μ, upper part brown; paraphyses filiform, 1–1.5 μ, lax, gelatin abundant; asci subclavate, 85–115 × 12.5–17.5 μ; spores 8, hyaline, simple, globose to ovoid, 16–23 × 11–14 μ.

This species grows on mosses and other lichens over limestone soils in moist habitats. It is known from Scandinavia, Novaya Zemlya, Iceland, Greenland, and in North America from Ellesmere Island to Alaska.

2. Leciophysma furfurascens (Nyl.) Gyeln.

Ann. Mus. Nat. Hungar. 32:176. 1939. *Pannaria furfurascens* Nyl., Flora 56:17. 1873. *Leciophysma occidentale* Dahl, Medd. om Grønland 150:44. 1950.

Thallus granulose, forming cushions or scattered, dark olive to blackish, to 2 cm broad, attached by tufts of rhizinae, the lobes globose, about 0.1 mm or to 0.2 mm long, sometimes forming a thin cortical parenchyma in older thalli, the interior hyphae reticulately arranged, the *Nostoc* through the thallus.

Leciophysma finmarkicum Th. Fr.

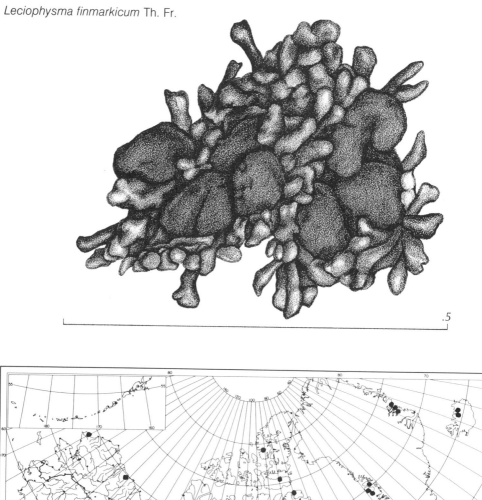

.5

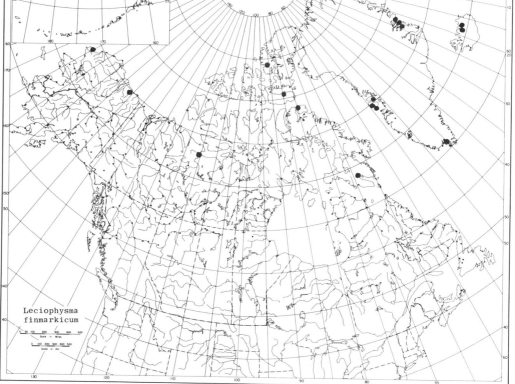

Leciophysma
finmarkicum

Apothecia brown or blackish brown, flat or convex, to 0.9 mm broad, proper margin distinct at first, then disappearing; subhymenium pale, 110–340 μ; hymenium 80–135 μ, upper part with brown gelatin; asci subclavate, 80–100 × 12–21.5 μ; spores uniseriate or biseriate, 8, hyaline, ellipsoid with slightly pointed tips, 11–20 × 7–9.5 μ.

This species grows over mosses and other lichens and also occasionally on barks. It is known from Scandinavia, Greenland, Canada, and South Dakota.

Leciophysma furfurascens (Nyl.) Gyel.

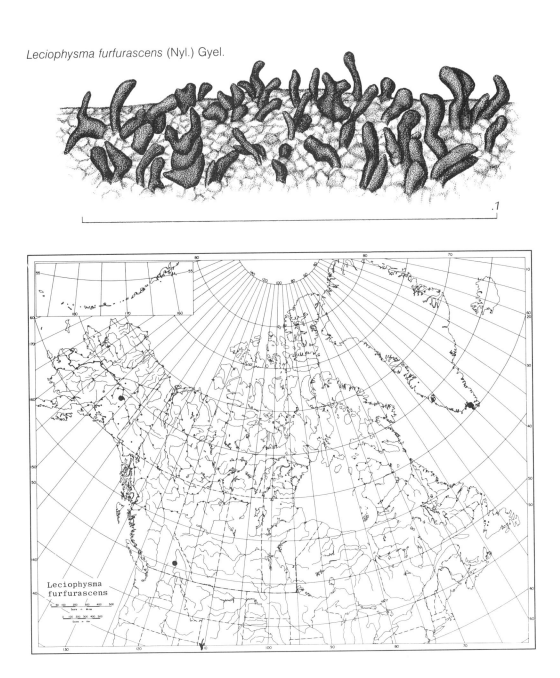

.1

Leciophysma
furfurascens

LEPROCAULON Nyl.

ex Lamy, Bull. Soc. Bot. France 25:352. 1878.

Thallus of small, slender, cartilaginous, more or less terete simple or branched, crowded pseudopodetia with leprose-sorediate, irregularly lumpy or minutely verrucose surface, glabrous or with slight arachnoid tomentum, in some species with root-like structures extending into the susbtratum. Lacking cephalodia.

Apothecia and pycnidia unknown.

Algae: *Trebouxia*.

Type species: *Leprocaulon nanum* (Ach.) Nyl. = *L. microscopicum* (Vill.) Gams ex Hawksw.

REFERENCE

Lamb, I. M. & Annmarie Ward. 1974. A preliminary conspectus of the species attributed to the imperfect lichen genus *Leprocaulon* Nyl. J. Hattori Bot. Lab. 38:499–553.

1. Leprocaulon subalbicans (Lamb) Lamb & Ward

J. Hattori Bot. Lab. 38:534. 1974. *Stereocaulon subalbicans* Lamb ex Imsh. Bryologist 60:220. 1957.

Pseudeopodetia forming more or less continuous crusts on soil, erect or partly decumbent, small, 2–4 mm tall, 0.25–0.50 mm thick; irregularly branched, terete or partly slightly flattened, white or ashy white, dull, with soft but not tomentose surface, base without root-like structures; pseudopodetia more or less uniformly covered with concolorous irregular poorly

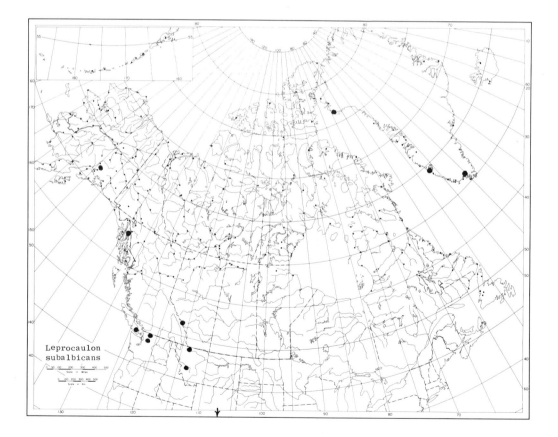

Leprocaulon
subalbicans

Leprocaulon subalbicans (Lamb) Lamb & Ward

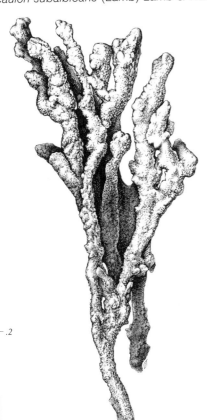

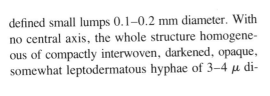

defined small lumps 0.1–0.2 mm diameter. With no central axis, the whole structure homogeneous of compactly interwoven, darkened, opaque, somewhat leptodermatous hyphae of 3–4 μ di-

ameter; those on the outside loose and floccose, the algae mostly in groups near the surface, chiefly in the small lumps.

Reactions: Very variable, dependent on the strain. Strain I K −, P + intense yellow; Strain II K+ brownish or brownish-yellow, P + red; Strain III K + indistinct yellow to strong yellow or sometimes orange-yellow to red; Strain IV K+ yellow, faint brownish yellow or K−, P + intense yellow.

Contents: Strain I, atranorin, psoromic, conpsoromic, and divaricatic acids + unidentified substance; Strain II, atranorin, protocetraric acid, unidentified fatty acids, and 2 other substances; Strain III, atranorin, thamnolic acid, and unidentified fatty acids; Strain IV, atranorin, squamatic, and baeomycesic acids and unidentified fatty acids.

The type specimen of the species from Chile belongs to Strain I. The strain which reaches the arctic is Strain IV.

This species grows on soil or on soil over rocks or in crevices of rocks. It is reported by Lamb and Ward from the Americas, Strains I and II from Chile, Peru, and Argentina, Strains III and IV in North America, the latter in Alaska and Greenland.

L. albicans (Th. Fr.) Nyl. differs in having distinctly developed phyllocladial granules, a well-developed central axis and a more chalky white color. In its Strain III with atranorin, squamatic and baeomycesic acids, and unidentified fatty acids it resembles Strain IV of *L. subalbicans*. It is reported in Strain III from Alaska, Mt. Roberts, Juneau, Krog, by Lamb and Ward. This suggests that any collections from further north should be carefully checked for each species.

LEPTOCHIDIUM Choisy

Bull. Mens. Soc. Linn. Lyon 21:165. 1952. *Polychidium* Sect. *Pseudoleptogium* (Jatta) Zahlbr. in Engler & Prantl, Naturl. Pflanzenfamil. 1:157. 1906.

Thallus foliose, similar to *Leptogium* or *Collema*, dark green or brownish black, with both upper and lower multicellular paraplectenchymatous cortices and usually with small hyaline cellular hairs projecting from the surface, especially on the margins and the exciple.

Leptochidium albociliatum (Desm.) Choisy

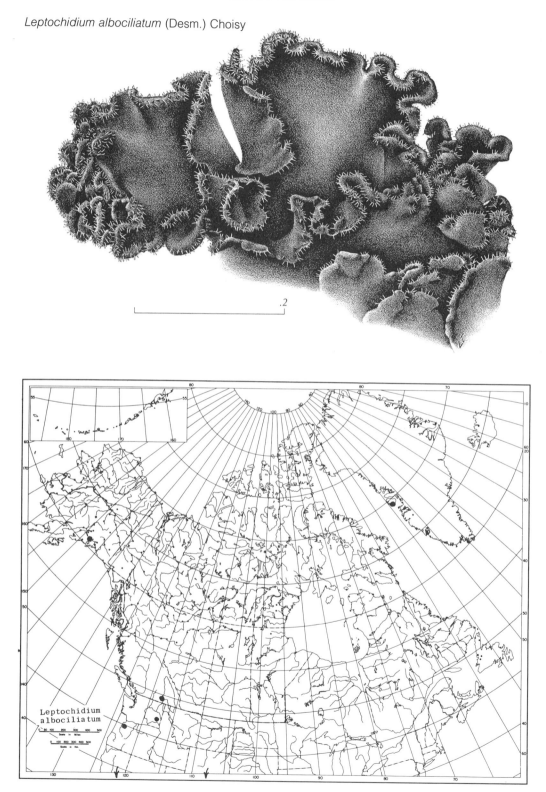

.2

Apothecia adnate, lecanorine with paraplectenchymatous margins; paraphyses more or less branched and capitate, septate; asci clavate; spores 8, hyaline, ellipsoid, 2-celled.

Alga: *Scytonema*.

Type species: *Leptochidium albociliatum* (Desm.) Choisy.

The genus is monotypic.

1. Leptochidium albociliatum (Desm.) Choisy

1952. *Leptogium albociliatum* Desm., Annal. Scienc. Nat. Bot. ser. 4, 4:132. 1855. *Polychidium albociliatum* (Desm.) Zahlbr., loc cit 1906. *Collema albociliatum* (Desm.) Nyl., Synops. Lich. 1:117. 1858.

Thallus dark greenish black, lobes 3–5 mm wide, the tips rounded or divided into smaller lobes or with crenulate margins; the upper surface smooth, usually with abundant small hyaline hairs projecting from the surface, especially marginally; the cortices multicellular, paraplectenchymatous; attached by fasciculate rhizines.

Apothecia to 1.2 mm broad, adnate; disk reddish brown, flat; spores hyaline, oblong ellipsoid to ellipsoid, 2-celled, 18–26 × 5–9 μ.

This species grows among mosses in moist habitats. It is rarely collected except in California where it appears to be frequent. It is known from Europe and North America, in the latter in the western states from Alaska to Washington, Oregon, and California, east to Colorado and with a disjunct occurrence on Disko Island in Greenland collected by P. Gelting (Lich. Groenl. Exsicc. 175). This disjunct type of range matches that of *Rhizocarpon bolanderi* and *Endocarpon tortuosum* (Thomson 1972).

Difficulty may be experienced in the western states in comparisons with *Leptogium burnetiae* var. *hirsutum* (Sierk) Jørg. The presence of isidia scattered over the thallus, the broader lobes 4–10 mm broad, the *Nostoc* rather than *Scytonema,* and the muriform spores may all be used to separate that species from this.

LEPTOGIUM S. Gray

Nat. Arr. Brit. Pl. 1: 400. 1821.

Thallus crustose, foliose, or fruticose; cortex of a single layer of irregularly isodiametric cells usually on both upper and lower surfaces; internally homiomerous with the algal cells scattered among loosely interwoven hyphae in most species but paraplectenchymatous throughout in one section.

Apothecia adnate, sessile or short stipitate, laminal or occasionally marginal; both thalloid and proper exciples present; disk brown to black; spores 8, few celled to muriform, fusiform to acicular, hyaline.

Algae: *Nostoc*.

Type species: *Leptogium tremelloides* (L. fil.) S. Gray

REFERENCES

Jørgensen, P. J. 1973. Ueber einige *Leptogium*-arten vom Mallotium-type. Herzogia 2:453–468.

Sierk, H. A. 1964. The genus *Leptogium* in North America, north of Mexico. Bryologist 67:245–317.

1. Thallus pseudoparenchymatous throughout.
   2. Thallus crustose, on soil, forming a broken areolate crust .... 2. *L. byssinum*
   2. Thallus foliose, sometimes minutely.
      3. Cortex poorly developed, cells small, indistinct, spores 3-septate .... 7. *L. parculum*
      3. Cortex well developed, cells distinct, spores muriform.
         4. Apothecia immersed to adnate, exciple lobulate, lobes of thallus non-isidiate ....
         .... 3. *L. crenatulum*
         4. Apothecia sessile, exciple sometimes granular but not lobulate.

1. Leptogium arcticum Jørg

Herzogia 2:454. 1973.

Thallus forming pulvinate masses to 8 cm broad, polyphyllus, the lobes irregular, to 8 mm broad, thin, translucent, 100–150 μ thick; the margins entire to irregularly cut, smooth above, black or brownish black, lacking isidia; the cells of the upper cortex poorly defined, in brown gelatinous matrix, those of the lower cortex very clear, large, 1–2 layers thick; the medulla loose with chains of algae scattered throughout; the hairs of the tomentum with cylindrical cells.

Apothecia unknown.

This species grows in low places in swales in lichen-rich tundras. It is known from Novaya Zemlya and from North America where it appears to be high arctic.

2. Leptogium byssinum (Hoffm.) Zw. in Nyl.

Act. Soc. Linn. Bordeaux 21:270. 1856. *Lichen byssinus* Hoffm., Enum. Lich. 46. 1784. *Leptogium sibiricum* Vain., Ann. Acad. Sci. Fenn. 27A:90. 1928.

Thallus forming a broken areolate crust to 2.5 cm broad, areolae to 3 mm broad, ashy brown to black, surface smooth to commonly granulose; paraplectenchymatous throughout, algal cells distributed through the thallus.

Apothecia common, immersed or adnate, to 1 mm broad; disk concave to flat, light brown to red-brown or black; thalloid exciple entire to granulose, concolorous with the thal-

lus, sometimes disappearing; proper exciple euparaplectenchymatous, 10 μ thick in center, to 45 μ at edges; subhymenium brownish; hymenium 95–150 μ, hyaline, with thin brown epithecium; paraphyses unbranched, 1 μ thick, little thickened at apices; asci cylindrico-clavate, 60–95 × 12–18 μ; spores 8, biseriate, ellipsoid to fusiform, apices rounded or pointed, 3–7 septate transversely, 0–1 septate longitudinally, 18–33 × 7–14 μ.

This species grows on soil, usually high in clay. It is a temperate circumpolar species, rarely collected perhaps because so inconspicuous, in North America ranging from New England to North Dakota and in the mountains in Colorado. It has been reported in Siberia under the name *Leptogium amphineum* Nyl. by Vainio and is included here under the possibility that it may be sought in the American Arctic.

3. Leptogium crenatulum (Nyl.) Vain.

Ark. Bot. 8(4):96. 1909. *Leptogium rivulare* var. *crenatulum* Nyl., Flora 58:106. 1875.

Thallus to 1.5 cm broad, of numerous spreading to erect lobes which are irregular, 0.1–0.8 mm broad, the margins entire to irregularly cut, light to dark brown; paraplectenchymatous throughout, the algae in chains through the thallus.

Apothecia common, adnate to immersed, to 0.5 mm broad; disk concave, brown to black; thalloid exciple with small lobules; proper exciple euparaplectenchymatous, 10–20

Leptogium arcticum Jørg.

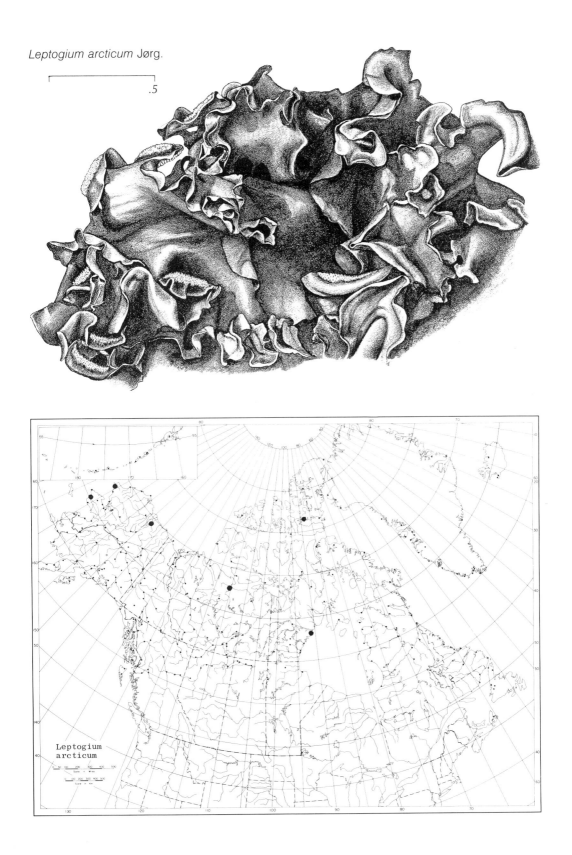

μ thick in center, thinning marginally; subhymenium yellowish; hymenium 115–135 μ, hyaline, with thin brown epithecium; paraphyses unbranched, 1.0–1.5 μ, slightly thicker at apices; asci cylindrico-clavate, 85–115 × 14–19 μ; spores 8, uniseriate, ellipsoid to subfusiform; 5 septate transversely, 1-septate longitudinally; apices rounded or pointed, 25–25 × 9–14 μ.

This species grows on soil. It is known only from collections made in Siberia and on Bering Island in the Bering Sea. It may be expected in Alaska.

4. Leptogium lichenoides (L) Zahlbr.

Cat. Lich. Univ. 3:136. 1924. *Tremella lichenoides* L., Sp. Pl. 1157. 1753. *Leptogium lacerum* (Retz.) Gray, Nat. Arr. Brit. Pl. 1:401. 1821. *Leptogium pulvinatum* (Hoffm.) Cromb., Mongr. Brit. Lich. 1:70. 1894.

Thallus to 6 cm broad, of numerous erect to semi-erect lobes sometimes forming distinct cushions, the lobes orbicular to elongate, 1–4 mm broad; the margins finely lobulate to fimbriate, the surface distinctly wrinkled, the heaviest wrinkles running the length of the lobes, lead gray to brown; underside attached by scattered tufts of hairs; cortex of 1 layer of isodiametric cells; medulla loosely interwoven, algal cells in chains.

Apothecia not uncommon, sessile, to 0.7 mm broad; disk brown to red-brown; thalloid exciple with numerous lobulate to coralloid outgrowths; proper exciple euparaplectenchymatous, 20–70 μ thick in center, thinning to 10–20 μ and then widening to 45–70 μ in margin; subhymenium yellowish; hymenium 130–200 μ, hyaline, with yellowish to brown epithecium; paraphyses unbranched, about 1.5 μ, thick, little thickened at apices; asci cylindrico-clavate, 125–185 × 16–21 μ; spores 8, biseriate or irregular, ellipsoid or subfusiform, apices rounded to occasionally pointed, 5–9 septate transversely, 1–3 septate logitudinally, 18–45 × 11–16 μ.

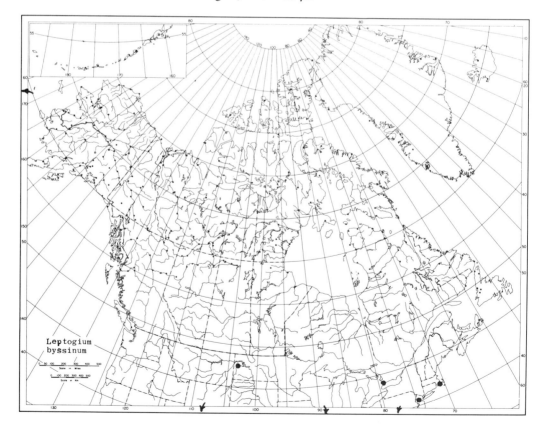

Leptogium
byssinum

Leptogium lichenoides (L.) Zahlbr.

.1

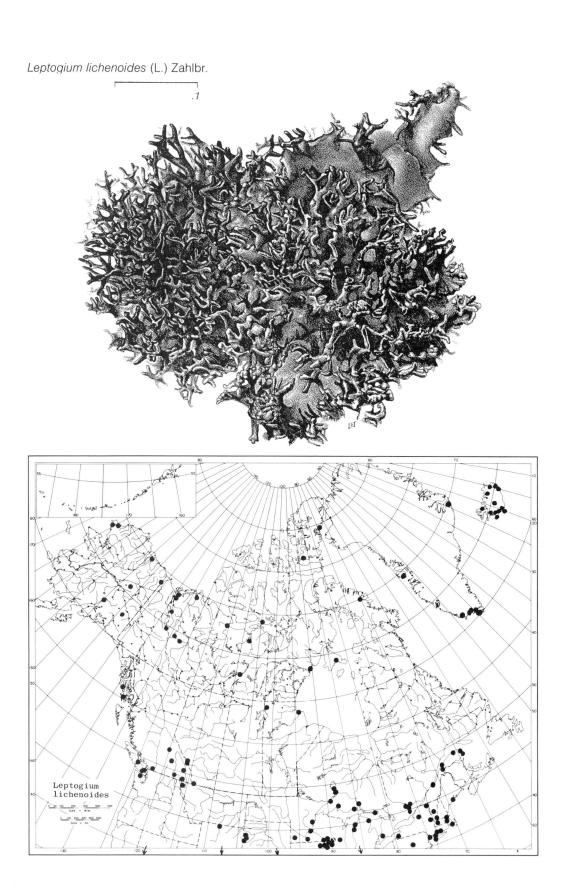

Leptogium
lichenoides

This species grows on soil and among mosses, commonly over rocks and in calcareous situations. It is a circumpolar species, arctic to temperate, and ranging over much of North America except the southeastern and Gulf coasts, the Central Plains, and the Great Basin.

5. Leptogium minutissimum (Flörke) Fr.
Sum. Veg. 122. 1846. *Collema minutissimum* Flörke, Deutsch. Lich. pt. 5:99. 1815.

Thallus to 2 cm broad, of numerous spreading to commonly erect and often imbricate lobes, the lobes to 2 mm broad; margins entire to irregularly lobulate, the surface smooth, occasionally with minute lobules, lead-gray to brown; underside with tufts of hair; thallus paraplectenchymatous throughout, algae in chains throughout the thallus; hairs 2–3 μ in diameter from lower cortex.

Apothecia sometimes abundant, sessile, to 1.5 mm broad; disk concave to flat, brown, red-brown, or dark chocolate brown; thalloid exciple entire, of same color as thallus or darker; proper exciple euparaplectenchymatous, about 10 μ broad but unclear; subhymenium hyaline to yellowish; hymenium 105–200 μ, hyaline, with thin yellow epithecium; paraphyses unbranched, 1–2 μ, slightly thicker at apices; asci cylindrico-clavate, 90–180 × 12–20 μ; spores uniseriate to irregular, ellipsoid with rounded tips, 3–5 septate transversely, 1-septate longitudinally, 21–35 × 7–14 μ.

This species grows on soil. It is circumpolar arctic to temperate, in North America ranging south to California, Colorado and West Virginia.

6. Leptogium palmatum (Huds.) Mont. in Webb & Berth.
Hist. Nat. Canar. 3(2):128. 1840. *Lichen palmatus* Huds., Fl. Angl. ed. 2:535. 1778.

Thallus to 7 cm broad, of tufted, mainly erect lobes, the lobes elongate, 5–20 mm long, 0.5–6.0 mm broad, the margins curling inward, especially at the apices to form antler-like tips; surface wrinkled, the wrinkles running the length

of the lobes, often shiny at the apices, lead-gray to brown; underside with tufts of hairs at times; cortex of 1 layer of isodiametric or elongate cells; medulla interwoven but compact, algal cells in chains.

Apothecia adnate to sessile, to 0.6 mm broad; disk concave to flat, brown to red-brown; thalloid exciple entire, of same color as thallus or cream colored, often wrinkled; proper exciple euparaplectenchymatous, 12–65 μ thick; subhymenium yellowish; hymenium 125–300 μ, hyaline, with thin brown epithecium; paraphyses unbranched, about 1 μ, slightly thickened at apices; asci cylindrico-clavate, 130–185 × 16–25 μ; spores 8, biseriate to irregular in ascus, ellipsoid to subfusiform, the apices rounded to pointed, 5–9 septate transversely, 1–2 septate longitudinally, 30–56 × 10–20 μ.

This species grows on soil, among mosses, on rocks, rarely on barks. It appears to be oceanic in requirements, growing along coastal regions in Europe, Japan, and the west of North America, in the latter from Alaska to California.

7. Leptogium parculum Nyl.
Flora 68:601. 1885.

Thallus minute, 1–2 mm broad, irregular or warty, but not distinctly lobed, numerous thalli forming a crust on the rock, dark brown to black; cortex of one layer of irregularly isodiametric to flattened cells; internal structure paraplectenchymatous throughout; algae distributed through the thallus.

Apothecia abundant, adnate to sessile; disk concave to flat, red-brown; proper exciple euparaplectenchymatous, 10–35 μ thick, indistinct in center; subhymenium hyaline; hymenium 115–125 μ, hyaline, with thin brown epithecium; paraphyses unbranched, 1–2 μ thick, slightly thicker at apices; asci cylindrico-clavate, 95–105 × 12–16 μ; spores 8, irregular in arrangement, subfusiform, 3-septate transversely, 18–24 × 6–8 μ.

This species grows on rocks. It is known only from the type locality in Alaska, near Port Clarence on the Seward Peninsula.

Leptogium minutissimum (Dicks.) Fr.

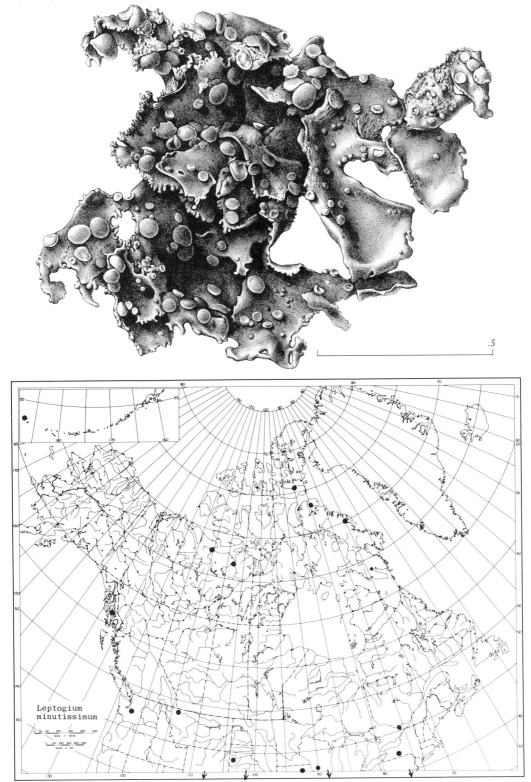

.5

Leptogium
minutissimum

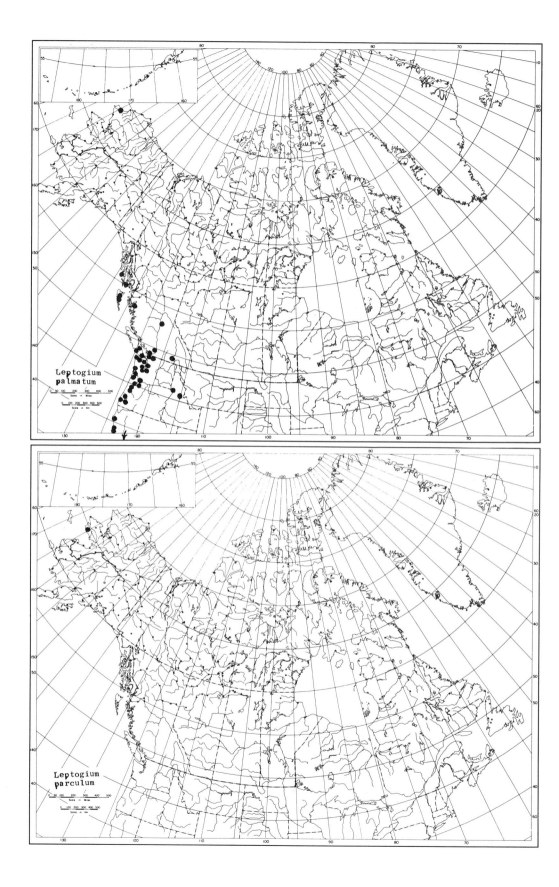

Leptogium
palmatum

Leptogium
parculum

Leptogium palmatum (Huds.) Mont. in Webb. & Berth.

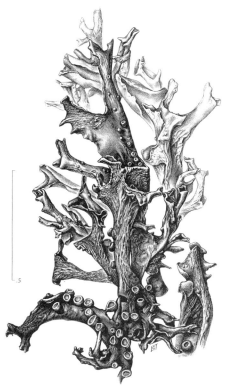

8. **Leptogium saturninum** (Dicks.) Nyl.

Act. Soc. Linn. Bordeaux 21:272. 1856. *Lichen saturninus* Dicks., Fasc. Pl. Crypt. Brit. 2:21. 1790.

Thallus to 6 cm broad, monophyllous to polyphyllous, the lobes rounded, to 10 mm broad; the margins entire to irregularly cut or isidiate, sometimes curling under the upper surface, smooth and with an abundance of granular, sometimes elongate and branched, isidia, lead-gray to olivaceous or black; underside covered with a dense tomentum of hairs with elongate cylindrical cells; cortex of 1 or 2 layers of iso-diametric cells, the cells of the lower cortex being larger than those of the upper, the medulla loosely interwoven, the algae in chains, more abundant near the cortices.

Apothecia rare, sessile, to 2.5 mm broad; disk brown to red-brown; the thalloid exciple entire to granulose, of same color as thallus or cream-colored; proper exciple euparaplecten-chymatous, 7–10 μ in center, widening to 20 μ marginally; subhymenium yellowish; hymenium 100–125 μ, hyaline, with thin brown ep-ithecium; paraphyses unbranched, 1–2 μ, slightly thicker at apices; asci cylindrico-clavate, 85–115 × 12–16 μ; spores 8, irregular in ascus, el-lipsoid, the apices rounded to pointed, 3–4 sep-tate transversely, 1 septate longitudinally, 20–25 × 7–10 μ.

This species grows on the bases of woody plants in the Arctic, especially on *Salix, Alnus, Be-tula, Ledum,* and also on rocks. It is a circumpolar species, arctic to temperate, ranging in North Amer-ica south to Georgia in the east, to California in the west.

9. **Leptogium sinuatum** (Huds.) Mass.

Mem. Lichenogr. 88. 1853. *Lichen sinuatus* Huds., Fl. Angl. 2 ed. 535. 1778. *Leptogium scotinum* (Ach.) Fr., Corp. Fl. Prov. Suec. 1:293. 1835.

Thallus to 5 cm broad, of erect to semi-erect lobes forming a cushion-like growth, the lobes orbicular, to 4 mm broad; margins entire to irregularly cut to lobulate, the upper surface roughened to distinctly wrinkled, dull or shiny, especially near the margins, lead-gray to dark brown; the cortex of 1 layer of isodiametric cells; the medulla loosely interwoven, algal cells in chains through the thallus but sometimes more near the upper cortex.

Apothecia common, adnate to sessile, occasionally on both sides of erect lobes; disk concave to flat, light to dark brown; thalloid margin entire, thick, sometimes concentrically wrinkled; proper exciple euparaplectenchyma-tous, 10–35 μ thick in center, widening to 35–45 μ in margin; subhymenium hyaline to yel-lowish; hymenium 115–170 μ, hyaline, with thin brown epithecium; paraphyses unbranched, 1 μ, slightly thickened at apices; asci cylindrico-cla-vate, 95–150 × 14–23 μ; spores 8, uniseriate to irregular, ellipsoid, the tips rounded to pointed, 7 septate transversely, 1–2 septate longitudi-nally, 25–35 × 12–14 μ.

This species grows on soil and among mosses, on bark and on rocks. It is circumpolar, mainly tem-

Leptogium saturninum (Dicks.) Nyl.

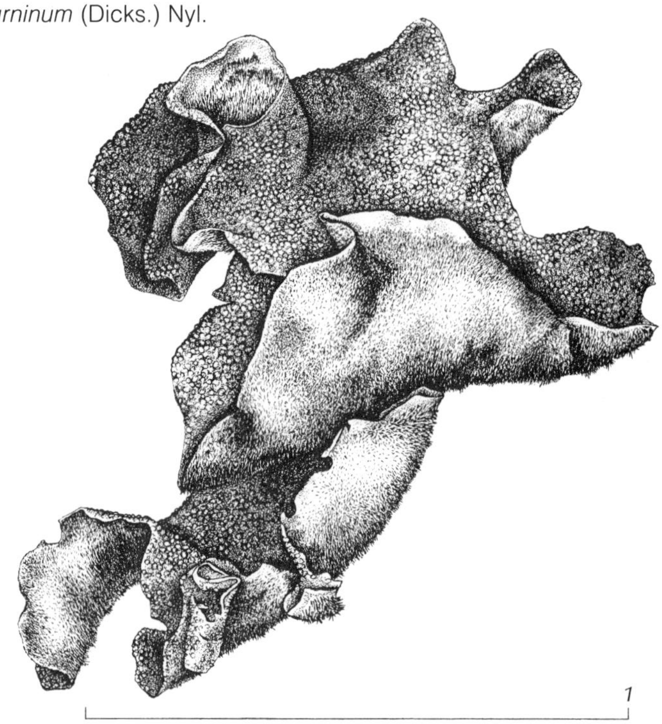

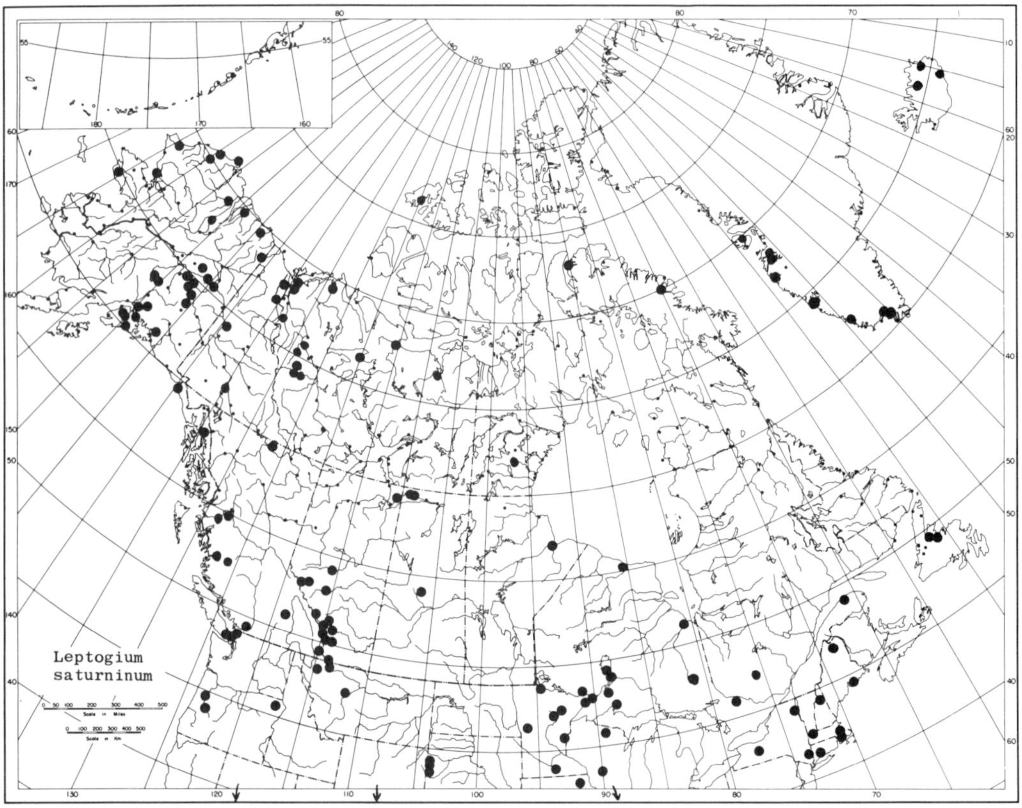

1

Leptogium
saturninum

Leptogium sinuatum (Huds.) Mass.

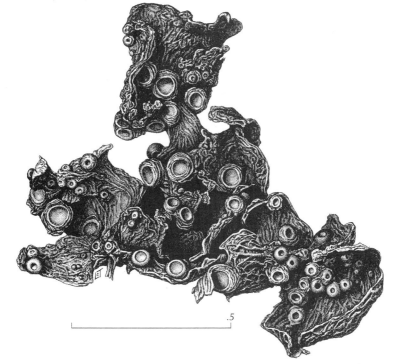

.5

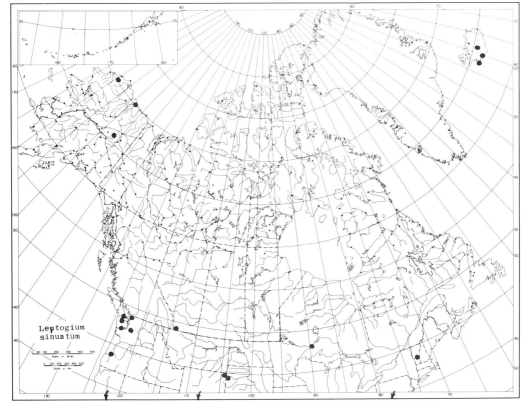

Leptogium
sinuatum

Leptogium tenuissimum (Dicks.) Fr.

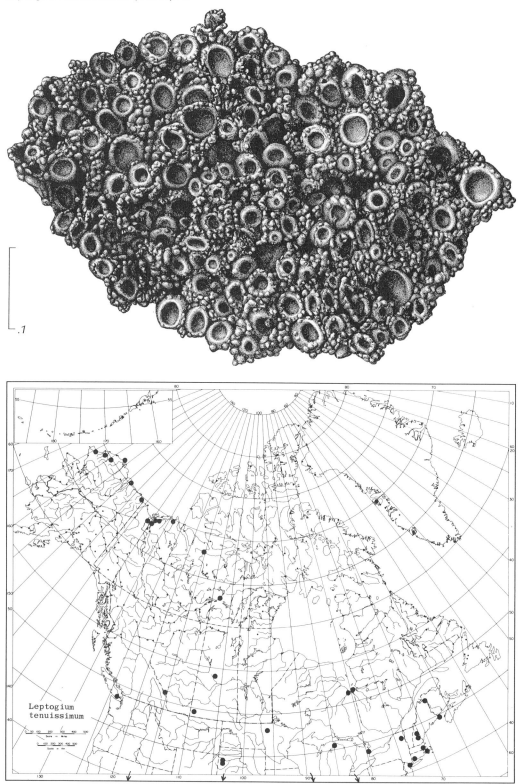

perate and boreal, but reaching into the Arctic in Alaska. It has not been much collected in North America but is known from Alaska to California, in the Rocky Mountains, and in the east as far south as Virginia.

10. Leptogium tenuissimum (Dicks.) Fr.

Corp. Fl. Prov. Suec. 1:293. 1835. *Lichen tenuissimus* Dicks., Pl. Crypt. Brit. 1:12. 1785. *Leptogium spongiosum* Nyl., Mem. Soc. Sci. Nat. Cherbourg 3:165. 1855. *Leptogium humosum* Nyl., Faun. Fl. Fenn., new ser. 1:8. 1858. *Leptogium manschorum* Vain., Ann. Acad. Sci. Fenn. 27A:90. 1928.

Thalli to 2 mm broad but coalescing to form a crust, lobes divided, to 0.2 mm broad, flattened to nearly terete, appearing as clustered coralloid growths, surface smooth, often with granular outgrowths; lead-gray or brownish to black; cortex of irregularly isodiametric cells; thallus paraplectenchymatous throughout, algae in chains or clustered; scattered hairs, 3–4 μ in diameter, growing from the lower cortex.

Apothecia common, adnate to sessile, to 0.8 mm broad; disk concave to flat, light to dark brown or black; thalloid exciple entire to lobulate; proper exciple euparaplectenchymatous but unclear; subhymenium hyaline to yellowish; hymenium 105–180 μ, hyaline, with thin yellow to brown epithecium; paraphyses unbranched, 1μ in diameter, slightly thickened at apices; asci cylindrico-clavate, 90–150 × 12–14 μ; spores 8, uniseriate to irregular, ovoid to ellipsoid or subfusiform, tips rounded or pointed, 3–7 septate transversely, 0–2 septate longitudinally, 17–37 × 9–14 μ.

This species usually grows on sandy soil but is also on sandstone and on tree barks. It is circumpolar, arctic to temperate, but so inconspicuous that it is rarely collected. It appears to be widely distributed across North America.

LOBARIA (Schreb.) Hoffm.

Deutschl. Fl. 2:138. 1796. *Lichen* (sect.) *Lobaria* Schreb., Linn. Gen. Pl. 8 ed. 768. 1791.

Thallus foliose, large, irregularly laciniate and polyphyllus; upper cortex several cells thick, paraplectenchymatous; medulla lax or compact; hyphae hyaline and pachydermatous; lower cortex thin, fibrous to paraplectenchymatous; lower surface tomentose, the tomental hyphae uniseriate, sometimes branched, light-colored to dark pigmented; rhizines simple to squarrose, sparse, to 3 mm long; cephalodia often present, internal or erumpent, containing *Nostoc*.

Apothecia hemiangiocarpic, subpedicellate, lecanorine; disk dark red; hymenium hyaline; paraphyses slender, simple or rarely branched near the apices; asci cylindrical to clavate; spores 8, 3-septate, hyaline, fusiform or acicular. Pycnidia immersed, conidia cylindrical, slightly swollen at tips.

Algae: *Myrmecia* in some species, *Nostoc* in others, *Nostoc* in the cephalodia when present.

Type species: *Lobaria pulmonaria* (L.) Bir.

REFERENCES
Hakulinen, R. 1964. Die Flechtengattung *Lobaria* Schreb. in Ostfennoskandien. Ann. Bot. Fenn. 1:202–213.
Jordan, W. P. 1973. The genus *Lobaria* in North America north of Mexico. Bryologist 76:225–251.
Yoshimura, I. 1971. The genus *Lobaria* of eastern Asia. J. Hattori Bot. Lab. 34:231–364.

1. Algal layer containing green algae, blue-green algae in cephalodia .... 3. *L. linita*
1. Algal layer containing blue-green algae, cephalodia absent.
    2. Isidia and soredia absent.
        3. Medulla P+, stictic or norstictic acids or both present .... 4. *L. pseudopulmonaria*
        3. Medulla P−, stictic and norstictic acids absent .... 2. *L. kurokawae*

2. Soredia present.
    4. Medulla P+, stictic acid present . . . . 5. *L. scrobiculata*
    4. Medulla P−, stictic acid absent . . . . 1. *L. hallii*

1. **Lobaria hallii** (Tuck.) Zahlbr.

Cat. Lich. Univ. 3:321. 1925. *Sticta hallii* Tuck., Proc. Amer. Acad. Arts 12:168. 1877.

Thallus to 11 cm broad, center firmly attached, leathery to cartilaginous, scrobiculate, sometimes smooth or ribbed, smoky gray to grayish olive when dry, much darker when wet; lobes large, broadly rounded, to 4 cm across, tips tending to be free and reflexed, pulverulent; soredia abundant on lamella, less on margins, punctiform when young, becoming confluent, dark; a fine pale tomentum present on upper side of young lobes, very variable in amount; lobules and isidia lacking; underside with flat naked areas separated by tomentose veins which often resemble those in *Peltigera;* lower side cream-colored to yellowish brown, the tomen-

tum darkening, with clusters of dark rhizines to 3 mm long. Upper cortex 18–45 μ, 4–8 cells thick; cephalodia absent.

Apothecia rare, to 2.2 mm broad, lamellar; disk flat to slightly convex, brown; margin entire, minutely warty and villose; hypothecium orange-brownish; hymenium 100–150 μ, hyaline; epithecium orange; paraphyses rarely branched, to 2.5 μ diameter; asci 125 19–24 μ; spores to 8 per ascus but some aborting, dark brown, 1-septate, fusiform with tips often mamillate containing large oil droplets, 23–41 × 9–14 μ.

Reactions: cortices K+ yellow, medulla K−, KC−, C−, P−.

Contents: unidentified yellow pigment plus unknowns.

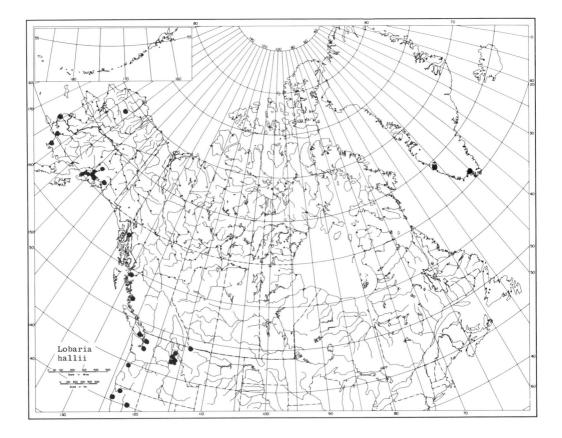

This species usually grows on twigs and trunks of woody plants. It has a disjunct distribution pattern, being known from western North America, Greenland, and Scandinavia. It appears to have strong oceanic tendencies.

2. Lobaria kurokawae Yoshim.
J. Hattori Bot. Lab. 34:297. 1971.

Thallus to 11 cm broad, loosely attached, leathery, strongly reticulately ridged and scrobiculate, brown to yellowish brown, darker when wet; lobes irregular in shape, overlapping, and more or less ascending, tips darkest and shiny; margins crenate to truncate-crenate; soredia and isidia lacking, lobules rare; lower surface with smooth irregular areas naked and separated by a dark tomental network; tomentum black; rhizines squarrose, to 4 mm long.

Upper cortex 33–70 μ, or 5–9 cell lay-ers, outer part reddish brown; algae blue-green; cephalodia absent.

Apothecia not found in North American material; in Asiatic material reported to be laminal, substipitate, flat, to 3 mm broad, the margin nearly entire, the excipulum verruculose, lacking algae, the hymenium 90 μ; spores hyaline to pale becoming 3-septate, fusiform, 20 × 8 μ.

Reactions: medulla K−, C−, KC−, P−; tomentum K+ blue-green.

Contents: lacking stictic and norstictic acids; retigeric acids A and B and several unknowns present; tomentum with thelephoric acid.

This species grows among mosses over soils and rocks. It is like the closely related *L. pseudopulmonaria* in being distributed in Asia and in the Alaskan region. It may well represent only a further chemical strain of that species.

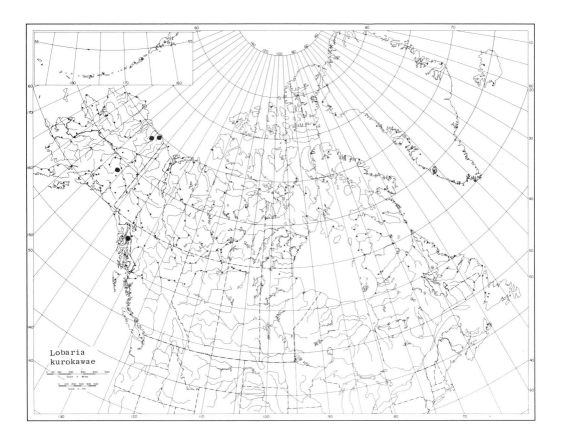

Lobaria kurokawae

3. Lobaria linita (Ach.) Rabenh.
Deutschl. Kryptogam. Fl. 2:65. 1845.

Thallus large, to 30 cm broad, irregularly lobate; lobes broad and rounded, rarely linear, tips often ascending, more or less shining, greenish gray to brown, greener when wet, reticulately ribbed or rugose, lacking isidia or soredia; lower side with pale naked swellings, separated by smooth, blackened veins, tomentum light-colored when young, rhizinae dark, to 4 mm long.

Cortex 30-50 μ, 5–6 layers of cells with large, 10 μ, lumina, algae green, cephalodia internal, numerous, appearing on the surfaces as swellings to 2 mm broad.

Apothecia common on upper surface and margins, to 4 mm broad, margin with numerous protuberances; disk brown; hypothecium 60–110 μ, brownish; hymenium 95–113 μ, hyaline or brownish; epithecium reddish brown; paraphyses 1.5–2.5 not capitate; asci clavate, 90–110 × 15–21 μ; spores 6–8 per ascus, hyaline, usually 1-septate (–3-septate) fusiform-elliptical, straight, 21–34 × 6–9 μ. Pycnidia laminar and marginal, conidia 4.5–5.5 × 1–1.5 μ.

Reactions: K–, medulla K–, KC–, P–.

Contents: tenuorin, methyl gyrophorate, methyl evernate plus other substances and several linitins irregularly present (Maass 1975).

This species grows on tree bases, rocks or earth, often among mosses. It is circumpolar arctic and alpine. In North America it is mainly in the western part of the continent, from Alaska south to Oregon and east as far as the west coast of Hudson Bay.

4. Lobaria pseudopulmonaria Gyeln.
Acta Fauna Fl. Universalis Ser. 2, Bot. (1(5–6):6. 1933.

Thallus to 9 cm broad, leathery, reticulate ridged and scrobiculate, brown or yellowish brown, darker when wet; lobes irregular, overlapping and more or less ascending, tips darkest and shining, margins crenate to truncate-crenate, lacking soredia or isidia, lobules rare; underside with smooth naked areas surrounded by black tomentum; rhizinae squarrose, to 3 mm long.

Upper cortex 28–38 μ, of 3–5 cell lay-

ers, outer part reddish-brown, algae blue-green; cephalodia absent.

Apothecia absent in North American material, on the upper surface, to 2 mm broad, margin entire, disk reddish brown; thalloid exciple verruculose; hymenium 110 μ; spores 3 septate, fusiform, 25 × 7–8 μ.

Reactions: thallus K–, medulla K+ yellow or red, P+ yellow or orange-red, tomentum K+ blue-green.

Contents: norstictic acid and triterpenoids always present, in Strain I stictic and constictic acids present, in Strain II the latter two substances are lacking. Thelephoric acid is present in the tomentum and rhizinae.

This species grows on soil and among mosses, sometimes over rocks. It is an amphi-Beringian species, common in eastern Asia and ranging into Alaska, to be expected eventually in British Columbia. North American material formerly referred to the isidiate species, *Lobaria retigera* (Bory) Trev., should be referred here.

5. Lobaria scrobiculata (Scop.) DC. in Lam. & DC.
Fl. France, 3 ed. 2:402. 1805.

Thallus to 14 cm broad, firmly attached, leathery, strongly scrobiculate, pale yellow to yellowish brown or dark yellow or olive-buff when dry, blackish when wet; lobes often imbricate to ascending, rounded to elongate to 2 cm wide, tips often strongly pulverulent; margins subentire or undulate crenate, rarely with scattered hairs, not thickened except where sorediate; soredia on lamella and margins, soralia punctiform, becoming confluent or linear on the margins, the soredia dark brown; lobules or isidia absent; lower surface smooth or rugose, not distinctly veined, densely tomentose with small scattered naked areas; tomentum often blackening, present to margins; rhizines often abundant in clusters, to 3 mm long.

Upper cortex irregular, outer part brownish, 34–37 μ thick, of 5–9 cell layers, algae blue-green, cephalodia absent.

Apothecia rare, sessile to short stipitate, to 1.5 mm broad; margin strongly papillate, rim

Lobaria linita (Ach.) Rabenh.

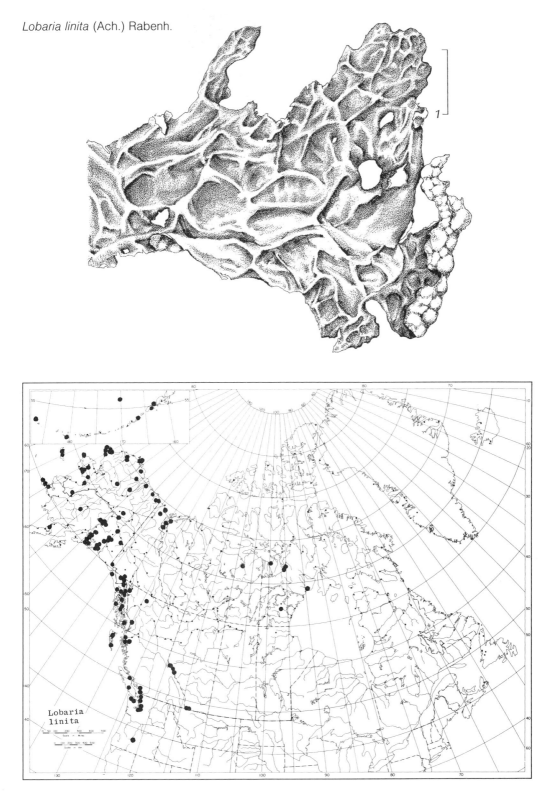

Lobaria pseudopulmonaria Gyeln. _____ *1*

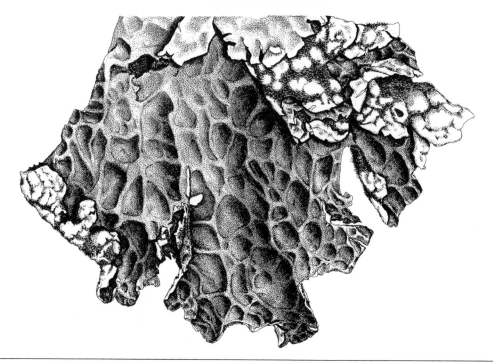

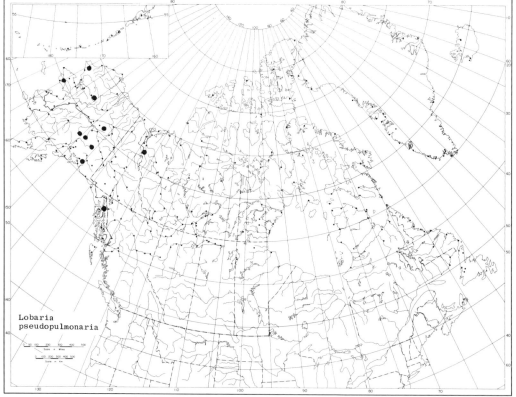

Lobaria
pseudopulmonaria

Lobaria scrobiculata (Scop.) DC.

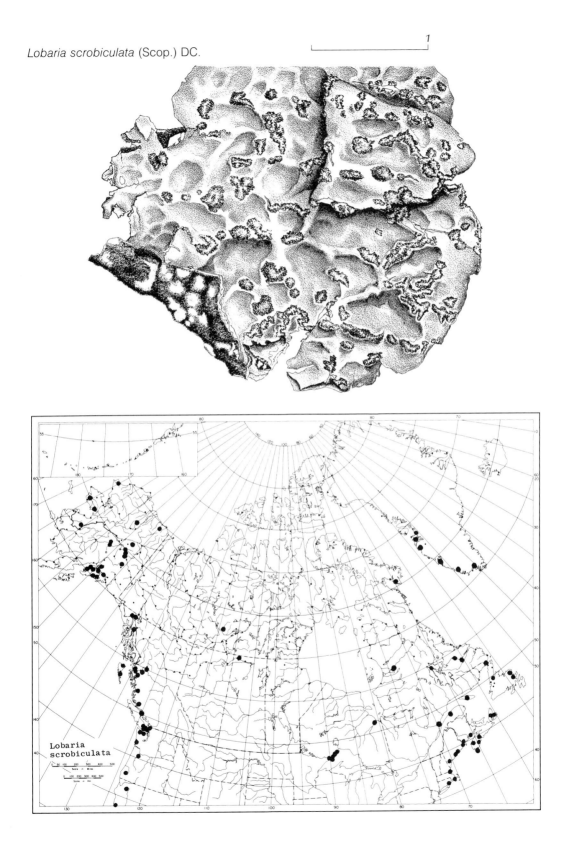

1

Lobaria
scrobiculata

smooth and undulate; hymenium 100–120 μ, pale brownish; hypothecium yellowish-brown; paraphyses 2 μ; asci 92–108 × 16–20 μ; spores 8 hyaline, 1–3 septate, acicular, straight or curved, ends often asymmetrical, 61–72 × 6–8 μ. Pycnidia unknown.

Reactions: medulla and lower cortex K+ orange-red, upper cortex K–, medulla P+ orange, medulla KC+ red, C–.

Contents: usnic acid, scrobiculin, nor-stictic acid, stictic acid, constictic acid and an unknown substance.

This species grows on trees at the bases, on rocks and on soil. It is a circumpolar boreal species. In North America it appears to show a strongly oceanic tendency with the main collections being near the coasts but with some collections from the mountains in Colorado and in the Appalachians.

The closely related *L. hallii* lacks scrobiculin norstictic and stictic acids and has fusiform spores.

MASONHALEA Kärnefelt

Bot. Not. 130:101. 1977.

Thallus prostrate, unattached, of major branches dividing irregularly or dichotomously, curling into ball-like masses when dry; upper surface dark brown, olive-brown when wet; lower surface concolorous on corticate portions but with large areas decorticate and whitish to bluish pruinose; upper cortex very thick and prosoplectenchymatous, composed of gelatinized and very thick-walled hyphae giving the plant a horny texture; medulla thick and dense, white.

Apothecia lateral, the margin thalloid; asci cylindrical to clavate; spores 8, hyaline, simple ellipsoid to subglobose. Pycnidia marginal; conidia cylindrical.

Algae: *Trebouxia*.

Type species: *Masonhalea richardsonii* (Hook.) Kärnef.

REFERENCE
Kärnefelt, I. 1977. *Masonhalea*, a new lichen genus in the Parmeliaceae. Bot. Not. 130:101–107.

The genus is monotypic, containing only one species.

1. Masonhalea richardsonii (Hook.) Kärnef.

Bot. Nat. 130:101. 1977. *Cetraria richardsonii* Hook. in Richardson, Bot. Appendix to J. Franklin, Narrative of a Journey to the shores of the Polar Sea 774. 1823. *Evernia richardsonii* (Hook.) Nyl., Mem. Soc. Imp. Sci. Nat. Cherbourg 5:99. 1857. *Parmelia richardsonii* (Hook.) Nyl., Flora 43:42. 1860. *Platysma richardsonii* (Hook.) Nyl., Synops. Lich. 1:306. 1860.

Thallus foliose, dichotomous to irregularly divided, when dry rolled up into ball-like masses, horny in texture; the lobes sharp pointed and antler-like; upper side chestnut-brown when dry, becoming olive-green to greenish brown when moist; underside paler brown with conspicuous pruinose patches, lacking rhizinae.

Outer cortex prosoplectenchymatous, 37–65 μ thick, the outer 15 μ brown pigmented; within this an inner cortex of longitudinally oriented but interwoven hyphae, about 100–125 μ in thickness, the hyphae slender, 2–2.5 μ, but quite densely arranged. Within this is a continuous layer of the algae and a loose medulla. The lower cortex is prosoplectenchymatous but lacks the inner fibrous layer of the upper cortex and is discontinuous, being lacking over the pruinose areas which appear to function as super-sized pseudocyphellae for gas exchanges. The dual upper cortex reminds one of that of some of the Ramalinas but is not on both surfaces as in that genus.

Apothecia rare, at the tips of lobes, the margin crenate, concolorous with the thallus, inflexed; disk 4–8 mm, brown, almost the same color as the thallus, shining; epithecium brownish, lower part brown; hypothecium pale brownish; hymenium 125 μ; paraphyses capitate, the tips 6 μ; spores 8, hyaline, thin-walled, simple, ellipsoid, 7.5–12.5 × 5–6 μ.

Masonhalea richardsonii (Hook.) Kärnef.

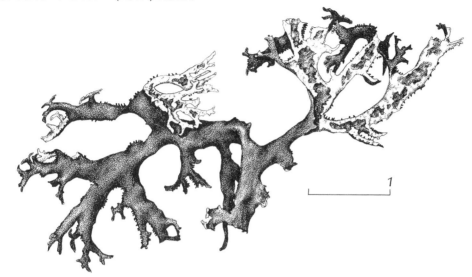

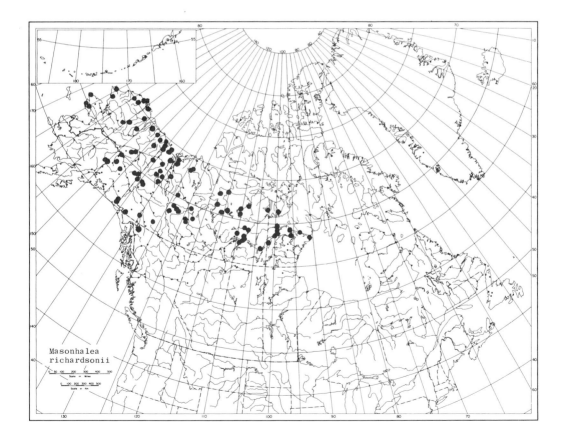

Reactions: medulla K−, C+, KC+ orange red to red, P−. UV+.

Contents: alectoronic acid.

This species is a *"wanderflechten,"* growing unattached and being blown across the tundras by the winds, gathering in low spots and wet places where it spreads out in the moisture. It is a Beringian element in the flora, ranging in Siberia and in North America east to Chesterfield Inlet on the west side of Hudson Bay and south in the mountains to British Columbia.

MASSALONGIA Körb.

Syst. Lich. Germ. 109. 1855.

Thallus small, foliose; lobes squamulose or elongated, more or less isidiose, esorediate; underside pale with brown or blackish rhizinae; upper cortex paraplectenchymatous; medulla with loosely interwoven hyphae, filled with globose *Nostoc* colonies; lower surface of densely interwoven hyphae.

Apothecia hemiangiocarpic, with proper margin, laminal or marginal; hymenium brown above, hyaline below; I+ blue; paraphyses septate, simple or branched; asci cylindrical with apically thickened walls; spores 8, hyaline, ellipsoid to fusiform, 2–4 celled. Pycnidia brown, conidiophores short celled, branched, producing the conidia terminally and laterally; conidia short cylindrical.

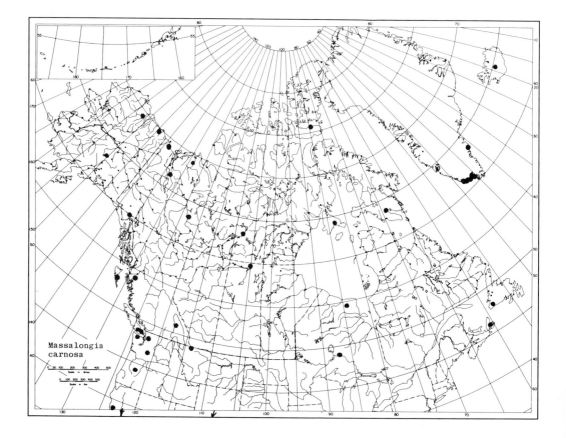

Massalongia
carnosa

Algae: *Nostoc*.

Type species: *Massalongia carnosa* (Dicks.) Körb.

REFERENCES

Gyelnik, V. 1940. Pannariaceae in Rabenhorst Krypt.-Fl. von Deutschl. u. d. Schweiz 9(2):135–272.

Henssen, A. 1963. The North American species of *Massalongia* and generic relationships. Can. J. Bot. 41:1331–1346.

1. Massalongia carnosa (Dicks.) Körb.

Syst. Lich. Germ. 109: 1855. *Lichen carnosus* Dicks., Fasc. Plant. Cryptog. Brit. 2:21. 1790. The extensive synonymy can be consulted in Gyelnik 1940.

Prothallus lacking. Thallus tiny foliose, of flattened lobes 0.5–2 mm broad, 0.5–3 (–10) mm long, irregularly branched and overlapping; upper side brown, smooth, or with round or globose isidia, the isidial tips sometimes breaking and simulating soredia, dull; underside whitish to brown with brown rhizinae; upper cortex of 3–8 cells covered by a gelatinous layer, pseudoparenchymatous; medulla of loosely interwoven mainly vertical hyphae, with numerous *Nostoc* colonies; lower cortex of longitudinal hyphae more or less loosely interwoven.

Apothecia often stipitate, brown, with proper margin composed of exciple plus pseudoexcipulum; hymenium 70–130 μ, upper part brown; paraphyses septate, unbranched, 2 μ thick, the tips thickened to 5–6 μ; subhymenial layer hyaline, of interwoven hyphae, 45–80 μ thick; exciple pseudoparenchymatous, pale; asci clavate, 57–79 × 11.5–14.5 μ; spores 8, hyaline, 2 (–3) celled, ellipsoid to fusiform, 11–27 × 4.5–8.5 μ. Conidia slightly dumbbell-shaped, 4–6 × 1 μ.

This species grows over mosses and humus, on boulders and rock faces, rarely on earth. It is an arctic and arctic-alpine species in Greenland and in North America known from Alaska to Baffin Island, south to Nova Scotia in the east, and Colorado and California in the west.

Henssen (1963) comments that M. *carnosa* and *Parmeliella lepidiota* are easily mistaken. The latter becomes sorediose and the upper cortex has angular cells with strongly gelatinous walls and connecting pores, the former lacks soredia and the upper cortex has rounded cells with thin walls.

NEPHROMA Ach.

Lich. Univ. 521. 1810.

Thallus foliose, greenish yellow or brown, upper and lower cortices paraplectenchymatous, heteromerous; medulla of loosely interwoven hyphae; lower surface glabrous, pubescent, or tomentose with thick-walled hairs.

Apothecia immersed on lower surface at tips of lobes; disks light to dark brown; exciple hyaline to dark brown; hymenium 60–90 μ thick; paraphyses unbranched; spores 8, 3-septate, light brown.

Algae: *Coccomyxa* or *Nostoc*.

Type species: *Nephroma arcticum* (L.)Torss.

REFERENCES

DuRietz, G. E. 1929. The lichens of the Swedish Kamtchatka expeditions. Ark. Bot. 22A(13):1–25.

Gyelnik, V. 1931. Nephromae novae et criticae. Ann. Crypt. Exot. 4:121–149.

Wetmore, C. M. 1960. The lichen genus *Nephroma* in North and Middle America. Publ. Mus. Mich. State Univ. 1(11):369–452.

1. Algal layer with green algae; blue-green algae restricted to cephalodia.
    2. Thallus yellow; cephalodia visible on upper surface; upper surface smooth . . . . 1. *N. arcticum*
    2. Thallus dull green or brownish; cephalodia internal, visible on lower surface; upper surface scaly or slightly pubescent . . . . 3. *N. expallidum*

1. Algal layer with blue-green algae; cephalodia lacking.
   3. Papillae present on lower surface; thallus usually tomentose below . . . . 7. *N. resupinatum*
   3. Papillae lacking on lower surface; underside tomentose or not.
      4. Thallus with soredia . . . . 6. *N. parile*
      4. Thallus lacking soredia.
         5. Thallus with isidia and or marginal teeth.
            6. Isidia branching, usually arising in clusters on ridges of lacunose upper surface
            . . . . 5. *N. isidiosum*
            6. Isidia and marginal teeth flat and short, unbranched, not in clusters, lacunose ridges
            lacking . . . . 4. *N. helveticum*
         5. Thallus lacking isidia or marginal teeth . . . . 2. *N. bellum*

1. Nephroma arcticum (L.) Torss.

Enum. Lich. Byssic. Scand. 7. 1843. *Lichen arcticus* L., Sp. Pl. 1148. 1753.

Thallus yellow-green, large foliose, forming large mats; lobes broad, crisped or entire, regeneration squamules common; upper side smooth; algal layer of green algae with the blue-green algae restricted to cephalodia under the upper surface; underside tomentose, light brown or white near the margins, black centrally.

Apothecia large, to 20 (–30) mm broad; disk light brown; exciple entire; spores subfusiform with thin walls, 23–27 × 4–6 μ.

Reactions: thallus K+ yellow, KC+ yellow, P+ yellow; medulla UV+.

Contents: nephroarctin, phenarctin, usnic acid, zeorin; the "carotenoid" of Wetmore is the nephroarctin.

This species grows on hummocks, over moist rock outcrops, and mossy banks. It is circumpolar boreal and low arctic, lacking in the higher latitudes.

2. Nephroma bellum (Spreng.) Tuck.

Boston J. Nat. Hist. 3:293. 1840. *Peltigera bella* Spreng., Syst. Veg. 16 ed. 4(1):306. 1827.

Thallus to 6 cm broad, the lobes to 1 cm broad; upper side glabrous or slightly pubescent, brown; lower surface rugulose, pale to dark brown, pubescent to tomentose; thallus lacking soredia or isidia but sometimes with small lobules or regeneration squamules; algae blue-green.

Apothecia to 10 (–15) mm broad; the margins entire to crenulate; disk light brown; spores ellipsoidal with thickened walls; 17–21 × 5–6 μ.

Reactions: all negative or P– or P+ yellow (substance unknown).

Contents: zeorin and nephrin usual, plus accessory neutral substances.

This species grows on mossy rocks in shady places and also on trees in similar habitats. It is circumpolar, hemiboreal to alpine, in North America ranging south to Arizona and to Virginia (in mountains). Perhaps its occurrence in south Greenland reflects a relic condition from the former presence of forest.

The closely related *N. laevigatum* which has a K+ pink to deep red violet reaction and P+ red in the medulla due to the presence of the anthraquinone nephromin in the medulla does not apparently reach the Arctic. It is also more oceanic in its requirements.

3. Nephroma expallidum (Nyl.) Nyl.

Flora 48:428. 1865. *Nephromium expallidum* Nyl., Oefvers. Kgl. Vetensk.-Akad. Förh. 17:295, 1860.

Thallus lobes to 1.8 cm broad, edges entire or crisped, often with regeneration lobules; upper side greenish brown to brown, greener when wet, glabrous, pruinose, or slightly pubescent; algal layer green, cephalodia internal and near the lower surface; underside light brown at margins, dark centrally and with varying amounts of tomentum.

Apothecia uncommon, to 10 mm broad, disks light to dark brown; spores with slightly thickened walls, occasionally abnormal in shape, 17–21 × 5–6 μ.

Reactions: all usual reagents negative.

Contents: nephrin, zeorin and neutral substance "B" (Wetmore 1960).

Nephroma arcticum (L.) Torss.

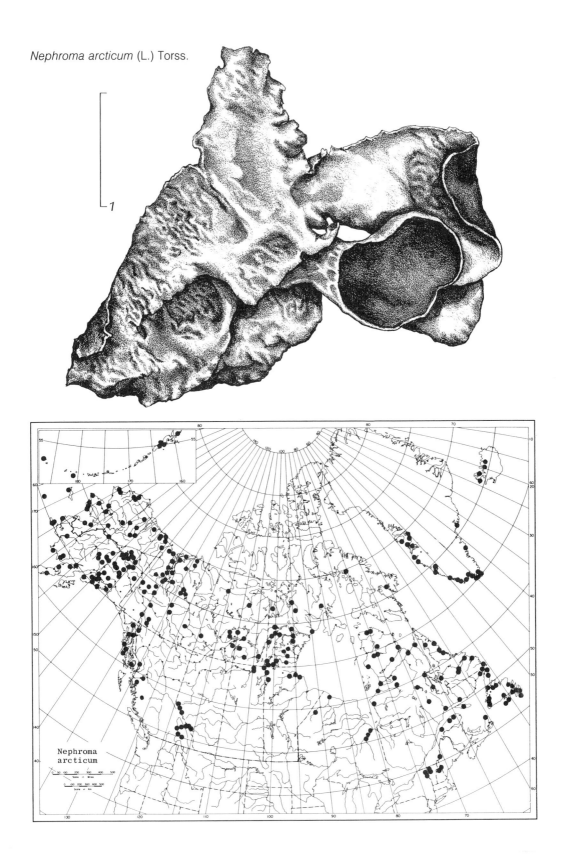

Nephroma bellum (Spreng.) Tuck.

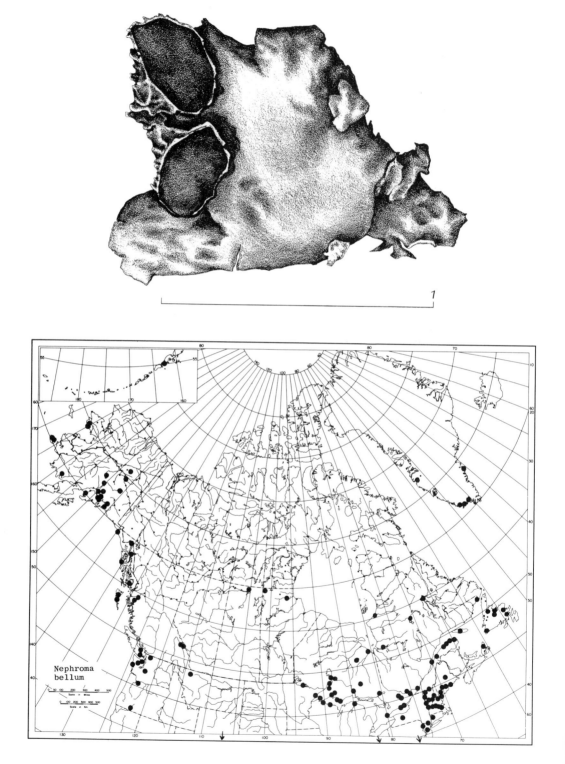

1

Nephroma
bellum

Nephroma expallidum (Nyl.) Nyl.

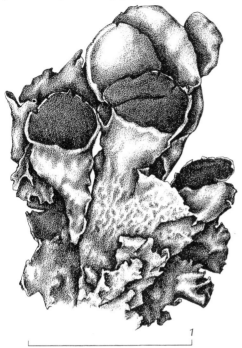

1

This species grows on tussocks, on hummocks, among mosses in crevices between rocks, commonly in *Cassiope* heaths. It is cirumpolar, arctic, and alpine, in North America ranging south only into northern Ontario and in the west into the Canadian Rockies.

4. **Nephroma helveticum** Ach.
Lich. Univ. 523. 1810.

Thallus to 8 cm broad, the lobes narrow, to ca. 5 mm broad; upper surface smooth with varying amounts of pubescence and small flat isidia, brown; lower surface tomentose or pubescent, dark brown to black; regeneration squamules common on the surface and edges of the thallus; algae of thallus blue-green.

Apothecia to 8 mm broad, common, margins often fringed with flat isidia, the reverse often scrobiculate; disks dark brown; spores ellipsoid, 20–23 x 6–7 μ.

Reactions: all negative.

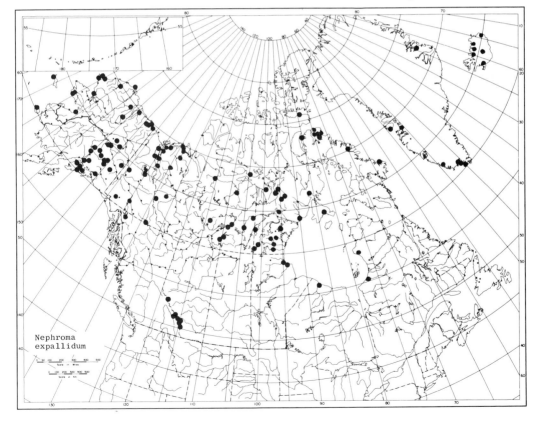

Nephroma
expallidum

Nephroma helveticum Ach.

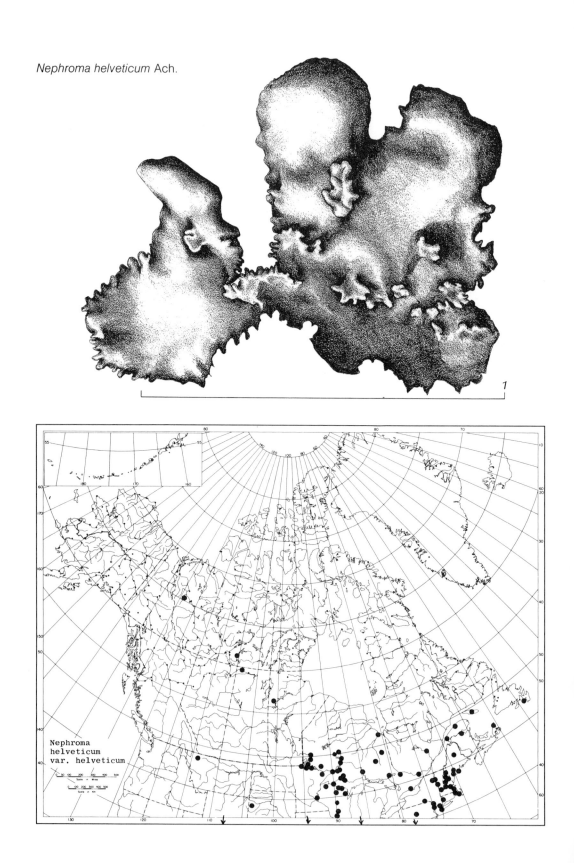

1

Nephroma
helveticum
var. helveticum

Nephroma isidiosum (Nyl.) Gyeln.

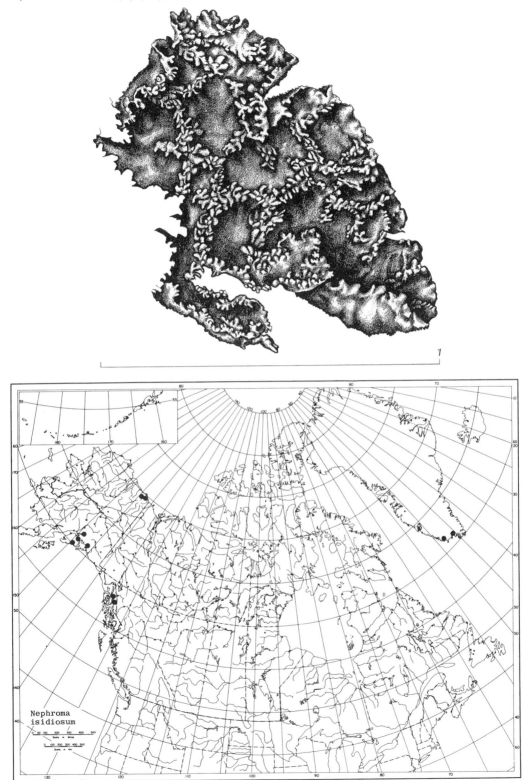

1

Nephroma
isidiosum

Nephroma parile (Ach.) Ach.

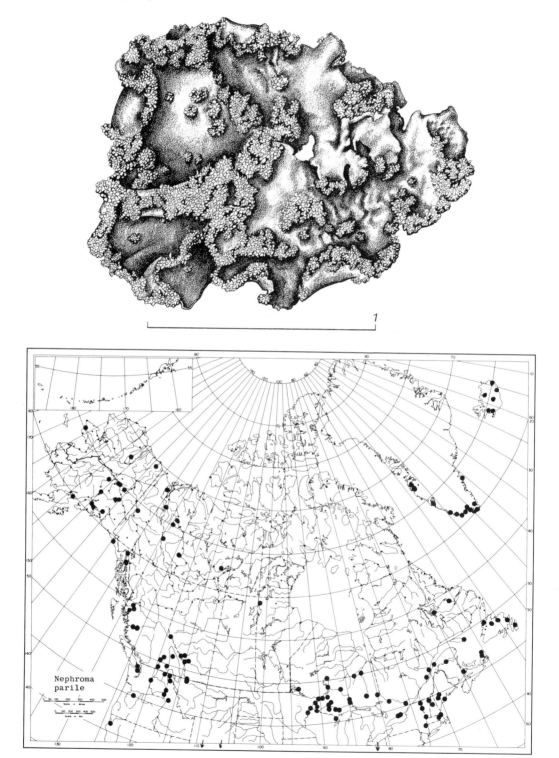

1

Nephroma resupinatum (L.) Ach.

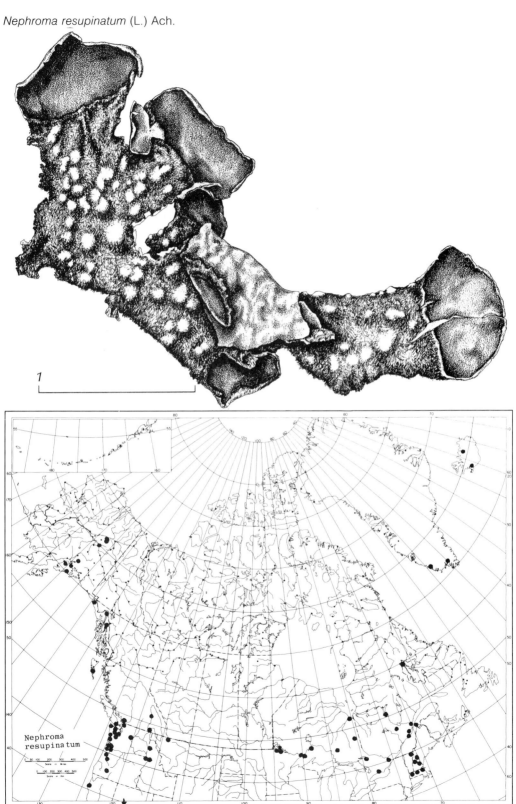

1

Contents: several neutral substances, + nephrin.

This species grows on mossy rocks and on trees in moist, cool, shady places and may cover large areas of the rocks.

It is circumpolar, boreal, and temperate in distribution; in North America it ranges east of the Rocky Mountains and northward to close to the forest-tundra border. In the west it is replaced by var. *sipeanum* (Gyeln.) Wetm. with a thicker thallus (250–290 μ vs 100–160 μ), and rounded and long isidia and teeth instead of short flat ones.

5. **Nephroma isidiosum** (Nyl.) Gyeln.

Ann. Crypt. Exot. 4:126. 1932. *Nephromium tomentosum* var. *isidiosum* Nyl., Notis. Sällsk. Faun. Fl. Fenn. Förh. II. 5:180. 1866.

Thallus brown, to about 5 cm broad; upper surface somewhat lacunose with many clusters of coralloid isidia 0.27 mm tall, usually arising from the ridges; lower surface covered with dense, short, dark brown tomentum and short scattered rhizinae; algae blue-green. No apothecia seen.

Reactions: none.
Contents: 3 neutral substances.

This species grows on rocks, among mosses and at the base of woody plants. It is probably a very rare circumpolar boreal to arctic species. Although it has been found in North America only in the northwest it does not appear to be amphi-Beringian as other collections were reported from Scandinavia, western Siberia, and Greenland.

6. **Nephroma parile** (Ach.) Ach.

Lich. Univ. 522. 1810. *Lichen parilis* Ach., Lich. Suec. Prodr. 164. 1798.

Thallus brown, about 8 cm broad, the lobes to 5 (−17) mm broad; upper surface smooth with round soralia forming patches of bluish-gray, soredia also on margins; lower side rugulose with variable amounts of tomentum; regeneration squamules often present; algae blue-green.

Apothecia rare; disks light brown; spores ellipsoidal, 16–18 × 6–7 μ.

Reactions: medulla K+ yellow or K−.

Contents, zeorin in most samples, nephrin (±), and several unknown substances.

This species grows on moist shaded rocks and on trees, in shade. It is circumpolar, boreal to temperate, in North America ranging south to Arizona in the west, and Tennessee in the east.

7. **Nephroma resupinatum** (L.) Ach.

Lich. Univ. 522. 1810. *Lichen resupinatus* L., Sp. Pl. 1148. 1753.

Thallus gray-brown to brown, small, to 9 cm broad, the lobes to 1 cm broad; upper side pubescent in varying amounts, regeneration lobes common; lower surface with papillae among the tomentum; algae of thallus blue-green. Apothecia to 11 mm broad, disks light brown; spores subfusiform with thin walls, 21–24 × 4–6 μ.

Reactions: none.
Contents: none.

This species grows on rocks and on tree barks in cool moist, shady places. It is circumpolar, hemiboreal to low alpine. In North America it is mainly in the boreal forest. Its occurrence in Greenland may reflect its persistence after the forest disappeared.

NEUROPOGON (Nees & Flot.) Nyl

Synops. Lich. 272. 1860.

Thallus fruticose, attached by a holdfast, branching with or without lateral fibrils; branches terete to angular, smooth or minutely scabrid, verruculose, plicate or foveolate, yellow with more or less variegation with black banding; cortex with yellowish granules and pachydermatous hyphae; medulla lax in arctic species, commonly compact in antarctic species; central axis ⅓ to ⅔ of the diameter of the branches in southern hemisphere species, ⅕ to ⅓ in northern, hard and horny in texture; sorediate in northern hemisphere species, the soredia yellow to blackish, borne in rounded soralia on the branches.

Apothecia lacking in northern, present in southern hemisphere species; black or brown-black disk; spores 8, simple, hyaline, ellipsoid to subglobose.

Algae: *Trebouxia*.

Type species: *Neuropogon antennarius* Nees & Flot.

Only one species occurs in the Arctic.

REFERENCES

Lamb, I. M. 1939. A review of the genus *Neuropogon* (Nees & Flot.) Nyl. with special reference to the Antarctic species. J. Linn. Soc. London Bot. 52:199–237. 1939.

—— 1964. Antarctic Lichens. 1. The genera *Usnea, Ramalina, Himantormia, Alectoria, Cornicularia*. British Antarctic Survey. Scientific Reports 38:1–34, + tables I–XIV and plates I–IX.

Lynge, B. 1941. On *Neuropogon sulphureus* (König) Elenk., a bipolar lichen. Skrift, norske Vidensk.-Akad., mat.-naturv. Kl. 10:1–35.

1. Neuropogon sulphureus (Koenig) Hellb.

Bih. svensk Vetensk.-Akad. Handl. 21(13):21. 1896. *Lichen sulphureus* Koenig, Olafsend Povelsen, Reise igion. Island. 16. 1772.

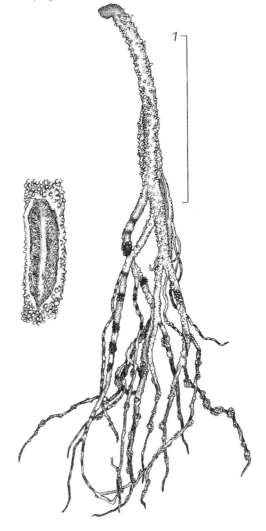

Neuropogon sulphureus (Koenig) Hellb.

Thallus small, 2–3 cm (–4 cm), erect to subdecumbent, caespitose branched from a small basal holdfast, the base yellow, the upper parts blackened, somewhat soft; main branches terete, 0.5–1.3 mm diameter, smooth or minutely scabrid, rarely faintly scorbiculate, dull or subshining, often fracturing to expose the central strand, the outer parts black ringed, the terminal branches entirely black; soredia small and in small punctiform soralia, blackened; axis ⅕ to ⅓ the diameter of the branches; medulla white, thick, lax, rarely compact.

Apothecia unknown.

Reactions: K−, P−; in a few Antarctic specimens K+ red, P+ yellow.

Contents: Usnic acid; in the Antarctic strain with the positive reactions there is a small amount of norstictic acid.

This species grows on granitic types of rocks, avoiding schists and calciferous types. It occurs from the coasts to considerable elevations, to 1700 msm in Greenland. Lynge (1941) speculates that it requires considerable atmospheric moisture.

The bipolar distribution of this lichen has caused considerable speculation. In the northern hemisphere it is on Greenland and Iceland and most of the arctic islands with the proper types of geology. In the southern hemisphere it occurs circumpolar on Antarctica and some adjacent islands. It also occurs on scattered high Andean stations in Equador, Peru,

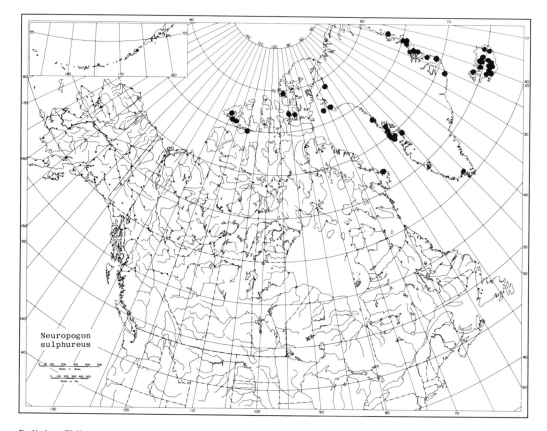

Neuropogon
sulphureus

Bolivia, Chile, and Argentina. Discussions center about the possible relictual nature of these stations and also about possible connections with continental drift. However a simpler possibility exists. Dr. David Parmelee of the University of Minnesota has informed me that Skua gulls which were banded in Antarctica have been recovered three months later in Greenland. These birds also happen to nest in the *Neuropogon* colonies. What could seem simpler than that these birds with bipolar migration habits could carry the soredia of this lichen on their bodies to later shed them in the Arctic? Very germane is Lamb's comment that the Antarctic population of the *Neuropogon* exhibits a different spectrum of variability than the Arctic. This would suggest selection of only some of the available variations present in the south to be carried to the north, a logical possibility with the chances inherent in such long-distance dispersal. Certainly the center of development of the species of *Neuropogon* lies in the southern hemisphere. Whether the skua gulls also visit the high Andes is still open to investigation. But they do seem to be the only possible carriers with the bipolar migration patterns.

OMPHALODISCUS Schol.

Nyt. Mag. Naturvid. 75:23. 1936.

Thallus foliose, umbilicate, mono- polyphyllus; smooth or reticulate-rugose; algal layer continuous, medulla loose, lower cortex sclerenplectenchymatous, rhizines present or absent. Apothecia adnate to slightly stipitate, lecideine with a common margin; the disk developing a central sterile column which usually changes into a slightly branched and angular fissure, secondary fissures then developing between these and the margins to form a very irregular but *not radiating* surface. It is sometimes necessary to section to see the central sterile column. Spores

8, hyaline and simple in the American Arctic species, brown and muriform in *O. crustulosus* (Ach.) Schol., a rare species reported from farther south in North America.

Algae: *Trebouxia*.

Type species: *Omphalodiscus decussatus* (Vill.) Schol.

REFERENCE
Llano, G. A. 1950. a monograph of the lichen family Umbilicariaceae in the Western Hemisphere. Navexos- P-831. 1-281.

1. Rhizinae present, cylindrical or flat; upper side coarsely wrinkled; lower side pink to brown . . . . 3. *O. virginis*
1. Rhizinae absent; upper side reticulate-rugose.
   2. Under side dark brown to black, rarely pruinose; margins often laciniate or perforate; apothecia uncommon . . . . 1. *O. decussatus*
   2. Underside light colored, dove gray with a dark area around the umbilicus, pruinose, smooth to finely chinky; margins entire; apothecia common . . . . 2. *O. krascheninnikovii*

1. Omphalodiscus decussatus (Vill.) Schol.

Nyt. Mag. Naturvid. 75:23. 1936. *Lichen decussatus* Vill., Hist. Plant. Dauphine 3:964. 1789. *Umbilicaria tessellata* var. *reticulata* Duby, Bot. Gall. 2:596. 1830. *Gyrophora discolor* Th. Fr., Svensk Vetensk. Akad. Handl. 7:31. 1867. *Umbilicaria ptychophora* Nyl., Flora 52:388. 1869. *Umbilicaria eximia* Hue, 2me. Exped. Antarctic. Franc. 55. 1915. *Gyrophora decussata* (Vill.) Zahlbr., Cat. Lich. Univ. 4:678. 1927. *Agyrophora decussata* (Vill.) Gyeln., Ann. Mycol. 30:44. 1932. *Umbilicaria rugosa* Dodge & Baker, Ann. Mo. Bot. Gard. 25:564. 1938. *Umbilicaria pateriforms* Dodge & Baker, ibid., p. 564. *Umbilicaria decussata* (Vill.) Zahlbr., Cat. Lich. Univ. 8:490. 1942. *Umbilicaria hunteri* Dodge, B.A.N.Z. Antarctic Research Exp. Rep. Ser. B7: 148 1948.

Thallus to 8 cm broad, monophyllus, umbilicate, rigid, the umbo raised, coarsely granulose, pruinose, reticulating with a raised well-developed pattern, the rugi lower toward the edge and becoming vermiform, occasionally with small lobules from the ridges or margins; upper surface dull, light brown or buff to dark brownish black; underside dark sooty black, the margins gray or light brown, occasionally spotted with dove gray blotches; smooth, the umbilicus flat, lacking rhizinae.

Thallus 300–600 μ thick, upper cortex 30–60 μ, palisadeplectenchymatous, brown; algal layer continuous, medulla very loose; lower cortex 99–115 μ, scleroplectenchymatous.

Apothecia to 3 mm broad, adnate, with sterile center button and fissures or covered with fissures, round with proper margin, becoming convex; hypothecium brown; hymenium 70 μ; paraphyses septate, branched or not, tips slightly thickened to 3 μ; spores 8, simple, hyaline, ellipsoid, 6.6–13.2 × 3.3–6.6 μ.

Reactions: K−, C+ red, KC+ red, P−.

Contents: gyrophoric acid (substances lacking in some, Dahl & Krog, 1973).

This is a nitrophilous species, preferring bird perches, etc. It is circumpolar arctic and alpine and also ranges south into South America.

2. Omphalodiscus krascheninnikovii (Sav.) Schol.

Nyt. Mag. Naturvid. 75:24. 1936. *Gyrophora reticulata* (Schaer.) Th. Fr., Lich. Scand. 1:166. 1871 as used in Herre, Contr. U.S. Nat. Herb. 13:316. 1910. *Gyrophora krascheninnikovii* Sav., Bull. Jard. Imp. Pierre Grand 14:117. 1914. *Gyrophora hultenii* DR., Ark. f. Bot. 22:14. 1929. *Gyrophora polaris* Schol. in Lynge & Schol., Skrifter Svalbard Ishavet 41:57. 1932. *Umbilicaria polaris* (Schol.) Zahlbr., Cat. Lich. Univ. 8:493. 1932. *Umbilicaria krascheninnikovii* (Sav.) Zahlbr., Cat. Lich. Univ. 10:405. 1939.

Thallus to 5 cm broad, monophyllus, rigid, umbo with a coarsely granulated crystal covering, with sharp reticulate ridges running to the margins, the ridges weakening toward the margins; upper surface dull (shining in var. *darrowii* in SW United States), ashy to ashy brown or ashy black, pruinose; underside with a dark spot around the umbilicus, dove gray or paler peripherally, the lower cortex nearly smooth or slightly granulose to rimulose; occasionally there may be a few cylindrical simple or branched rhizinae near the margins.

Thallus 200–500 μ thick; with upper dead material zone 50–100 μ, a thin upper cortex 18–30 μ thick, a discontinuous algal zone,

Omphalodiscus decussatus (Vill.) Schol.

1

Omphalodiscus krascheninnikovii (Sav.) Schol.

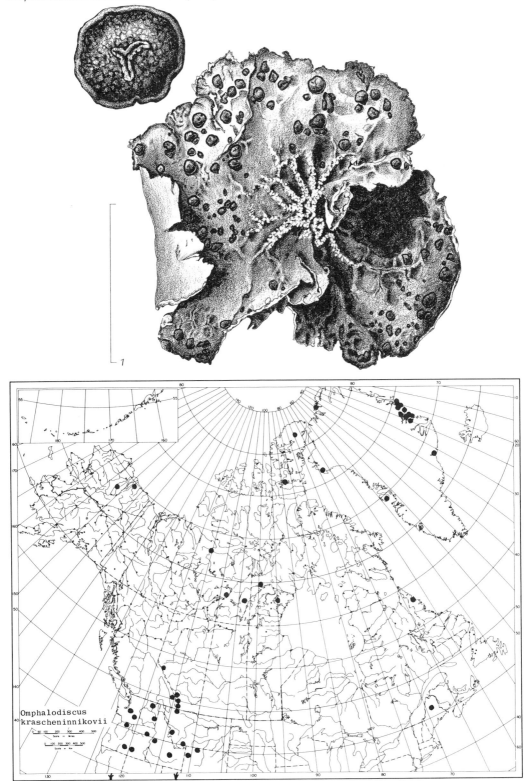

1

Omphalodiscus
krascheninnikovii

Omphalodiscus virginis (Schaer.) Schol.

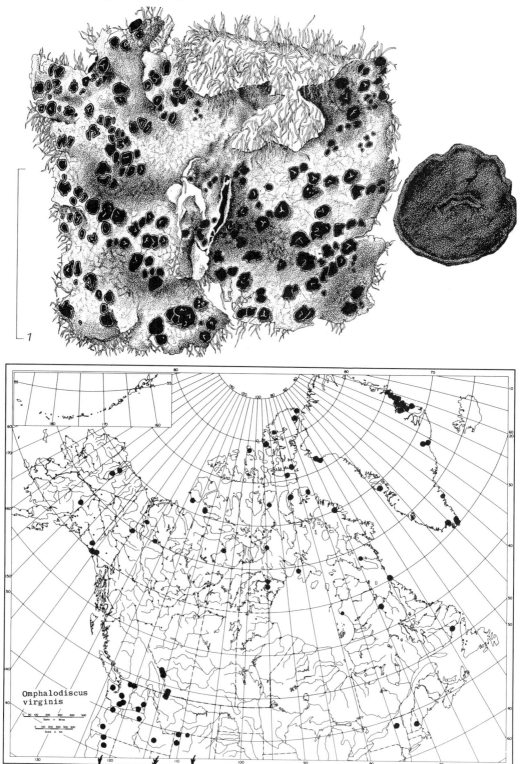

1

a very loose medulla, a lower cortex 66–100 μ, scleroplectenchymatous.

Apothecia to 3 mm, sessile to substipitate, common; disk flat, black, with central sterile button and many secondary fissures; hypothecium brown, 40–50 μ; hymenium 55–65 μ; paraphyses simple, septate, tips to 3.4 μ; spores 8, simple, hyaline, ellipsoid, 5–7 × 3.8–4 μ.

Reactions: K−, C+ red, KC+ red, P−.

Contents: gyrophoric acid (Krog, 1968).

Growing on acid rocks, this is a circumpolar arctic and alpine species, in North America growing south to California and Arizona in the west and Mt. Kataadn in Maine in the east.

3. Omphalodiscus virginis (Schaer.) Schol.

Nyt. Mag. Naturvid. 75:25. 1936. *Umbilicaria virginis* Schaer., Biblioth. Univ. Genève 36:153. 1841. *Umbilicaria lecanocarpoides* Nyl., Flora 43:418. 1860. *Umbilicaria rugifera* Nyl., Lich. Scand. 117. 1861. *Umbilicaria stipitata* Nyl., Lich. Scand. 289. 1861. *Gyrophora virginis* (Schaer.) Frey, Hedwigia 69:222. 1929.

Thallus to 10 cm broad, monophyllus, umbilicate, leathery; umbo not elevated, often wrinkled with weak reticulating irregular ridges which fade marginally, margins irregular, crenate to torn; surface smooth, dull, deep olive-buff to yellowish-brown, sometimes darker brown; underside with well developed umbilicus which extends outward as buttresses or veins, smooth or granular, pink or salmon to orange-buff, with rhizinae thicker toward the margins, the rhizinae cylindrical or flat, simple or slightly branched, darker than the lower surface.

Thallus ca. 266 μ thick; upper cortex 33 μ; algal layer continuous; medulla loose; lower cortex 66 μ, scleroplectenchymatous.

Apothecia to 4.5 mm broad, numerous, sessile; disk round and flat, becoming convex and irregular, central sterile button and fissures often absent and only seen in sections; hypothecium light brown; hymenium to 55 μ, paraphyses septate, branched, light brown, the tips to 2.4 μ; spores simple, hyaline, 10–17 × 6–10 μ.

Reactions: K−, C+, red, KC+ red, P−.

Contents: gyrophoric acid.

This species grows on acid rocks. It is circumpolar arctic-alpine, in North America occurring in the Arctic and in the western mountains south into Mexico but lacking in the eastern United States, except for the Adirondack and White Mountains.

PANNARIA Del.

in Bory, Diction. Class. Hist. Nat. 13:20. 1828.

Thallus foliose or squamulose, heteromerous, often with a well developed hypothallus; lobes contiguous or separate, small, to 4 mm broad, when contiguous forming an almost crustose appearance; flat, convex or concave, smooth, glabrous or pruinose; upper side gray or brown; underside pale or black; margins often effigurate; upper cortex paraplectenchymatous, hyaline or brown, 15–46 μ thick or in *P. hookeri* 57–76 μ thick; medulla loose, lower cortex lacking.

Apothecia usually common, to 2.5 mm broad, margin thalloid, entire or crenulate, sometimes disappearing; disk concave to flat, becoming convex, reddish brown to black; epithecium not granular, hyaline or brownish, greenish in *P. hookeri,* hymenium hyaline, 60–135 μ; paraphyses 1–2 μ, unbranched, tips only slightly thickened, septate; asci clavate; spores 8, single celled, hyaline, often with rough surface. Pycnidia immersed, conidia straight, cylindrical.

Algae: *Nostoc,* usually in glomerules.

Type species: *Pannaria rubiginosa* (Ach.) Bory.

REFERENCES
Gyelnik, K. Pannariaceae. in Rabenhorst Kryptogamen-Flora 9(2):135–272. 1940.
Jørgensen, P. M. 1978. The lichen family Pannariaceae in Europe. Opera Botanica 45:1–124.
Weber, W. A. 1965. The lichen flora of Colorado. 2. Pannariaceae. Univ. Colo. Studies, Biology. 16:1–10.

1. Thallus with coarse soredia to isidioid soredia . . . . 1. *P. conoplea*
1. Thallus lacking soredia.
  2. Thallus lobate at margins, on rocks, disk black . . . . 2. *P. hookeri*
  2. Thallus of small squamules, not effigurate at margins; on soil, disk rust-brown . . 3. *P. pezizoides*

1. Pannaria conoplea (Ach.) Bory

Dict. Class. Hist. Nat. B: 20. 1828. *Parmelia conoplea* Ach., Lich. Univ. 467. 1810. *Pannaria rubiginosa* var. *lanuginosa* Zahlbr., Cat. Lich. Univ. 3:258. 1925. *Pannaria pityrea* Sensu Degel., Bot. Not. 104. 1929.

Plants with well developed hypothallus, blue, greenish blue or green, forming a broad fimbriate margin; thallus small foliose, squamulose, the marginal lobes to 4 mm broad, 7 mm long, crisped and with edges turned up, inwards from the edges the margins of the lobes become sorediate with coarse almost isidioid soredia, the isidia bluish pruinose, the center of the thallus often becoming covered with confluent soredia; the upper side light brown to slaty brown, smooth, dull, usually pruinose toward the tips; underside covered with the dense blue to blue-black hyphae; upper cortex 38–48 μ thick, lower cortex lacking.

Apothecia sessile, to 1.5 mm broad, the margin light brown, more or less entire but sorediate; proper exciple not visible; disk chestnut brown, smooth, epruinose; epithecium red brown, granular; hymenium 75–95 μ, hyaline; paraphyses hyaline, 1.5 μ; asci cylindrico-clavate; spores 8, hyaline, ellipsoid, 15–16×9–10 μ.

Reactions: The hymenium I+ blue, the asci then becoming red; thallus PD+ orange-red.

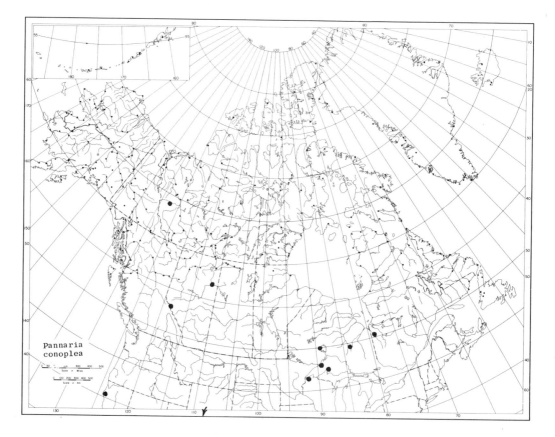

Pannaria conoplea

Contents: atranorin and pannarin.

This species grows on soil and among mosses, in more southerly portions of the range also on tree trunks. It is circumpolar, boreal. In North America it scarcely touches the northern edge of the boreal forest.

2. Pannaria hookeri (Borr.) Nyl.

Mem. Soc. Sci. Nat. Cherbourg 5:109. 1857. *Lichen hookeri* Borr. in Sm. & Sowerb., Engl. Bot. 32: plate 2283. 1811.

Hypothallus thin, black or blackish; thallus small squamulose, the center almost areolate granulose, 250–500 μ thick or even to 2 mm thick, the squamules irregular or fan-shaped, scattered or imbricate, 0.5–0.8 mm long, 0.2–0.5 mm broad, the thallus margin effigurate-lobate with the marginal lobes to 3 mm long and 1.5 mm broad; upper side whitish gray, gray or brownish gray to brown; underside black; upper cortex thick, 57–76 μ, paraplectenchymatous. lower cortex lacking.

Apothecia sessile, to 2 mm broad, the margin crenulate, whitish gray or dark gray, persistent; proper exciple lacking; disk black, dark brown when wet, smooth, epruinose, flat; epithecium dark green to blackish green; hymenium hyaline, 100–125 μ, asci cylindrico-clavate, lacking I+ apical apparatus; spores 8, hyaline, ellipsoid, not rough, 11.5–17 × 7.5–11 μ.

Reactions: hymenium I+ blue.

Contents: Small amounts of pannarin difficult to show with PD+ tests according to Jørgensen (1978).

This species grows on acid rocks. It is circumpolar, arctic and alpine, but in North America appears to be limited to the eastern Arctic. It is a rarely collected species.

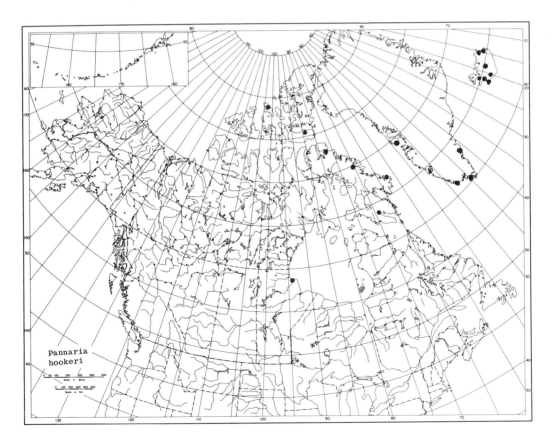

Pannaria
hookeri

3. Pannaria pezizoides (Web.) Trev.

Lichenotheca Veneta no. 98. 1869. *Lichen pezizoides* Web.,
Spicil. Flor. Goettingens 200. 1778. *Pannaria brunnea* (Sw.)
Mass., Ricerch. Auton. Lich. 113. 1852.

Hypothallus usually weakly developed, black to blue-black; thallus of small squamules less than 1 mm broad, toward the center forming a granulose thallus, upper side pale brown, underside white; upper cortex 38–45 μ thick, the medullary layer very compact but not paraplectenchymatous; thallus lacking soredia or isidia.

Apothecia sessile, to 2 mm broad, the margin crenulate and also granular, persistent; proper exciple not visible; disk light reddish brown, smooth, epruinose, slightly convex; epithecium light brown, inspersed; hymenium 95–135 μ, hyaline; asci cylindrico-clavate; spores 8, hyaline, ovate, somewhat pointed, the walls rough with a mamillose episporium, 26.5–30.5 × 9.5–12 μ.

Reactions: hymenium I+ blue, in concentrated I becoming red.

Contents: no lichen substances.

This species grows on soils, seldom on old wood or on mosses; if on rocks it is on soil over the rocks. It is circumpolar, arctic-alpine to boreal, but lacking in the northernmost or high arctic in North America.

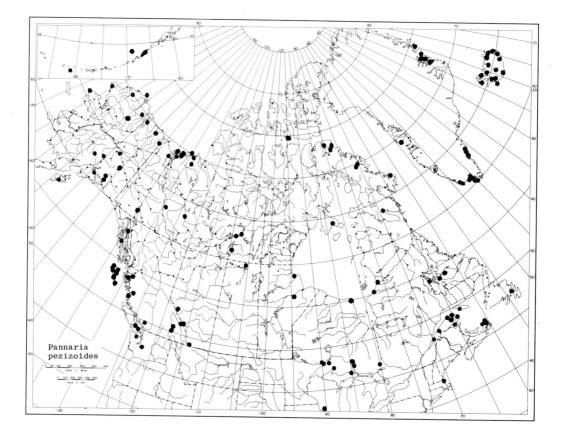

Pannaria
pezizoides

PARMELIA Ach.

Meth. Lich. 153. 1803.

Thallus foliose, dorsiventral, flat or nearly terete, upper cortex paraplectenchymatous or prosoplectenchymatous; lower cortex also prosoplectenchymatous; medulla dense or loose, white.

Apothecia laminal, round, adnate to sessile; margin thalloid; epithecium pale; hypothecium hyaline; asci clavate; spores 8, hyaline, simple, ellipsoid. Pycnidia laminal, innate or partly projecting, often on margin of apothecia; fulcra endobasidial; conidia straight, cylindrical or spindle-shaped.

Algae: *Trebouxia.*

Type species: *Parmelia saxatilis* (L.) Ach.

REFERENCES

Ahti, T. 1969. Notes on brown species of *Parmelia* in North America. Bryologist 72:233–239.

—— 1966. *Parmelia olivacea* and the allied non-isidiate and non-sorediate corticolous lichens in the northern hemisphere. Acta. Bot. Fenn.70:1–68.

Bird, C. D. & A. H. Marsh. 1973. Phytogeography and ecology of the lichen family Parmeliaceae in southwestern Alberta. Can. J. Bot. 51:261–288.

Esslinger, T. L. 1977. A chemosystematic revision of the brown Parmeliae. J. Hattori Bot. Lab. 42:1–211.

Hale, M. E., Jr. 1971. *Parmelia squarrosa,* a new species in section Parmelia. Phytologia 22:29.

Hillmann, J. 1936. Parmeliaceae in Rabenhorsts Kryptogamen-Flora, 2 ed. 9 5(3):1–309.

Krog, H. 1951. Microchemical studies on *Parmelia.* Nytt Mag. Naturvidensk. 88:57–85.

—— 1963. *Parmelia almquistii* and its distribution. Bryologist 66:28–31.

—— 1967. On the identity of *Parmelia substygia* Räs. Blyttia 25:33.

1. Thallus gray or blue-gray.
   2. Thallus sorediate or isidiate.
      3. Thallus sorediate.
         4. The soredia on a raised network on the upper side of the lobes; lobes gray; containing salazinic acid and atranorin . . . . 21. *P. sulcata*
         4. The soredia along the edges of the lobes; lobes yellowish gray; containing usnic acid and protolichesterinic acid as well as salazinic acid and atranorin . . . . 6. *P. fraudans*
      3. Thallus isidiate, the isidia scattered over the surface; containing salazinic acid and atranorin.
         5. Rhizines simple, unbranched; thallus often brownish gray; cortex shiny; lobaric acid sometimes present . . . . 13. *P. saxatilis*
         5. Rhizines squarrosely branched; thallus light mineral gray, not shiny; lobaric acid never present . . . . 17. *P. squarrosa*
   2. Thallus without either soredia or isidia; containing salazinic and lobaric acids and atranorin . . . . 11. *P. omphalodes*
1. Thallus brown.
   6. Thallus with soredia or isidia or both.
      7. Thallus sorediate or with both soredia and isidia.
         8. Medulla C+ rose or red.
            9. Upper surface with distinct pseudocyphellae; on rocks; with soredia only; containing gyrophoric acid . . . . 20. *P. substygia*
            9. Upper surface lacking pseudocyphellae or with only rough spots on the cortex; on woody plants, rarely rocks; with soredia and often also isidia; containing lecanoric acid . . . . 19. *P. subaurifera*
         8. Medulla C−.
            10. Corticolous; lacking substances or with fumarprotocetraric and protocetraric acids; soralia laminal, punctiform; lobes lacking pseudocyphellae . . . . 10. *P. olivaceoides*
            10. Saxicolous; with perlatolic and stenosporic acids.
               11. Lobes shiny, with obscure to distinct submarginal pseudocyphellae; soralia laminal

and submarginal, punctiform to strongly capitate and stipitate, partly arising
from pseudocyphellae . . . . 3. *P. disjuncta*

    11. Lobes dull, lacking pseudocyphellae; soralia mainly terminal on primary lobes
or on tips of secondary more or less erect lobes, arising from dissolution of
cortex. . . . . 16. *P. sorediosa*

7. Thallus isidiate.

    12. Isidia flattened, club-shaped, hollow, no substances present . . . . 5. *P. exasperatula*

    12. Isidia not flattened or club-shaped, cylindrical and solid.

        13. "Isidia" granular, dark clustered isidioid soredia; + or − fumarprotocetraric
acid and protocetraric acid . . . . 10. *P. olivaceoides*

        13. Isidia are true isidia, cylindrical, more or less branched.

            14. Medulla C+ rose or red.

                15. Isidia fine, 0.4×0.02–0.06 mm diameter, seldom branched; soredia
usually also present; containing lecanoric acid . . . . 19. *P. subaurifera*

                15. Isidia coarse, 0.2–1.0×0.05–0.1 mm, often branched; no soredia pres-
ent; containing lecanoric acid + usually also rhodophysin . . . .
. . . . 7. *P. glabratula*

            14. Medulla C−.

                16. Isidia arising as small conical to hemispherical papillae with pseudocy-
phellae at the tip; when mature with anisotomic branching . . . .
. . . . 4. *P. elegantula*

                16. Isidia arising as small spherical to hemispherical papillae lacking pseu-
docyphellae at tips, when mature with isotomic branching . . . .
. . . . 8. *P. infumata*

6. Thallus lacking soredia or isidia.

    17. Medulla P+, yellow, yellow-orange, or red.

        18. Lower surface lacking rhizinae; medulla P+ bright yellow; containing alectoralic and
barbatolic acids . . . . 2. *P. alpicola*

        18. Lower surface with rhizinae; medulla P+ orange, yellow-orange, or red-orange.

            19. On rocks; lobes with distinctive pseudocyphellae; cortex thick, 30–50 μ . . . . . .
. . . . 18. *P. stygia*

            19. On bark; lobes lacking pseudocyphellae or with tiny fleck-like pseudocyphellae;
cortex thin, less than 30 μ.

                20. Thallus large, 2–9 cm; lobes 2–6 mm; pseudocyphellae often numerous,
surface dull to slightly shiny; apothecia in center of thallus; spores 11–16×6–
9μ . . . . 9. *P. olivacea*

                20. Thallus smaller, 1–5 cm; lobes 1–3 mm; pseudocyphellae absent or few; lobes
shiny, smooth; apothecia reaching edge of thallus; spores 9–13×5.5–8.5μ
. . . . 14. *P. septentrionalis*

    17. Medulla P−.

        21. Medulla C+ rose or red.

            22. Upper surface lacking pseudocyphellae; underside without rhizines; containing
olivetoric acid . . . . 1. *P. almquistii*

            22. Upper surface with pseudocyphellae; underside with rhizines; containing gyro-
phoric acid . . . esorediate specimens of . . . . 20. *P. substygia*

        21. Medulla C−.

            23. Thallus small, blackish, subcrustose rhizines absent; medulla KC+ rose or reddish;
containing alectoronic and α-collatolic acids; in eastern Siberia, not yet in North
America . . . . 15. (*P. sibirica* Zahlbr.)

            23. Thallus small foliose; rhizines present; medulla KC−.

                24. Upper surface with distinctive pseudocyphellae . . . . 18. *P. stygia*

                24. Upper surface lacking pseudocyphellae; containing perlatolic and stenosporic
acids. . . . . 12. *P. panniformis*

1. Parmelia almquistii Vain.

Ark. Bot. 8(4):32. 1909. *Allantoparmelia almquistii* (Vain.) Essl.,
Mycotaxon 7:46. 1978.

Thallus appressed to pulvinate or panniform, loosely adnate, lobes to 1 mm broad,
more or less convex, linear or nearly so, rather
torulose, discrete to entangled; upper surface
olive brown to reddish brown or blackening,
more or less smooth or often torulose or contorted, dull or slightly shiny; lacking soredia,
isidia, or pseudocyphellae but with the center of
the thallus becoming covered by tiny finger-like
0.05–0.3 mm broad lobules which may become
imbricate and panniform; lower surface pale
brown or tan, occasionally partly dark brown or
black, irregularly plicate and pitted, dull, lacking rhizinae but attached by holdfasts.

Apothecia not common, sessile, concave to flat, then convex, to 5 mm broad, margin entire to coarsely crenate or with lobules like
those of the thallus; hymenium 45–57 μ; spores
ellipsoid, 10–11.5 × 4.5 − 6.5 μ. Conidia bifusiform to enlarged at one end, 4–5 × 1 μ.

Reactions: K− , HNO$_3$− : medulla K− ,
KC+ and C+ rose-red to red-orange, P− .

Contents: olivetoric acid.

This species grows on rocks. It appears to be
an amphi-Beringian species with its range from north
central Siberia across Arctic North America to Baffin
Island and eastern Greenland. As with some other
arctic species it also ranges south along the coast in
British Columbia.

2. Parmelia alpicola Th. Fr.

Lichenes Arctoi 57. 1860. *Allantoparmelia alpicola* (Th. Fr.)
Essl., Mycotaxon 7:46. 1978.

Thallus appressed to pulvinate, loosely
adnate to somewhat crustose, lobes to 1.5 mm
broad, convex, elongate, rather strongly torulose, continuous to much imbricated; upper surface dark brown to black or grayish black,

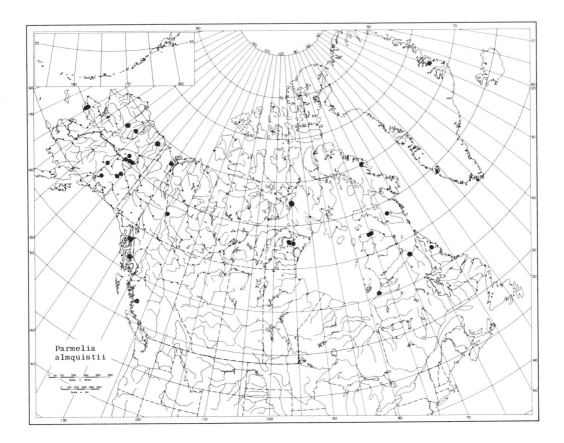

Parmelia
almquistii

sometimes with a reddish brown tinge, smooth toward the periphery, becoming torulose and fissured areolate centrally, dull, or slightly shiny at lobe tips, lacking soredia, isidia or pseudocyphellae; lower surface black, more or less longitudinally plicate, sometimes slightly pitted, lacking rhizines but occasionally with sparse lobate holdfasts.

Apothecia common, sessile, concave to flat, to 7 mm broad, the margin entire to coarsely crenate, occasionally lobulate; hymenium 45–63 μ; spores ellipsoid to subglobose, 7.5–10 × 5–7 μ. Conidia 4.5–6 × 1 μ.

Reactions: cortex K−, HNO$_3$−; medulla P+ strong yellow, K+ pale to dingy yellow, C− or C+ rose, KC+ rose-red.

Contents: alectoralic acid, barbatolic acid with unknowns according to Esslinger (1977), one of them a fatty acid; with protolichesterinic acid in addition according to Krog (1968).

This is a species of rocks and outcrops, mainly acidic. It is an alpine and boreal to arctic circumpolar species ranging south in North America to Labrador, Quebec, Colorado, and Washington state.

3. Parmelia disjuncta Erichs.

Ann. Mycol. 37:78. 1939. *Parmelia sorediata* var. *coralloidea* Lynge, Lichens from Novaya Zemlya 200. 1928. *Parmelia granulosa* Lynge in Lynge & Scholander, Skr. om Svalbard og Ishavet 41:74. 1932. *Parmelia denali* Krog, Norsk Polarinst. Skr. 144:105. 1968. *Melanelia disjuncta* (Erichs.) Essl., Mycotaxon 7:46. 1978.

Thallus appressed to occasionally pulvinate, moderately to loosely adnate, the lobes 0.8–1.5 (–3) mm broad, flat to slightly concave or convex, short to elongate, often flabellate; upper surface dark olive brown, dark brown, or blackish brown, infrequently paler; smooth or pitted toward the tips, centrally rugulose or warted, dull, rarely slightly pruinose, lacking isidia, with small and obscure to distinct pseudocyphellae; sorediate, the soredia laminal and submarginal, punctiform and capitate often short stipitate, sometimes eroded, the soredia dark gray to black, granular to isidioid; lower surface dark brown to black, smooth to rugulose, rhizinate.

Apothecia infrequent, sessile or short stipitate, to 3 mm broad; the margin partly to

completely sorediate, occasionally entire, pseudocyphellate; hymenium 49–66 μ; spores ellipsoid, 9–12.5 × 5–7 μ. Conidia 6–7.5 × 1 μ.

Reactions: cortex K−, HNO$_3$−; medulla K−, C−, KC− or rarely faint rose, P−.

Contents: perlatolic and stenosporic acids plus accessory unknowns.

This species grows on rocks, rarely on old wood or bark. It is circumpolar, arctic, and boreal, ranging in North America into Maine, the Great Lakes, and into Arizona and southern California in the west.

4. Parmelia elegantula (Zahlbr.) Szat.

Magyar Bot. Lapok 28:77. 1930. *Parmelia olivacea* * *Parmelia aspidota* var. *elegantula* Zahlbr., Verh. Vereins. Natur.- und Heilk. Pressburg 8:39. 1894. *Melanelia elegantula* (Zahlbr.) Essl., Mycotaxon 7:47. 1978.

Thallus appressed, moderately adnate, lobes 1–4(–7) mm broad, flat, short and rounded to slightly elongate; upper surface pale olive green to dark olive brown, red-brown or blackening, smooth to slightly pitted or wrinkled, dull, frequently lightly to heavily pruinose, lacking soredia; isidiate with isidia which commence as conical to hemispherical papillae with obscure to distinct pseudocyphellae at the tips and then elongate to cylindrical isidia to 1.5 mm tall and 0.4 mm broad, often branched anisotomically; lower surface pale tan to dark brown or black, smooth to wrinkled or trabeculate, rhizinate.

Apothecia infrequent, sessile to short stipitate, to 3.5 mm broad, the margin entire to becoming papillate or isidiate; hymenium 45–75 μ; spores ellipsoid to broadly ovoid or subglobose, 8–11.5 × 4.5–7 μ. Conidia 7 × 1 μ.

Reactions: cortex K−, HNO$_3$−; medulla K−, C−, KC−, P−.

Contents: none shown.

This species grows on rocks, on soil over rocks, on bark, and over mosses. It is apparently circumpolar arctic, boreal, and montane. In the Americas it is practically bipolar, occurring in western North America from California to Alaska and in South America in Chile and Argentina, but it extends eastward into the Canadian eastern Arctic and Greenland and Iceland. Its lack in the eastern provinces of Canada and New England appears anomalous with such a broad world distribution.

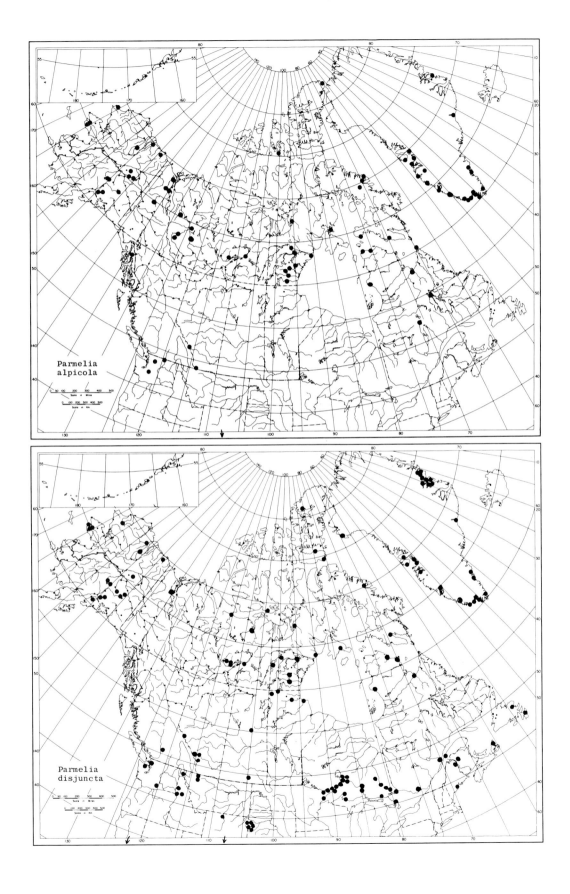

Parmelia
alpicola

Parmelia
disjuncta

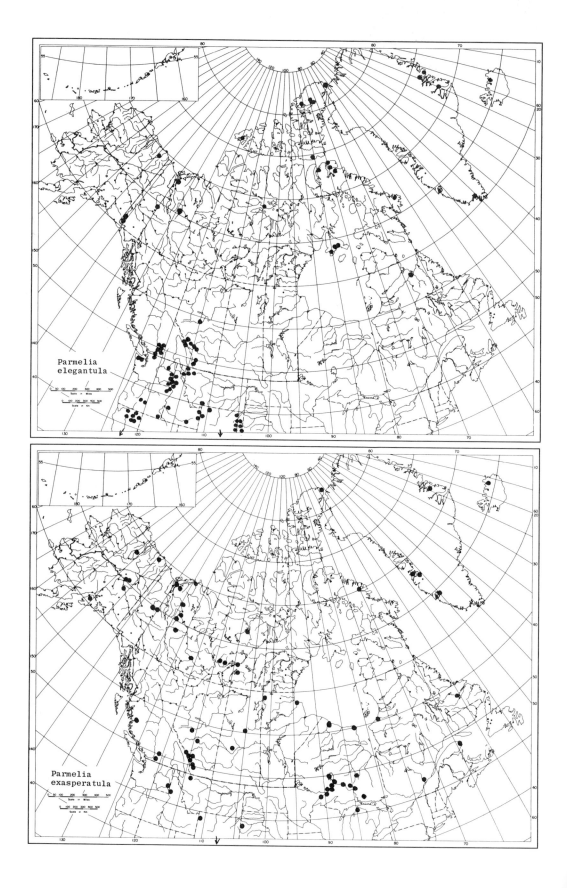

Parmelia
elegantula

Parmelia
exasperatula

Parmelia exasperatula Nyl.

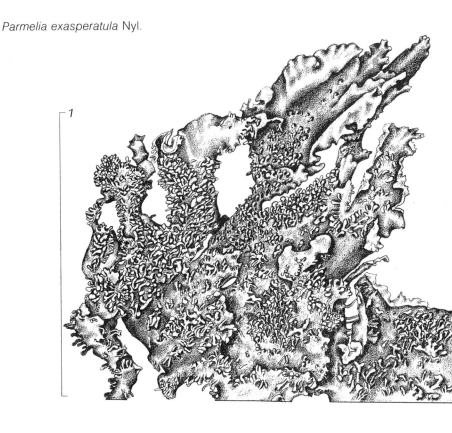

5. Parmelia exasperatula Nyl.

Flora 56:299. 1873. *Melanelia exasperatula* (Nyl.) Essl., Mycotaxon 7:47. 1978.

Thallus appressed or raised at periphery, loosely to moderately adnate, lobes 2–5 mm broad, more or less flat but often with reflexed margins, broadly rounded; upper surface pale olive green to olive brown or red-brown; smooth to slightly wrinkled or sometimes with shallow reticulations, often very oily shiny, centrally dull, occasionally with slight scattered pruina; lacking soredia or pseudocyphellae, with isidia which arise as small hemispherical to spherical papillae and enlarge to form inflated club-shaped or spatulate isidia to 2.0 mm long, simple or occasionally slightly forked at the tip; under surface pale to dark tan, rarely blackening, smooth or wrinkled, rhizinate.

Apothecia uncommon, sessile, to 3 mm broad, the disk flat or concave; margin entire to crenate; hymenium 47–60 μ; spores broadly ellipsoid or subglobose, 8–10.5 × 5.5–8 μ.

Reactions: cortex K−, HNO$_3$−; medulla K−, C−, KC−, P−.

Contents: no substances known.

This species grows mainly on bark and twigs, rarely on rocks. It is circumpolar, arctic, boreal, and montane. In North America it ranges south to Quebec, the Great Lakes area, and in the west to New Mexico and California.

6. Parmelia fraudans Nyl.

Lich. Scand. 100. 1861.

Thallus dichotomously or irregularly branched, slightly channeled above, pale brownish or yellowish gray with coarse yellowish or gray soredia partly scattered in rounded groups on the upper surface but mainly along the margins in abundance; underside black with abundant black, simple to irregularly dichotomously branched rhizinae; upper cortex with pored epicortex, and prosoplectenchymatous;

Parmelia fraudans Nyl.

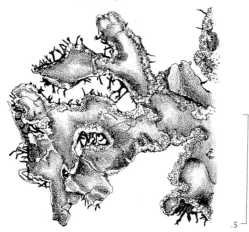

.5

medulla white; upper cortex 15–20 μ thick, lower the same.

Apothecia rare, described by Lynge in Norway; to 14 mm broad; margin thin, isidiate-sorediate; disk brown, rough, dull; hypothecium hyaline; hymenium 65–80 μ, hyaline with upper part yellowish brown; paraphyses conglutinate, branched, septate, the tips capitate-thickened; asci saccate, 55–70 × 10–18 μ; spores uniseriate to biseriate, thick walled, ellipsoid, 13–16 × 7.8–9.2 μ.

Reactions: cortex K+ yellow, medulla K+ yellow becoming red.

Contents: usnic acid, protolichesterinic acid, salazinic acid in medulla and atranorin in the cortex.

This species grows on rocks.

It is circumpolar arctic and boreal to alpine. In North America it reaches southward into the United States only in the west in the Rocky Mountains where it occurs from Colorado to New Mexico.

7. Parmelia glabratula (Lamy) Nyl.
Flora 66:532. 1883. *Parmelia fuliginosa* * *P. glabratula* Lamy, Bull. Soc. Bot. France 30:353. 1883. *Melanelia glabratula* (Lamy) Essl., Mycotaxon 7:47. 1978.

Thallus appressed, loosely adnate, lobes 1–3 (–5) mm broad, short and rounded to more

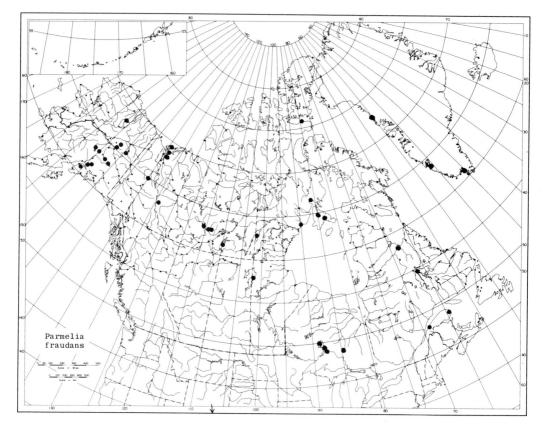

Parmelia
fraudans

or less elongate; upper surface pale olive green, olive brown to reddish brown or blackening, smooth to wrinkled or pitted, lacking soredia or pseudocyphellae, sparsely or densely isidiate, the isidia more or less cylindrical, often branched, attenuate or knobby at the tips, easily broken off at the base to leave white spots, to 1.5 mm tall; lower surface dark brown to black, paler toward the edges, rhizinate.

Apothecia frequent, sessile, to 6 mm broad, flat or concave; margin entire when young, soon papillate to isidiate; hymenium 60–75 μ; spores ellipsoid to subglobose, 10–14 × 5.5–9.5 μ. Conidia 6–7.5 × 1 μ.

Reactions: cortex K−, HNO$_3$−; medulla K− or K+ violet where orange pigmented, C+ red, P−.

Contents: lecanoric acid, and where the medulla is orange pigmented it contains rhodophysin (skyrin).

This species grows on bark on or rocks. It is temperate circumpolar in distribution, in North America ranging south to North Carolina in the east and southern California in the west, northward extending to the Arctic in the Mackenzie River valley.

8. Parmelia infumata Nyl.

Flora 58:359. 1975. *Melanelia infumata* (Nyl.) Essl., Mycotaxon 7:48. 1978.

Thallus appressed, loosely adnate, lobes 1–4 (−6) mm broad, flat to slightly convex, short and rounded or more elongate; upper surface usually dark red-brown to blackish brown, smooth or shallowly pitted, usually dull, frequently slightly pruinose, lacking soredia or pseudocyphellae; sparsely to densely isidiate, the isidia arising as small hemispherical papillae, then enlarging to form cylindrical simple to isotomically branched isidia to 2 mm tall and 0.12 mm thick, sometimes hollow; lower surface dark brown to black, paler toward the periphery, smooth to wrinkled or weakly trabeculate, sometimes appearing channelled below with the downturned lobe edges, rhizinate.

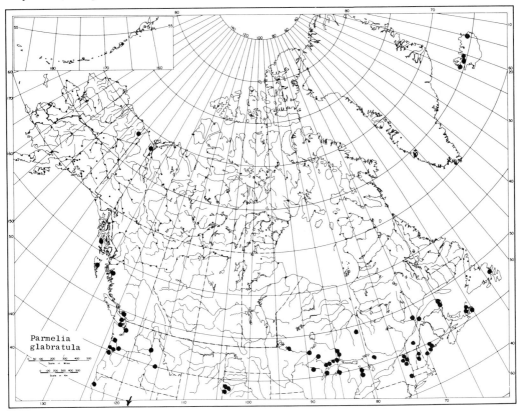

Parmelia
glabratula

Parmelia infumata Nyl.

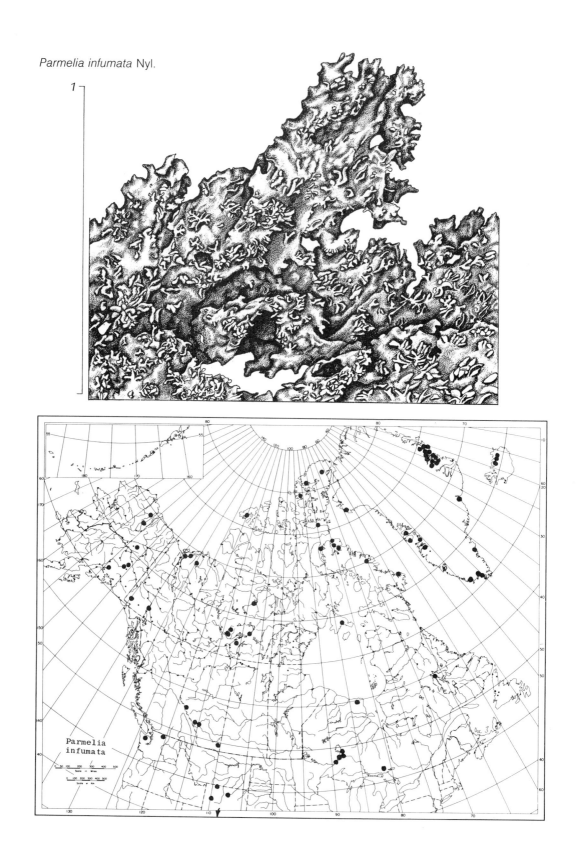

Parmelia
infumata

Apothecia unknown.

Reactions: cortex K−, HNO₃−; medulla K−, C−, KC−, P−.

Contents: none.

This species grows on rocks, usually in the open but with moist surroundings. It is European and North American, in the latter continent it is arctic, boreal, and montane ranging south to Colorado, the Great Lakes region and New England.

9. Parmelia olivacea (L.) Ach.
Meth. Lich. 213. 1803. *Lichen olivaceus* L., Sp. Pl. 1143. 1753. *Melanelia olivacea* (L.) Essl., Mycotaxon 7:48. 1978.

Thallus appressed, moderately adnate, lobes 2–6 (–8) mm broad, flat, short and rounded to elongate, discrete to strongly imbricate; upper surface olive-brown to dark-brown, smooth to strongly rugose and pitted near the periphery, inward more rugose and sometimes tuberculate; dull throughout or sometimes shiny at the periphery, sometimes partly pruinose; lacking soredia or isidia, with numerous pseudocyphellae; lower surface dark brown to black, smooth to rugose, rhizinate.

Apothecia common, sessile to short stipitate, more or less concave, to 9 mm broad; margin entire or crenate to tuberculate, with numerous pseudocyphellae; hymenium 52–72 μ; spores ellipsoid to ovoid 11–16×5–9 μ. Conidia 5.5–7×1 μ, needle-shaped to bifusiform.

Reactions: cortex K−, HNO₃−; medulla K− or rarely K+ dingy orange, KC−, P+ yellow-range to red-orange or rarely P−.

Contents: fumarprotocetraric and protocetraric acids or none.

This species grows on bark and twigs, rarely on rocks. It is circumpolar, boreal, in North America ranging south to New England, the Great Lakes area, and Colorado.

10. Parmelia olivaceoides Krog
Norsk Polarinst. Skr. 144:109. 1968. *Melanelia olivaceoides* (Krog) Essl., Mycotaxon 7:48. 1978.

Thallus appressed or reflexed toward the periphery, loosely adnate, lobes 1–4 (–6) mm broad, flat, short and rounded or somewhat elongate; upper surface olive-brown to darker reddish brown, smooth to weakly pitted or wrinkled, dull or shiny, commonly pruinose, sorediate with the soralia laminal, punctiform to becoming confluent, the soredia granular to strongly isidioid, dark brown to black-brown, if abraded then whitish, occasionally as single strongly isidioid soredia, the soredia arising by disintegration of the upper cortex; lower surface dark brown to black-brown, rhizinate.

Apothecia rare, sessile, 0.6 mm broad; margin with much soredia; hymenium 52 μ; spores ovoid or ellipsoid, 9–11.5×4–5 μ.

Reactions: cortex K−, HNO₃−; medulla K−, C−, KC−, P− or P+ red.

Contents: fumarprotocetraric and protocetraric acids or no substances.

This species grows on bark, rarely on rocks. It is a boreal austral species with a bipolar distribution pattern, reaching the Arctic at the Firth River on the Alaska-Yukon boundary.

11. Parmelia omphalodes (L.) Ach.
Meth. Lich. 204. 1803. *Lichen omphalodes* L., Sp. Pl. 1153. 1753.

Thallus of elongate lobes 5–10 mm broad, irregularly branched, the axils rounded; upper side bluish gray to brownish or blackening over much of the thallus, with slightly raised network of whitish markings, sulcate between, lacking soredia or isidia; underside black with narrow peripheral brown zone, with black simple to dichotomously branched rhizinae; medulla white; epicortex pored, upper cortex prosoplectenchymatous.

Apothecia adnate on the upper surface; margin of same color as thallus, disk to 8 mm, flat or concave, dark brown, dull; epithecium brown; hypothecium hyaline; hymenium 62–74 μ, hyaline; asci saccate-clavate, 54–64×22–24 μ; spores ellipsoid with thickened (1–2 μ) wall, 14–20×8–12 μ.

Reactions: cortex K+ yellow, medulla K+ yellow turning red-orange.

Parmelia olivacea (L.) Ach.

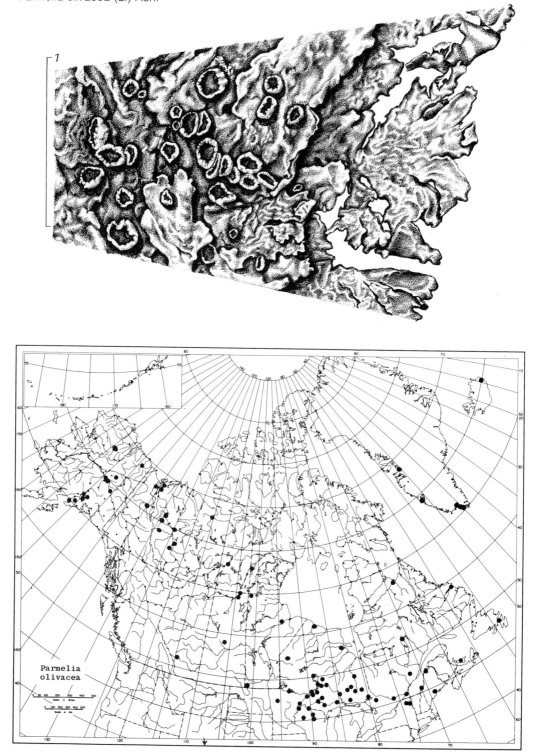

Parmelia
olivacea

Contents: atranorin in the cortex, salazinic and lobaric acids in the medulla.

This species grows on boulders and rock outcrops, usually in full sun. It is circumpolar arctic and boreal, in North America ranging south in the Appalachian Mountains and to Washington and Montana in the west.

12. **Parmelia panniformis** (Nyl.) Vain.

Medd. Soc. Fauna Fl. Fenn. 6:124. 1881. *Parmelia prolixa* f. *panniformis* Nyl., Synops. Meth. Lich. 397. 1860. *Melanelia panniformis* (Nyl.) Essl., Mycotaxon 7:46. 1978.

Thallus appressed to pulvinate, more or less panniform, loosely adnate; lobes to 1.5 mm broad, more or less flat, short and rounded to more elongate, discrete to imbricate; upper surface olive-brown to dark brown or reddish brown, smooth to weakly pitted at the periphery; inward the main lobes hidden by the thickly set imbricate small lobules which are dorsiventral, short to elongate, 0.5 mm broad, simple or branched, pale tan to black below, occasionally rhizinate below, rarely sorediate, the soredia possibly pathogenic; pseudocyphellae occasionally present but obscure at the tips of lobules; lower surface black, smooth to plicate or rugose, dull or slightly shiny, rhizinate.

Apothecia uncommon, sessile, flat to concave, to 3 mm broad; margin entire to papillate to tuberculate or lobulate; hymenium 40–59 μ; spores ellipsoid, 9–11.5 × 4.5–7 μ. Conidia cylindrical 5–7 × 1 μ.

Reactions: cortex K−, HNO$_3$−; medulla K−, C−, KC−, or rarely KC+ dingy rose, P−.

Contents: perlatolic and stenosporic acids.

This species grows on rocks in the open. It is circumpolar in the northern hemisphere and is also in the Andes in South America. In North America it is low arctic and boreal, ranging southward to New York, the Great Lakes region, and in the west to Oregon and Colorado.

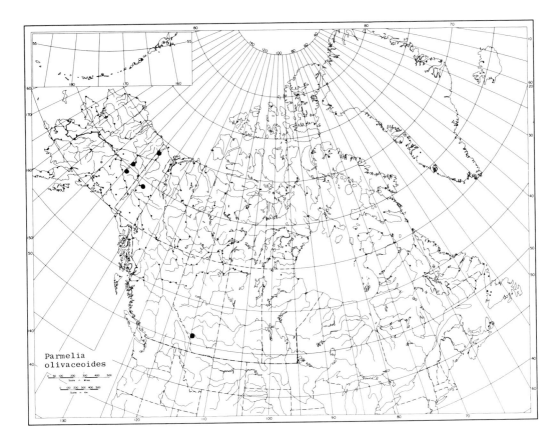

Parmelia olivaceoides

Parmelia omphalodes (L.) Ach.

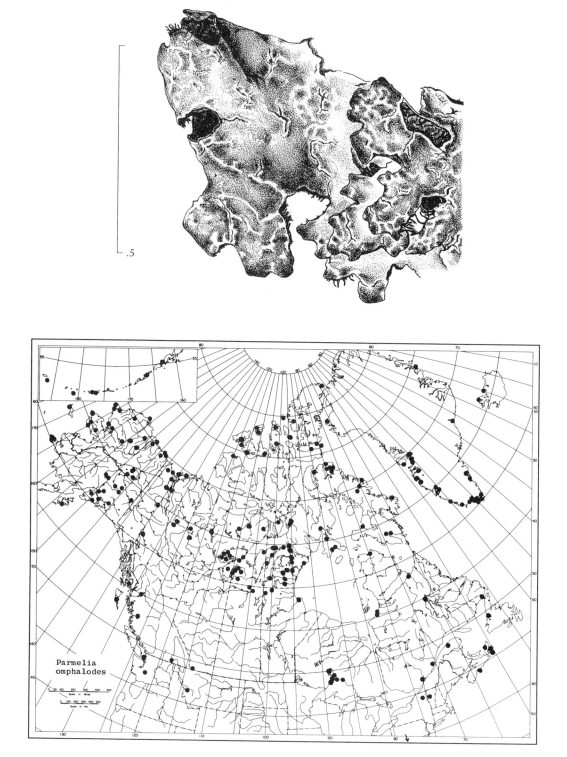

13. **Parmelia saxatilis** (L.) Ach.

Meth. Lich. 204. 1803. *Lichen saxatilis* L., Sp. Pl. 1142. 1753.

Thallus loosely attached, the lobes radiating, elongate, usually to 2–6 mm (–10) broad, the sinuses rounded between the lobes, sometimes the lobes very narrow and much imbricated in heaps, irregularly branching; upper side ashy gray or bluish gray, with slight netting and sulcation, lacking soredia but with short to cylindrical and sometimes branching isidia scattered over the surface, the tips of the isidia browned; underside black, the margins brown and shining, with simple or slightly branched but not squarrose rhizinae abundant and loosely attaching the plant to the substratum; upper cortex prosoplectenchymatous, medulla white.

Apothecia adnate to short pedicellate on the upper surface; margin entire to isidiate, concolorous with the thallus or browning; disk concave, brown or reddish brown, dull, epruinose; epithecium brown, smooth; hymenium 75–100 μ, hyaline; paraphyses conglutinate, septate, slightly branched; asci clavate, $70-72 \times 24-30$ μ; spores ellipsoid, $13-19 \times 8-12$ μ.

Reactions: cortex K+ yellow; medulla K+ yellow turning red-orange, KC−, P+ red.

Contents: atranorin in cortex, salazinic and lobaric acids in the medulla.

This species grows on acid rocks, less often on trees, and over mosses and on old wood. It is circumpolar and arctic to boreal, ranging south in the Appalachian Mountains and in the Rocky Mountains.

14. **Parmelia septentrionalis** (Lynge) Ahti

Acta Bot. Fenn. 70:22. 1966. *Parmelia olivacea* var. *septentrionalis* Lynge, Bergens Mus. Arbok 1912 (10):4. 1912. *Melanelia septentrionalis* (Lynge) Essl., Mycotaxon 7:48, 1978.

Thallus appressed, moderately adnate, lobes to 4 mm broad, flat to slightly convex, short and rounded to elongate, discrete to imbricate; upper surface olive-brown to dark brown, smooth to slightly pitted or rugose, dull

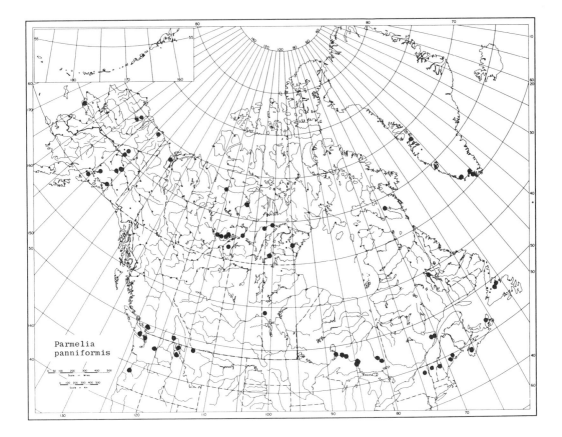

Parmelia
panniformis

Parmelia saxatilis (L.) Ach.

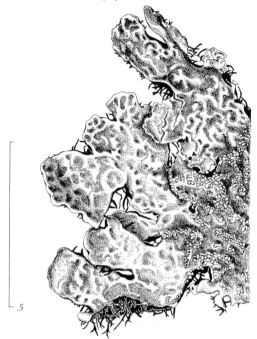

.5

throughout or strongly shiny, especially at the periphery, occasionally slightly pruinose, lacking soredia or isidia, with pseudocyphellae sparse or absent; lower surface dark brown or black, smooth or rugose, rhizinate.

Apothecia common, crowded in the thallus center but extending closer to the margin than in *P. olivacea,* concave when young but becoming convex with age, to 3 (–5) mm broad, margin entire to crenulate, with many pseudocyphellae; hymenium 52–78 μ; spores ellipsoid 9–13 (–15) × 5.5–8.5 μ. Conidia 5–6.5 × 1 μ, needle shaped to bifusiform.

Reactions: cortex K–, HNO_3–; medulla K– or K+ slightly yellow or yellow-range, KC–, P+ yellow-orange to red-orange or P–.

Contents: fumarprotocetraric acid and protocetraric acid or rarely no substances.

This species grows on bark and twigs. It is circumpolar and boreal, in North America growing south to Alberta, the Great Lakes area and New England.

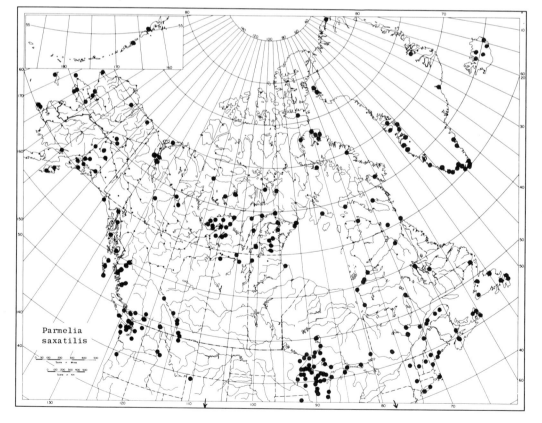

Parmelia
saxatilis

Parmelia septentrionalis (Lynge) Ahti

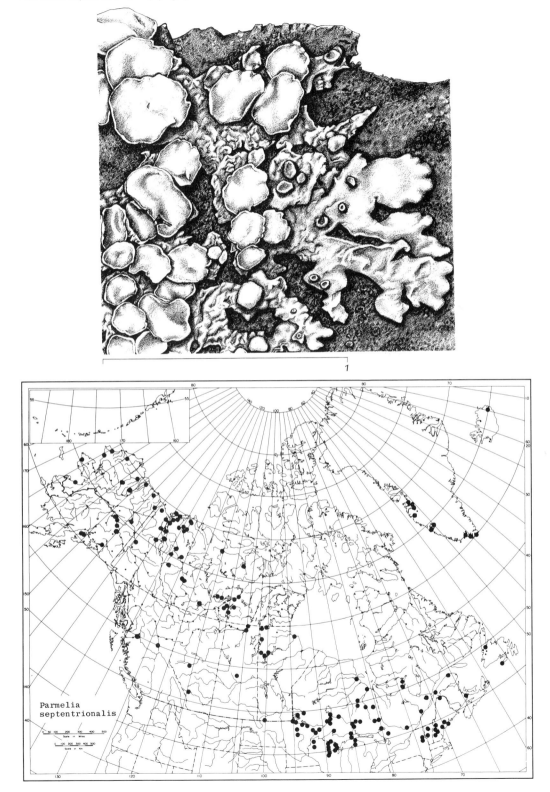

1

Parmelia
septentrionalis

15. **Parmelia sibirica** Zahlbr.

Cat. Lich. Univ. 6:47. 1930 nom. nov. for *Parmelia nigra* Vain. Ark. Bot. 8(4):31. 1909. *Allantoparmelia sibirica* (Zahlbr.) Essl. Mycotaxon 7:46. 1978.

Thallus crustose or subcrustose, appressed and tightly adnate, lobes to 0.4 mm broad, short convex and torulose, discrete; upper surface dark blackish brown to black, smooth at the periphery, soon fissured and areolate; shiny, lacking pseudocyphellae, soredia and isidia; lower surface black, lacking rhizines.

Apothecia sessile, flat, to 1.5 mm broad, margin entire, hymenium 50–58 μ; spores ellipsoid to ovoid, $8–10 \times 4.5–5.5\ \mu$.

Reactions: cortex K−, HNO$_3$−; medulla K−, KC+ rose-red, C−, P−,

Contents: alectoronic and α-collatolic acids plus unidentified substances.

This species grows on rocks. It is known only from eastern Siberia. But as it is known from just across the Bering Straits at Lawrence Bay and Kon-yam Bay it should be expected in Alaska. Hence it is included in this treatment.

16. **Parmelia sorediosa** Almb.

in Krok & Almquist, Svensk Flora för Skolor 2, Kryptogamer, 6th ed. 134. 1947 (nom. nov. for *P. sorediata* (Ach.) Th. Fr. *Melanelia sorediosa* (Almb.) Essl. Mycotaxon 7:47. 1978.

Thallus appressed to loosely adnate, lobes 0.5–1.5 (−3) mm broad, flat, elongate and linear; upper surface olive-brown to gray-brown, occasionally approaching black; smooth or often reticulately ridged and pitted, especially at the tips, usually dull, sometimes shiny, rarely slightly pruinose, lacking isidia or pseudocyphellae, sorediate, the soredia terminal on small lateral branches or sometimes on primary lobes by dissolution of the cortex, the soralia more or less capitate, the soredia brownish gray to darker, granular to isidioid; lower surface very dark brown to black, smooth to slightly rugose, rhizinate.

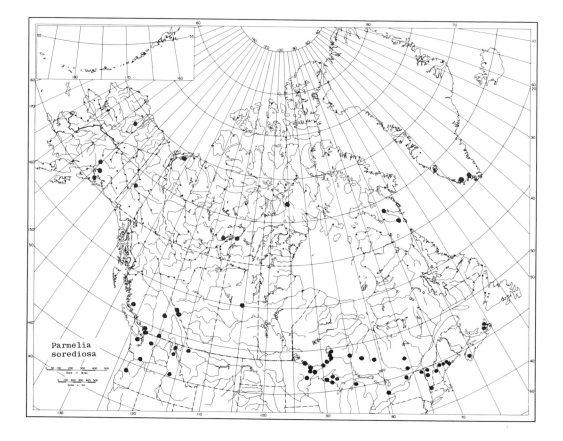

Parmelia
sorediosa

Apothecia rare, sessile or short stipitate, to 2.5 mm, margin entire to slightly crenate; hymenium 49–55 μ; spores ellipsoid to slightly ovoid, 9–11 × 4.5–6 μ.

Reactions: cortex K−, HNO$_3$−; medulla K−, C−, KC− or faint rose, P−.

Contents: perlatolic and stenosporic acids plus accessory unknowns.

This species grows on acid rocks, rarely bark. It is circumpolar arctic and boreal. In North America it occurs southward to New England, the Great Lakes area and, in the west, to Colorado and Oregon.

17. Parmelia squarrosa Hale
Phytologia 22:29. 1971.

Parmelia squarrosa is exceedingly similar to *P. saxatilis* with more elongate lobes, a lighter mineral gray color, slightly greater formation of the isidia toward the margins and richly squarrosely branched rhizines.

Reactions: cortex K+ yellow, medulla K+ yellow turning red orange.

Contents: atranorin in the cortex, salazinic acid in the medulla. The lobaric acid which is present in *P. saxatilis* is lacking in this species.

This species grows on rocks and among mosses. It is known from North America where it appears to be boreal and temperate but reaches the Arctic on Southhampton Island. It is also known from eastern Asia and Nepal.

18. Parmelia stygia (L.) Ach.
Meth. Lich. 203. 1803. *Lichen stygius* L., Sp. Pl. 1143. 1753. *Parmelia stygia* var. *septentrionalis* Lynge, Skr. Norske Vidensk.-Akad. Oslo, Mat.-Naturv. Kl. 6:81. 1938. *Melanelia stygia* (L.) Essl., Mycotaxon 7:47. 1978.

Thallus appressed to subpanniform, lobes to 3 mm broad, flat to slightly concave or convex, sometimes terete, usually elongate to lin-

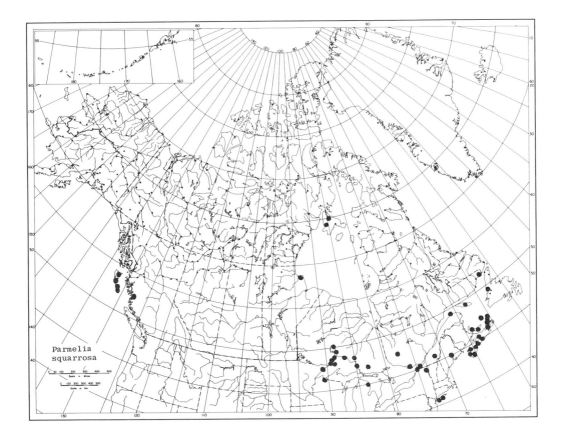

Parmelia squarrosa

Parmelia stygia (L.) Ach.

1

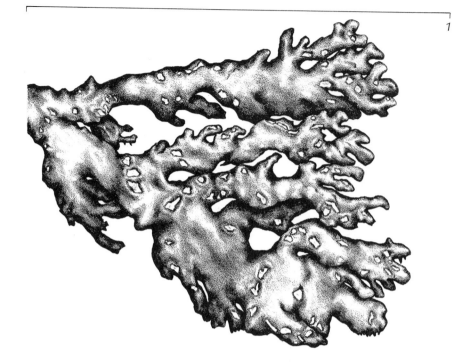

Parmelia stygia (L.) Ach. with apothecia

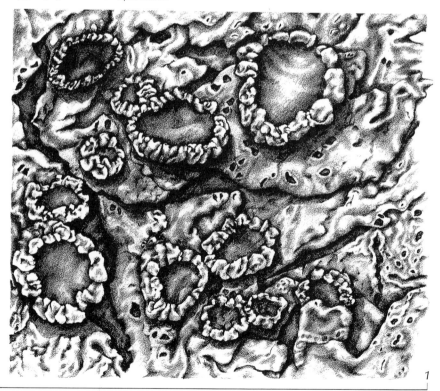

1

Parmelia stygia var. *conturbata* (Arn.) Dalla Torre & Sarnth. with erect, branching lobes.

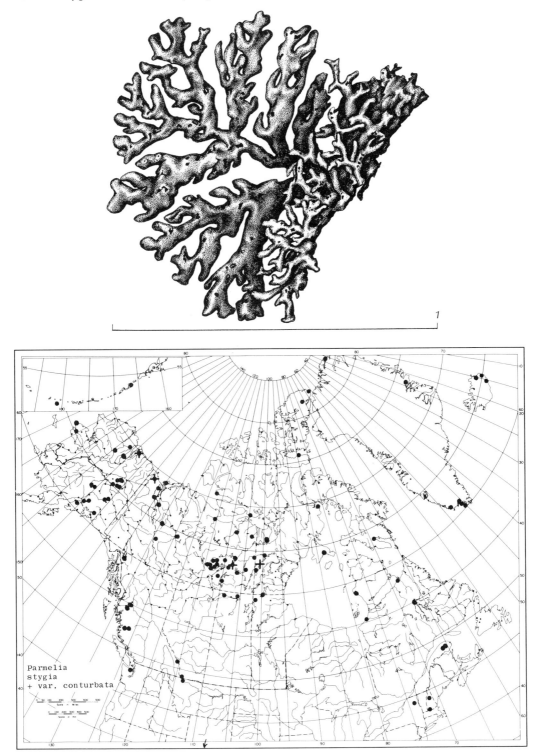

Parmelia
stygia
+ var. conturbata

ear-elongate; upper surface dark olive-brown to black, smooth peripherally, inward becoming coarsely plicate or rugose, usually shiny but sometimes dull, without true isidia but sometimes with small spherical to clavate papillae which usually elongate to form dorsiventral lobules or erect branched upright lobes, usually with abundant laminal whitish or darkening pseudocyphellae; lower surface black, sometimes paler at periphery, smooth to slightly rugose or pitted, rhizinate.

Apothecia common, sessile to short stipitate, concave to flat, to 8 mm broad; margin coarsely warted and crenate, pseudocyphellate; hymenium 42–55 μ; spores ellipsoid to subglobose, 7–10.5 × 4.5–6 μ. Conidia bifusiform, 3.5–5.5 × 1 μ.

Reactions: K−, HNO$_3$−; medulla K− or K+ orange, KC−, P− or P+ orange-red.

Contents: Race 1. fumarprotocetraric and protocetraric acids; Race 2. caperatic acid; Race 3. fumarprotocetraric acid (rarely absent), protocetraric acid, and caperatic acid; Race 4. fumarprotocetraric acid, protocetraric acid and two unknown fatty acids; Race 5. caperatic acid and two fatty acids; Race 6. no substances.

This species grows on rocks. It is a circumpolar arctic, boreal, and montane species which in North America ranges south to Washington, Montana, and North Carolina.

19. Parmelia subaurifera Nyl.
Flora 56:22. 1873. *Melanelia subaurifera* (Nyl.) Essl., Mycotaxon 7:48. 1978.

Thallus appressed or slightly reflexed at the periphery; lobes 1–4 (–6) mm broad, more or less flat, short and rounded; upper surface olive green to olive brown or reddish brown, smooth to rugose, especially centrally, dull or occasionally shiny; sorediate or isidiate or more

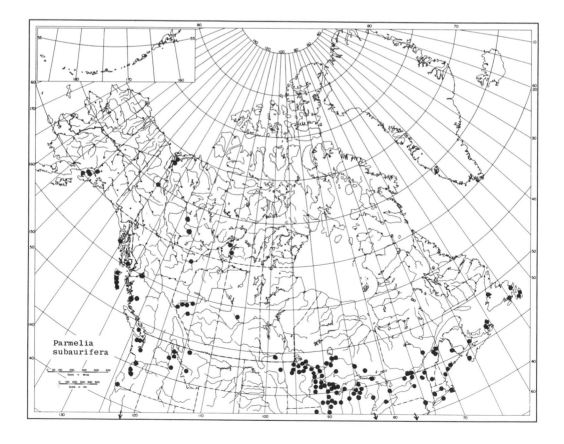

Parmelia
subaurifera

usually both; the soralia laminal, punctiform but spreading and becoming confluent; the soredia white, pale green, yellow, or brown, the soredia granular to becoming isidioid; the isidia arising between the soredia; cylindrical, sometimes branched, sometimes with very small, obscure pseudocyphellae on the lobes; lower surface pale to dark brown or black, paler toward the edges, smooth to rugose, moderately rhizinate.

Apothecia rare, sessile, flat, to 2.5 mm broad; margin entire when young, soon sorediate and isidiate; hymenium 47–65 μ; spores ellipsoid, 10–13 × 5.5–7 μ. Conidia rare, weakly fusiform, 5.5–7 × 1 μ.

Reactions: cortex K−, NNO$_3$−; medulla K−, C+ rose-red to red, KC+ red, P−.

Contents: lecanoric acid.

This species is circumpolar, boreal, and temperate, very common in North America and reaching the Arctic at the mouth of the Mackenzie River.

20. **Parmelia substygia** Räs.

Lichenes Fenniae Exs. 51. 1935. *Parmelia saximontana* R. Anderson & W. A. Weber, Bryologist 65:236. 1962. *Melanelia substygia* (Räs.) Essl., Mycotaxon 7:47. 1978.

Thallus appressed to pulvinate, moderately to loosely adnate; lobes 1–3 (−4) mm broad, flat to slightly convex, short and rounded to flabellate, sometimes elongate; upper surface pale olive brown to yellowish brown or reddish brown or blackening, smooth to weakly pitted, inward tending to be fissured; dull or shiny, with laminal pseudocyphellae, with laminal and marginal soredia, the soralia punctiform to capitate, the soredia pale to dark brown, granular, occasionally lacking soredia; lower surface dark brown to black, paler toward the tips, rhizinate.

Apothecia frequent, sessile to short stipitate, concave to flat, to 6 mm broad; margin entire to slightly crenate, pseudocyphellate and becoming sorediate; hymenium 49–60 μ; spores ellipsoid, 8.5–11 × 4.5–7 μ. Conidia 5–7 × 1 μ, cylindric.

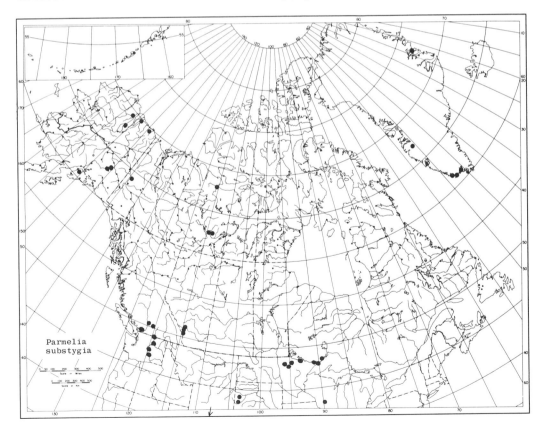

Parmelia substygia

Reactions; cortex K−, HNO$_3$−; medulla K−, C+ rose or reddish, KC+ red, P−.

Contents: gyrophoric acid plus 2 unknowns.

This species grows on acid rocks. It is a circumpolar arctic, boreal, and montane species which in North America ranges southward to southern California, Arizona, and the western Great Lakes area.

21. Parmelia sulcata Tayl.

In Mack., Flora Hibern. 2:145. 1836.

Thallus loosely attached to the substratum, irregularly branching, the lobes elongate, to 6 mm broad, the edges entire or notched; upper side blue-gray, ashy gray, or brownish, with a raised network of fine ridges, sulcate between, pruinose occasionally in patches or toward the tips, the ridges with round or elongate soralia with lighter soredia than the general color; underside black with the edges brown, with numerous simple or irregularly branched black rhizinae; the upper cortex has effigurate maculae and is prosoplectenchymatous.

Apothecia uncommon, adnate or on short pedicels, to 9 mm broad; margin entire or often sorediate; disk flat, dull, brown, epruinose; epithecium brownish; hypothecium yellowish; hymenium 60–74 μ, hyaline; asci clavate, 40–50 × 20–22 μ; spores 8, ovate or ellipsoid, poorly developed, 12–18 × 5–8 μ.

Reactions: cortex K+ yellow, medulla K+ yellow becoming red.

Contents: atranorin in cortex, salazinic acid in medulla.

This species grows on rocks, bark, humus, and soil. It is circumpolar, arctic to temperate, and very common across North America.

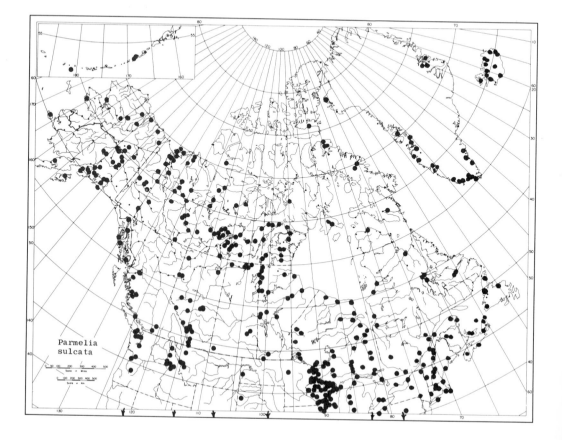

Parmelia sulcata

PARMELIELLA Müll. Arg.

Mem. Soc. Phys. Hist. Nat. Genève 16. 376. 1863.

Thallus foliose, squamulose or crustose-granular, heteromerous; upper side glabrous; upper cortex paraplectenchymatous; lower cortex lacking; rhizinae lacking.

Apothecia with open disk; margin lecideine (lacking algae); proper exciple present, darker or lighter than the disk; epithecium yellowish brown, reddish brown, or brown; hypothecium hyaline or in *P. tryptophylla* red-brown; hymenium 70–122 μ thick, hyaline; paraphyses 2 μ thick, septate, unbranched, the tips not or only slightly thickened; asci clavate; spores 8, single-celled, hyaline, the outer wall often rough.

Algae: *Nostoc,* in glomerules.

Type species: *Parmeliella tryptophylla* (Ach.) Müll. Arg.

REFERENCES

Gyelnik, V. 1940. Pannariaceae in Rabenhorsts Kryptogamen-Flora 9(2):135–272.
Jørgensen, P. M. 1978. The lichen family Pannariaceae in Europe. Opera Botanica 45:1–124.
Weber, W. A. 1965. The lichen flora of Colorado. 2. Pannariaceae. Univ. Colo. Studies No. 16: 1–10.

1. Thallus isidiate, hypothecium red-brown, hymenium I+ blue . . . . 3. *P. tryptophylla*
1. Thallus not isidiate; hypothecium hyaline; hymenium I+ blue turning wine red.
   2. Thallus entirely granular, lacking squamules even at the margins, spores 15–20 × 7–12 μ . . .
   . . . . 1. *P. arctophila*
   2. Thallus squamulose, brown with blue pruina; spores 13–23 × 8–11.5 μ . . . . . . 2. *P. praetermissa*

1. Parmeliella arctophila (Th. Fr.) Malme

Svensk Bot. Tids. 1915:121. 1915. *Pannaria arctophila* Th. Fr., Bot. Not. 8. 1863.

Thallus granulose, the granules 0.12–0.3 mm thick, spherical, not branched but coalescing to form a continuous crust, swelling when wet and drying into wrinkled clumps, gray with brown suffusion; upper cortex 8–12 μ, paraplectenchymatous; algae glomerulate in a dense medulla; underside attached by hyphae; although resembling the thallus of *Pannaria pezizoides* in structure, this lacks the bordering squamules of that species.

Apothecia adnate, to 0.7 mm broad; margin lecideine, narrow, entire, paler than the disk, not gelatinous; disk flesh-colored, flat to convex, smooth, epruinose; epithecium brown, not inspersed; hymenium hyaline, 75–95 μ; with compact plug-like apical apparatus; paraphyses 1.5 μ thick, unbranched; spores ellipsoid, the outer wall smooth both ends round, 15–50 × 7–12 μ.

Reactions: hymenium I+ blue, turning wine-red.

Contents: unknown.

This species grows over mosses on soil. It is arctic and alpine, probably circumpolar. In North America it is exceedingly rare. It was first reported from the Mt. McKinley area in Alaska by W. A. Weber (1965). Jørgensen (1978:111) expressed doubt that this species is in the Pannariaceae, suggesting it may be possibly referred to the Collemataceae or Placynthiaceae on account of the thallus texture.

2. Parmeliella praetermissa (Nyl.) P. James

Lichenologist 3:98. 1965. *Pannaria praetermissa* Nyl. in Chyd. & Furuhj., Not. Sällsk, Fauna Flora Förh. 1:97. 1858. *Lecidea carnosa* var. *lepidiota* Sommerf., Suppl. Flora Lappon. 174. 1826. *Pannaria lepidiota* (Sommerf.) Th. Fr., Nova Acta Soc. Sci. Upsal. ser 3, 3:174. 1861. *Parmeliella lepidiota* (Sommerf.) Vain., Termeszetr. Füzetek 22:308. 1899.

Thallus squamulose, the squamules 1–3 mm long, 0.75–1.5 mm broad, centrally small and thickly beset with erect lobules which are variously called soredia, isidia, or isidioid soredia in the manuals; these lobules are usually white pruinose; they also produce bluish-gray soredia; the thallus is brown to reddish brown when dry, bluish gray when wet; the underside is pale to white, lacking rhizinae; the upper cortex is hyaline to yellowish brown, paraplecten-

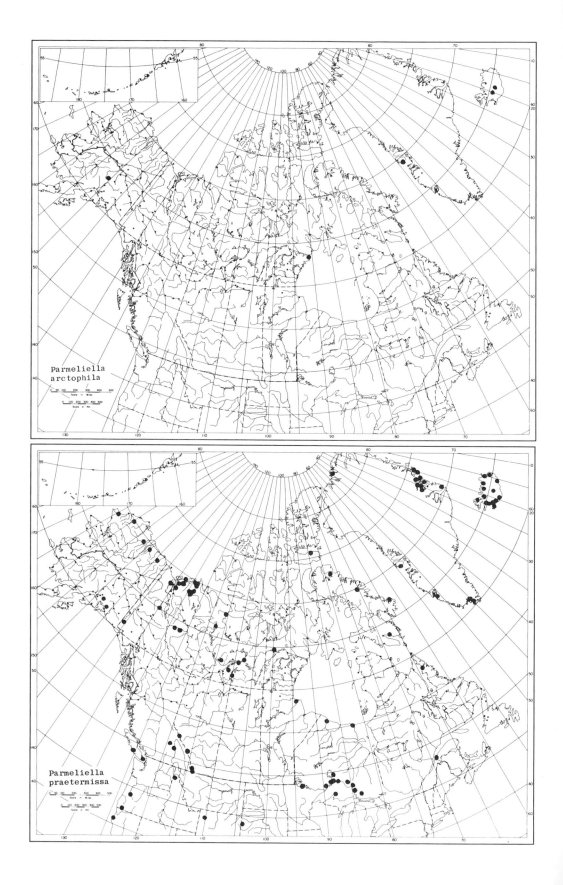

Parmeliella
arctophila

Parmeliella
praetermissa

chymatous, 38–45 μ thick; the algae in glomerules in the algal layer; lower cortex lacking.

Apothecia adnate, lecideine, to 1 mm broad, the proper exciple narrow, entire, much paler than the disk; occasionally the margin thalline and then granular; the disk reddish brown to brownish black, flat to usually convex, epruinose; epithecium brownish yellow, not inspersed; hymenium hyaline, 110–120 μ; paraphyses 2 μ; spores ellipsoid, both ends rounded, containing oil droplets, 13–23 × 8–11.5 μ.

Reactions: hymenium I+ blue-green turning wine-red.

Contents: unknown.

This species grows on soil and among mosses. It is arctic, boreal, and alpine. It ranges into California in the west, and is in the Great Lakes region and in Quebec in the east in North America.

Two other names are associated with *Parmeliella* in North American Arctic and subarctic. *Parmeliella oblongata* Lynge, Lichens of West Greenland 27 (1937) has been examined by A. H.

Magnusson (Ark. Bot. ser 2 Bot. 2:143. 1952); and the type specimen determined as belonging to *Bacidia (Bilimbia) melaena* f. *turfosa. Parmeliella waghornei* (Eckf.) Arn. type specimen at PH has been examined by me. It is in very poor condition with a fraction of an apothecium left by insect ravages and no spores, but the pink color of the apothecium and the resemblance of the thallus to *Lecidea vernalis* as well as the reported size of the spores suggest that species. The name should be rejected.

3. **Parmeliella tryptophylla** (Ach.) Müll. Arg.

Mem. Soc. Phys. Hist. Nat. Genève 16. 376. 1862. *Lecidea thriptophylla (sic)* Ach., K. Vetensk. Akad. nya Handl. 29:272. 1808. *Parmeliella corallinoides* sensu Zahlbr., Annal. Naturhist. Hofmuseum Wien 13:462. 1899. Non *Stereocaulon corallinoides* Hoffm., Deutsch. Flora 129. 1796.

Hypothallus well developed, blue-green to blue-black, extending beyond the thallus 1–1.5 mm; thallus squamulose, the squamules apparent at the thallus margins; the lobes to 1.5 mm long and 0.5 mm broad, the tips rounded;

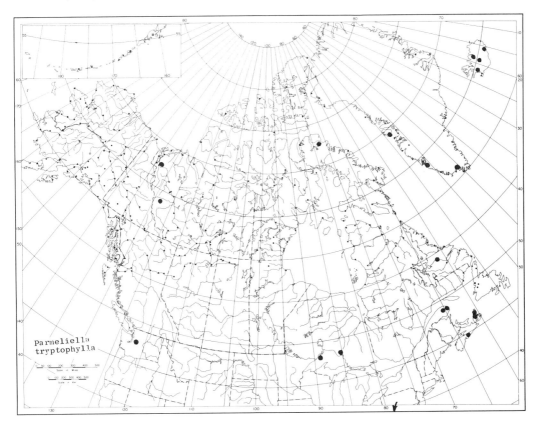

Parmeliella tryptophylla

centrally the ridges and lamina isidiate with cylindrical, branched isidia 0.04–0.12 mm thick, commencing as papillae on the upper surface and eventually covering the center of the thallus; upper surface cherry red to brown, blueblack when wet, dull, epruinose, esorediate; underside black or blue-black and with fine black or blue-black tomentum; upper cortex hyaline or yellowish brown, paraplectenchymatous, 22.5–30.5 μ thick, algae glomerate in the algal layer; lower cortex lacking.

Apothecia adnate, round, with proper exciple; thalloid exciple lacking; margin narrow, entire, dull, of the same color as the disk or paler; disk rust red, smooth, epruinose; epithecium hyaline or reddish; hymenium hyaline, 70–100 μ; asci with compact plug-like apical apparatus; paraphyses hyaline, 2 μ thick; spores ellipsoid, one end more pointed than the other, 13–17 × 5–7.5 μ.

Reactions: hymenium I+ blue, other reactions all −.

Contents: unknown, Dey (Bryologist 81:1–93. 1978) records traces of atranorin.

This species grows on mosses and soil in the arctic but is mainly on trees further south. It is a circumpolar, temperate to subalpine or arctic, species but is apparently seldom collected in North America. In the latter it reaches South Carolina and Louisiana. It is also reported from Australia and New Zealand.

PARMELIOPSIS (Nyl.) Nyl.

Not. Sallsk. F. Fl. Fenn. Forh. 5:121. 1866. *Parmelia* Sect. *Parmeliopsis* Nyl., Lich. Scand. 105. 1861.

Thallus foliose, orbicular, dorsiventral, both sides corticate; the cortices not paraplectenchymatous but of more or less vertically oriented hyphae.

Apothecia on the upper side, discoid, with thalloid margin; asci 8-spored; spores hyaline, 1-celled, ellipsoid or rod-shaped, straight or curved, thin walled; paraphyses simple or slightly branched, generally becoming quite gelatinous, the tips with cross walls.

Pycnidia immersed in the surface or margins of the lobes. Conidia borne terminally from the joints of the sterigmata (exobasidial) instead of laterally (endobasidial) as in *Parmelia;* short and straight or long and curved.

Algae: *Cystococcus.*

Type species: *Parmeliopsis ambigua* (Wulf.) Nyl.

REFERENCES
Culberson, W. L. 1955. Note sur la nomenclature, repartition et phytosociologie du *Parmeliopsis placorodia* (Ach.) Nyl. Rev. Bryol. Lich. 24:334–337.
Hillmann, J. 1934. Zur benennung einiger *Parmeliopsis*-arten. Fedde, Repert. Spec. nov. reg. veg. 33:168–176.
Riddle, L. W. 1917. The genus *Parmeliopsis* of Nylander. Bryol. 20:69–76.

1. Thallus capitate sorediate but not isidiate; underside black, brown toward margins; containing divaricatic acid and either atranorin or usnic acid.
    2. Thallus yellow; K− containing divaricatic and usnic acids . . . . 2. *P. ambigua*
    2. Thallus gray; K+ yellow; containing divaricatic acid plus atranorin. . . . . 3. *P. hyperopta*
1. Thallus isidiate or isidiate and diffuse sorediate; underside pale; containing atranorin and thamnolic acid . . . . 1. *P. aleurites*

1. Parmeliopsis aleurites (Ach.) Nyl. em. Lettau

Hedw. 52: 228. 1912. *Lichen pallescens* Hoffm., Enum. Lich. 66. 1784 (non L., non Necker) *Lichen aleurites* Ach., Prodr. 117. 1798. *Placodium diffusum* Hoffm., Descr. Adumb. Pl. III: 12. 1801. *Parmelia aleurites* (Ach.) Ach., Meth. Lich. 208. 1803.

Parmelia placorodia Nyl., Lich. Scand. 1:106. 1861 (non Ach.). *Parmeliopsis placorodia* (Nyl.) Nyl., Syn. Lich. 2:55. 1863. *Cetraria aleurites* (Ach.) Th. Fr., Lich. Scand. 109. 1871. *Platysma diffusum* (Hoffm.) Leight., Lich. Flor. G. Br., 3 ed. 96. 1879. *Parmeliopsis aleurites* (Ach.) Nyl., Synops. Lich. 2:54. 1863 (*P. hyperopta* (Ach.) erroneously attributed to this name).

Parmeliopsis aleurites (Ach.) Nyl.

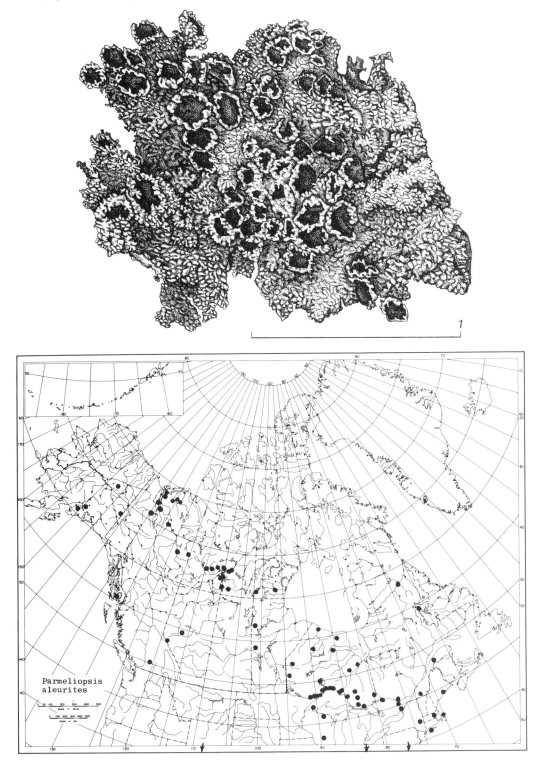

Cetraria diffusa (Hoffm.) Sandst., Abh. naturw. Ver. Bremen 21:206. 1912. *Parmeliopsis pallescens* (Hoffm.) Zahlbr., Cat. Lich. Univ. 6:15. 1929.

Thallus foliose, orbicular, reaching 10 cm. in diameter but usually small when on twigs; lobes usually 1–4 mm. long, up to 2 mm broad, flat, irregularly pinnately branched, the tips rounded; upper side pale ashy gray or whitish, slightly arachnoid under strong magnification, dull or slightly shiny; with short, granulose or cylindrical isidia which are gray or brownish and may merge into paler gray or greenish masses of soredia to the center of the thallus; underside pale, white or brownish, with pale brownish rhizinae or warts distributed to the margin.

Upper cortex 15–20 μ, brownish, composed of more or less vertically oriented hyphae; algal layer continuous, 40–70 μ thick; medulla loose or dense, 80–100 μ, the hyphae 3–3.5 μ thick; lower cortex pale or brownish, 10–18 μ thick.

Apothecia small, 2–6 mm broad, flat, adnate; the disk pale gray or brownish or reddish brown, dull, epruinose; the thalloid exciple persistent, entire or breaking up into soredia or isidia, concolorous with the thallus; epithecium yellowish or brownish with a gray or colorless covering; hymenium hyaline or yellowish, 50–70 μ thick; hypothecium hyaline; paraphyses unbranched or slightly branched; asci 8-spored; spores 1-celled, hyaline, thin-walled, ellpitical, 5.5–9 × 3.5–5 μ. Pycnidia small, black, attached to the margins of the lobes, 0.1 mm in diameter; conidia straight, short, 3.5–4.5 × 0.8 μ.

Reactions: K+ yellow, C−, P+ yellow then red.

Contents: atranorin and thamnolic acid.

This species grows on the twigs and trunks of spruce and pine, rarely on broad leaved trees and shrubs, sometimes on old wood, occasionally on rocks. It grows in Europe and North America. It is distributed from the tree line in Canada in the Great Slave Lake region south, in the east to Pennsylvania and in the west to Colorado. It is reported from Alaska (Nylander 1888; Stair 1947) but this may possibly be based on material of *P. hyperopta*.

2. Parmeliopsis ambigua (Wulf.) Nyl.

Syn. Lich. 2:54. 1863. *Lichen ambiguus* Wulf., in Jacq., Coll. Austriac. 4:239. 1790. *Parmelia ambigua* Ach., Meth. Lich. 207. 1803. *Parmelia diffusa* var. *ochromatica* Wallr., Flor. Crypt. Germ. 3:497. 1831. *Parmelia ambigua* var. *diffusa* Schaer., Enum. Crit. Lich. 47. 1850. *Imbricaria diffusa* (Schaer.) Koerb., Syst. Lich. Germ. 83. 1855. *Parmeliopsis diffusa* (Schaer.) Poetsch, in Poetsch & Schiedermayr, Aufzähl. Erzharzogt. Osterr. Pflanzen 250. 1872. *Parmelia diffusa* (Schaer.) Kummer, Führer 1. d. Flechtenk. 56. 1874.

Thallus foliose, orbicular, usually 1–2 cm., occasionally up to 5 cm., broad, closely adhering to substrate, irregularly dichotomous, the marginal lobes quite discrete; lobes flat, the tips somewhat upturned; upperside yellow, straw-colored, yellowish green or gray green, dull or shining, the dead central parts becoming black, with capitate soralia which are either surficial or pedicillate at the tips of short lobes, the soredia yellowish green; underside black to the center, brown to the margins, shining, with pale or blackening rhizinae to the margins.

Upper cortex smooth, brown or greenish brown, 10–20 μ thick; algal layer continuous, medullary layer loose, sometimes poorly developed, 10–20 μ thick, of 2–4 μ thick hyphae; lower cortex well developed, of hyphae more or less parallel to the surface.

Apothecia adnate on the upper surface or short pedicillate up to 2 mm. broad; the thalloid exciple yellow or brownish, entire or becoming sorediate; the disk chestnut brown or red-brown, dull, at first concave, becoming flat or convex, epruinose; epithecium 8–12 μ thick, pale brown or yellowish, smooth, with a covering of amorphous material; hymenium hyaline, 50–60 μ, quite gelatinous; hypothecium hyaline; paraphyses unclear; asci clavate, 8-spored, spores hyaline, 1-celled, elliptical or one end more pointed, 7–11 × 2.5–3 μ. Pycnidia sunken in the thallus; conidia sickle-shaped, 12–18 × 0.5–1.0 μ.

Reactions: K−, KC+ yellow, P−.

Contents: divaricatic and usnic acids.

This species is very abundant on the twigs and bark and wood of coniferous trees and beyond the forest on birch and willow. It also grows on broad-leaved trees and occasionally on rocks and mosses

Parmeliopsis ambigua (Wulf.) Nyl.

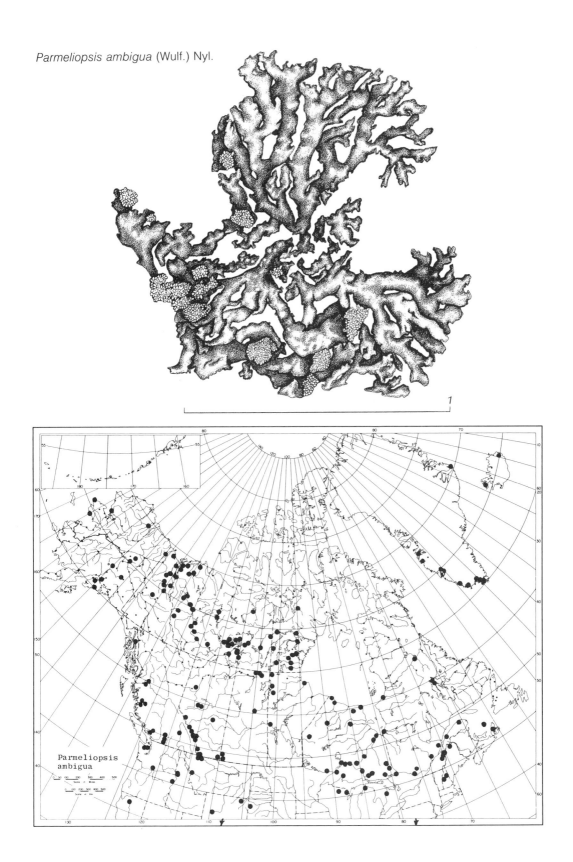

1

Parmeliopsis
ambigua

Parmeliopsis hyperopta (Ach.) Arn.

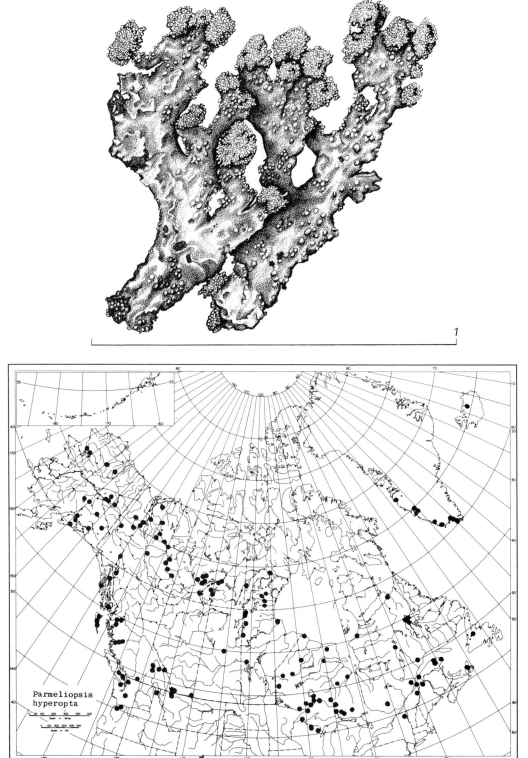

1

Parmeliopsis
hyperopta

over rocks. Its distribution is circumpolar in the boreal forest. In North America it grows from Alaska to Baffin Island, south in the east to S. Carolina, and in the west to Colorado and New Mexico.

3. Parmeliopsis hyperopta (Ach.) Arn.

Verh. Zool. Bot. Gesell. Wien 30:117. 1880. *Parmelia hyperopta* Ach., Syn. Lich. 208. 1814. *Parmelia diffusa* var. *leucochroa* Wallr., Flora Crypt. 3:497. 1831. *Parmelia diffusa* var. *albescens* Rabh., Deutsch. Krypt. Flora 2:56. 1845. *Parmelia ambigua* var. *albescens* (Rabh.) Schaer., Enum. Crit. Lich. 47. 1850. *Imbricaria hyperopta* (Ach.) Körb., Syst. Lich. Germ. 73. 1855. *Parmeliopsis aleurites* Nyl., Syn. Lich. 2:54. 1863 (non *Parmelia aleurites* Ach.).

Thallus foliose, orbicular, up to 6 cm. broad, deeply lobed, irregularly dichotomous; lobes 5–15 mm long, 1–2 mm broad, the tips broadened, rounded or crenulate; upper side whitish or ashy gray, brownish toward the tips, dull or shining, lacking isidia but with capitate soralia on the surface or at the tips of short lobes, the soredia blue-gray or whitish gray; underside black, dull, the margins brown and shining, bearing dark rhizinae over the surface up to the margins.

Upper cortex brownish, 16–20 μ thick; algal layer continuous or interrupted; medulla hyaline, loose, of 4 μ thick hyphae; lower cortex brown to black, 16–20 μ thick.

Apothecia surficial or marginal, slightly pedicillate, up to 2 mm broad, thalloid margin whitish or brown, crenulate, lacking soredia or becoming sorediate; the disk pale or dark brown, shining, flat or becoming convex, epruinose; epithecium brown or yellow brown, 14–30 μ thick; hymenium hyaline or yellowish, very gelatinous, 52–62 μ thick; hypothecium hyaline; paraphyses scarcely distinguishable; asci clavate, 8-spored; spores hyaline, 1-celled, short elliptical or rod-shaped with the ends rounded, 8.5–12 × 2–4 μ. Pycnidia immersed in the thallus; conidia sickle-shaped, seldom straight, 16–22 × 0.5–0.8 μ.

Reactions: cortex K+ yellow, medulla K−, C−, P−.

Contents: atranorin and divaricatic acid.

This species is common on both coniferous and broad-leaved trees, on old stumps, burned logs, and old wood, and on acid rocks. It is a circumpolar northern forest species which grows north of the tree line on birches, *Ledum,* willows, etc., and on rocks. Its range is from Alaska to Ellesmere Island and Greenland, south to New Hampshire and Wisconsin, and to Colorado and Oregon in the west.

PELTIGERA Willd. em. L. Rabenh.

Deutsch. Crypt. Flor. II. 1845. *Peltigera* Willd., Flor. Berolinens. Prod. 347. 1787.

Thallus foliose, more or less lobed, loosely attached to the substrate by rhizinae; differentiated into a plectenchymatous upper cortex, a continuous algal layer, and a loosely interwoven layer of medullary hyphae to which the rhizinae are attached, lower cortex lacking; the lower surface usually more or less veined; upper surface smooth, scabrid or with superficial hyphae, especially at the margins.

Apothecia round, marginal or on marginal lobules, horizontal or vertical, often with the sides reflexed; exciple lacking, a cortical rim around the apothecium which is sunken in the thallus, the margins usually becoming crenate; hypothecium hyaline to brown; hymenium hyaline to brownish; upper part of the paraphyses red-brown, paraphyses simple; asci cylindrico-clavate, the upper part considerably gelatinized; spores 8, fusiform or acicular, sometimes slightly curved, 3–9 septate.

Algae: *Nostoc* or *Dactylococcus.*

Type species: *Peltigera canina* (L.)Willd.

REFERENCES

Ahti, T. & O. Vitikainen. 1977. Notes on the lichens of Newfoundland. 5. Peltigeraceae. Ann. Bot. Fenn. 14:89–94.

Kristinsson, H. 1968. *Peltigera occidentalis* in Iceland. Bryologist 71:38–40.

Kurokawa, S., Y. Jinzenji, S. Shibata, & H. Chiang. 1966. Chemistry of Japanese *Peltigera* with some taxonomic notes. Bull. Nat. Sci. Museum, Tokyo 9:101–114.

Lindahl, P. 1953. The taxonomy and ecology of some *Peltigera* species, *P. canina* (L.) Willd., *P. rufescens* (Weis.) Humb., *P. praetextata* (Flk.) Vain. Svensk Bot. Tids. 47:94–106.

Lindahl, P. 1962. Taxonomical aspects of some *Peltigera* species. *P. scutata* (Dicks.) Duby, *P. scabrosa* Th. Fr. and *P. pulverulenta* (Tayl.) Nyl. Svensk Bot. Tids. 56:471–476.

Ohlsson, K. E. 1973. New and interesting lichens of British Columbia. Bryologist 76:366–387.

Thomson, J. W. 1950. The species of *Peltigera* of North America north of Mexico. Amer. Midl. Nat. 44:1–68.

—— 1948. Experiments upon the regeneration of certain species of *Peltigera* and their relationship to the taxonomy of this genus. Bull. Torrey Bot. Club. 75:486–491.

—— 1955. *Peltigera pulverulenta* (Tayl.) Nyl. takes precedence over *Peltigera scabrosa* Th. Fr. and becomes of considerable phytogeographic interest. Bryologist 58:45–49.

1. Algae of the thallus green, warty cephalodia containing blue-green algae on the upper or lower surface of the thallus.
  2. Upper side of the thallus with warty cephalodia and often marginally erect tomentum; under surface with scattered fasciculate rhizinae, thallus and lobes large, up to 6 cm broad; apothecia vertical.
    3. Under surface with nearly continuous nap of tomentum, no definite veins . . . . 1. *P. aphthosa* var. *aphthosa*
    3. Under surface with well-developed vein structure . . . . 1. *P. aphthosa* var. *leucophlebia*
  2. Upper side of the thallus lacking cephalodia and etomentose; underside with small black cephalodia on the veins; thallus small, to 2 cm across; apothecia horizontal . . . . 11. *P. venosa*
1. Algae of thallus blue-green, thallus lacking cephalodia.
  4. Upper side of thallus smooth or verruculose scabrid, lacking tomentum.
    5. Thallus margins sorediate . . . . 3. *P. collina*
    5. Thallus margins esorediate.
      6. Upper side verruculose scabrid . . . . 10. *P. scabrosa*
      6. Upper side smooth, lacking tomentum.
        7. Upper side sorediate in spots . . . . 2. *P. canina* f. *sorediata*
        7. Upper side esorediate.
          8. Apothecia horizontal; spores fusiform 4-celled, $25–45 \times 3.5–6$ μ . . 4. *P. horizontalis*
          8. Apothecia vertical; spores acicular, 4–10 celled, $48–71 \times 3–4.5$ μ . 9. *P. polydactyla*
  4. Thallus with tomentum above, at least toward the margins.
    9. Upper surface at tips with a combination of tomentum plus scabrid surface with some erect hairs in the scabrous areas; veins below of the flattened polydactyliform type. . . . . 8. *P. occidentalis*
    9. Upper surface with either erect or flattened tomentum but not both, not scabrid; veins below lacking or caninaeform (very narrow and well defined) but not broad, flat polydactyliform.
      10. Tomentum at margins erect, veins lacking below . . . . 6. *P. malacea*
      10. Tomentum more or less appressed to upper surface, thallus with veins below.
        11. Upper side with scattered peltate isidia . . . . 5. *P. lepidophora*
        11. Upper side lacking isidia or if present they are marginally attached not peltate, and associated with cracks or the margins of the thallus.
          12. Rhizinae simple or fibrillose; medulla 300–500 μ thick. . . . . 2. *P. canina*
          12. Rhizinae and veins penicillate; medulla 70–110 μ thick. . . . . 7. *P. membranacea*

1. Peltigera aphthosa (L.) Willd.
Flora Berolinensis 347. 1787. *Lichen aphthosus* L., Sp. Pl. 1148. 1753.

Thallus large, sometimes over a meter across, the lobes broad, up to 10 cm. long and 6 cm. broad, although usually smaller, edges ascending. Upper surface smooth in the center with an erect tomentum of hairs 6–8 μ in diameter and up to 65 μ long at the very margin, warty cephalodia containing nostocoid algae 6–9 μ in diameter scattered over the upper surface; pale to apple green when moist, leaden

Peltigera aphthosa (L.) Willd. var. *aphthosa*

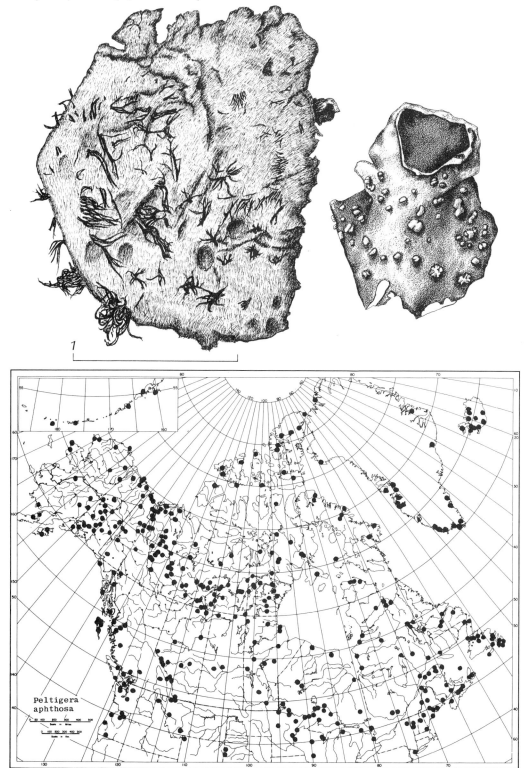

1

Peltigera aphthosa var. *leucophlebia* Nyl. underside

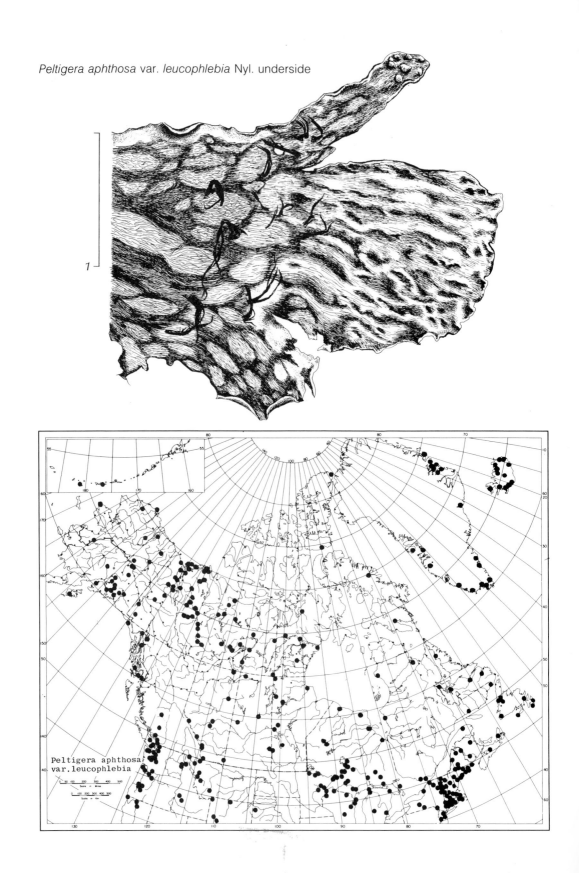

Peltigera aphthosa
var.leucophlebia

green when dry; under surface pale at the margins, blackening toward the center of the thallus, either veinless and covered with a fine even nap of tomentum or with white or dark veins, rhizinae scattered, fasciculate.

Algal layer continuous, 25–35 μ thick; upper part of the cortex pseudoparenchymatous, hyaline to pale brown, 25–55 μ thick; medulla of very loosely interwoven hyphae 10 μ in diameter and with a lumen of 6 μ, hyphae of the veins in var. *leucophlebia* darker and more thick walled, 12 μ in diameter with a lumen of 4–5 μ, oriented in the direction of the vein; medulla 460–1350 μ thick; rhizoids composed of bundles of hyphae continuous with the medullary hyphae but becoming dark brown, 7μ thick and very thin walled, cellular and anastomosing; hymenium 880–120 μ thick; hypothecium about 100 μ thick. Algae of thallus 2.5–6 μ diameter; algae of cephalodia 6–9 μ diameter.

Apothecia large, up to 16 mm. across, on extended lobules, the sides reflexed, the margins becoming crenate, the reverse with a cortex at first, leaving scaly patches of cortex on the back of the larger apothecia; disk reddish brown to blackish brown; hypothecium brown; hymenium hyaline to pale brown; paraphyses simple, thickened and darkened at the tips to a red-brown; asci cylindrico-clavate 80×21 μ, 8 spored; spores acicular, 3–9 septate, $46–63 \times 5$ μ (Tuckerman reports $48–70 \times 4–7$; Fink reports $45–75 \times 4–7$).

Reactions: K−, C−, KC−, P−, I stains the young asci bright blue, no other part being affected.

Contents: zeorin, tenuiorin, phlebin A and B and several unknowns according to Kurokawa et al (1966) and tenuiorin, methyl gyrophorate, methyl lecanorate, methyl evernate, methyl orsellinate, 4–0 methylgryophorate (trace), gyrophoric acid (trace) and evernic acid (trace) according to Maass (1975).

This species grows on earth, rotting logs and stumps, on humus, and on rocks, usually in moist places. It is circumpolar, arctic, boreal, and temperate. It ranges in North America south to California, Arizona, New Mexico in the west, and to North Carolina in the east.

The typical material of this species lacks veins and is covered with an even tomentum. On the underside it is bluish or brownish black at the center and pale white or brownish at the margins. When dealing with a large sequence of specimens one can arrange them from this extreme along a gradient with slight to increasing amounts of more cottony and white interspaces exposing the medullary hyphae and narrowing of the veins which appear until at the other extreme one has specimens which have very definite vein patterns. By some lichenologists the specimens with the very strong vein patterns on the underside are called *Peltigera leucophlebia* (Nyl.) Gyeln. Such specimens seem to be more common southward in the range of the species and the var. *aphthosa* or typical extreme is more common in the northern part of the range. There is also a slight difference in habitat preferences with the typical specimens coming from moister habitats than the *leucophlebia* type of specimens. Therefore I have retained the opinion that these can at most be treated as varieties with the extremes characterized by the veined under sides being called var. *leucophlebia* Nyl. In my paper on the genus in North America (1950) I had used the name var. *variolosa* (Mass.) Thoms. being misled by the description of var. *leucophlebia* by Nylander supposedly having an entirely white underside. However the type specimen of Nylander at H does have the black or dark veins of this extreme and the two names belong to the same taxonomic entity, with Nylander's taking precedence.

2. Peltigera canina (L.) Willd.

Flora Berolinens. Prodrom. 347. 1787. *Lichen caninus* L., Sp. Pl. 1149. 1753.

Thallus very variable; size up to 30 cm. or more across; lobes up to 8 cm. across; upper surface slate gray, greenish brown, or brown above with an arachnoid or dense tomentum toward the margins, the center dull or shining; in one form with coarse gray soredia in rounded spots on the upper surface; another form with the flat platelets or isidia of regeneration following injury; under surface with a strong network of raised white to brown veins, with white to brown, simple, fasciculate, or fibrillose rhizinae which sometimes form a confluent spongy mat below.

Algal layer continuous, 40–80 μ thick; upper part of the cortex plectenchymatous, hya-

Peltigera canina (L.) Willd. f. *sorediata* Schaer.

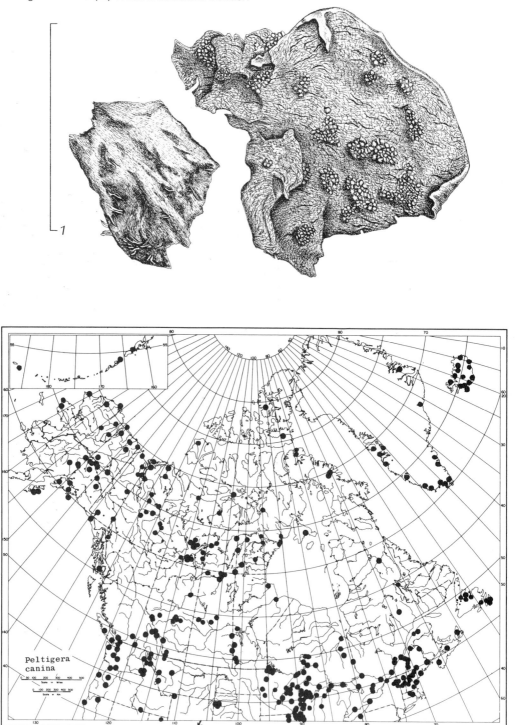

Peltigera canina (L.) Willd. var. *rufescens* (Weis.) Mudd

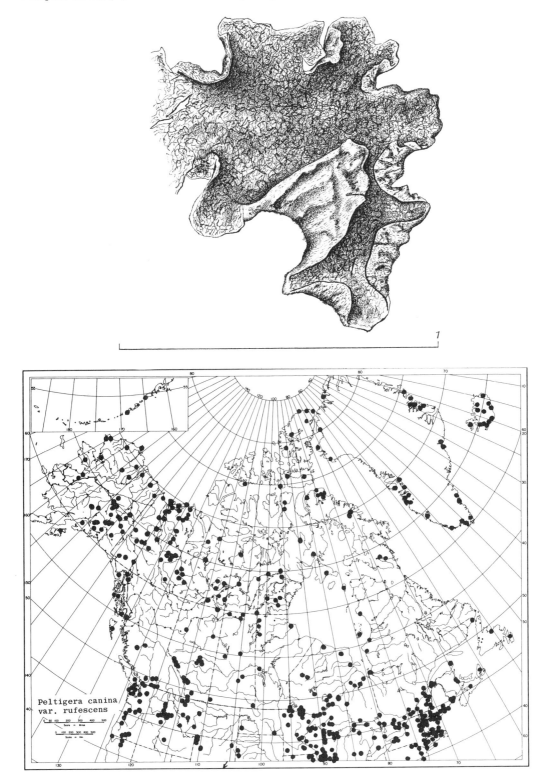

1

Peltigera canina
var. rufescens

line to pale brown, 40–70 μ thick, with tri-chomatic hyphae; medulla of rather compact hyphae, more or less arranged parallel to the surfaces, 5–8 μ in diameter with a lumen of 3–5 μ, medulla 300–500 μ thick; hymenium 70–110 μ thick; hypothecium 35–60 μ thick. Algae 4–8 μ in diameter.

Apothecia vertical on extended lobules 6–10 mm. across, the sides usually reflexed, the disk light brown to reddish brown or chestnut brown or black, the margin crenate; hypothecium brown; hymenium hyaline to brown; paraphyses thickened and red-brown at the tips; asci cylindrico-clavate 90 × 10 μ, 8 spored; spores acicular, hyaline to yellowish, 23–67 × 3–6.5 μ (Tuckerman reports 44–70 × 3–6, Fink reports 38–72 × 3–5).

Reactions: K−, C−, KC−, P−, I stains the asci blue, other parts are negative.

Contents: tenuiorin or no substances.

This species grows on soil, rocks, rotten logs, and bases of trees. It is a cosmopolitan species, ranging over much of the world and in North America over most of the continent.

It is a very variable species. Juvenile thalli are often characerized by the presence of soredia in definite rounded soralia over the upper side of the thallus and are often identified as f. *sorediata* Schaer. or even as a species (*P. erumpens* Vain.). Specimens which form small flat isidia at the margins of the lobes and around cracks in the thallus as regeneration phenomena have been classified as f. *innovans* (Korb.) Thoms. by me but are recognized as a separate species, *Peltigera praetextata* (Flörke) Vain. by many authors. The experimental evidences for these points of view are discussed in the papers by Thomson and Lindahl cited here. Also controversial is the relationship between *Peltigera canina* and what I recognize as its var. *rufescens* (Weis.) Mudd but which many lichenologists recognize as *Peltigera rufescens* (Weis.) Humb. The latter has smaller, narrower lobes which are more involute on the margins and with the cells of the upper cortex being more thick-walled, 2–4 μ rather than 1–2 μ. It is also more brittle. But these characteristics could just as easily be vegetative responses to environmental stress and the "*rufescens*" types of thalli occupy the drier more highly insolated habitats than the "*canina*" types which occur in moist, more shaded habitats. Further experimental observations are needed on these problems. The maps in this volume retain these as separate entities for this reason.

3. Peltigera collina (Ach.) Schrad.

J. für die Bot. 1:78. 1801. *Lichen collinus* Ach., Lichenogr. Suec. Prodr. 162. 1798. *Peltigera scutata* (Dicks.) Duby, Bot. Gallic. 2:599. 1830.

Thallus medium-sized, several cm across, the lobes about a cm. broad but up to 2 cm. broad, curled and crisped at the edges; upper side leaden-gray or more commonly brown to chestnut brown, usually slightly scabrid or pebbly but often smooth, often pruinose-scabrid at the tips of the lobes; the margins usually with large round gray soredia in masses, but these may occasionally be lacking; under surface malacea-like, lacking veins, or polydactyla-like, the veins tan to chestnut brown or black, the interspaces yellowish-white and cottony; rhizinae short, fasciculate or fibrillose, sometimes forming a dense mass toward the center of the thallus.

Algal layer continuous, 25–45 μ thick; upper part of the cortex plectenchymatous; lower part of this hyaline, the upper part very thick and brown, 40–45 μ thick; medulla of loosely interwoven hyphae 9–10 μ in diameter with a lumen of 5–6 μ, medulla 75–300 μ thick; hymenium 150 μ thick in hyaline part, 60 μ thick in brown part; hypothecium 50–60 μ thick. Algae 4–6 μ in diameter.

Apothecia small, up to 3mm. across, borne at the tips of the lobes, horizontal, or erect and with the sides reflexed; disk chestnut-brown to black, the margins crenate, with a pseudo-parenchymatous cortex on the reverse; hypothecium brown; hymenium hyaline below, with a brown pseudoparenchymatous layer above formed by the tips of the paraphyses; paraphyses simple, thickened, darkened and cellular at the tips; asci cylindrico-clavate 80 × 9 μ, 8 spored; spores often degenerate, acicular, 3–7 septate, 27–64 × 2.5–5.5 μ (Tuckerman reports 50–70 × 3–4, Fink reports 48–68 × 3–4.5).

Reactions: K−, C−, KC−, P−, I stains the young asci blue; other parts are negative in reaction.

Peltigera collina (Ach.) Schrad.

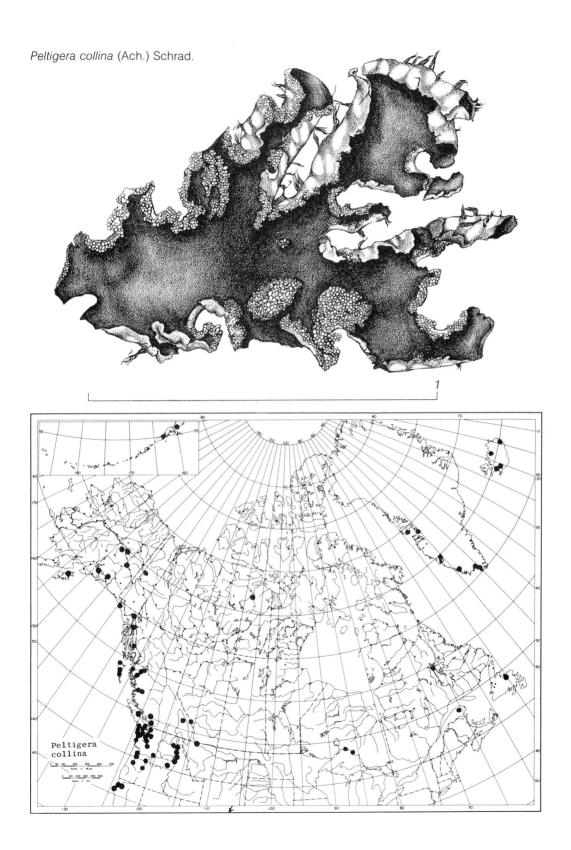

Peltigera
collina

Contents: tenuiorin, zeorin and 4 unidentified substances. Kurokawa, et al. (1966).

This species grows over mosses, on bark and on rocks and rotting logs. Although circumpolar this species does not have a continuous range. It appears to be strongly oceanic. It is common in the western parts of North America, very rare in the east, and on the north shore of Lake Superior. It also occurs in South America.

4. Peltigera horizontalis (Huds.) Baumg.

Flora Lipsiens. 562. 1790. *Lichen horizontalis* Huds., Flora Anglica 543. 1762.

Thallus large, the rounded lobes up to about 3 cm. broad and several cm. long, the margins usually ascending; upper surface smooth, shining, bluish gray or greenish gray when moist, lead gray or brownish gray or brown when dry, lacking tomentum or cylindrical isidia but sometimes producing tiny flat squamules or "isidia" when regenerating following injury; under surface with broad flat veins forming an almost confluent tomentum, the veins brown to black, the interspaces white; rhizinae black, sparse, fasciculate.

Algal layer continuous, 60–80 μ thick; upper part of the cortex plectenchymatous, 40–60 μ thick, hyaline below, grading into pale brown at top; medulla of very loosely interwoven hyphae 12 μ in diameter with a lumen of 9 μ, in the veins the hyphae are dark brown with a diameter of 9 μ, and a lumen of 7 μ, medulla 300–600 μ thick. Hymenium about 80 μ thick; hypothecium about 50 μ thick. Algae 5–9 μ in diameter.

Apothecia horizontal, at the edge of the lobes or on short lobules usually 3–6 mm. across but up to 12 mm., the margin crenate; the disk chestnut brown rarely pruinose, the reverse with scattered patches of cortex; hypothecium pale brown; hymenium hyaline below, brown above; paraphyses simple, thickened and darkened at the tips; asci cylindrico-clavate, 60×10 μ, 8 spored; spores fusiforn. 3 septate, 24–45×3.5–6 μ (Tuckerman reports 33–46×6–8 μ, Fink reports 30–48×5–7.5 μ).

Reactions: K−, C−, KC−, P−, I turns the young asci blue; other parts are negative.

Contents: tenuiorin, zeorin, dolichorrhizin, scabrosin A and B, plus 4 unidentified substances.

This species grows on soil, among mosses, on fallen logs and over rocks in moist woods. It is circumpolar, south boreal to temperate, reaching barely to the arctic in Iceland, Alaska, and the Mackenzie District of the Northwest Territories.

5. Peltigera lepidophora (Nyl.) Vain.

Medd. Soc. Fauna Flora Fenn. 6:130. 1881. *Peltigera canina* var. *lepidophora* Nyl. in Vain., Medd. Soc. Fauna Flora Fenn. 2:49. 1878.

Thallus small, of small lobes up to 2 cm. across, the edges ascending, sometimes making the lobes cochleate; upper surface tomentose and with peltate isidia resembling cephalodia but containing the same alga as the thallus scattered over the surface; brown to grayish blue; under surface pale to brown with irregular raised venation; rhizinae simple to fibrillose, pale to brown.

Algal layer continuous, 40–75 μ thick; upper parts of the cortex plectenchymatous, hyaline to pale brown, 40–50 μ thick; medulla of loosely interwoven hyphae of 5 μ diameter with a lumen of 4 μ; medulla 75–150 μ thick. Algae 3–6 μ in diameter.

Apothecia small, 1.5–4.5 mm. across, on extended lobules, the margin crenate; disk brown; hypothecium color not mentioned; hymenium hyaline to pale brown; paraphyses thickened and red-brown at the tips; asci sparse; spores acicular, 55–85×4–4.5 μ.

Reactions: K−, C−, KC−, P−, I−.
Contents: tenuiorin.

This species grows on soil and among mosses. It is a circumpolar arctic to temperate species in North America ranging south to New England, the Great Lakes area, and in the west to Colorado. Recent work in the American Arctic has greatly extended its known range over that mapped in Thomson 1950.

6. Peltigera malacea (Ach.) Funck.

Crypt. Gewächse 33 heft: 5. 1827. *Peltidea malacea* Ach., Synops. Lich. 240. 1814.

Thallus middle-sized, up to about 10 cm. in diameter, the lobes ascending, up to about 4

Peltigera horizontalis (Huds.) Baumg.

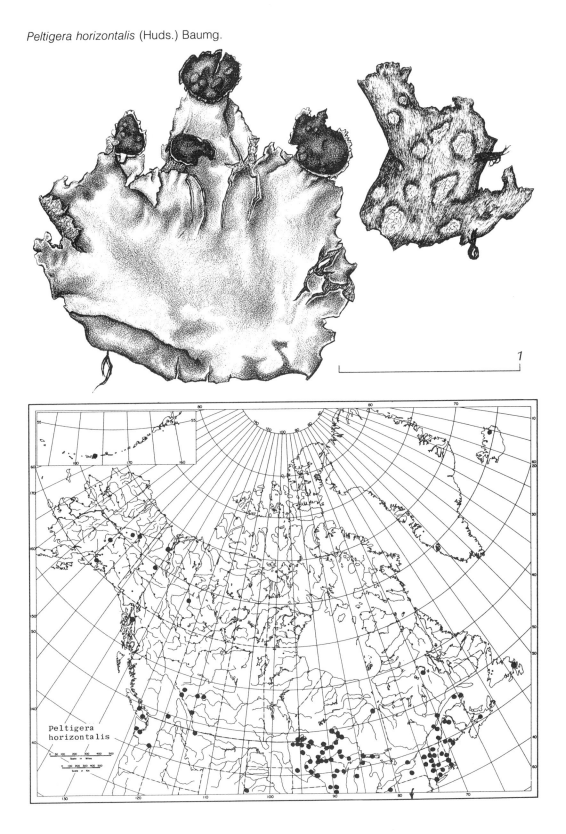

1

Peltigera
horizontalis

Peltigera lepidophora (Nyl.) Vain.

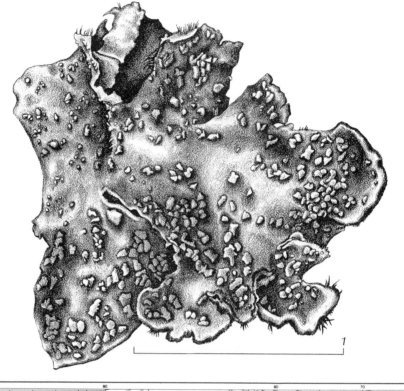

1

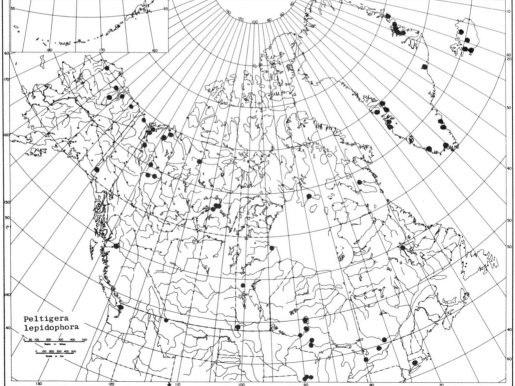

Peltigera
lepidophora

Peltigera malacea (Ach.) Funck.

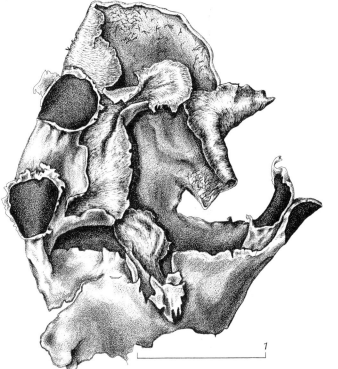

1

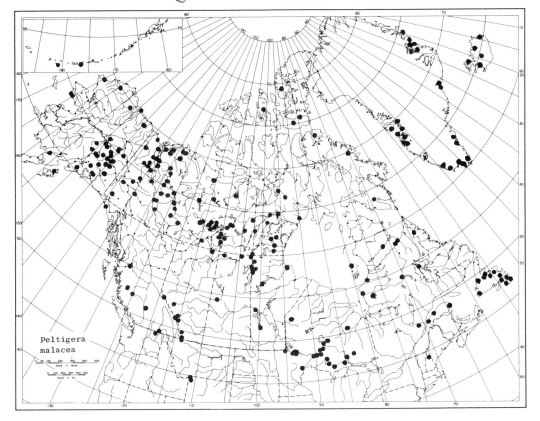

Peltigera
malacea

cm. long and 1–2 cm. broad; upper surface brown to brownish green or apple green, shining toward the center, with a sparse erect tomentum and sometimes also whitish pulverulent toward the apices of the lobes; the margins are inrolled and form a light border to the lobes; under surface brownish at the margins, blackish brown to the center, lacking veins and covered with a fine even tomentum, sometimes with a few whitish interspaces; rhizinae sparse, black, fasciculate.

Algal layer continuous, 35–50 μ thick; upper part of the cortex plectenchymatous, hyaline to pale brown, 25–50 μ thick, with trichomatic hyphae of 3–12 μ diameter and up to 150 μ long; medulla of exceedingly loosely interwoven hyphae, 10 μ in diameter with a lumen of 7 μ, medulla 450–1,800 μ thick; hymenium about 110 μ thick; hypothecium about 60 μ thick. Algae 4–8 μ in diameter.

Apothecia vertical, borne on extended lobules, the sides reflexed; disk chestnut to blackish brown; 4–8 mm. diameter, margin crenate; hypothecium pale to dark brown; hymenium pale brown to brown; paraphyses simple, thickened and darkened at the tip to a red-brown; asci cylindrico-clavate, 80×12 μ, 8 spored; spores acicular, 3–5 septate, 46–72×3.5–6 μ (Tuckerman reports 52–72×4–6; Fink reports 50–72×4–6).

Reactions: K−, C−, KC−, P−, I stains the asci blue, other parts being negative.

Contents: tenuiorin, zeorin, and 3 unknown substances (Nuno 1966).

This species grows on soil and among mosses, particularly amid shrubby vegetation. It is a circumpolar arctic and boreal species ranging south to the mountains in New England, the region of the Great Lakes, and into Colorado in the west. It is rare in the Arctic Islands.

lightly arachnoid-tomentose at the margins, the tomentose zone sometimes being extremely narrow, light yellow-brown to chestnut brown when dry, greenish brown to grayish brown when moist; under surface light, cream to pale brown with strongly raised veins which have a finely penicillate surface, veins pale to brown; rhizinae sparse, simple but with a penicillate surface, pale to black, long, up to a cm. in length.

Algal layer continuous, 25–40 μ thick; upper part of the cortex plectenchymatous, hyaline to pale brown below, brown above, 30–60 μ thick; medulla of thickly interwoven hyphae 8–10 μ in diameter with a lumen of 4–8 μ; hyphae of veins smaller and thicker walled, about 7 μ in diameter with a lumen of 2–3 μ; medulla 70–110 μ thick. Hymenium 90–100 μ thick; hypothecium 20–60 μ thick. Algae 3–5 μ in diameter.

Apothecia small, 3–6 mm. across, on extended lobules, the sides often reflexed, the margins thin, crenate, the reverse usually corticate and smooth; disk reddish to chestnut brown; hypothecium brown; hymenium hyaline to pale brown below, red-brown above; paraphyses simple, thickened and darkened at the tips to a red-brown; asci cylindrico-clavate, 9–15×60–80 μ; spores acicular, 3–5 septate, 3–4×44–75 μ (Herre reports 3.5–5×40–73.5 μ and 3–7 septate).

Reactions: K−, C−, KC−, P−, I stains the asci blue, no other part being affected.

Contents: tenuiorin and zeorin (MC test).

This species grows on earth, rotting logs, and stumps, over humus and rocks. It is probably a circumpolar temperate to boreal species with oceanic tendencies. It occurs commonly in the western parts of North America, less common in New England to Newfoundland and in a few scattered spots in the Arctic.

7. Peltigera membranacea (Ach.) Nyl.

Bull. Soc. Linn. Normand., Ser. 4, 1:74. 1887. *Peltidea canina* var. *membranacea* Ach., Lich. Univ. 518. 1810.

Thallus large as in *P. canina*, the lobes broad, usually 1–4 cm. broad, margins somewhat rolled under, lobes often overlapping, very thin; upper surface smooth and usually shining,

8. Peltigera occidentalis (Dahl) Krist.

Bryologist 71:38. 1968. *Peltigera scabrosa* var. *occidentalis* Dahl., Medd. om. Grønl. 150(2):68. 1950.

Lobes to 1–2 cm broad, margin slightly elevated or revolute; upper side gray, gray-brown, or brown sometimes with yellowish tint, dull or slightly shining, usually tomentose along

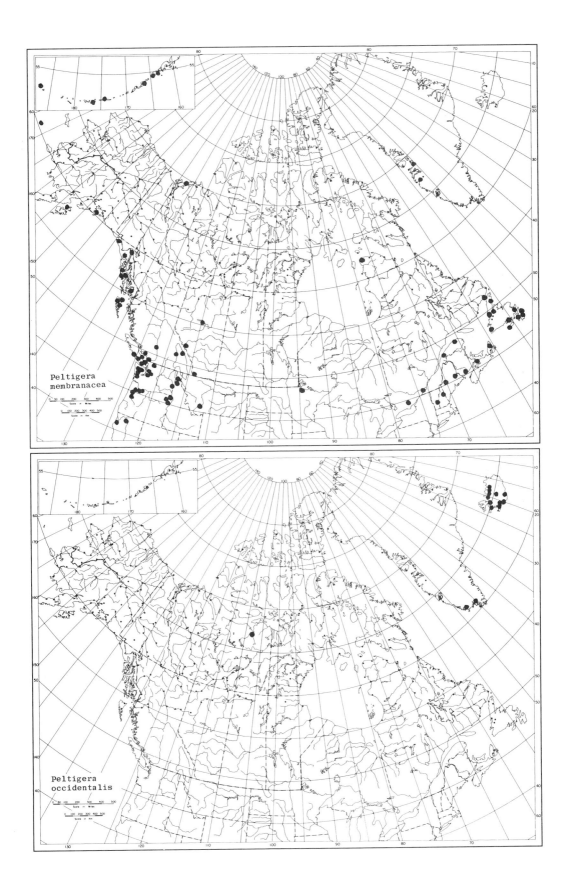

Peltigera
membranacea

Peltigera
occidentalis

the margin, the margins sometimes verruculose-scabrid, isidia-like lobules at margins or along cracks, often absent; underside pale brown to white with a fine network of usually very distinct, flattened dark brown veins like those of *P. polydactyla;* rhizines brown to pale brown at the margins, simple or fasciculate; upper cortex 22–45 μ thick, the cells isodiametric, the verruculose areas with warts of thick-walled cells and the warts also with a few erect hairs; algal layer 50–95 μ; medulla 190–450 μ.

Apothecia 4–10 mm broad, on slightly ascending lobes, nearly vertical or almost horizontal, the sides sometimes recurved, brown above, lacking cortex or algal layer below or rarely with scattered bits of cortex near the margin; hymenium 105–132 μ, hyaline to light brown; paraphyses with brown tips; hypothecium pale brown; asci 8-spored; spores 4-celled 50–73 × 2.6–5 μ.

Reactions: K−, C−, KC−, P−,

Contents: tenuorin, zeorin, methyl gyrophorate, triterpine I, 3 unnamed substances (Ohlsson 1973).

Growing on soil among mosses, this species has been reported from Iceland, Greenland, and Alberta. The latter specimens from Alberta however are of *Peltigera canina* f. *spongiosa* and the report consequently incorrect. It has recently been discovered at Bathurst Inlet, N.W.T.

9. Peltigera polydactyla (Neck.) Hoffm.
Descript. et Adumbr. Plant. Lich. 1:19. 1790. *Lichen polydactylon* Neck., Meth. Muscorum 85. 1771.

Thallus variable, ranging in size from single, pusilloid erect lobes to large thalli almost a meter across; the lobes imbricate, up to 4 cm. across and several cm. long, the margins ascending, some forms crisped or proliferate; upper surface dark greenish blue when moist, slaty or bluish or greenish gray or brownish when dry; smooth, shining; under surface with broad, flat, white, or usually, brown or black, reticulated veins with very small white interspaces, rhizoids sparse, short and fasciculate or in one variety over 5 mm. long and simple.

Algal layer continuous, 20–30 μ thick; upper part of the cortex plectenchymatous 30 μ

thick, hyaline to pale brown, medulla of very loosely interwoven hyphae 8 μ in diameter with a lumen of 5 μ; medulla 200–700 μ thick. Hymenium 100–110 μ thick; hypothecium about 60 μ thick. Algae 3–6 μ in diameter.

Apothecia medium sized, 2–5 mm. in diameter, on erect narrow lobules at the ascendent tips of the lobes, sometimes borne in pairs, the sides usually reflexed, the disk reddish or chestnut brown to black, the margin crenate, the reverse corticate; hypothecium brown; hymenium hyaline; asci cylindrico-clavate, 80–105 × 10 μ, 8 spored; the spores acicular, slightly curved, 48–105 × 3–4 μ (Tuckerman reports 69–90 × 3–4; Fink reports 60–100 × 3–4).

Reactions: K−, C−, KC−, P−, I stains the asci blue, no other part being affected.

Contents: tenuiorin, dolichorrhizin, zeorin, unidentified substances.

This species grows on soil, moss, logs, rocks, and bases of trees in moist woods and is reported to be cosmopolitan. It ranges throughout North America but appears to be lacking in the Arctic Islands other than Baffin Island.

10. Peltigera scabrosa Th. Fr.
Nova Acta Reg. Soc. Scient. Upsal. ser. 3, 3:145. 1861.

Thallus medium sized to large, lobes often large, up to 5 cm. across, although usually smaller; upper surface greenish gray or yellowish brown to brown, dull, scabrid; under surface with broad flat veins like *P. polydactyla,* the veins brown to blackish brown, the interspaces white to pale brown; rhizinae fasciculate, sometimes forming a thick mat to the center of the thallus.

Algal layer continuous, 40–75 μ thick; upper part of the cortex plectenchymatous, hyaline with the upper part brown, 50–70 μ thick; medulla of loosely interwoven hyphae 11 μ in diameter with a lumen of 4 μ, medulla 50–600 μ thick; hymenium 110–140 μ thick in the hyaline part, 20–25 μ thick in the brown part; hypothecium about 60 μ thick. Algae 6–10 μ diameter.

Apothecia large, up to 6 mm. across, borne on extended lobules, vertical, often round or with the sides reflexed; disk dark reddish

Peltigera polydactyla (Neck.) Hoffm.

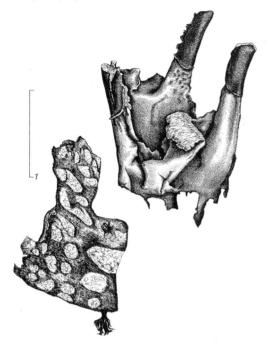

brown, the margin entire to slightly irregular, the reverse with a cortex; hypothecium hyaline to pale brown; hymenium hyaline below, deep brown above; paraphyses simple, thickened and dark red-brown at the tips; asci cylindrico-clavate, $10 \times 105 \ \mu$, 8 spored; 5–7 septate, spores acicular $63–95 \times 2–3 \ \mu$. (Tuckerman reports 60–94 $\times$ 3–4; Fink reports 60–100 $\times$ 3–4).

Reactions: K−, KC− or KC+ or −, P−.

Contents: tenuiorin, zeorin, scabrosin B, gyrophoric acid + or −, an unknown triterpine, and substance V. (Ohlsson 1973). This species lacks a second unknown triterpene which is present in *P. pulverulenta* according to Ohlsson.

This species grows on soil and among mosses, occasionally at the bases of woody plants. It is circumpolar, arctic, boreal, and alpine. In North America it ranges south to New England, Montana, and Washington.

The relationship of this species to *P. pulverulenta* is a controversial question. Ohlsson has added

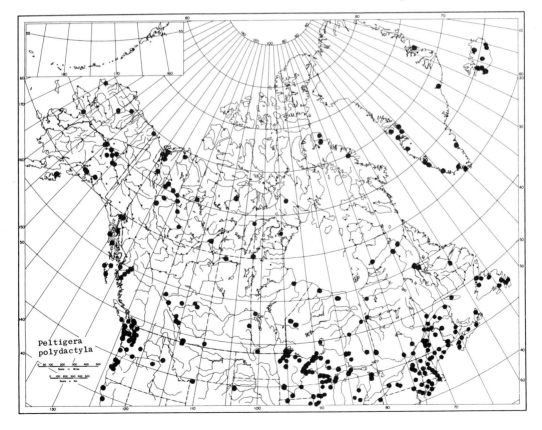

Peltigera
polydactyla

Peltigera scabrosa Th. Fr.

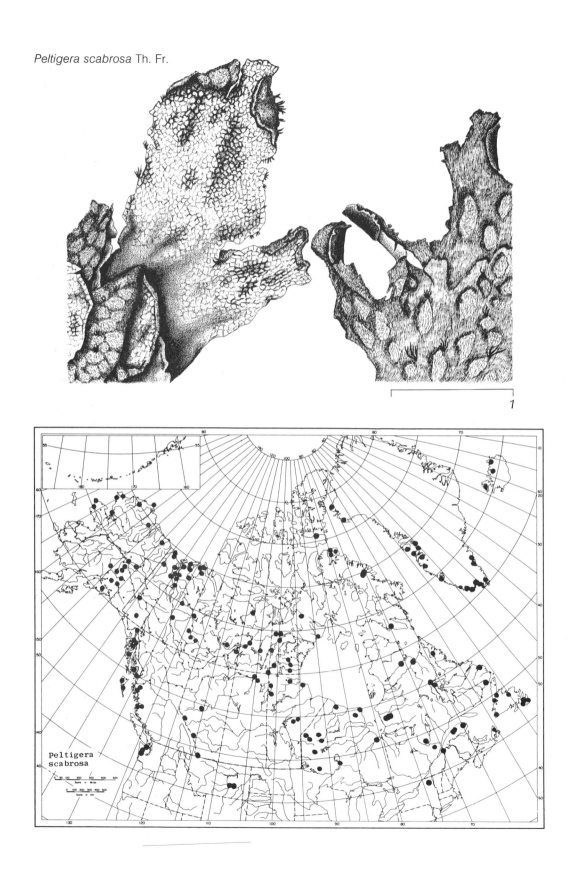

1

Peltigera venosa (L.) Baumg.

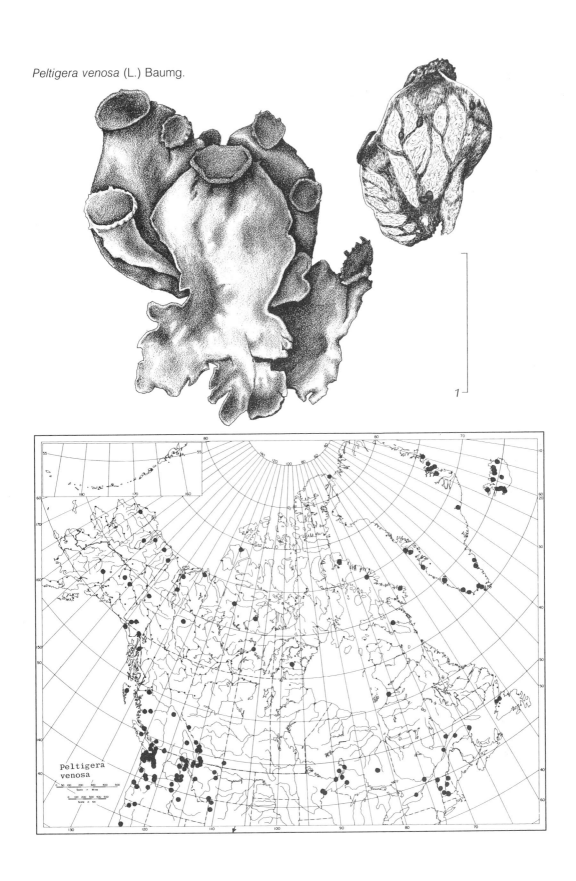

1

a new dimension with reference to the presence or absence of a second triterpene but a thorough study of the constancy of these substances is yet to be performed. Those interested in the problem should consult the papers cited above in the generic listing.

11. Peltigera venosa (L.) Baumg.
Flora Lipsiens. 581, 1790. *Lichen venosus* L., Sp. Pl. 1148, 1793.

Thallus small, of single, rounded, fan-shaped lobes, sometimes somewhat divided, up to 2 cm. across, flattened; upper surface smooth, shining, apple green when moist, grayish or brownish green when dry; under surface white or light brown with dark brown or black veins; small dark green or brown to black, warty cephalodia located on the veins and containing nostocoid algae; a single group of rhizoids at one side attaches the thallus to the substrate.

Algal layer continuous, 50–65 μ thick; upper part of the cortex pseudoparenchymatous, hyaline to pale brownish, 40–60 μ thick; medulla of very loosely interwoven hyphae 8 μ in diameter with a lumen of 4.5 μ; hyphae of the veins very dark brown, more thickly interwoven, 6 μ in diameter with a lumen of 2.5 μ, medulla 150–300 μ thick; hymenium 100–110 μ thick; hypothecium about 150 μ thick. Algae of thallus 2–3 μ in diameter; algae of cephalodia 4–6 μ diameter.

Apothecia small, horizontal, up to 5 mm. in diameter, round, borne at the margin of the thallus, the margin crenate; the disk reddish to blackish brown, the reverse with a thick, dark pseudoparenchymatous cortex; hypothecium pale to dark brown with a layer of hyphae inspersed with air immediately below; hymenium hyaline to pale brown; paraphyses simple, coherent thickened and darkened to a red-brown at the tips; asci cylindrico-clavate, 75 × 11 μ, 8 spored; spores fusiform, (1−)3 septate (rarely 5 septate), 24–40 × 6–8 μ (Tuckerman reports 30–46 × 7–10; Fink reports 28–45 × 6–9).

Reactions: K−, C−, KC−, P−, I stains algae of thallus blue, other parts negative.

Contents: tenuiorin, zeorin, phlebin A and B and unidentified substance (Kurokawa et al 1966).

This species grows on moist soil and rocks with calcareous seepage, on moist cliffs, talus slopes, edges of tussocks, edges of animal burrows and like microhabitats. It is circumpolar, arctic, boreal, and alpine. In North America it ranges south to Pennsylvania and California.

PHAEOPHYSCIA Moberg

Symboliae Bot. Upsal. 22(1):29. 1977. *Physcia* sect. *Euphyscia* b. *Sordulenta* 1. *Brachysperma* Vain., Acta Soc. Fauna Flora Fenn. 7:144. 1890. *Physcia* subgen. *Euphyscia* sect. *Brachysperma* subsect. *Sordulenta* group *Obscura* (DR.) Thoms., Nova Hedw. 7:102. 1963.

Thallus foliose, lobate to fruticose, gray-brown to brown or greenish brown, epruinose; underside black or dark brown, not white, with simple rhizinae of the same color; upper cortex paraplectenchymatous; medulla of loosely interwoven hyphae, white or with red pigments (skyrin); lower cortex brownish or blackish, paraplectenchymatous with isodiametric cells or rarely prosoplectenchymatous.

Apothecia lecanorine, laminal, sessile, often with rhizinae or lower borders forming a corona; disk brown to black; hymenium hyaline; hypothecium hyaline; asci cylindrical, spores 8, brown, thick-walled, 1 septate. Pycnidia laminal immersed, the tips black, conidia ellipsoid and small, less than 4 μ. Atranorin is absent in this genus.

Algae: *Trebouxia*

Type species: *Phaeophyscia orbicularis* (Neck.) Moberg.

REFERENCES

Esslinger, T. L. 1978. Studies in the lichen family Physciaceae. 2. The genus *Phaeophyscia* in North America. Mycotaxon 7:283–320.
Moberg, R. 1977. The lichen genus *Physcia* and allied genera in Fennoscandia. Symbol. Bot. Upsal. 22(1):1–108.

1. Lacking soredia or isidia.
   2. Lobes ascending, nearly erect, long and narrow, much branched with numerous marginal cilia, among mosses . . . . 1. *P. constipata*
   2. Lobes appressed, rarely ascending, lacking marginal cilia, on rocks . . . . 2. *P. endococcinea*
1. Thallus isidiate or isidiate-sorediate, on rocks . . . . 3. *P. sciastra*

1. Phaeophyscia constipata (Norrl. & Nyl.) Moberg

Symbol. Bot. Upsal. 22:33. 1977. *Physcia constipata* Norrl. & Nyl., Herb. Lich. Fenn. Exs. 218. 1882.

Thallus irregular, loosely attached, sometimes forming tufts, the lobes slender, to 0.3 mm broad; branches irregular, some running out into pointed tips, marginally ciliate, lacking soredia, isidia or pruina, thin, to 300 μ thick, gray to gray-brown or darker brown, apple green when moist; underside pale whitish with pale rhizinae; upper and lower cortices both paraplectenchymatous, occasionally with yellowish pigment on the lower surface, the pigment K+ violet.

Apothecia rare, laminal; disk dark brown, epruinose; hymenium hyaline, the upper part brown; paraphyses capitate; asci clavate; spores brown, 1-septate, $20-22 \times 9-10$ μ. Conidia straight, $3-4$ μ.

Reactions: all negative except when yellowish pigment is on lower side and this then K+ violet.

Contents: no substances identified, the yellowish pigment probably an anthraquinone.

This species grows on soil and among mosses, preferably in limestone regions. It is very rare, arctic-alpine, and known from Europe and North America.

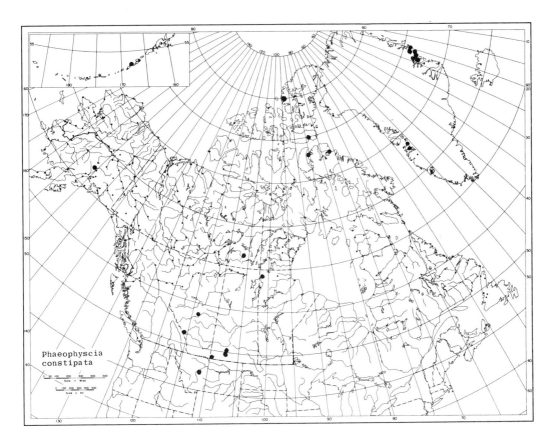

Phaeophyscia
constipata

2. Phaeophyscia endococcinea (Körb.) Moberg

Symb. Bot. Upsal. 22(1):35. 1977. *Parmelia endococcinea* Körb., Parerga Lich. 36. 1859. *Physcia endococcinea* (Körb.) Th. Fr., Bot. Not. 150. 1866. Physcia lithotodes Nyl., Flora 58:360. 1875. *Phaeophyscia decolor* (Kash.) Essl., Mycotaxon 7:299. 1978.

Thallus small, fairly closely attached to the substratum, lobes 0.2–0.3 (–0.5) mm broad, discrete or more or less imbricated, dichotomous or irregularly pinnately divided, pale to dark greenish brown above, greener when moistened; underside black with dark rhizinae; upper cortex 25–30 μ thick; medulla white or red (skyrin); lower cortex dark.

Apothecia common, to 2 mm broad; disk brown-black, margin entire to crenulate; underside of exciple paraplectenchymatous; hypothecium yellowish or hyaline; hymenium 100–120 μ; spores 17–26 × 8–13 μ.

Reactions: cortex and medulla K− except medulla K+ red-violet when red with skyrin.

Contents: zeorin and skyrin in the medulla.

This species grows on rocks and among mosses on rocks, rarely on tree bases. It is a circumpolar arctic-alpine and boreal species but is rarely collected. In North America this species ranges south to New Mexico in the west and into New England in the east. The northern material lacks the skyrin and has a white medulla. It would be classed as *Phaeophyscia decolor* (Kash.) Essl. by Esslinger.

Specimens of *Phaeophyscia sciastra* which lack the isidia are occasionally encountered in arctic material and would key to this species but lack the zeorin as pointed out by Moberg (1977).

3. Phaeophyscia sciastra (Ach.) Moberg

Symbol. Bot. Upsal. 22(1):47. 1977. *Parmelia sciastra* Ach., Meth. Lich. 68. 1803. *Physcia sciastra* (Ach.)DR., Svensk. Bot. Tids. 15:168. 1921.

Thallus small, coalescing to form larger mats; lobes 0.2–0.5 mm broad, linear to fan-shaped at tips; upper side gray-black or dark

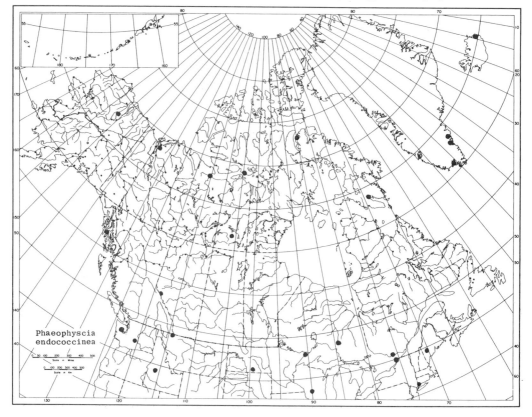

Phaeophyscia
endococcinea

Phaeophyscia sciastra (Ach.) Moberg

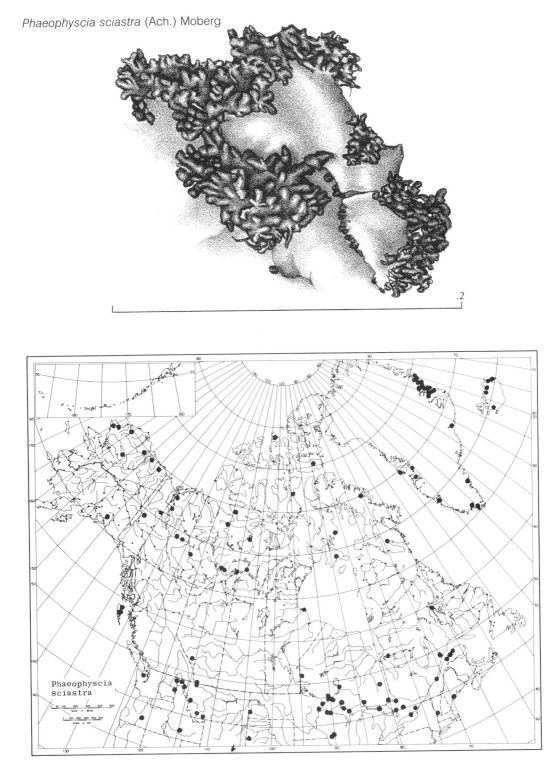

.2

Phaeophyscia
sciastra

brown to black, sometimes bluish pruinose, isidiate on the margins or lamina, sometimes with soredia-like structures formed at breaks but not a truly sorediate species, the isidia darker than the thallus, sometimes quite black; underside black, with dark rhizinae; upper cortex with amorphous covering 6–10 μ thick.

Apothecia rare, sessile, to 2 mm broad; margins entire to deeply crenate; disk red-brown to black, dull, epruinose; hymenium 100 μ; spores brown to nearly black, 15–24 × 8–12 μ.

Reactions: all negative.

Contents: no lichen substances found.

This species grows on acidic rocks, on mosses over rocks, occasionally over mosses on soil in the arctic. It is a circumpolar arctic, boreal, and temperate species, in North America ranging southward to the northern states and in the west to Arizona.

PHYSCIA (Schreb.) Michx.

Flora boreali-americana 2:326. 1803. *Lichen* sect. *Physcia* Schreb. in Linn., Gen. Plant. 2:798. 1791.

Thallus small foliose, appressed to ascending, deeply incised, margins sometimes ciliated or fibrillose; upper cortex paraplectenchymatous; medulla loosely interwoven, white, containing always atranorin and sometimes zeorin; underside white to pale brown; lower cortex prosoplectenchymatous or rarely paraplectenchymatous; underside attached by rhizinae.

Apothecia laminal, sessile, or short-stalked, with lecanorine margin; disk brown to black, naked or with white pruina (reddish pruinose in some extraterritorial species); hymenium hyaline, I+ blue; hypothecium hyaline; paraphyses conglutinate, septate, more or less branched, thickened and brown at apices; asci cylindrical; spores 8, 2-celled, thick-walled, the walls thickened at the ends and the center making the lumen more or less hourglass-shaped, brown, the surface smooth. Pycnidia immersed in the lamina, the conidia terminal and lateral on short-celled and branched conidiophores, the conidia subcylindrical and 4–7 × 1–1.5 μ.

Algae: *Trebouxia*.

Type species: *Physica tenella* (Scop.)DC.

REFERENCES

Kashiwadani, H. 1975. The Genera *Physcia*, *Physconia*, and *Dirinaria* (Lichens) of Japan. Ginkgoana 3:1–77.
Moberg, R. 1977. The lichen genus *Physcia* and allied genera in Fennoscandia. Symbol. Bot. Upsals. 22(1):1–108.
Poelt, J. 1965. Zur Systematik der Flechtenfamilie Physciaceae. Nova Hedwigia 9:21–32.
—— 1974. Die Gattungen *Physcia*, *Physciopsis*, und *Physconia*. Flechten des Himalaya 6. Khumbu Himal. 6(2):57–100.
Thomson, J. W. 1963. The lichen genus *Physcia* in North America. Beih. Nova Hedwigia 7:1–172.

1. Lobes ascending with long marginal cilia.
    2. Tips of lobes with helmet-shaped inflated tips which are sorediate on the interior of the underside; thallus white or gray-white . . . . 1. *P. adscendens*
    2. Tips of lobes with terminal labriform soralia; thallus dark ashy gray to blackish .   6. *P. subobscura*
1. Lobes lacking long marginal cilia.
    3. Thallus lacking soredia or isidia; upper side white spotted, medulla as well as cortex K+ yellow.
        4. Thallus white to gray-white, on bark, spores 20–25 × 8.5–11 μ . . . . 2. *P. aipolia*
        4. Thallus blue-gray to ash-gray, on rocks, spores 17–20 × 8–9 μ . . . . 5. *P. phaea*
    3. Thallus with soredia.
        5. Soredia laminal, capitate, medulla as well as cortex K+ yellow, atranorin and zeorin present . . . . 3. *P. caesia*
        5. Soredia labriform, terminal or also with laminal crateriform soralia; medulla K−, atranorin in cortex only . . . . 4. *P. dubia*

1. Physcia adscendens (Fr.) Oliv.

Lich. Fl. Orne 1:79. 1882. *Parmelia stellaris* var. *adscendens* Fr., Summa Veg. Scand. sect. 1:105. 1845.

Thallus of small rosettes or isolated lobes, more or less free and erect, attached by rhizinae; lobes narrow and up to 1.5 cm long, irregularly pinnately branched, flat to convex, light gray-white with a bluish cast, greener when moist, epruinose, lacking insidia but the lobe tips cupule- or helmet-shaped and containing greenish soredia, the margins ciliate with pale cilia with darkening tips; underside white, the pale rhizinae blackening; upper cortex paraplectenchymatous, lower cortex prosoplectenchymatous.

Apothecia stalked, to 2 mm broad, laminal; disk red-brown to black-brown; hypothecium pale yellowish brown; hymenium hyaline with the upper part brownish; paraphyses simple or branched, capitate; asci cylindrical to clavate; spores 8, brown, little constricted, the ends and septa thicker than the sides, the lumina hourglass-shaped, $15-23 \times 7-10$ μ. Pycnidia laminal, immersed, the conidia cylindrical, $3.5-6 \times 1$ μ.

Reactions: upper cortex K + yellow, medulla K −.

Contents: atranorin in the cortex.

This species grows mainly on the trunks and branches of trees and shrubs, both deciduous and coniferous. It occasionally also grows on old wood, rocks, and tombstones. It is circumpolar, mainly boreal to temperate, but is occasional in the low arctic.

2. Physcia aipolia (Humb.) Fürnrohr

Flora Ratisbonensis. 249. 1839. *Lichen aipolius* Ehrh. ex Humb., Fl. Friburg. Specim. 19. 1793.

Thallus of rosettes or separate lobes, irregularly pinnate, sometimes with the center becoming a mass of conglomerate lobes, the lobes to 3 mm broad, upper side whitish-gray to bluish gray, densely white spotted, sometimes slightly pruinose, lacking soredia or isidia but sometimes with bullate warty lobes; underside white to pale brown with pale to brown rhizinae; upper cortex paraplectenchymatous, the algae very conglomerated and often extending into the cortex, the intervening medullary tissues and clearer tissues causing the white spotting; lower cortex prosoplectenchymatous.

Apothecia usually common, to 2 mm broad, the margin smooth to crenulate, the disk red-brown to black and usually heavily pruinose; hypothecium hyaline; hymenium hyaline, the upper part brown-yellow; paraphyses usually unbranched to branched at apices; asci cylindrical to clavate; spores 8, 2-celled, brown, the walls much thickened at the septum and apices, $16-29 \times 7-11$ μ. Conidia $4-6 \times 1$ μ.

Reactions: cortex and medulla K + yellow, P −, C −, hymenium I + blue turning reddish.

Contents: atranorin in cortex and medulla, zeorin in medulla.

This species grows on deciduous trees and shrubs in the Arctic, especially on *Salix* and *Betula,* but also occurs on conifers, more rarely on rocks. It is circumpolar boreal to temperate, in the Arctic ranging beyond the tree line to some extent on shrubs.

3. Physcia caesia (Hoffm.) Fürnrohr

Flora Ratisbonensis 250. 1839. *Lichen caesius* Hoffm., Enum. Lich. 65. 1784.

Thallus small, to 3 cm broad, often coalescing with others; the lobes appressed, sometimes narrow, less than 1 mm, sometimes broad and to 3 mm broad, convex, with laminal or terminal usually capitate but sometimes labriform soralia, these sometimes emptying and becoming crater-like; upper side ashy gray to bluish gray, white spotted; underside white to brownish with pale to dark rhizinae; upper cortex paraplectenchymatous; algae glomerulate and penetrating into the upper cortex; lower cortex prosoplectenchymatous.

Apothecia uncommon, laminal, sessile, to 1.5 mm broad; disk brown black, sometimes pruinose; margins slightly crenulate or entire; hypothecium hyaline or yellowish; hymenium hyaline, the upper part brownish; paraphyses simple or rarely branched, capitate; asci cylindrical to clavate; spores brown, 1-septate, the apical and septal walls thickened and the lumen hourglass shaped, $15-24 \times 7.5-11$. Conidia $4-5 \times 1$ μ.

Physcia adscendens (Fr.) Oliv.

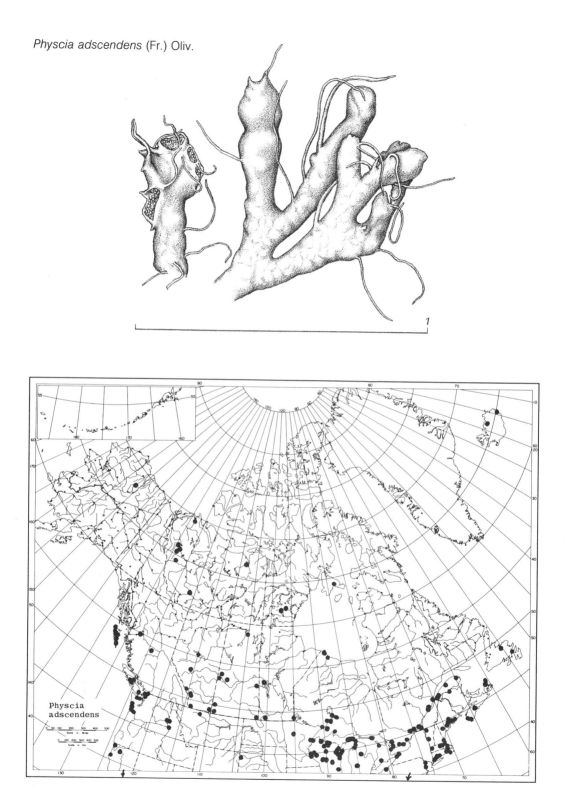

Physcia aipolia (Humb.) Fürnrohr.

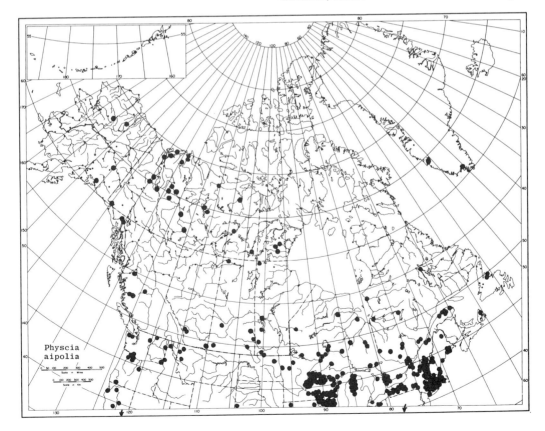

Reactions: cortex and medulla K+ yellow, P−, C−.

Contents: atranorin in cortex and medulla, zeorin in medulla.

This species grows usually on rocks, frequently those well manured by birds, as perch rocks, but also grows on tree bases, old wood, concrete, and mortar. It is circumpolar, arctic, and boreal, ranging southward in North America to New England, across the northern states and south in the Rocky Mts.

4. Physcia dubia (Hoffm.) Lettau

Hedwigia 52:254. 1912. *Lobaria dubia* Hoffm., Deutschl. Flora 156. 1796. Incl. *Physcia intermedia* Vain., Medd. Soc. Fauna Fl. Fenn. 2:51. 1878, and *Physcia teretiuscula* (Ach.) Lynge, Videns. Selsk. Skr. Oslo 8:96. 1916. (*vide* Moberg 1977: 77).

Thallus usually coalescent with others, the lobes soon raised to reflexed, more or less convex, to 2 mm broad but often narrow, broadening toward the tips, with labriform soralia at the tips of the lobes or on short side branches, sometimes laminal and crater-shaped

Physcia aipolia

Physcia caesia (Hoffm.) Fürnrohr.

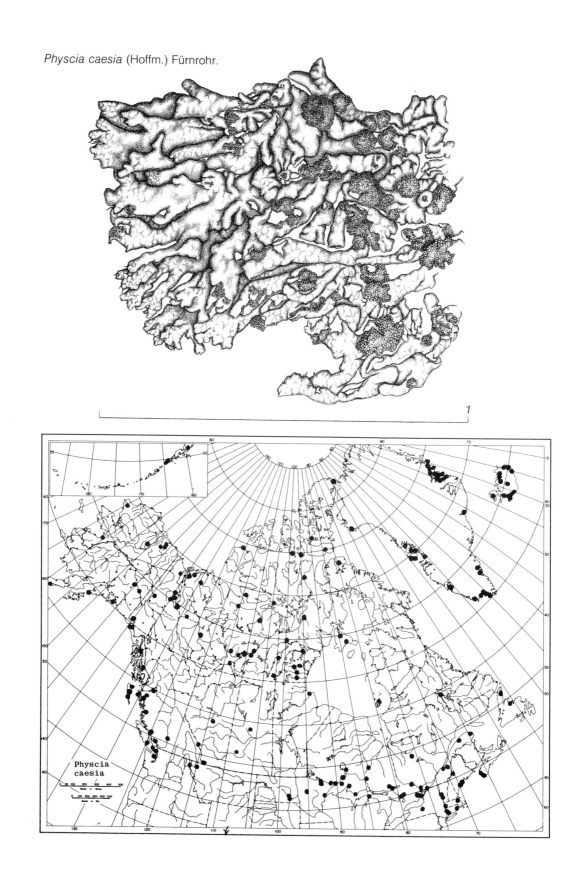

1

Physcia
caesia

Physcia dubia (Hoffm.) Lettau

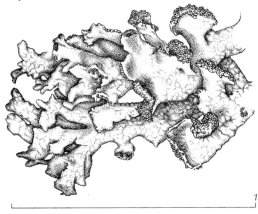

in part, ashy gray to dark gray, more or less glossy; underside white or brownish, with pale or brownish, sometimes black, rhizinae; upper cortex paraplectenchymatous; lower cortex prosoplectenchymatous.

Apothecia uncommon, to 2 mm broad, the margin entire to crenulate or sorediate; disk black-brown, epruinose; hypothecium hyaline to yellowish; hymenium hyaline, the upper part brown; paraphyses simple or branched, capitate; spores 8, brown, the apices and septal walls thickened, straight or slightly bean-shaped, 15–28 × 6–11 μ. Conidia 4–6 × 1 μ.

Reactions: cortex K+ yellow, P−, C−, medulla K−, P−, C−.

Contents: atranorin present only in the cortex.

This species grows on rocks, old bones, weathered boards, and characteristically in places manured by birds. It is circumpolar arctic, boreal, and temperate, in North America ranging into the northern states and south in the Rocky Mountains.

5. **Physcia phaea** (Tuck.) Thoms.

Beih. Nova Hedw. 7:54. 1963. *Parmelia phaea* Tuck. in Darlington, Flora Cestria 3d ed. 440. 1853. *Physcia melops* Duf. ex Nyl., Flora 57:16. 1867.

Thallus rosette forming, sometimes several coalescent, the lobes irregular, convex, ra-

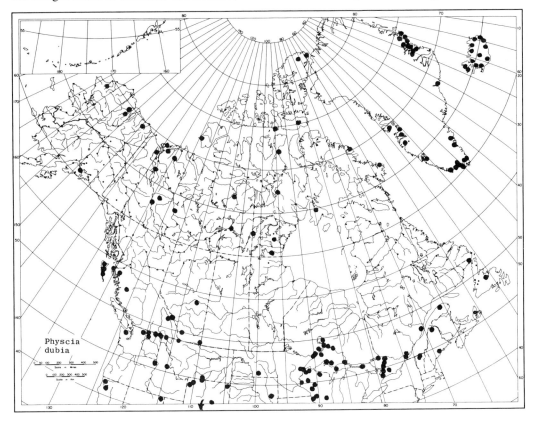

Physcia dubia

diating, usually about 1 mm broad; upper side very blue-gray to violet-gray and white spotted as in *P. aipolia,* lacking soredia or isidia; underside pale brown to brown with rhizinae of the same color or darker; upper cortex paraplectenchymatous, the algae glomerate and distributed as in *P. aipolia;* lower cortex prosoplectenchymatous.

Apothecia numerous, adnate to appressed, to 2 mm broad, margins thick, entire to crenulate; disk brown-black, epruinose; hypothecium hyaline; hymenium hyaline, the upper part brown-yellow; paraphyses unbranched or slightly so above; asci clavate; spores 8, brown, 1-septate, the end walls and septum thickened; the lumina angular; 15–22 × 7–10 μ. Conidia 4–6 × 1 μ.

Reactions: cortex and medulla K+ yellow, P–, C–.

Contents: atranorin in both cortex and medulla, zeorin in medulla.

This species is easily confused with rock-living specimens of *P. aipolia* but is distinguished by the more blue color, the smaller spores, and the more appressed habit.

This species grows on acid rocks. It is perhaps amphi-Atlantic, currently known from Europe and North America. It appears to be boreal and temperate in distribution.

6. Physcia subobscura Nyl.

Flora 52:389. 1869. (Art. 72. Note 1). *Physcia stellaris* var. *marina* Nyl., Not. Sallsk. F. F. Fenn. Förh. 3:86. 1857. *Physcia stellaris* var. *subobscura* Nyl. Not. Sällsk. F. Fl. Fenn. Förh. 1858. 59:239. *Physcia marina* (Nyl.) Lynge in Zahlbr., Annal. Hofmus. Wien 30:217. 1916 (illeg.). *Physcia nigricans* var. *groenlandica* Dahl, Medd. om Grønl. 150(2):153. 1950.

Thallus small to middle-sized, the marginal lobes low lying, the central forming small tufts or cushions; lobes narrow, to 0.3 mm broad, the tips sorediate with labriform soralia, the tips then erect to reflexed, the margins ciliate with blackening cilia; upper side light gray to very

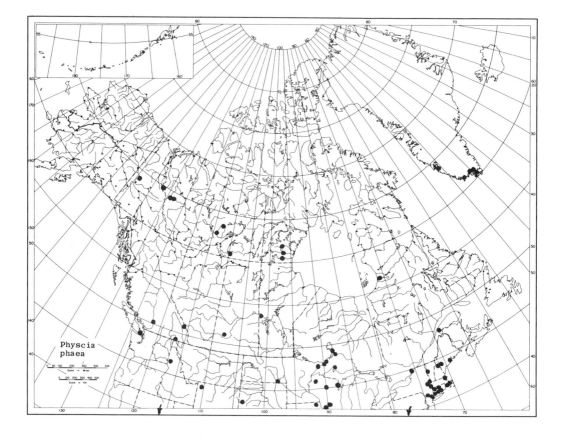

Physcia
phaea

dark, smooth, epruinose; underside white to brownish with black rhizinae; upper cortex paraplectenchymatous; lower cortex prosoplectenchymatous.

Apothecia laminal, sessile to short pedicillate, to 2 mm broad, disk flat, brown, bluish pruinose; epithecium brown; hymenium low, 60–70 μ, hyaline; paraphyses unbranched or branched at tips, capitate; asci clavate; spores brown, 1 septate, "physica type" with hourglass lumina, $15–18 \times 9–10$ μ. Conidia cylindrical 3–4 μ.

Reactions: cortex K+ yellow, medulla K−.

Contents: atranorin in the upper cortex.

This species grows on rocks in the vicinity of seashores and in the vicinity of rainwater depressions. It is probably circumpolar but reports are from Europe and North America in arctic and boreal localities. It is very rare in North America.

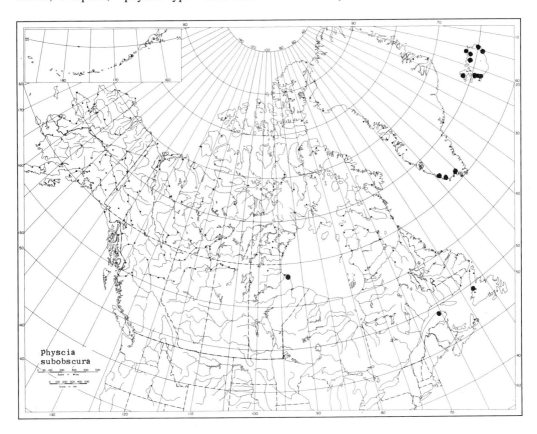

Physcia subobscura

PHYSCONIA Poelt

Nova Hedw. 9:30. 1965.

Thallus foliose, gray, gray-brown to brown, usually whitish pruinose, loosely attached, rhizines usually squarrose. Upper cortex thick, scleroplectenchymatous or paraplectenchymatous; lower cortex prosoplectenchymatous, white or black.

Apothecia lecanorine, often lobulate on the margins; hymenium hyaline, I+ blue; hypothecium pale brown; asci cylindrical or clavate; spores 8, 1-septate, brown, usually more than

$27\mu \times 15\mu$, the walls of the "Physconia type" nearly uniform in thickness and the surface verrucose with shallow depressions visible under exceedingly high magnifications. Pycnidia immersed in the thallus, flask-shaped, the conidia subcylindrical and $4-7 \times 1-1.5$ μ. Lacking atranorin, several unknown substances present.

Algae: *Trebouxia*.

Type species: *Physconia pulveracea* Moberg

REFERENCES

Gunnerbeck, E. and R. Moberg. 1979. Lectotypification of *Physconia*, a generic name based on a misnamed type species—a new solution to an old problem. Mycotaxon 8:307–317.
Moberg, R. 1977. The lichen genus *Physcia* and allied genera in Fennoscandia. Symbol. Bot. Upsalienses 22:1–108.
Poelt, J. 1965. Zur Systematik der Flechtenfamilie Physciaceae. Nova Hedw. 9:21–32.
—— 1974. Die Gattungen *Physcia*, *Physciopsis*, und *Physconia*. Flechten des Himalaya 6. Khumbu Himal. 6(2):57–100.
Thomson, J. W. 1963. The lichen genus *Physcia* in North America. Beih. Nova Hedw. 7:1–172.

1. Lobes esorediate . . . . 2. *P. muscigena*
1. Lobes sorediate . . . . 1. *P. detersa*

1. Physconia detersa (Nyl.) Poelt

Nova Hedw. 9:30. 1965. *Parmelia pulverulenta* var. *detersa* Nyl., Synops. Lich. 1:420. 1860.

Thallus foliose, irregular, to 8 cm broad, the lobes often overlapping, to 3 mm broad, the tips often concave; upper side light gray-brown or gray-green to dark brown or chestnut brown, smooth or corrugated, usually pruinose with white or blue-white pruina; sorediate, the soredia marginal or less commonly laminal, coarse to isidiate; underside whitish marginally, usually quickly becoming black, with the marginal rhizines often simple to slightly fasciculate but soon becoming squarrose with fine hairs; the rhizines usually black; upper cortex scleroplectenchymatous, medulla loose, white, the lower cortex prosoplectenchymatous.

Apothecia uncommon, laminal, to 3 mm broad, the margins entire to sorediate; the disk naked to strongly bluish pruinose; hymenium 120–160 μ; asci 78–105 × 27–30 μ; spores biseriate, 1-septate, brown, straight or slightly curved, the center constricted, the walls even in thickness, the surface minutely verrucose, 24–33 × 13–17 μ. Conidia cylindrical, straight, 3.5–7 × 1–2 μ.

Reactions: all negative with usual reagents.

Contents: no substances found (Moberg 1977). Differing from *P. grisea* in which Moberg found four unidentified substances in TLC.

This species grows mainly on the trunks of broad-leaved and coniferous trees but also occurs on rocks and over moss. It is circumpolar, boreal, and temperate, reaching the tundras only at their borders with the forest. Its occurrence in Greenland may reflect a relict from a more mild climatic period.

The broader delimitation of *Physcia grisea* (Lam.) Zahlbr. in my study of 1963 was altered by Poelt (1963), such that those plants with simple rhizines and predominately pale underside were retained in a narrower concept as *Physconia grisea* (Lam.) Poelt. Other plants with squarrose rhizines and a soon blackening underside were placed in *Physconia detersa* (Nyl.) Poelt. Moberg (1977) followed these separations, including the separation of these two species into two different sections of the genus *Physconia*. Examination of North American specimens in a considerable collection in WIS shows that they mainly fall into the category of *Physconia detersa* as treated by Poelt and Moberg. A few specimens could possibly be placed as *Physconia grisea* and are mainly western. A large problem with many North American specimens of *P. detersa* lies in the presence of a fairly broad outer white zone on the underside with only simple to fasciculate rhizines, and only toward the center does the underside blacken and the rhizines become squarrose. Moberg states that *P. detersa* is black or brown to the tips of the lobe undersides and mentions only squarrose rhizines. Poelt keys out *P. detersa* as having a whitish to deep black underside with rhizines soon black and squarrose. The North American specimens of *Physconia detersa* do not seem to have as clear-cut species distinctions as European material. The Arctic specimens in North America have

Physconia detersa (Nyl.) Poelt

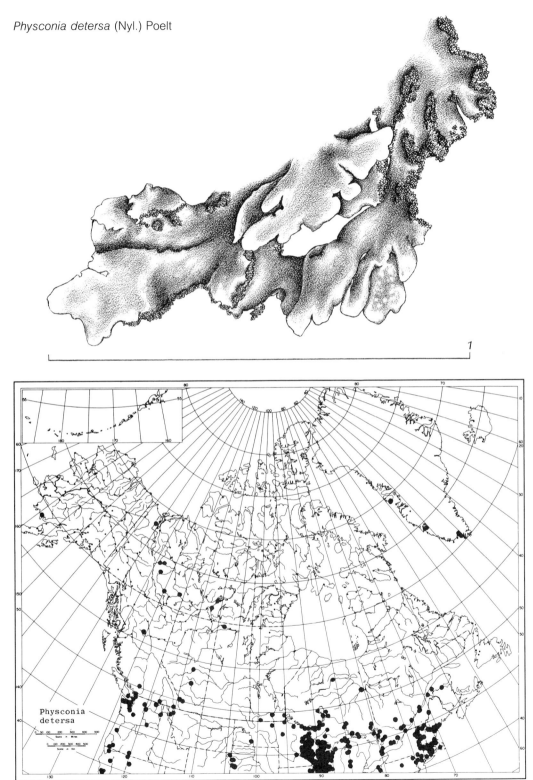

1

Physconia
detersa

Physconia muscigena (Ach.) Poelt

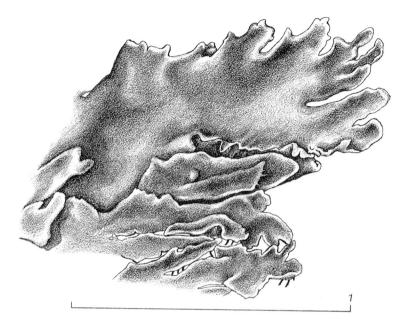

1

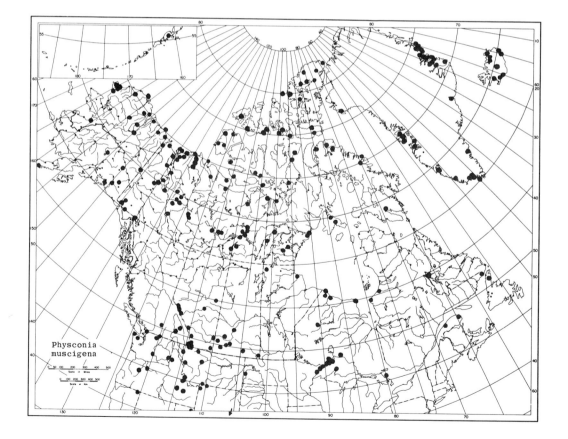

Physconia
muscigena

been placed in *P. detersa* as they become black below, at least centrally, and the rhizines also become squarrose centrally.

2. Physconia muscigena (Ach.) Poelt
Nova Hedw. 9:30. 1965. *Parmelia muscigena* Ach., Lich. Univ. 472. 1810. *Physcia muscigena* (Ach.) Nyl., Acta Soc. Linn. Bordeaux 21:308. 1856.

Thallus foliose, to 12 cm across, loosely attached, pale brown to dark brown, usually more or less covered with a whitish pruina, in some forms densely so, sometimes more or less erect squarrose; the upper side more or less concave; underside black with black squarrose rhizinae; upper cortex paraplectenchymatous; medulla white; lower cortex prosoplectenchymatous.

Apothecia to 5 mm broad, margin often with lobules; hypothecium hyaline, 25–35 μ, upper part brown; paraphyses coherent, capitate, to 5–6 μ, occasionally branched toward the tips; asci 77–107 μ; spores 8, biseriate, brown, the walls uniform in thickness, the lumena rounded, 22–30 × 12–17 μ. Conidia cylindrical, straight, 3.5–6 × 1 μ.

Reactions: all usual reactions negative.
Contents: no lichen substances.

This species is well known as a calciphile and also for growing in well-dunged places such as bird perches. It grows over rocks as well as on humus soils and is particularly common on limestones. It is circumpolar, arctic-alpine, and boreal.

PILOPHORUS Th. Fr.
De Stereocaulis et Pilophoris Commentatio. 1–42. 1857.

Primary thallus crustose, granulose, or areolate, with cephalodia. Pseudopodetia simple or branched with hollow or loose medullary layer, a glomerulate algal layer, covered with squamules or soredia, in some species with a boundary layer of dense tissue at the base of the apothecium.

Apothecia terminal on the pseudopodetia or their branches; black, globose but clavate in 1 species, lacking epithecium but with black substance between the tips of the paraphyses; paraphyses simple, strongly gelatinous; asci narrowly clavate; spores 8, hyaline, simple, spherical to ellipsoid. Cephalodia on the primary thallus or base of the pseudopodetia, sacculate, containing *Nostoc* or *Stigonema* (in 2 European species). Pycnidia black, flask-shaped, at apices of sterile pseudopodetia; fulcra exobasidial; conidia curved, 1 × 5 μ.

Algae: *Protococcus* and with *Nostoc* or *Stigonema* in the cephalodia.

Type species: *Philophorus robustus* Th. Fr.

REFERENCES

Alstrup, V. 1976. Two species of *Pilophorus* new to Greenland. Lichenologist 8:96–97.

Jahns, H. M. 1970. Remarks on the taxonomy of the European and North American species of *Philophorus* Th. Fr. Lichenologist 4:199–213.

———— 1981. The genus *Pilophorus*. Mycotaxon 13:289–330.

1. Pseudopodetia branching, esorediate, a boundary pigmented layer absent at the base of the apothecium . . . . 2. *P. robustus*
1. Pseudopodetia unbranched, short, thickest at the middle, sorediate especially above, apothecia rare but with boundary of pigment at base when present . . . . 1. *P. cereolus*

1. Pilophorus cereolus (Ach.) Th. Fr.
Lich. Scand. 1:55. 1871. *Lichen cereolus* Ach., Lich. Suec. Prodr. 89. 1789.

Primary thallus persistent, of sorediate granules 0.1 mm broad; gray-green and with scattered brown cephalodia. Pseudopodetia short, to 5 mm tall, sorediate, the sterile stalks thickest at the middle, irregularly curved; interior compact, strongly gelatinized; algal layer sorediate.

Philophorus cereolus (Ach.) Th. Fr.

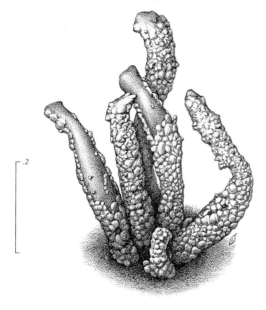

.2

Apothecia rare, solitary at tips of pseudopodetia, black, spherical, to 1 mm broad, a
dark boundary between the stalk and the base of
the apothecium; hymenium 80 μ; spores 8, hyaline, round when young, becoming spindle-
shaped, $5.5–6.5 \times 14.5–21$ μ.

Reactions: K+ yellow.

Contents: atranorin and zeorin.

This species grows on rock outcrops and
stones. Its known range is confused by problems in
nomenclature but it is common in central Europe and
Scandinavia and is reported from Disko Island,
Greenland, Newfoundland, New York, and Minnesota.

2. Pilophorus robustus Th. Fr.
De Stereocaulis et Pilophorus Commentatio. 41. 1857.

Primary thallus granular, gray-green with
the brown cephalodia scattered and also on base
of podetia; the thallus evanescent. Pseudopodetia stout, irregularly to unbellately branched

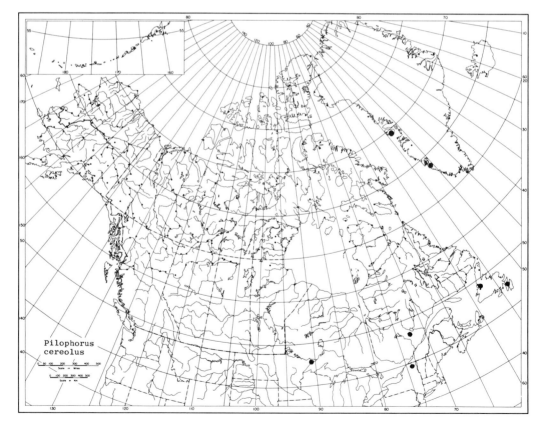

Pilophorus
cereolus

Pilophorus robustus Th. Fr.

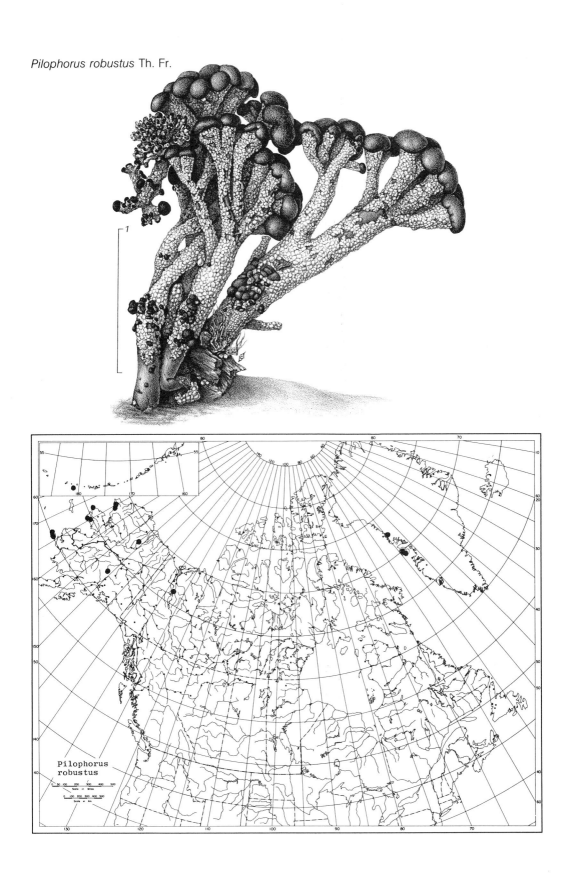

1

Pilophorus
robustus

above, to 2.5 cm tall, not becoming hollow; the hyphae strongly gelatinized, pachydermatous, the cells vertically oriented, algal layer glomerulate.

Apothecia numerous, crowded at the tips of the branches, to 2.5 mm broad, globose; margin of the hymenium incurved and reversing upward before uniting with the pseudopodetium surface, the center with the vegetative tissue extending up into the center of the apothecium and lacking a boundary layer at the base of the apothecium; hymenium 200 μ; asci 8 spored; spores simple, hyaline, spherical to becoming spindle-shaped, 4.0–6.5 × 18–24 μ.

Reactions: K + yellow.

Contents: atranorin and zeorin.

This species grows on rocks and gravels and on outcrops. It is circumpolar but apparently rather disjunct, occurring in Alaska and Greenland in our area.

PLACYNTHIUM (Ach.) S. F. Gray

A Natural Arrangement of British Plants 395. 1821. *Collema* * *Placynthium* Ach., Lich. Univ. 628. 1810. *Pterygium* Nyl., Bull. Soc. Bot. Fr. 1:328. 1854. *Wilmsia* Körb., Parerga Lich. 406. 1865. *Anziella* Gyeln. in Rabenh., Kryptog. Fl. Deutschl. Oesterr. und d. Schweitz 9(2):31. 1940.

Thallus squamulose or stellate radiating, flattened or filiform, a dark prothallus present in some species; with or without isidia; attached by rhizinae. Internal anatomy radial or dorsiventral, the hyphae reticulate or pseudoparenchymatous.

Apothecia hemiangiocarpic with thalloid or proper margin; hymenium brown or green in upper part, blue with I; asci cylindrical with apically thickened wall; spores 8, hyaline, 2-, 3-, or 4-celled, ellipsoid or fusiform; paraphyses septate, ends pointed, thickened or not. Pycnidia with dark or brown ostiole; conidiophores short celled, branched, producing conidia terminally and laterally, the conidia rod- or dumbell-shaped.

Alga: Rivulariaceae.

Type species: *Placynthium nigrum* (Huds.) S. F. Gray.

REFERENCES
Gyelnik. V. 1940. Lichinaceae, Heppiaceae in Rabenh. Krypt. Fl. Deutschl. u. d. Schweiz. 9(2)1:1–134.
Henssen, A. 1963. The North American species of *Placynthium*. Can. J. Bot. 41:1687–1724.

1. Lobes narrow, less than 0.2 mm broad.
    2. Blue-black prothallus usually present, lobes squamulose, not effigurate . . . . 2. *P. nigrum*
    2. Prothallus lacking, lobes filiform or canaliculate, effigurate at the margin . . . . 1. *P. aspratile*
1. Lobes flat, broader, 0.5–1 mm broad, the tips fan-shaped broadened.
    3. Dark prothallus present, thallus dwarf fruticose, central lobes cylindrical or flattened; epithecium greenish blue, blue or brown; hymenium 80–115 μ thick . . . . 3. *P. pannariellum*
    3. Dark prothallus not noticeable, thallus an areolate crust, the central part squamulose; epithecium brown; hymenium 83–95 μ thick . . . . 4. *P. rosulans*

1. Placynthium aspratile (Ach.) Henss.

Can. J. Bot. 41:1699. 1963. *Parmelia aspratilis* Ach. in Weber, *Beitrage zur Naturkunde* 152. 1810. *Collema asperellum* Ach., Lich. Univ. 629. 1810. *Lichen asperellus* (Ach.) Wahlenb., Flora Lapp. 441. 1812. *Pterygium asperellum* (Ach.) Nyl., Not. Sällsk. Fauna Fl. Fenn. Förh. 4:236. 1858. *Placynthium asperellum* (Ach.) Trev., Lichenotheca Veneta 98. 1869. *Placynthium vrangianum* Gyeln. in Rabenh., Kryptog. Fl. Deutschl. 9(2)58. 1940.

Prothallus seldom present. Thallus of more or less aggregated canaliculate lobes, effigurate at the border, the center becoming granulose or isidiate and breaking up into areoles of 1–2 mm diameter; upper side shining, olive or blackish; underside dark blue-green, rhizinose; isidia erect, cylindrical, to 3 mm tall;

hyphae of thallus mainly parallel, more or less connected to an upper pseudoparenchyma, the lower layers blue-green.

Apothecia to 1 mm, with dark proper exciple; the exciple dark, radiate; disk black, urceolate, becoming flat or convex; hymenium 80–100 μ, upper part dark green; paraphyses septate, branched, 2.5 μ thick, the end cells pointed or thickened; subhymenium brown, 100–130 μ, the hyphae interwoven; asci cylindrical, 44–55 × 12.5–18 μ; spores 4 or 8 per ascus, broad or narrowly ellipsoid, hyaline, 2–4 celled, 11.5–18 × 4.5–7 μ.

Algae: Rivulariaceae or Scytonemataceae, mainly in upper part of thallus.

This species is very common on both calcareous and acidic rocks. A circumpolar species, it is widely distributed in arctic Canada from Alert to the north shore of Lake Superior and to Colorado in the western mountains.

The canaliculate shining marginal lobes and the granulose or isidiose center aggregated into areolate masses are characteristic of this species. A dendroid prothallus is mentioned by Henssen as sometimes being developed in arctic specimens.

2. Placynthium nigrum (Huds.) S. F. Gray

A Natural Arrangement of British Plants 395. 1821. *Lichen niger* Huds., Flora Anglica 2:524. 1778. Numerous synonyms can be consulted in Gyelnik (1940).

Prothallus usually present, blue-black, sometimes quite fimbriate. Thallus squamulose, not effigurate, lobes flat, 0.4–1.5 mm broad with crenate or digitate margin, the squamules sometimes scattered over the prothallus, crowded toward the center and forming areoles 1–4 mm broad; cylindrical isidia sometimes present and mainly horizontal; upper side olive or blackish; lower side blue-green and with dark rhizinae; hyphae forming a reticulum in most of the thallus with a few strands in the lower part of the thallus being oriented parallel to the surface and forming a lower cortex.

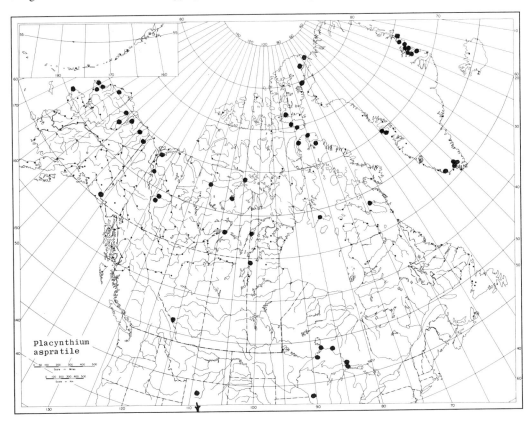

Placynthium
aspratile

Apothecia to 1 mm broad, with dark proper margin; disk brown or black, urceolate, becoming flat or convex; hymenium 70–175 μ, the upper part violet brown or dark green; paraphyses septate, branched, 2.5 μ thick, end cells pointed or thickened; subhymenium brown, 90–280 μ thick, of interwoven or pseudoparenchymatic hyphae; cells at lower part of the dark violet or green exciple producing hairs, the cells of the exciple radiating; asci cylindrical, 39–54 × 8–15.5 μ; spores 8, hyaline, narrowly ellipsoid, 2-, 3-, or 4-celled, 7–17 × 3.5–5.5 μ.

Alga: *Dicothrix* in the Rivulariaceae.

Characteristically this species grows on dry calcareous rocks but may occur on other rocks in more moist places and rarely on other substrata. It is circumpolar, arctic, and temperate. It is easily distinguished by its conspicuous blue-black prothallus and the flat squamulose thallus.

3. Placynthium pannariellum (Nyl.) H. Magn.

Fort. Över Skand. Växter 24: 1936. *Pterygium pannariellum* Nyl., Notis. Sallsk. Fauna Fl. Fenn. Förh. 1:236. 1858. *Leucothecium pannariellum* (Nyl.) Blomb. & Forss., Enum. Plant. Scand. 108. 1880.

Prothallus green, greenish blue, or bluish black. Thallus effigurate, the marginal lobes flat, the tips fan-shaped, more or less branched, to 0.7–1.0 mm broad; the central lobes dwarf fruticose, brown, blackish brown or black, dull, cylindrical or more or less squamulose, to 0.5 mm broad, sometimes cushion-like or forming a thickened areolate crust; lower side blue-green with dark rhizinae; interior hyphae longitudinally extended but forming a pseudoparenchyma, the cells rounded or angular.

Apothecia to 1 mm broad with dark proper margins; disk dull, urceolate to convex, blackish brown to black; hymenium 80–115 μ

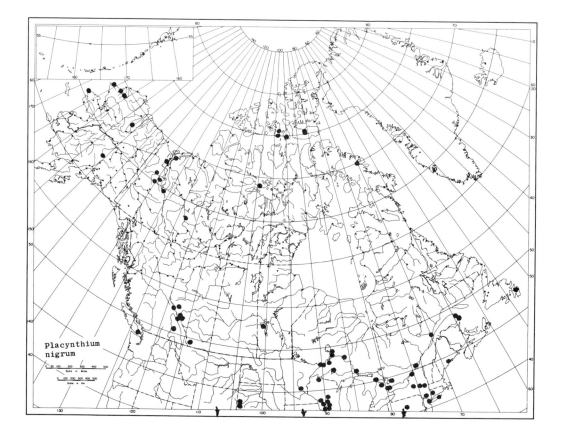

Placynthium
nigrum

thick, hyaline; epithecium greenish blue, blue, or brown; paraphyses septate, unbranched, 3 μ thick, the tips thickened; subhymenium brown or reddish brown; asci clavate, 10–16 μ broad; spores 8, hyaline, ellipsoid, 3–4 celled, 19–20 × 5.5–6.5 μ.

Algae: in Scytonemataceae.

This species grows on siliceous rocks. It is arctic, subarctic, and alpine, growing in Europe and in Greenland. It has not yet been found in North America (Henssen 1963). A report from Maine by Merrill (1914) was based on *P. flabellosum*.

4. Placynthium rosulans (Th. Fr.) Zahlbr.

Cat. Lich. Univ. 3:235. 1925. *Leucothecium rosulans* Th. Fr., Bot. Not. 12. 1863. *Leucothecium coralloides* var. *rosulans* (Th. Fr.) Hellb., Nerikes Lafflora 25. 1871.

Prothallus lacking. Thallus effigurate; the marginal lobes flat, the tips fan-shaped, little branched, about 0.5 mm broad; the central lobes coralloid and forming an areolate crust, dull, chestnut brown or blackish brown; lower side blue-green with dark rhizinae; hyphae of thallus longitudinally extended but forming a pseudo-parenchyma, the cells rounded or angular.

Apothecia adnate, to 0.5 mm broad, with dark proper margin; disk dull, a little concave, becoming flat, blackish brown to black; hymenium 83–95 μ thick, hyaline, the epithecium brown; paraphyses septate, unbranched, 3.5 μ thick, the tips thickened; subhymenium brownish black; exciple paraplectenchymatous, brown or brownish black; asci clavate, 15–17 μ thick; spores 8, hyaline, ellipsoid, 4-celled, 18–19 × 5.5–6 μ.

Algae: Scytonemataceae.

This species grows on siliceous rocks which are occasionally covered by flowing water. It is found in Sweden and has been reported from Greenland (locality unstated) by Henssen (1963). It has not yet been found in North America.

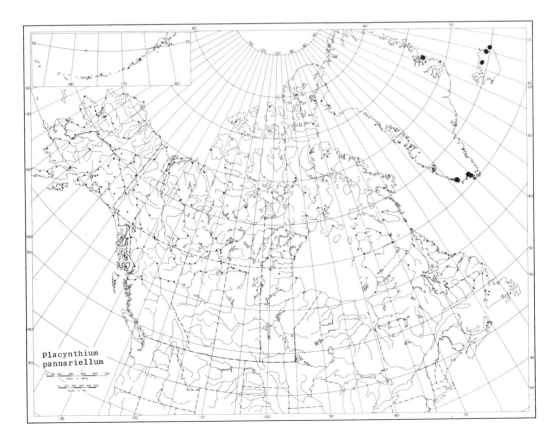

Placynthium
pannariellum

PLATISMATIA W. Culb. & C. Culb.

Contr. U.S. Nat'l. Herb. 34(7):524. 1968.

Thallus foliose, of medium size, lobes to 2.5 cm broad; upper side usually rugose, commonly pseudocyphellate or isidate, ashy or blue-gray; underside with rhizines, black, punctate; upper and lower cortex prosoplectenchymatous.

Apothecia large, to 4 cm broad, marginal or submarginal, commonly perforate; epithecium pale; subhymenium I+ blue; asci clavate; spores 8, hyaline, simple, ellipsoid or subglobose.

Contents: atranorin and caperatic acid, fumarprotocetraric acid in one species.

Type species: *Platismatia glauca* (L.) W. Culb. & C. Culb.

Only one species reaches into the American arctic tundras.

1. Platismatia glauca (L.) W. Culb. & C. Culb.

Contr. U.S. Nat'l. Herb. 34(7):530. 1968. *Lichen glaucus* L., Sp. Pl. 1148. 1753. *Lobaria glauca* (L.) Hoffm., Deutschl. Flor. 149. 1795 (1796). *Cetraria glauca* (L.) Ach., Meth. Lich. 296. 1803. *Physcia glauca* (L.) DC. in Lam. & DC., Fl. Franc. ed. 3. 2:401. 1805. *Parmelia glauca* (L.) Hepp, Lichenen-Fl. Würzburg 23. 1824. *Platysma glauca* (L.) Frege, Deutschl. Bot. Taschenb. 2. Theil für das Jahr 1812:167. 1812.

Thallus medium sized; the lobes to 2 cm broad, margins with soredia, simple or coralloid isidia, or both, or branched and divided into fruticose lobes which in turn may have soredia or isidia or both; upper surface whitish or whitish green, or tinged with yellow, smooth or becoming reticulately wrinkled, not pseudocyphellate; smooth or with coralloid or sorediose isidia in patches or scattered mostly on margins; lower surface jet black, margins brown, colored like the upper surface, or white, smooth or reticulately wrinkled and pitted; rhizines few

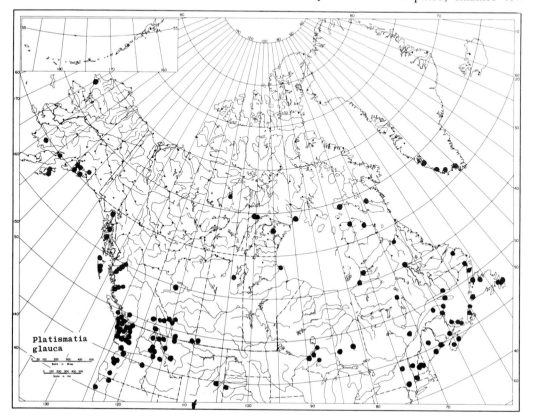

Platismatia glauca

or many, brown or black, simple or branched.

Upper cortex prosoplectenchymatous, 16–26 μ thick; medulla white, 62–200 μ thick, I + lavender; lower cortex prosoplectenchymatous, 16–25 μ thick, I + lavender; lower cortex prosoplectenchymatous, 16–25 μ thick.

Apothecia very rare, to 1 cm broad, marginal, perforate or not; hymenium 34–56 μ; subhymenium 16–52 μ, I + lavender to bright purple; asci clavate, I + blue or blue-green; spores 8, hyaline, simple, ellipsoid to ovoid, 3.5–8.5 × 3–5 μ.

Reactions: medulla K −, C −, KC −, P −, UV −; cortex K + yellow.

Contents: caperatic acid and atranorin.

This species grows on trees and on rocks and rarely soil. It is circumpolar in the northern hemisphere and also is in Patagonia and in the mountains of Kenya and Tanzania. It is in southern Greenland and ranges from Alaska to Baffin Island in North America, south in the boreal forest and in the mountains to California, Colorado, and North Carolina.

POLYCHIDIUM (Ach.) S. F. Gray

A Natural Arrangement of British Plants. 401. 1821. *Collema* subgen. *Polychidium* Ach., Lich. Univ. 658. 1810. *Leptogidium* Nyl., Flora 56:195. 1873. *Dendriscocaulon* Nyl., Lichenes Novae Zeylandica 10. 1888. *Leptodendriscum* Vain., Acta Soc. Flora Fauna Fenn. 7:219. 1890.

Thallus fruticose, dichotomously branched, the branching progressively smaller toward the tips, forming small dendroid masses; the interior hyphae longitudinally oriented in the outer parts, paraplectenchymatous in the older, basal parts; thallus with a single cortical layer of cells or several layers.

Apothecia lateral; disk brown, lacking thalloid margin, with a thick proper exciple and sunken disk; hymenium brown above; subhymenium of loose hyphae; exciple paraplectenchymatous; paraphyses unbranched, septate, capitate; asci cylindrical, the end walls thickened, the walls I + blue; spores 8, 1- or 2-celled, hyaline, ellipsoid or spindle-shaped, thin or thick walled.

Algae: *Nostoc* and *Scytonema*.

Type species: *Polychidium muscicola* (Sw.) S. F. Gray.

REFERENCE
Henssen, A. 1963. Eine revision der flechtenfamilien Lichinaceae und Ephebaceae. Symbol. Bot. Upsal. 18(1):1–123.

1. Thallus tips shining, dichotomous, not coralloid, usually with apothecia . . . . 1. *P. muscicola*
1. Thallus tips dull, coralloid branched, apothecia unknown . . . . 2. *P. umhausense*

1. Polychidium muscicola (Sw.) S. F. Gray

A Natural Arrangement of British Plants. 402. 1821. *Lichen muscicola* Sw., Nova Acta Reg. Soc. Upsal. 4:248. 1784. *Polychidium kalkuense* Räs., Ann. Bot. Soc. Vanamo 18:59. 1943.

Thallus fruticose, chestnut brown or blackish, tips shining, interior of masses dull, forming small cushion-like masses, lobes dichotomous or palmately divided, the branches of equal size or with a main axis and smaller side branches, to 4 mm tall, lobes 0.2 mm thick, terete at the tips; cortex of 1–2 (–3) cells thick with short cilia which toward the base become rhizinae; interior hyphae loosely interwoven, more or less parallel to the surface, toward the base forming a paraplectenchyma.

Algae: *Nostoc*.

Apothecia to 2 mm broad; disk red-brown; exciple of the same color as the thallus; subhymenium 70–110 μ; hymenium 90–100 μ; paraphyses septate, unbranched, 1–2 μ thick; asci clavate to broadly cylindrical 45–62 × 9–22 μ; spores 8, 2-celled, uniseriate or biseriate, spindle-shaped, 19–25.5 × 4.5–6.5 μ. Conidia 1.5–3.5 × 1 μ.

This species grows among mosses usually in moist situations and over acidic rocks. It is known

Polychidium muscicola (Sw.) S. F. Gray

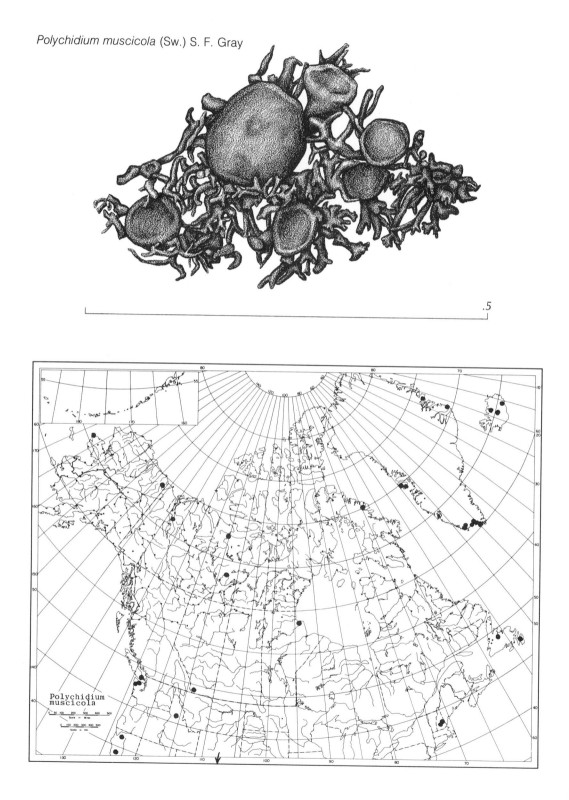

.5

Polychidium
muscicola

from Europe, Iceland, Greenland, and in North America where its range is from Alaska to Ellesmere Island south to New England, Colorado, and California.

2. Polychidium umhausense (Auersw.) Henss.

Symbol. Bot. Upsal. 18(1):102. 1963. *Cornicularia* (?) *umhausensis* Auersw., Hedwigia 8:113. 1869.

Thallus small, fruticose, brown, paler in shaded interior of the cushion-like mass; the tips

coralloid, the branching irregular; sides with many fine white hairs; cortex 3–4 cells thick; interior hyphae longitudinally oriented and radiating out into the branches.

Apothecia unknown.

This species grows among mosses in moist situations. It is known from Europe and from North America in North Carolina, Massachusetts, and Alaska.

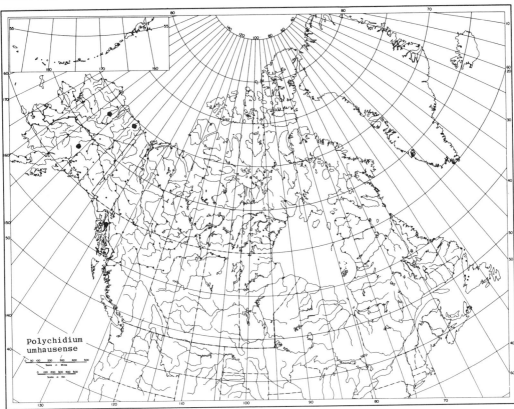

Polychidium
umhausense

PSEUDEPHEBE Choisy

Icon. Lich. Univ. ser. 2, fasc. 1, unpaged. 1930. *Alectoria* sect. *Pseudephebe* (Choisy) Choisy, Bull. Mens. Soc. Linn. Lyon 24:26. 1955. *Parmelia* sect. *Teretiuscula* Hillm. Rabenh. Krypt. Fl. 9, 5(3):104. 1936. *Alectoria* sect. *Subparmelia* Degel., Nyt. Mag. Naturvid. 70:286. 1938. *Alectoria* sect. *Teretiuscula* (Hillm.) Lamb, Br. Antarct. Surv. Sci. Rep. 38:26. 1964. *Alectoria* subgen. *Teretiuscula* (Hillm.) Hawksw. Lichenologist 5:201. 1972.

Thallus fruticose, prostrate, partly subcrustose in the center, attached by hapters; branching isotomic dichotomous, terete or dorsiventrally compressed, dull or shiny, lacking spinules, isidia, soredia or pseudocyphellae; cortex of longitudinally oriented hyphae which are prosoplectenchymatous in a single layer toward the surface; medullary hyphae without projections.

Apothecia lateral, thalloid margin of same color as the thallus, sometimes ciliate; asci clavate, thick walled; spores 8, lacking distinct epispore, hyaline, ellipsoid, simple. Pycnidia common.

Type species: *Pseudephebe pubescens* (L.) Choisy.

REFERENCES

Brodo, I. M. & D. L. Hawksworth. 1977. *Alectoria* and allied genera in North America. Opera Botanica 42:1–164.
Degelius, G. 1937. Lichens from southern Alaska and the Aleutian Islands, collected by Dr. E. Hulten. Medd. Göteborgs Bot. Trad. 12:105–144.
DuRietz, G. E. 1926. Vorarbeiten zu einer "Synopsis Lichenum" I. Die Gattungen *Alectoria, Oropogon* und *Cornicularia*. Ark. f. Bot. 20A(11):1–43.
Hakulinen, R. 1965. Ueber die Verbreitung und das Vorkommen einiger Nördlichen Erd- und Steinflechten in Ostfennoskandien. Aquilo, Ser. Bot. 3:22–66.
Hawksworth, D. L. 1972. Regional studies in *Alectoria* (Lichenes) 2. The British Species. Lichenologist 5:181–261.

1. Branches terete, evenly thickened, the apices not attached, the internodes elongate, giving a loose hairy appearance to the plant . . . . 2. *P. pubescens*
1. Branches becoming more or less flattened, unevenly thickened, the tips becoming adnate, the internodes short, giving the plant a denser appearance than the previous . . . . 1. *P. minuscula*

1. Pseudephebe minuscula (Nyl. ex Arn.) Brodo & Hawksw.

Opera Botanica 42:140. 1977. *Imbricaria lanata* var. *minuscula* Nyl. ex Arn., Ver. Zool.-Bot. Ges. Wien 28:293. 1878. *Parmelia minuscula* (Nyl. ex Arn.) Nyl., Bull. Soc. Linn. Normand. ser. 4, 1:205. 1887. *Alectoria minuscula* (Nyl. ex Arn.) Degel., Nyt. Mag. Naturvid. 78:286. 1938. Additional nomenclatural notes may be sought in Brodo & Hawksworth.

Thallus fruticose, very closely adpressed to the substratum, becoming subcrustose; branches flattened, to 1 mm broad, more or less terete at tips, irregularly isotomic dichotomous with numerous short lateral branchlets, internodal distance short, less than 1 mm; dark brown to black, shiny or dull, the base sometimes pale.

Apothecia frequent, lateral; the disk to 3 mm broad, dark red-brown to black, margins concolorous with the thallus; spores 8, hyaline, ellipsoid, 7–8 × 6.7–9 μ. Pycnidia abundant, imbedded in the thallus; conidia straight, 6–8 × 1–2 μ.

Reactions: all negative.

Contents: no lichen substances known.

This species is common on rock faces, boulders, and on windswept gravels of eskers and ridges. It is circumpolar arctic-alpine in both the northern and southern hemispheres. It seems to prefer drier locations than *P. pubescens,* being inland along the Pacific coast as compared with that species (see maps). In the east it occurs south to Newfoundland, in the west south to Arizona in the mountains.

2. Pseudephebe pubescens (L.) Choisy

Icon. Lich. Univ. ser. 2, 1: no paging. 1930. *Lichen pubescens* L., Sp. Plant. 2:1155. 1753. *Cornicularia pubescens* (L.) Ach., Meth. Lich. 305. 1803. *Usnea lanea* Ehrh. ex Hoffm., Deutsch. Fl. 2:135. 1796. *Cornicularia lanata* var. *nitida* Ach., Lich. Univ. 616. 1810. *Parmelia pubescens* (L.) Vain., Medd. Soc. Fauna Flora Fenn. 14:22. 1886. *Alectoria lanea* (Ehrh. ex Hoffm.) Vain., Medd. Soc. Fauna Flora Fenn. 14:21. 1886. *Alectoria pacifica* Stiz., Proc. Calif. Acad. Sci. 5:537. 1895. *Bryopogon pubescens* (L.) Choisy, Bull. Mens. Soc. Linn. Lyon 21:172. 1952.

Thallus fruticose, prostrate, closely appressed to the substrate, usually less than 1 cm tall; branching isotomic dichotomous, interwoven, terete to slighly compressed to 0.2 mm diameter; usually concolorous but sometimes paler at the base, dark brown to black, shining; hapters attaching the plant to the substratum; spinules, soredia, isidia, pseudocyphellae all lacking.

Apothecia lateral, the margin concolorous with the thallus; the disk gray-black or brown, flat to convex, to 5.5 mm broad; spores 8, simple hyaline, ellipsoid 7–12 × 6–8 μ.

Reactions all negative.

Content: no lichen substances known.

This species grows on rock faces, boulders, and gravels. It is an arctic-alpine species which tends to be more abundant northward. In North America it ranges southward to Labrador in the east and to Guadalupe in the west. It is also circumpolar in the southern as well as the northern hemisphere.

Pseudephebe minuscula (Nyl. ex Arn.) Brodo & Hawksw.

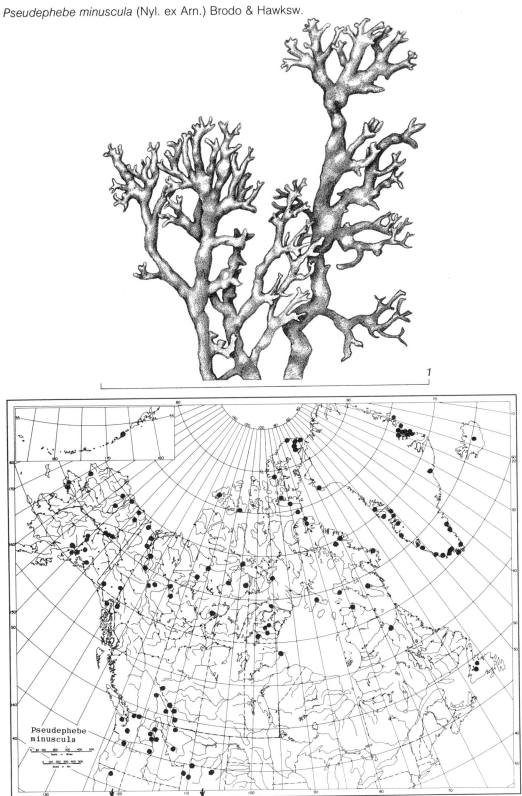

1

Pseudephebe
minuscula

Pseudephebe pubescens (L.) Choisy

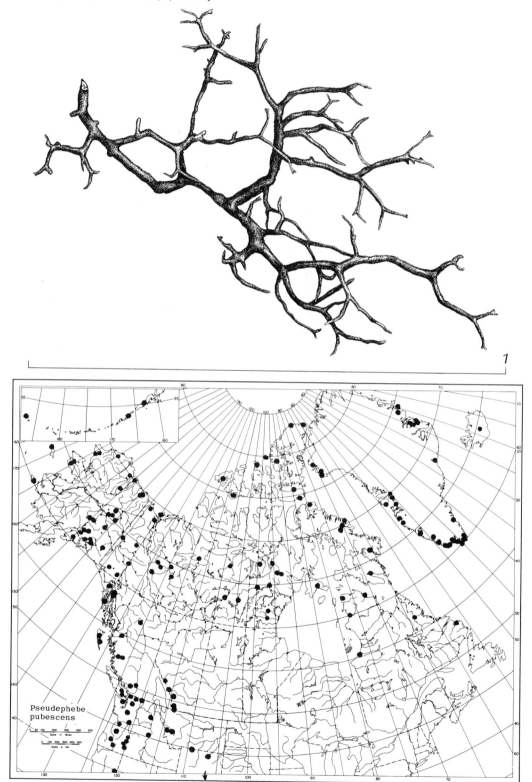

1

Pseudephebe
pubescens

PSOROMA Michx.

Flora Bor.-Amer. 321. 1803.

Thallus squamulose, granulose-squamulose, close to substratum or raised; upper cortex paraplectenchymatous of several cells thickness; algal and medullary layers indistinct, lower cortex of interwoven hyphae, attached by rhizinae.

Apothecia adnate to sessile; disk concave or flat, red to red-brown; thalloid exciple colored like the thallus, the reverse with short hairs, crenate to granulose; subhymenium yellowish brown; hymenium hyaline, I+ blue; paraphyses simple, coherent; asci clavate, the apical apparatus with a distinct manubrium (cuff-like); spores 8, hyaline, 1-celled with a thin but warty wall, ellipsoid to ovoid.

Algae: *Myrmecia*.

Type species: *Psoroma hypnorum* (Vahl.) S. F. Gray.

REFERENCE
Jørgensen, P. M. 1978. The lichen family Pannariaceae in Europe. Opera Bot. 45:1–123.

1. Psoroma hypnorum (Vahl) S. F. Gray

Nat. Arr. Brit. Plants 1:445. 1821. *Lichen hypnorum* Vahl. Flora Danica 6, fasc. 16:8. 1787.

Thallus of coarsely granular squamules, to 1 mm long, 0.5 mm broad; yellowish brown to brown, greener when wet and with a golden luster, over a pale hypothallus, often mixed with cephalodia containing *Nostoc;* upper cortex 30–40 μ, paraplectenchymatous; medullary layer usually with the yellow-green *Myrmecia* algae but some squamules may have clusters of *Nostoc.*

Apothecia sessile to slightly immersed among the squamules, to 5 mm broad; disk concave to flat, red-brown, dull; margin thick, inflexed, with short irregular hairs, concolorous with the thallus, a poorly developed proper margin within; subhymenium yellowish brown; hymenium 100–120 μ, upper part brown, lower hyaline, I+ deep blue; paraphyses simple, 1–2 μ, slightly clavate; asci clavate to subcylindric with an I+ blue apical apparatus with a distinct manubrium; spores 8, hyaline, simple, ellipsoid to spindle-shaped or ovoid, rough walled, 8–12 × 19–34 μ.

Reactions: all negative except those of the hymenium I+ blue.

Contents: none known.

This species grows among mosses and on soil, occasionally on the bases of woody plants, usually on much humus, rarely on barks higher up. It is common in boreal and arctic regions and is circumpolar.

PYCNOTHELIA (Ach.) Duf.

Rev. Clad. 5. 1817. *Baeomyces β Pycnothele* Ach., Meth. Lich. XXXVI & 323. 1803. *Cenomyce* †††† *Pycnothelia* (Ach.) Ach., Lich. Univ. 571. 1810. *Cenomyce* II. *Retiporae* Del. in Duby, Bot. Gall. 620. 1830. *Cladonia* sect. II. *Pycnothele* (Ach.) Fr., Lich. Eur. Ref. 242. 1831. *Cladonia* subgen. *Pycnothelia* (Ach.) Vain., Acta Soc. Fauna Flora Fenn. 4:47. 1887.

Primary thallus long persistent crustose, verruculose, lacking a true cortex. Psuedopodetia simple or sparingly branched, persistent at base, neither cup-forming nor squamulose or sorediate; true cortex lacking, the outer medullary layer with the algal cells forming a persistent and subcontinuous pseudocortex; inner medullary or cartilaginous layer variable in thickness and not clearly defined.

Psoroma hypnorum (Vahl) S. F. Gray

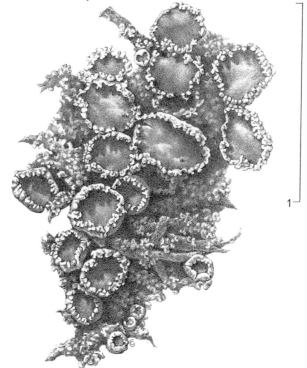

1

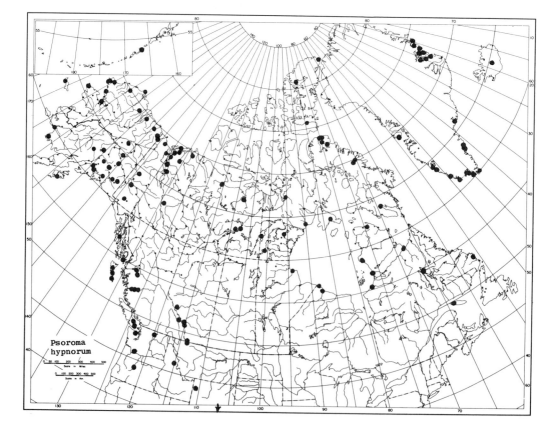

Psoroma
hypnorum

Apothecia small, sub-peltate and forming crowded apical clusters, often lacking. Spores simple or becoming 1 septate.

Only one species occurs in this genus which is often classified in the genus *Cladonia*.

1. Pycnothelia papillaria (Ehrh.) Duf.

Rev. Clad. 5. 1817. *Lichen papillaria* Ehrh., Hannov. Magaz. 218. 1783. *Cladonia papillaria* (Ehrh.) Hoffm., Deutschl. Flora 2:117. 1796.

Primary thallus persistent, crustose, greenish gray, of low minute contiguous verruculae. Pseudopodetia simple wart-like projections, papilliform, simple or sparingly or densely branched, cupless; smooth or verruculose but without true cortex; very fragile when dry.

Reactions: K + yellow, KC −, C −, P −.

Contents: atranorin and d-protolichesterinic acid.

This species grows on sandy soil in open fields, old woods, and the edges of rock outcrops. It ranges in Europe and North America, in the latter from Greenland and Nova Scotia to Georgia, Alabama, and Arkansas. It reaches the Arctic only in Greenland where it is known only from Quqertatsiaq in the extreme south collected by Eberlin in the period 1883–1885.

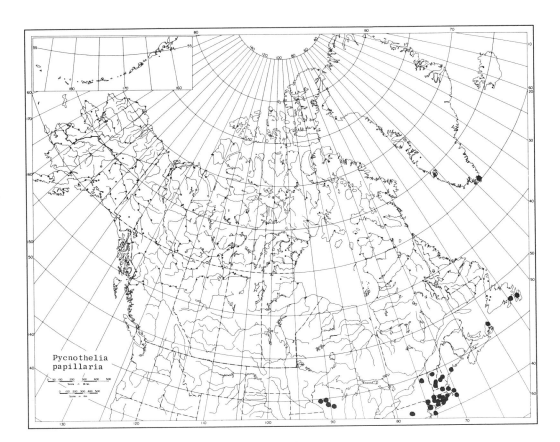

Pycnothelia
papillaria

RAMALINA Ach.

Meth. Lich. 122. 1810.

Thallus fruticose, erect to hanging, attached to the substrate by a holdfast; flattened or cylindrical, the entire surface corticate with more or less longitudinally oriented hyphae or else hyphae at right angles to the length; an inner mechanical layer of cartilaginous strands close under the outer cortex; the medulla of arachnoid hyphae; soredia common; pseudocyphellae common; the surface often longitudinally rugose.

Apothecia terminal or lateral, flat or concave; margin thalloid, concolorous with the thallus; disk pale, greenish, pruinose or not; epithecium pale; hypothecium hyaline; paraphyses unbranched, clavate, coherent; asci clavate; spores 8, hyaline, 2- or 4-celled, straight or curved, ellipsoid to spindle-shaped.

Algae: *Trebouxia*.

Type species: *Ramalina calicaris* (L.) Fr.

REFERENCES

Asahina, Y. 1938. *Ramalina* Arten aus Japan. J. Jap. Bot. 14:721–730; 15:205–223. 1939.
Bowler, P. A. & P. W. Rundel. 1974. The *Ramalina intermedia* complex in North America. Bryologist 77:617–623.
Bowler, P. A. 1977. *Ramalina thrausta* in North America. Bryologist 80:529–532.
Howe, R. H., Jr. 1913–1919. North American species of the genus *Ramalina*. Bryologist 16:65–74, 81–89; 17:1–7, 17–27, 33–40, 49–52, 65–69, 81–87.
Keissler, K. von. 1960. Usneaceae in Rabenhorsts Kryptogamen-Flora IX. (5)4.
Krog, H. & P. W. James. 1977. The genus *Ramalina* in Fennoscandia and the British Isles. Norw. J. Bot. 24:15–43.
Nylander, W. 1870. Recognito monographica Ramalinarum. Bull. Soc. Linn. Normand. Ser. 2, 4:101–181.

1. Cortex well developed, distinctly two layered and chondroid, tips of branches not hooked.
    2. Soralia cupuliform, mainly on tips; plants partly hollow with few fenestrations; containing evernic and obtusatic acids . . . . 2. *R. obtusata*
    2. Soralia maculiform, lateral and laminal to terminal.
        3. Containing evernic and obtusatic acids . . . . 3. *R. pollinaria*
        3. Containing sekikaic and homosekikaic acids . . . . 1. *R. intermedia*
1. Cortex poorly developed with a thin layer of fibrous periclinal hyphae longitudinally oriented over the chondroid inner layer; branches often hooked fiddle-head-like around a few coarse soredia . .
    . . . . 4. *R. thrausta*

1. Ramalina intermedia (Del. ex Nyl.) Nyl.

Flora 56:66. 1873. *Ramalina minuscula* ssp. *intermedia* Del. ex Nyl., Bull. Soc. Linn. Normandie 4:166. 1870.

Thallus fruticose, small, 1–5 mm tall; much branched, more or less rigid, the branches flattened to subcylindrical, the apices tapering to isidioid tips, basal attachment plate-like; cortex greenish-yellow, smooth and shiny; soralia marginal and terminal; the soredia whitish, occasionally bearing isidioid spinules; medulla lax, white.

Apothecia rare, marginal, subterminal; disk flat to concave or convex, to 3 mm broad, disk pale yellowish green; spores uniseptate, slightly curved, 4×10–$12 \ \mu$. Pycnidia not seen.

Reactions: K$-$, C$-$, KC$-$, P$-$.

Contents: sekikaic, homosekikaic, and usnic acids, $\pm$ atranorin and an unknown (Bowler & Rundel 1974).

Usually this species grows on moist cliffs and rocks in shaded conditions, rarely on bark. It is boreal and temperate ranging from the vicinity of the mouth of the Mackenzie River south to Arizona and New Mexico, eastward to Quebec and Nova Scotia and is common in the Appalachian Mountains. It barely reaches the Arctic. As it occurs rarely in Europe it may be considered to be amphi-Atlantic.

2. Ramalina obtusata (Arn.) Bitt.

Jahrb. Wiss. Bot. 36: 435. 1901. *Ramalina minuscula* var. *obtusata* Arn., Verh. Zool. Bot. Ges. Wien 25:472. 1875.

Ramalina intermedia (Del. ex Nyl.) Nyl.

Thallus fruticose, tufted, short, to 3 cm tall, the lobes 2–4 mm broad, fistulose; cortex shining smooth or obscurely spotted, pale green to straw-colored, pellucid when wet, with few fenestrations; soralia common, in cupule shaped tips or occasionally labriform; the soredia farinose.

Apothecia not known.
Reactions: K−, C−, KC−, P−.
Contents: usnic and evernic acids.

This species grows on twigs of coniferous and deciduous trees in the boreal forest, rarely on rocks. It has seldom been collected so that its range is uncertain but appears to be similar to that of *R. intermedia* in appearing amphi-Atlantic.

3. **Ramalina pollinaria** (Westr.) Ach.

Lich. Univ. 608. 1810. *Lichen pollinarius* Westr., Kgl. Vitensk. Acad. Nya Handl. 16:56. 1795.

Thallus fruticose, separate or in swards, tufted; small but reported to reach 6 cm length;

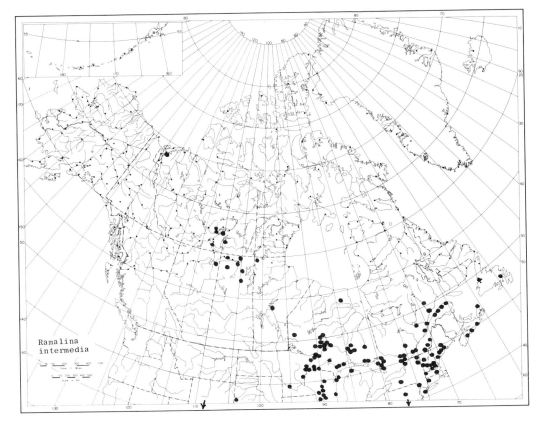

Ramalina
intermedia

Ramalina obtusata (Arn.) Bitt.

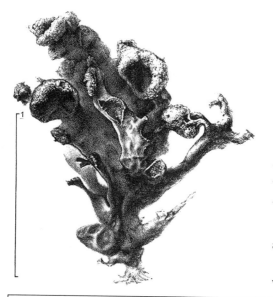

usually much branched, the lobes flat to terete, especially toward the apices, with numerous nodulose proliferations, sometimes deeply notched or incised; surface smooth, shining, appearing cartilaginous, the larger lobes slightly foveolate; soralia irregularly spreading, not well defined, mainly laminal, subterminal to terminal, rarely marginal, often seemingly sublabriform at the tips; the soredia somewhat granular.

Apothecia rare, described as being mainly terminal, to 10 mm broad but usually less; margin concolorous with the thallus, the reverse smooth or furrowed; disk concave or flat, brown, usually pruinose; epithecium brownish; hypothecium brownish; spores 8, hyaline, straight, elongate, spindle-shaped, 2-celled, $4–5 \times 10–15$ μ.

Reactions: $K-$, $C-$, $KC+$ yellow, $P-$.
Contents: usnic, evernic, and obtusatic acids.

This species grows upon bark and twigs, dead wood and rocks. It is circumpolar, boreal, and tem-

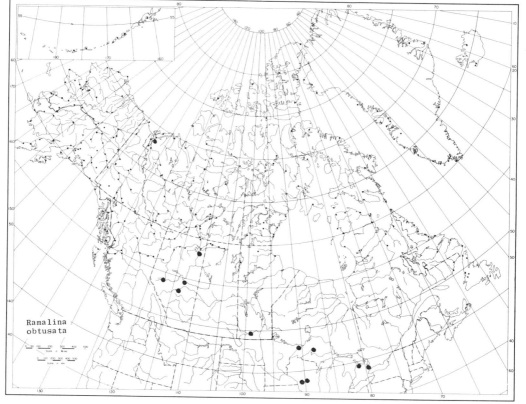

Ramalina
obtusata

Ramalina pollinaria (Westr.) Ach.

perate with a poorly defined North American range.

This species and *R. polymorpha* may be mistaken but the latter is reported by Krog & James (1977) as having a rough cortex and a lack of medullary substances. The chemistry will also separate this from *R. intermedia* in cases of doubt although usually the wider branches, more than 1 mm, and irregular formation of soredia in *R. polymorpha* versus more slender, less than 1 mm, branches and small oval soralia in *R. intermedia* differentiate them.

4. **Ramalina thrausta** (Ach.) Nyl.

Syn. Lich. 1:296. 1860. *Alectoria thrausta* Ach., Lich. Univ. 596. 1810. *Alectoria crinalis* Ach., Lich. Univ. 594. 1810. *Ramalina crinalis* (Ach.) Gyeln., Fedde Repert. 38:251. 1935.

Thallus thread-like, pendent or prostrate, the branches terete, interwoven, divergent or closely angled, the tips thin, filamentous, straight or curled like a fiddle-head; the surface smooth, glossy rarely with a few oval pseudocyphellae; the cortex a thin layer of increased cellular branching over an inner prosoplecten-

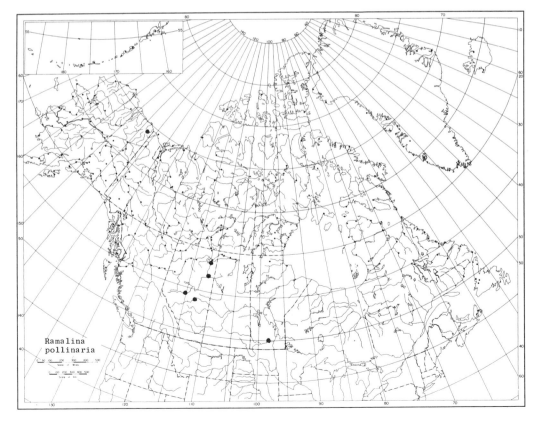

Ramalina
pollinaria

Ramalina thrausta (Ach.) Nyl.

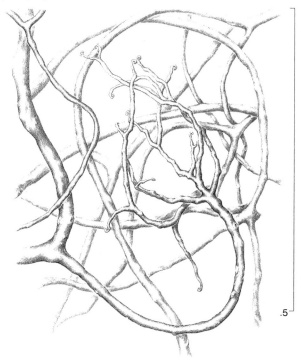

.5

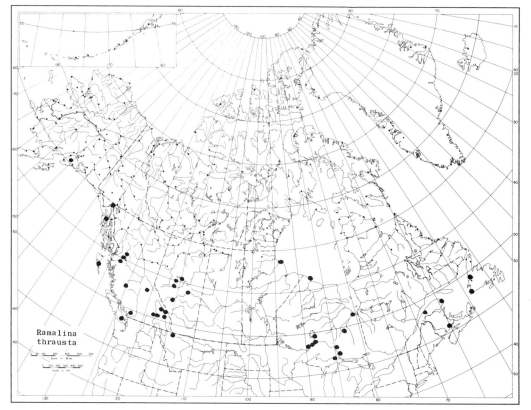

Ramalina
thrausta

chymatous chondroid sheath; soredia present or absent, from well-formed soralia and may be accompanied by isidioid spinules; medulla arachnoid, white.

Apothecia unknown.

Reactions: K −, KC + yellow, P −.

Contents: an unknown close to divaricatic acid and usnic acid, a rare trace unknown, or sometimes no medullary substance (Bowler, 1977). Usnic acid in the cortex.

This species is corticolous on conifers and broad-leaved trees and occasionally is saxicolous or terricolous, occurring on gravel in tundra in the northern Manitoba collection cited by Bowler (1977). It is circumpolar in the boreal forest, reaching only the southern edge of the tundra. In North America it occurs from Alaska to Newfoundland, occurring in the tundra near Churchill, Manitoba.

SIPHULA Fr.

Syst. Orb. Veget. 1:238. 1825.

Thallus fruticose, erect, sparingly branched, sparingly attached at the base to the substratum by rhizinae or the rhizinae lacking by death of the lower part as the podetia grow up through mosses; the podetia longitudinally sulcate, swollen at the nodes, simple or sparingly branched to caespitose; a cortex of paraplectenchymatous hyphae, the medullary layer of compact hyphae more or less longitudinally oriented, the algae in a glomerulate layer in the outer part of the medullary layer.

Apothecia known only in *S. patagonica* Vain. of lecanorine type with thalloid margin, 2 mm. broad, at the tip of branches; hypothecium and hymenium hyaline; asci with 4–6 spores; the spores 32 septate, spindle-shaped, hyaline, $16-20 \times 6-7 \ \mu$.

Algae: *Trebouxia.*

Type species: *Siphula ceratites* (Wahlenb.) Fr.

REFERENCES

Arwidsson, T. 1926. Die Verbreitung von *Siphula ceratites* (Wg.) E. Fr. Anlässlich der Auffindung der Art in Schweden. Bot. Not. 1926:379–392.

Faegri, K. 1952. Om utbredelsen av *Siphula ceratites* (Wbg.) E. Fr. i Norden. Blyttia 10:77–87.

Hakulinen, R. 1965. Über die Verbreitung und das Vorkommen einiger nördlichen Erd- und Steinflechten in Ostfennoskandien. Aquilo, Ser. Botanica 3:22–66.

Mathey, A. 1971. Contribution à l'étude du genre *Siphula* (Lichens) en Afrique. Nova Hedw. 22:795–878.

Räsänen, V. 1937. Das Rätsel der Flechtengattung *Siphula* Fr. gelöst. *Siphula patagonica* Vain. mit Apothecien und Sporen gefunden. Annal. Bot. Soc. Zool. Bot. Fenn. 'Vanamo' 8(7):3–8.

1. Siphula ceratites (Wahlenb.) Fr.

Syst. Orb. Veg. 1:238. 1825. *Baeomyces ceratites* Wahlenb., Flora Lapponica 459. 1812.

Podetia to 7 cm. tall, 2 mm. broad, sparingly dichotomously branched, the branches ascending and swollen at the nodes, white when fresh, darkening in the herbarium to grayish or yellowish-gray, longitudinally furrowed, lacking soredia or isidia, internal structure as above.

Apothecia not known.

Reactions: C −, K + orange, P −.

Contents: siphulin and siphulitol (Culberson 1969).

This species grows in cold seepages, sometimes under the water, below permanent or nearly permanent snowbanks. Along the Pacific coast it is also reported to grow in seepages. It is also found beside pools in the tundra.

This species ranges as a circumpolar high arctic species from northern Norway, Sweden, Finland, across Siberia, to North America where it occurs in Alaska, south to British Columbia, in the Northwest Territories to Philpots Island east of Devon Island, in Labrador and is known from three localities in western Greenland.

Siphula ceratites (Wahlenb.) Fr.

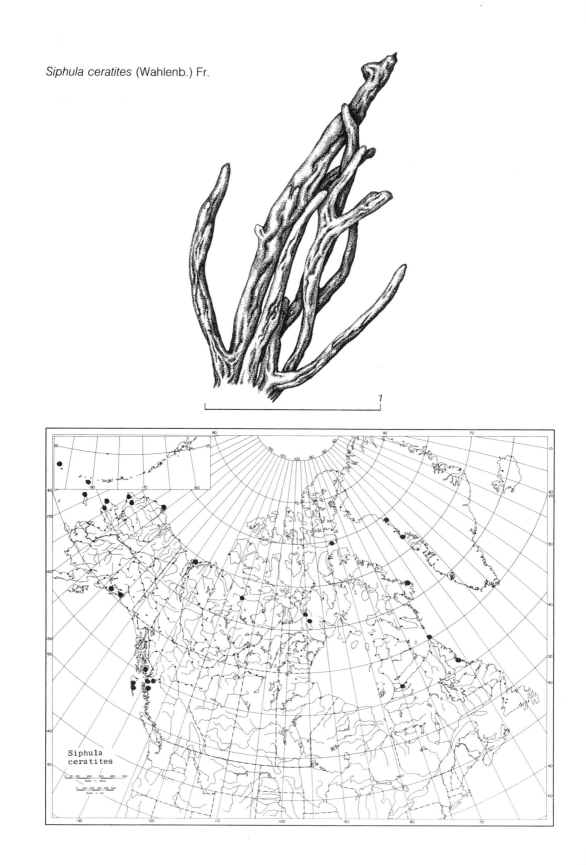

1

Siphula
ceratites

SOLORINA Ach.

Vet. Akad. Nya Handl. 228. 1808.

Thallus foliose or, in one species, reduced to a collar around the apothecium, the lobes elongate or short; upper cortex paraplectenchymatous; lower cortex paraplectenchymatous in the vicinity of the apothecia; heteromerous, the algae *Nostoc* and *Coccomyxa*, the two in separated layers in the same thallus in *S. crocea*, the *Nostoc* in internal or external cephalodia in the other species.

Apothecia round, immersed in the upper surface; the disk slightly to very deeply concave, red-brown; proper exciple lacking; subhymenium shallow; hymenium hyaline to brownish; paraphyses unbranched, the tips red-brown; asci clavate; spores 2–4–8, brown, ellipsoid with a constriction between the two cells.

Algae: *Nostoc* and *Coccomyxa*.

Type species: *Solorina crocea* (L.) Ach.

REFERENCES
Gilbert, O. L. 1975. Distribution maps of lichens in Britain. Lichenologist 7:180–192.
Thomson, N. F. & J. W. Thomson. Spore ornamentation in the lichen genus *Solorina*. Ms to Bryologist.

1. Underside of thallus red-orange or yellow . . . . 2. *S. crocea*
1. Underside of thallus gray to brown.
   2. Ascus containing 4 spores.
      3. Thallus broad, foliose lobed, lacking external cephalodia . . . . 4. *S. saccata*
      3. Thallus reduced, consisting of granules, squamules and external cephalodia around the
         apothecium . . . . 5. *S. spongiosa*
   2. Ascus containing 2 or 8 spores, thallus sometimes of small squamules with deep disk of apothecium in center.
      4. Ascus containing 2 large spores 60–104 × 34–40 μ, thallus usually pruinose . . . . . 1. *S. bispora*
      4. Ascus containing 8 smaller spores, 34–40 × 18–21 μ, thallus usually not pruinose . . . .
       . . . . 3. *S. octospora*

1. Solorina bispora Nyl.

Synops. Lich. 1:331. 1860.

Thallus foliose, sometimes reduced to a small area around an apothecium, sometimes slightly lobate, slaty or reddish brown, white pruinose to becoming scabrid above, the margins slightly raised; underside pale, attached centrally by rhizinae; cortex paraplectenchymatous; algae *Coccomyxa* (green).

Apothecia deeply sunken in the upper surface of the thallus; disk red to red-brown or blackish, to 2 mm broad, sometimes surrounded by remnants of the cortex through which the apothecium broke; subhymenium hyaline; hymenium hyaline below, reddish in upper part, 220–260 μ; paraphyses slightly branched, septate, the tips little thickened, sometimes darkened; asci clavate, 180 × 54 μ; spores 2, becoming red-brown, 1-septate, ellipsoid, the walls sometimes roughened, 60–104 × 34–40 μ.

This species grows on calcareous soils, sometimes on humus, at edges of frost boils, on earth edges, usually in somewhat moist sites. It is circumpolar arctic-alpine. Its range is poorly known in North America, especially in the east where it has not yet been collected south of the Arctic Islands. It is known from the Belcher Islands in Hudson Bay, from Churchill, and south to Colorado in the Rocky Mountains.

2. Solorina crocea (L.) Ach.

Vet. Akad. Nya Handl. 228. 1808. *Lichen croceus* L., Sp. Pl. 1149. 1753.

Thallus foliose, forming rosettes, dark green when moist, ashy to reddish brown when dry; lobes large, to 3 cm broad in large exam-

Solorina bispora Nyl.

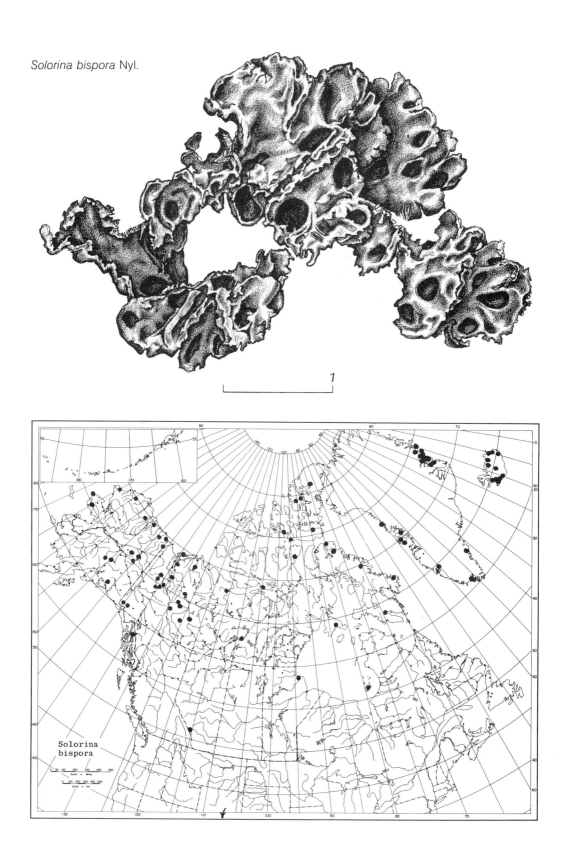

1

Solorina
bispora

Solorina crocea (L.) Ach.

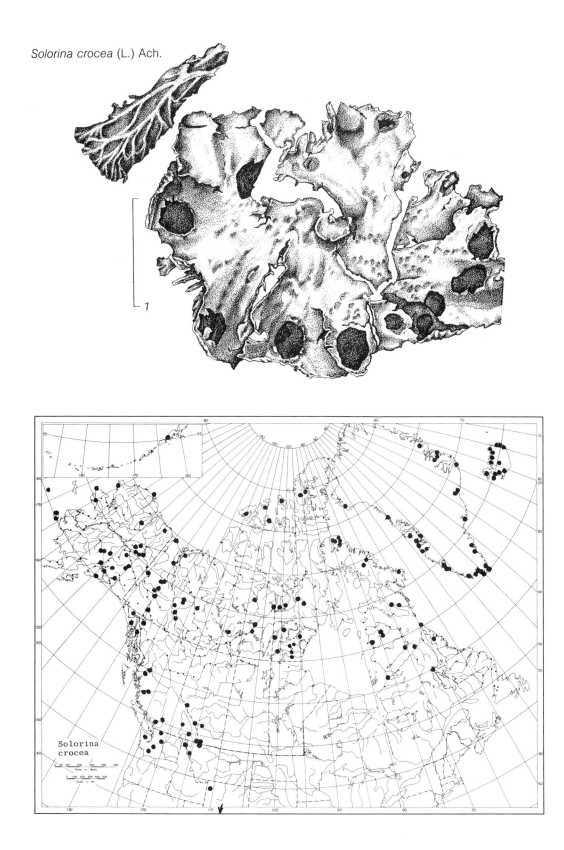

1

Solorina
crocea

ples, flat to slightly concave; upper surface smooth to roughened; lower side with conspicuous orange flabellate veins and yellow to orange interspaces; rhizinae black; upper cortex thick, paraplectenchymatous, algae in two distinct layers, *Coccomyxa* above, *Nostoc* below; lower cortex only under apothecia and there paraplectenchymatous; medullary hyphae orange.

Apothecia round to slightly angular, immersed in the upper surface, to 1 cm broad, redbrown to chestnut brown; subhymenium hyaline, 40–70 μ; hymenium 140–180 μ, hyaline below, yellow above; paraphyses coherent, unbranched, septate, the tips thickened and redbrown; asci 110–130 × 22–24 μ; spores 6 or 8, brown, 2-celled not constricted between the cells, 34–42 × 10–14 μ.

This species prefers moist seepage areas, below late snowbanks, in spring seepages, over clay soils, but may also be found on frost boils and in drier limestone barrens. It is circumpolar arctic-alpine, ranging in North America south to Labrador, Quebec, New Mexico, and Washington.

3. Solorina octospora (Arn.) Arn.

Verhandl. zool. bot. Ges. Wien 26:371. 1876. *Solorina saccata* var. *octospora* Arn., Verhandl. zool. bot. Ges. Wien 23:103. 1873.

Thallus foliose, several lobes or reduced to a single lobe surrounding the apothecium; thin, reddish brown, epruinose, somewhat concave with raised margins, upper surface roughened; underside pale to reddish brown, with pale to brown rhizinae; upper cortex paraplectenchymatous, algae *Coccomyxa* (green); KC + red, containing methylgyrophorate.

Apothecia in the upper surface of the thallus, not quite as sunken as in *S. bispora* but concave, often surrounded by remnants of cortex; disk reddish brown; subhymenium pale brownish; hymenium hyaline below, yellowing above, 210–260 μ; paraphyses coherent, septate, scarcely branched, the tips darkening, not thickened; asci clavate, 180 × 26 μ; spores 8, 2-celled, ellipsoid, becoming red-brown, 35–40 × 18–21 μ.

This species grows on soils and humus, sometimes on soil over rocks. Probably circumpolar arctic-alpine, it is known from Europe, Greenland, Spitzbergen, Iceland, and in North America from Baffin Island to Alaska, south in the Rocky Mountains to Colorado and New Mexico. It is a rarely collected species.

4. Solorina saccata (L.) Ach.

Vet. Akad. Nya Handl. 228. 1808.

Thallus foliose, membranaceous, thin; lobes to 2 cm broad, seldom broader, forming larger groupings of lobes, loosely attached, flat or slightly convex, sometimes slightly scrobiculate, apple green to slaty when fresh, becoming reddish in the herbarium, occasionally slightly pulverulent above; underside pale tan with scattered pale rhizinae; upper cortex paraplectenchymatous; medullary layer with only 1 alga, *Coccomyxa* but *Nostoc* in internal cephalodia; lower cortex lacking.

Apothecia rounded, 2–5 mm broad, immersed in upper side of thallus, sometimes several per lobe; lacking proper exciple, breaking through the upper cortex and with remnants of this sometimes around the periphery of the disk, which is deep in the thallus; the disk dark brownred; hymenium 120–280 μ, hyaline to reddish; paraphyses 5–6 μ, unbranched, septate, the tips red-brown, moniliform; asci 180–210 × 24–32 μ; spores 4, brown, 2-celled, 34–60 × 16–26 μ.

This calciphile species grows best in very moist places as in spray from waterfalls, in moist microhabitats as hummock sides, sides of animal burrows, edges of solifluction lobes, seepages, and moist shaded cliff sides. It is circumpolar, arctic, and boreal, in North America ranging south to Vermont, Wisconsin, South Dakota, Alberta, and British Columbia.

5. Solorina spongiosa (Sm.) Anzi

Comm. Soc. Critt. Ital. 1:136. 1862. *Lichen spongiosus* J. E. Smith, Engl. Bot. 20: pl. 1374. 1805.

Thallus scant and fragile, of small squamules and granules around the very urceolate apothecia, bluish gray to reddish gray, mixed with *Nostoc* colonies forming cephalodia; upper

Solorina octospora (Arn.) Arn.

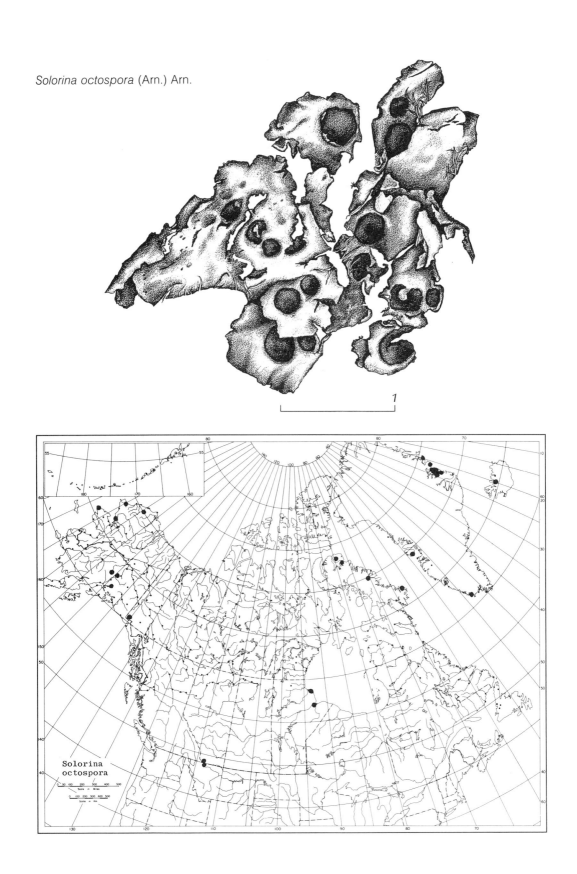

1

Solorina
octospora

Solorina saccata (L.) Ach.

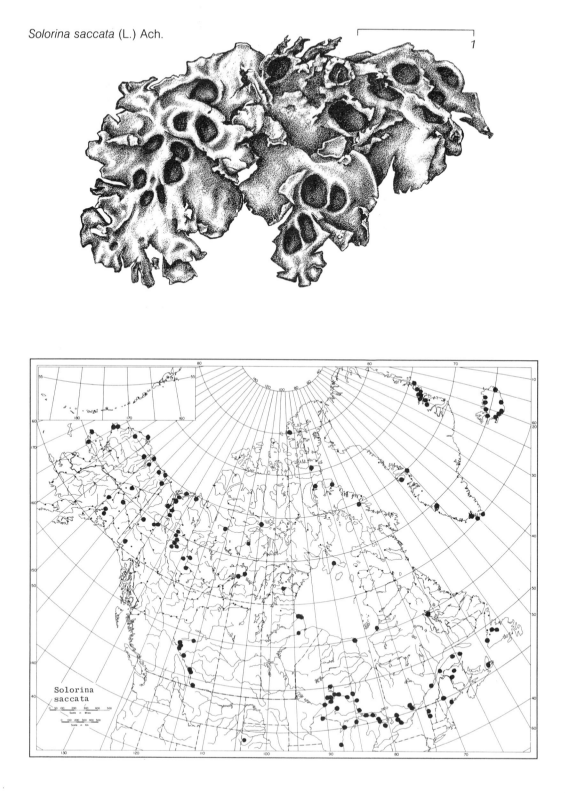

Solorina spongiosa (Sm.) Anzi

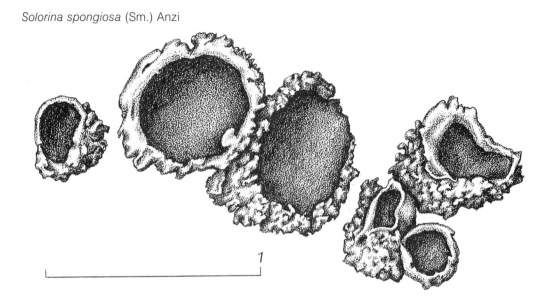

1

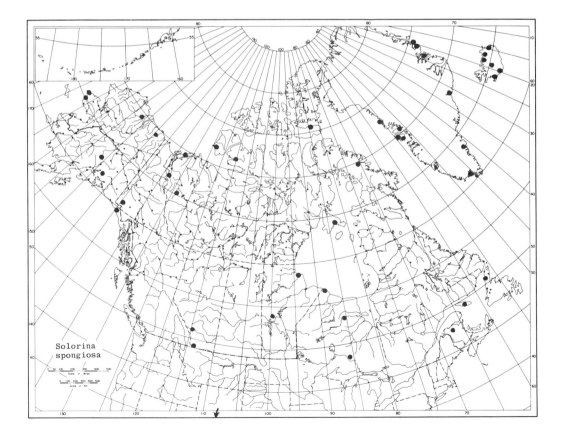

Solorina
spongiosa

cortex paraplectenchymatous, the algal layer containing *Coccomyxa;* lower cortex lacking, the lower side of the apothecia and squamules attached to the substratum by rhizines.

Apothecia deeply urceolate; the disk dark brownish red to blackening; hymenium hyaline, 160–200 μ; paraphyses unbranched, hyaline with the tips red-brown, 6–8 μ, coherent, little

thickened; asci 90–120 μ; spores 4, becoming brown, 2-celled, ellipsoid, 30–50 × 18–22 μ.

This species grows in moist, calcareous habitats. It is probably circumpolar arctic-alpine, reported from Europe, Novaya Zemlya, Bear Island, Iceland, Greenland, and in North America from Baffin Island to Alaska and south to Newfoundland, Ontario, Colorado, and Washington.

SPHAEROPHORUS Pers.

Ann. Bot. Usteri 7:23. 1794.

Thallus fruticose, terete or flattened, erect or procumbent, heteromerous, solid to subfistulose, sometimes with a hypothallus which forms small erect papillae, or hypothallus absent; cortex chondroid, the cells with small lumena.

Apothecia in the tips of the branches, the covering a thalloid globose head which bursts without a definite ostiole; lacking proper exciple; the mazaedium and spores exposed as a black mass at the lacerate tips; asci clavate, yielding a lavender pigment which dries blue from acetone extracts; spores 8, uniseriate, globose or nearly so, simple, brown.

Algae: *Cystococcus*.

Type species: *Sphaerophorus globosus* (Huds.) Vain.

REFERENCES
Keissler, K. 1938. Pyrenulaceae bis Mycoporaceae, Coniocarpineae. in Rabenh. Krypt. Flora Deutschl. Osterr. Schweiz 9(1):1–846.
Lindsay, D. C. 1972. Notes on Antarctic Lichens. 6. The genus *Sphaerophorus*. British Antarct. Surv. Bull. 28:43–48.
Mituno, M. 1938. *Sphaerophorus* arten aus Japan. J. Jap. Bot. 14:659–669.
Ohlsson, K. E. 1973. New and interesting lichens of British Columbia. Bryologist 76:366–387.
Rehm, A. 1971. A chemical study of *Sphaerophorus globosus* and *S. fragilis*. Bryologist 74:199–202.
Sato, M. 1968. Revision of the New Zealand lichens. Miscel. Bryol. Lich. 4:150–152.

1. Medulla I+ blue, thallus partly foveolate, not very fragile . . . . 2. *S. globosus*
1. Medulla I−, thallus not foveolate, fragile . . . . 1. *S. fragilis*

1. Sphaerophorus fragilis (L.) Pers.
Ann. Bot. Usteri 7:23. 1974. Lichen fragilis L., Sp. Pl. 1154. 1753.

Thallus erect, in close tussocks or tufts on the rock substratum; branching sympodial or dichotomous, the branches terete, bluish gray or ashy gray, brown in strong light; medulla I−.

Apothecia rare, in the globose tips of branches; hypothecium brown; upper part of hymenium blue black; paraphyses slender, the tips branched; asci cylindrical; spores 8, uniseriate, turning from blue into black, the walls rugulose papillate and very dark, spherical, 7–16 μ.

Reactions: Medulla I−, UV+ blue, in chemical strains with hypothamnolic acid K+ purple, KC+ red.

Contents: Sphaerophorin and fragilin in all specimens, squamatic acid accessory and thamnolic acid accessory in European and Icelandic material.

This species grows on rocks and rock outcrops. It is circumpolar and arctic-alpine occurring south to New England and Washington state.

2. Sphaerophorus globosus (Huds.) Vain.
Results Voyage S.Y. Belgica Bot. 35. 1903. *Lichen globosus* Huds., Flora Anglica 1:460. 1762. *Sphaerophorus coralloides* Pers., Annal. Bot. 1:23. 1794.

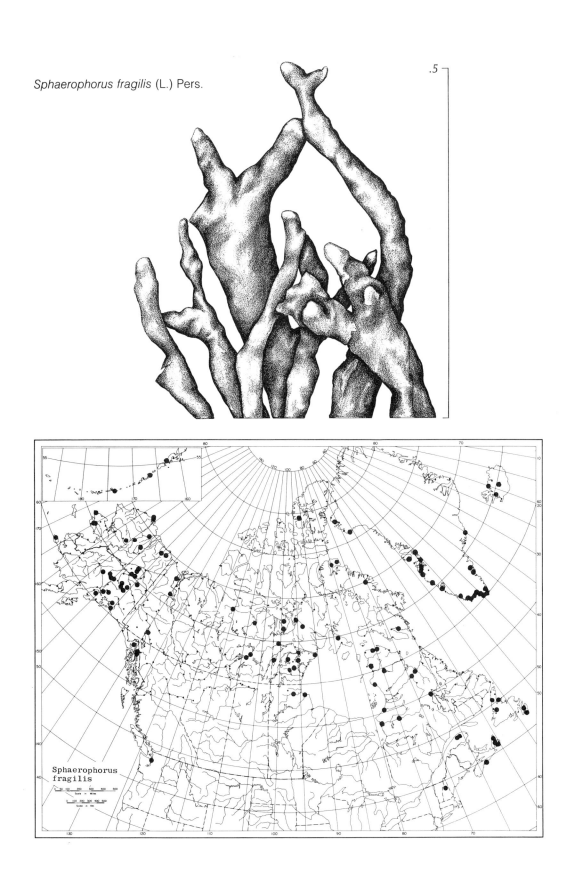

Sphaerophorus fragilis (L.) Pers.

.5

Sphaerophorus
fragilis

Sphaerophorus globosus (Huds.) Vain.

1

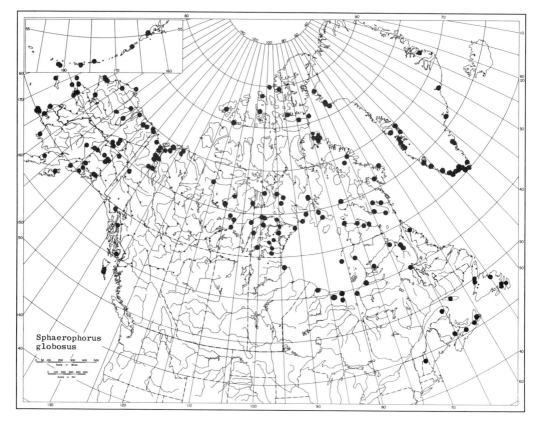

Sphaerophorus
globosus

Thallus erect, branching sympodially or dichotomously, the branches terete, partly impressed foveolate, smooth, the bases brown stained, the upper parts bluish gray to sunbrowned; medullary layer loose, I+ blue.

Apothecia rare, within the globose tips of branches, dehiscing lacerately; hypothecium reddish above, pale below; paraphyses poorly developed, thin; asci cylindrical, purplish; spores 8, uniseriate, simple, globose or nearly so, the walls rugulose papillate, 7–15 μ.

Reactions: Medulla I+ blue, UV+; K+ yellow, P+ yellow in race containing thamnolic acid; P−, KOH+ reddish purple (fading) when with hypothamnolic acid.

Contents: sphaerophorin and fragilin in all specimens, accessory squamatic and thamnolic acids, hypothamnolic acid accessory in European specimens (Rehm 1971).

This species grows on humus, soil, and over rocks. It is a common, circumpolar, arctic-alpine species ranging south to New England in the east and to British Columbia in the west.

Although there are reports of this species south in the Pacific coastal rain forest from Alaska to California on tree trunks, these reports appear to be based on the much more laterally branched material classified as *S. globosus* var. *gracilis* (Mull. Arg.) Zahlbr. or *S. tuckermanii* Räs. Ohlsson (1977) in a later study than that of Rehm found that the tundra species contained only sphaerophorin as a constant and that the rain forest material contained in addition accessory squamatic, and thamnolic or hypothamnolic acids. The differences in habit, habitat, distribution (Japan and west coast of North America), and chemistry appear to support the separation of *S. tuckermanii* as a distinct species from the arctic-alpine *S. globosus*.

STEREOCAULON (Schreb.) Hoffm.

Deutschl. Flora 2:128. 1796. nom. conserv. *Lichen* sect. *Stereocaulon* Schreb. in L., Genera Plant. ed. 8, 2:768. 1791.

Primary thallus crustose, areolate or warty, granulose-sorediate, or squamulose; with pseudopodetia arising from the primary thallus, usually branched, the center solid with a central medullary layer surrounded by a more or less cartilaginous outer layer, often with a tomentum covering it; algae in phyllocladia which vary from granulose to squamulose, digitate, subcoralloid, or peltate; with cephalodia on the primary thallus or pseudopodetia containing *Nostoc* or *Stigonema*.

Apothecia pale to dark brown, at first with a light margin, usually becoming convex and immarginate, the outer side with a palisade-like plectenchyma, with a central loose medullary layer; hypothecium hyaline or yellowish; hymenium mostly I+ blue, hyaline; paraphyses simple or slightly branched, usually capitate; asci with thickened tips; spores 8, acicular or fusiform, 4− many celled with only transverse walls. Pycnidia mostly imbedded at the tips of the phyllocladia; fulcra exobasidial; conidia bacillar to filiform, straight or bent, usually $8–12 \times 0.5$ μ.

Algae: *Trebouxia* in the thallus, *Nostoc* or *Stigonema* in the cephalodia.

Type species: *Stereocaulon paschale* (L.) Hoffm.

REFERENCES

Dodge, C. W. 1929. A synopsis of *Stereocaulon* with notes on some exotic species. Ann. Crypt. Exot. 2:95–153.

DuRietz, G. I. 1926. Skandinaviens *Stereocaulon*-arter. Svensk. Bot. Tids. 20:95.

Frey, E. 1933. Cladoniaceae. In Rabenh. Kryptog. Flora ed. 2 Abt. 4 halfte 1:61–202.

Lamb, I. M. 1951. On the morphology, phylogeny, and taxonomy of the lichen genus *Stereocaulon*. Can. J. Bot. 29:522–584.

——— 1961. Two new species of *Stereocaulon* occurring in Scandinavia. Bot. Not. 114:265–275.

——— 1972. *Stereocaulon arenarium* (Sav.) M. Lamb, a hitherto overlooked Boreal-Arctic lichen. Occ. Papers Farlow Herb. 2:1–11.

——— 1973. *Stereocaulon sterile* (Sav.) Lamb and *Stereocaulon groenlandicum* (Dahl) Lamb, two more hitherto overlooked lichen species. Occ. Papers Farlow Herb. 5:1–17.

——— 1977. A conspectus of the lichen genus *Stereocaulon* (Schreb.) Hoffm. J. Hattori Bot. Lab. 43:191–355.

——— 1978. Keys to the species of the lichen genus *Stereocaulon* (Schreb.) Hoffm. J. Hattori Bot. Lab. 44:209–250.

1. Primary thallus persisting, the short pseudopodetia growing between the well-developed primary squamules.
    2. With lateral phyllocladia.
        3. With tubular phyllocladia... 18. *S. saviczii*
        3. With granular phyllocladia or becoming sorediate.
            4. On sand and earth; cephalodia frequent on the thallus, smaller and more rare on the pseudopodetia, red-brown to black, hemispherical, the surface rough, containing *Stigonema;* pseudopodetia with scanty granular structures or containing apothecia; containing only atranorin or atranorin and lobaric acid .... 7. *S. condensatum*
            4. On rocks and clay; cephalodia infrequent, gray-brown or violet, granular; the sterile pseudopodetia mostly capitate-sorediate; containing atranorin and lobaric acid........
            .... 16. *S. pileatum*
    2. With terminal leaf-like apical expansions, esorediate; containing atranorin and lobaric acid ..
    .... 2. *S. apocalyptum*
1. Primary thallus more or less disappearing.
        5. Primary thallus at first well developed, becoming scant between the pseudopodetia, which arise from squamules similar to the phyllocladia; pseudopodetia capitate-sorediate.
            6. UV−; pseudopodetia fragile, dying at base; containing atranorin and lobaric acid .... 24. *S. uliginosum*
            6. UV+; pseudopodetia not fragile, base persistent; containing atranorin, miriquidic, perlatolic, and anziaic acids .... 6. *S. capitellatum*
        5. Primary thallus mainly lacking, at the most forming a granular or sorediate crust; pseudopodetia conspicuous, mainly over 1 cm tall, lacking soredia in most species.
            7. Phyllocladia peltate, the center dark green, the margins pale; growing on rocks.
                8. Cephalodia containing *Stigonema.*
                    9. Pseudopodetia erect; P+ red, containing atranorin and stictic acid; esorediate .... 25. *S. vesuvianum*
                    9. Pseudopodetia dorsiventral; sorediate; P+ yellow, containing atranorin and lobaric acid .... 22. *S. symphycheilum*
                8. Cephalodia containing *Nostoc;* pseudopodetia in dense tufts.
                    10. P+ red, containing atranorin, stictic acid and small amounts of norstictic acid .... 3. *S. arcticum*
                    10. P−, containing atranorin and porphyrilic acid (dendroidin) ......
                    .... 4. *S. arenarium*
            7. Phyllocladia not peltate, the centers not darker than the margins.
                11. On rocks; pseudopodetia tightly attached by holdfasts, little or not tomentose.
                    12. Phyllocladia not coralloid but warty, granular, or squamulose, broadened or more or less spherical and in the latter case sorediate.
                    13. Pseudopodetia little branched; phyllocladia dissolving into granular soredia.
                        14. Soredia capitate; pseudopodetia more branched; containing atranorin and porphyrilic acid............. [*S. leprocephalum*]
                        14. Soredia diffuse; pseudopodetia little or not branched.
                            15. P+ yellow, containing atranorin and lobaric acid; pseudopodetia rigid and woody .... 8. *S. coniophyllum*
                            15. P+ red, containing atranorin, stictic and norstictic acids and unidentified substances; pseudopodetia not woody
                            .... 20. *S. spathuliferum*
                    13. Pseudopodetia branching; phyllocladia not sorediate.
                        16. Pseudopodetia whitish, woody, the base lacking tomentum.

17. Phyllocladia irregularly grain-like, the tips often cauliflower-like, with granular small phyllocladia, often soredia-like; UV − , containing atranorin and porphyrilic
acid (dendroidin) . . . . 5. *S. botryosum*

17. Phyllocladia crenate to digitate, squamulose, never soredia-like; UV + , containing atranorin, miriquidic acid,
perlatolic and anziaic acids. . . . . 12. *S. groenlandicum*

16. Pseudopodetia in low, flat, cushion-like masses, the marginal pseudopodetia decumbent, gray in color, finally somewhat tomentose and light-colored, or bare and then blackish; phyllocladia a little broadened, crenulate or incised,
squamulose-warty, not in cauliflower-like masses; apothecia
small, convex, dark.

18. Pseudopodetia ascending, more or less tomentose, the
horizontal basal portion firmly attached to the substratum; phyllocladia warty, confluent with whitish tips that
occasionally burst into soredia . . . .
. . . . 9. *S. dactylophyllum* var. *occidentale*

18. Pseudopodetia in the middle of the cushion erect, the
base whitish, becoming tomentose upwards, the tips
much branched and covered with flat, squamulose, and
granulose white phyllocladia . . . . 19. *S. saxatile*

12. Phyllocladia coralloid, long cylindrical, simple or pinnately or irregularly
branched or coalescing into squamules.

19. Small species; pseudopodetia 1.5–2.5 cm tall; phyllocladia minute, 0.2–0.4 mm long, congestedly branched with coralloid divisions 0.1–0.15 mm diameter . . . . 21. *S. subcoralloides*

19. Taller species, phyllocladia coarser.

20. Pseudopodetia 2–8 cm tall, 1 mm thick; apothecia 2–5 mm
broad; hymenium 50 μ high, central cylinder of pseudopodetium K + yellow; eastern arctic, amphi-Atlantic . . . .
. . . . 9. *S. dactylophyllum*

20. Pseudopodetia over 5 cm tall, 2–3 mm thick; apothecia 1–
4 mm broad; hymenium 55–65 μ high; central cylinder K − ;
Beringian bilateral . . . . 14. *S. intermedium*

11. On earth, sand, and clay, between gravel, only occasionally on rocks and
in such cases on cracks containing detritus; pseudopodetia easily lifted
(except for *S. glareosum*), the base often thinner than above, always with
a distinct, often very thick, tomentum (except for *S. rivulorum* which has
thickly entangled pseudopodetia).

21. Phyllocladia at least in part rounded foliose or rounded squamulose,
margins more or less crenate or incised, in part becoming granulose;
pseudopodetia thickly arachnoid tomentose, distinctly dorsiventral even
when appearing partly erect; cephalodia small; apothecia very small,
numerous, sunken; containing atranorin, stictic and norstictic acids
. . . . 23. *S. tomentosum*

21. Phyllocladia neither foliose nor squamulose, mainly dactyliform, attached together at the base, granulose, warty, or small squamulose
but in the latter case not rounded; apothecia reaching over 1 mm diameter.

22. Pseudopodetia either very fragile or clearly dorsiventral.

23. Pseudopodetia forming confused entangled cushions, occasional pseudopodetia scattered, very fragile, with very thin

or no tomentum, not clearly dorsiventral, white with a rosy
tinge; phyllocladia coarse in comparison with the pseudopodetia, running together, whitish; apothecia flat, large, frequent; containing atranorin and lobaric acid ... 17. *S. rivulorum*

23. Pseudopodetia prostrate, forming low, flat cushions, underside only exceptionally with phyllocladia (the prostrate forms
of *S. paschale* with small granular or long finger-shaped
phyllocladia).

    24. Pseudopodetia always closely pressed to the substrate,
partly intergrown, with blackish tomentum; phylloclla-
dia squamulose, forming an almost continuous blue or
slate gray crust; apothecia small, high convex, blackish
brown; cephalodia rare .... 19. *S. saxatile*

    24. Pseudopodetia somewhat ascending with distinct main
axis; with whitish or pale rosy tomentum; phyllocladia
likewise whitish, at most whitish gray, not blue-gray or
dark gray; thick warty, always clearly swollen; apothecia finally flat, clear red-brown; cephalodia white,
spherical........................... 1. *S. alpinum*

22. Pseudopodetia not exceeding fragile, mainly erect or plainly
ascending, at least the fruiting pseudopodetia not decumbent;
phyllocladia more or less equally distributed around the pseudopodetia at least on the erect fruiting stalks.

    25. Cephalodia abundant and distinct, dark olive-brown, flat,
scabrid, containing *Stigonema;* pseudopodetia not rigid, bases
dying, the mats easily lifted, the base thinner than above;
phyllocladia botryose clustered, mostly small granular, little
more than 0.1 mm broad, seldom somewhat elongate and
short irregularly finger-shaped; containing atranorin and lobaric acid .... 15. *S. paschale*

    25. Cephalodia indistinct or smooth and nearly spherical to fis-
sured brown to dark brown; pseudopodetia strong, rigid, in
S. grande thicker at the base than higher; phyllocladia larger,
not botryose clustered.

        26. With abundant phyllocladia projecting beyond the to-
mentum.

            27. Pseudopodetia mainly over 4–5 (8) cm tall, with
distinct main axis which tapers up to the apothecia;
phyllocladia a mix of *S. alpinum* types with verrucose to thickened squamuliform with incised
margins and of *S. paschale* types with clustered
granulose to clustered digitate shapes usually massed
at regular intervals on fruiting podetia; apothecia
large, to 3 mm broad, soon emarginate and fissured; cephalodia small, pale brown smooth to
pulvinate and divided, containing *Nostoc* or *Scytonema* .... 11. *S. grande*

            27. Pseudopodetia strong, erect, but short, only 1–2.5
cm tall, almost adherent; phyllocladia elongate,
cylindrical or the tips somewhat thicker, clavate,
mainly irregularly massed: apothecia finally flat with
a distinct, thin clear margin; cephalodia large 2–3

mm broad, often fissured, brown to red-brown, containing *Nostoc* .... 10. *S. glareosum*
26. Phyllocladia smaller, granulose, in part almost hidden by the dense tomentum; apothecia terminal soon swollen and split; cephalodia dark brown, smooth, containing *Nostoc* .... 13. *S. incrustatum*

1. Stereocaulon alpinum Laur.

ex Funck, Cryptog.-Gewächse 33:6. 1827. *Stereocaulon tomentosum* γ *alpinum* Th. Fr., Comm. 30. 1857. *Stereocaulon paschale* var. *alpinum* Mudd, Manual Brit. Lich. 66. 1861.

Primary thallus disappearing; pseudopodetia dorsiventral or erect, caespitose, firmly attached to the ground, 1–4 cm tall, 0.5–1 (–2) mm thick at the base; the base pale brownish or blackish, sparingly branched; lower part without phyllocladia but with a grayish, whitish, or roseate tomentum; phyllocladia crowded toward the apices, whitish-gray with a bluish tinge or whitish, turning pale dirty yellowish in herbarium, granuliform, or united, somewhat flattened, lobate, 0.5 mm broad, indistinctly stalked, appressed to the pseudopodetium; cephalodia on lower side, hemispherical, grayish-white, 0.3–0.8 mm, partly concealed in the tomentum, containing *Nostoc.*

Central medullary layer of pachydermatous, 3–4 μ, loosely conglutinate, strictly parallel hyphae; exterior hyphae dirty yellowish, gelatinous, pachydermatous, 6–8 μ thick; cortex of phyllocladia 15–25 μ, grayish.

Apothecia terminal on the podetia or grouped on the upper branches, immarginate; disk dark brown, 1–1.5 mm broad, convex or irregularly swollen; lower side with the center tomentose, the margin bare; exterior hyphae brownish yellow, gelatinous, to the edge becoming parallel and perpendicular to the surface; hypothecium 35–40 μ, hyaline; hymenium 40–50 μ (60–70 μ acc. Frey), yellowish, I + blue; paraphyses slightly coherent, 1.7–2.0 μ thick, the tips brownish, capitate, 3–4 μ; asci elongate clavate 50–55 × 9–11 μ; spores 8, cylindrical, with one end narrower, slightly bent or straight, 3 septate 27–38 × 2.5–3 μ. Conidia straight, bacilliform, 5 × 1.5.

Reactions: Central axis KOH −, phyllocladia KOH + yellow, P + yellow, containing atranorin and lobaric acid, β-sitosterin (Lamb 1977).

This species grows on the ground, especially under a late snow cover. It appears to be circumpolar, arctic-alpine, ranging south to New Hampshire, Colorado, and Washington in North America. It is also reported from the southern hemisphere.

2. Stereocaulon apocalyptum Nyl.

ex Middendorff, Reise in den aussersten Norden und Osten Siberiens, 4, anhang 6:1v, footnote. 1867. *Stereocladium apocalyptum* (Nyl.) Nyl. Flora 58:302. 1875.

Primary thallus disappearing; pseudopodetia white to pale white, tufted, to 4 cm tall; the apices expanded to flattened foliaceous, stiff, wrinkled, the margins crisped and much paler, the bases darkening to black, about 1 mm thick, bare; the underside of the foliose part tomentose; cephalodia blackish-brown, verrucose, on the phyllocladia, containing *Stigonema.*

Sections of the phyllocladia 0.4–0.5 mm thick, with a thick gelatinous transparent cortex 50–150 μ thick, the cells indistinct; algal layer continuous, of small 5–10 μ cells; medulla 200–300 μ thick of densely intricate gelatinous hyphae 10–18 μ thick with a 1 μ lumen; lower cortex lacking; the blackish pseudopodetia with the outer 30–40 μ dark brown, of 4–5 μ hyphae; the interior white, of densely packed conglutinate pachydermatous hyphae 2–3 μ thick.

Apothecia described by Savicz: marginal on the phyllocladia, 0.25–0.5 mm diameter, at first concave, becoming subglobose, reddish brown; paraphyses distinct, the apices darkened; asci 20–30 × 7–12 μ; spores 3–5 septate, acute, 12–22 × 3.5–4.5 μ. Conidia straight, 4.5–6.5 μ.

Stereocaulon alpinum Laur.

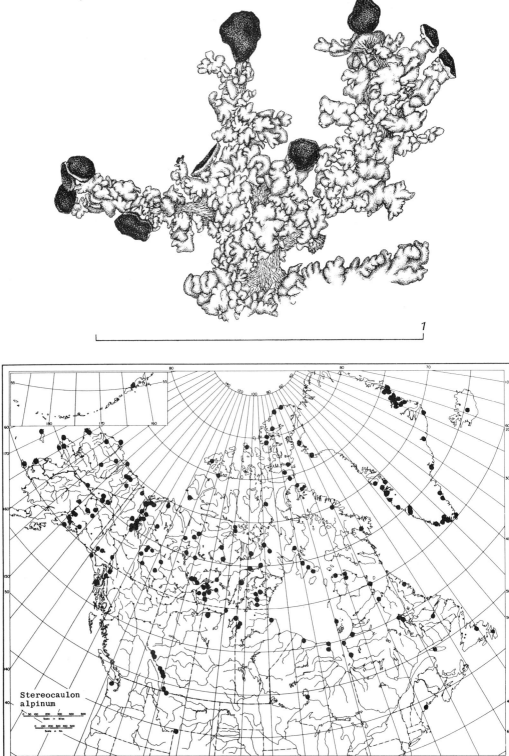

1

Stereocaulon
alpinum

Stereocaulon apocalyptum Nyl.

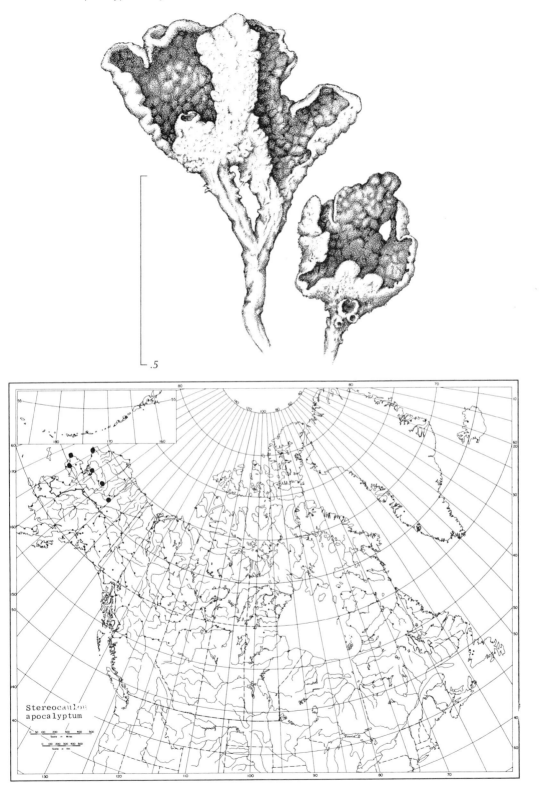

.5

Stereocaulon
apocalyptum

Reactions: K+ yellow, P−.

Containing: atranorin and lobaric acid.

This species grows over rocks in the Arctic. It is amphi-Beringian, being known from Japan, eastern Asia, and Alaska where it has been reported from St. Michel (Riddle 1910), the Ogotoruk Creek drainage (Krog 1962), and Mt. Harper in the Yukon district (Krog 1962).

This is extremely similar in morphology to *Stereocaulon wrightii* Tuck. which differs only in chemistry. *S. wrightii* contains atranorin and stictic acid and according to Lamb (1977) occurs only in Japan and the USSR, including Arakamchene Island on the west side of Bering Strait.

3. Stereocaulon arcticum Lynge

Norske Vid.-Akad. Skr. I. Math.-Nat. Kl. I. Oslo 6:69. 1938.

Primary thallus disappearing, of small granules; pseudopodetia in dense tufts 1–3 cm tall, thickly branched toward the tips, ashy to blackened, covered with discontinuous coarse gray tomentum dispersed in patches, decaying and blackening toward the base; very fragile as compared with *S. vesuvianum;* phyllocladia dense toward the tops, when young distinctly peltate with greenish centers and pale margins, concave, slightly more granulose than in *S. vesuvianum;* cephalodia lacking cortex, botryose, small, dark red-brown, containing *Nostoc*.

Central medullary hyphae dense, pachydermatous, 8–10 μ thick, parallel, exterior of loosely tomentose pachydermatous hyphae 6–12 μ thick; cortex of phyllocladia to 50 μ thick, gelatinous, the hyphae indistinct but almost paraplectenchymatous, the cells to 12 μ broad.

Apothecia rare, lateral, sessile or short-stalked, flat when young, becoming convex, small, to 1 mm broad; disk epruinose, dark chestnut, dull, becoming emarginate; cortical hyphae in perpendicular layer 25–65 μ thick; the hyphae 5–6 μ, pachydermatous; the hypothecium grading into the cortex; hymenium thin, 50 μ, upper part yellow-brown; para-

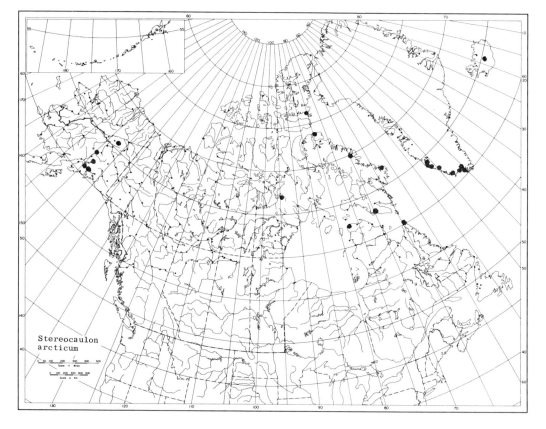

Stereocaulon arcticum

Stereocaulon arenarium (Sav.) Lamb

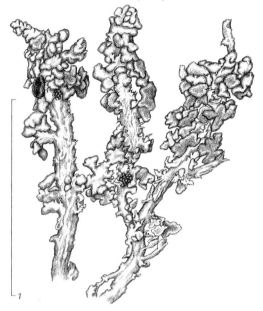

physes scarcely united, tips clavate, 3–3.5 μ thick; fertile asci rare; spores straight, subcylindrical to fusiform, 1–3(–5) septate, 18–30 × 2.5–3 μ. (Fertile characters based on Lynge 1938.)

Reactions: K+ yellow, P+ yellow to orange. Hymenium I+ blue turning dirty yellowish brown, apices of asci persistently blue.

Contents: atranorin, stictic acid, small amounts norstictic acid (Lamb 1977).

This species grows upon soil and in meadows and on gravel. It is boreal-arctic and circumpolar according to Lamb (1977).

4. **Stereocaulon arenarium** (Sav.) Lamb

Occ. Papers Farlow Herb. 2:1–11. 1972. *Stereocaulon denudatum* var. *pulvinatum* f. *arenarium* Savicz, Bot. Materialy, Notul. System. ex Inst. Cryptog. Hort. Bot. Petropol. 2:171. 1923.

Pseudopodetia in compact clumps, densely caespitose, loosely or firmly attached, 1–3 cm tall, brownish, bare and dead below,

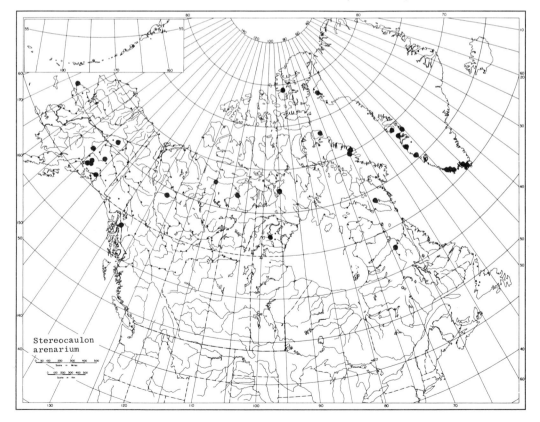

Stereocaulon
arenarium

tomentose above simple to sparingly irregularly branched; phyllocladia nodulose or verrucose to flattened peltate, olive bluish or yellowish centrally, paler at the margins, esorediate; cephalodia scarce to frequent, on the pseudopodetia between the phyllocladia, pale to dark brown, verrucose or subbotryose, containing *Nostoc* or *Stigonema*.

Apothecia rare, lateral, small, 0.5–1.0 mm, blackish; exciple at sides 30–60 μ thick of conglutinated pachydermatous hyphae; central cone hyaline to pale brownish; hypothecium 30–70 μ, pale brownish; hymenium 50–60 μ, upper part reddish brown; paraphyses simple or branched, 1–1.5 μ; ripe spores not known.

Reactions: K+ yellow, P+ pale yellowish.

Contents: atranorin and porphyrilic acid (dendroidin).

This species grows on moss and *Dryas* heaths on hummocks or gravelly soil. It is known from Greenland across westward into Kamtchatka. Morphologically it resembles *S. arcticum* or *S. vesuvianum* but differs in the chemistry.

5. Stereocaulon botryosum Ach. em. Frey

Lich. Univ. 581. 1810. em. Frey in Rabenh., Krypt. Fl. 9(4)1:120. 1933. *Stereocaulon fastigiatum* Anzi, Catal. Lich. Sond. 11. 1860. *Stereocaulon paschale β pulvinatum* Schaer., Lich. Helvet. Spicil. 6:274. 1833. *Stereocaulon corallinum β pulvinatum* (Schaer.) Schaer., Enum. Crit.Lich. Europ. 180. 1850. *Stereocaulon evolutum* var. *fastigiatum* (Anzi) Th. Fr., Lich. Scand. 1:45. 1874. *Stereocaulon tomentosum* var. *botryosum* (Ach.) Nyl., Scand. 64. 1861. *Stereocaulon evolutum* f. *fastigiatum* (Anzi) Anders, Strauch und Laubflechten Mitteleurop. 124. 1928.

Primary thallus farinose, disappearing; pseudopodetia firmly attached to rocks, in dense tufts, mainly radiating from a center, the base 0.5–1.5 mm thick, ochraceous, branching from the base upwards, the tips becoming fastigiate and forming cauliflower-like masses; phyllocladia glaucous to grayish-white, very dense on upper parts, granular or verrucose, 0.2–0.3 mm broad, rarely flattened; usually naked below and tomentose above; cephalodia rare, inconspicuous, tuberculose, containing *Nostoc;*

Central medullary hyphae 2–5 μ thick, parallel, densely conglutinate, exterior forming an indistinct cylinder; outermost hyphae gelatinous, 7–10 μ, pachydermatous.

Apothecia terminal on elongated pseudopodetia or level with the surface of the tuft, sometimes corymbose, dark brown, 1.5–3 mm, flat to becoming convex and immarginate, lower side covered with tomentum; exciple brownish yellow, its hyphae interwoven, more or less parallel; hypothecium dense, 40 μ; hymenium grayish-white; asci I+ blue, hymenial gelatin I−; upper part yellow-brown; paraphyses more or less free, 1–1.5 μ, tips clavate, 3–4.5 μ; asci cylindrico-clavate, 50–55 × 8–10 μ; spores 4–6, fusiform, one end narrower, 3-septate, 20–31 × 4–5 μ. Conidia straight, 5–7 × 0.6–0.8 μ.

Reactions :K+ yellow, P−.

Contents: atranorin and porphyrilic acid.

This circumpolar species is arctic-alpine or subalpine and grows on rocks.

6. Stereocaulon capitellatum Magn.

Göteborgs K. Vet. O. Vitterh. Samh. Handl. 30:39. 1926. *Stereocaulon farinaceum* Magn., ibid. p. 72.

Primary thallus disappearing; pseudopodetia strongly attached to the ground forming cushions 1–2 cm broad, 5–10 (–11) mm high, the bases 0.3–0.6 mm thick, white with a rosy tinge, the upper parts abundantly branched and the tips becoming capitate sorediate; phyllocladia on upper side of decumbent pseudopodetia, flattened verruciform, glaucous white, confluent, soon disappearing into farinose soredia; cephalodia inconspicuous, imbedded in the tomentum, containing *Nostoc*.

Central medullary layer densely conglutinate, of pachydermatous hyphae 2–3 μ thick, the lumen 1–1.5 μ, KOH+ yellow, gray or dark yellow-brown in color (with the same KOH reaction); outer hyphae white, forming a thick tomentum 50–90 μ thick of white gelatinous thick-walled hyphae 5–7 μ diameter, on the upper side the algae are in glomerules or more or less continuous; phyllocladia with a cortex 15–25 μ thick, dissolving into soredia, the soredia with a distinct 5 μ thick layer of thick-walled cells.

Apothecia and pycnidia not known.

Stereocaulon botryosum Ach. em. Frey

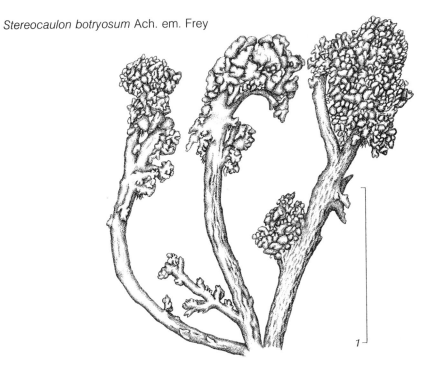

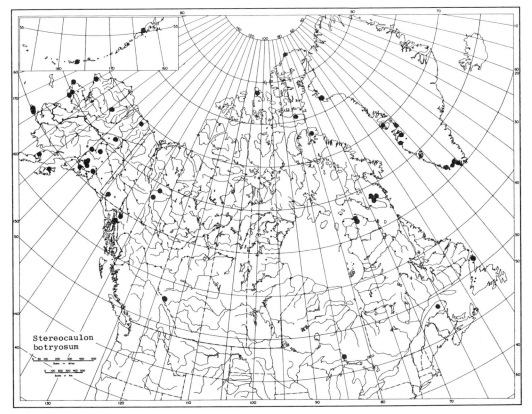

Stereocaulon
botryosum

Stereocaulon capitellatum Magn.

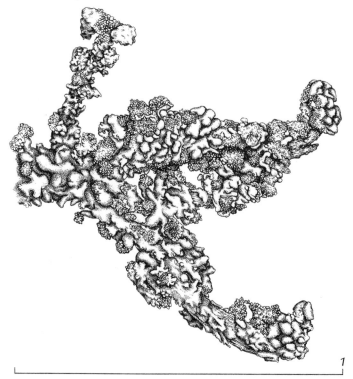

Reactions: axis and phyllocladia K+ yellow.

Contents: atranorin, miriquidic acid, perlatolic acid, and anziaic acid (Lamb 1973).

This species grows on earth. It has been reported from Europe, Iceland, and Greenland (locality not specified) by Lamb (1977) but not yet from the American continent.

7. Stereocaulon condensatum Hoffm.

Deutschl. Flora 2:130. 1796. *Lichen condensatus* (Hoffm.) Ach., Lich. Suec. Prodr. 209. 1798. *Patellaria pileata* var. *thamnodes* Wallr., Flora Crypt. Germ. 30:441. 1831. *Cereolus condensatus* (Hoffm.) Boist., Nouv. Flore Lich. 2:33. 1903.

Primary thallus persistent, of warty or elongated and incised blue-green phyllocladia with abundant cephalodia interspersed among them; pseudopodetia 2–3 (–15) mm tall, 0.8–1.2 mm thick, erect, simple or little branched, the upper part white or grayish tomentose; phyllocladia small, verruciform to incised-squamuliform, horizontal, grayish-white; cephalodia abundant, dark reddish brown to black, spherical with rough surface, those on the pseudopodetia smaller (0.2–0.5 mm) than those on the primary thallus (0.5–1.0 mm); algae usually *Stigonema*.

Central medullary hyphae densely conglutinate with only a few interstices toward the center, pachydermatous, 4–6 μ in diameter, the lumen 1.8–2 μ, outer more gelatinous and 8–11 μ in diameter, tomentum 50–70 μ thick. Phyllocladia with cortex 18–25 μ thick of thick-walled hyphae 3–4 μ in diameter, medullary hyphae thick-walled, looser in structure, 4–9 μ in diameter; algae in glomerules or continuous, oval or spherical, 7–15 μ, *Trebouxia*.

Apothecia common, terminal, 1–2 mm, with thin, pale margin when young, soon convex, often breaking into smaller parts; disk dark brown, the central cone dense but not gelatinous; hypothecium 50 μ, gray; hymenium 40–50 μ, upper part brown; paraphyses contiguous, 1–1.5 μ, the tips to 4 μ; asci elongate-clavate,

Stereocaulon condensatum Hoffm.

1

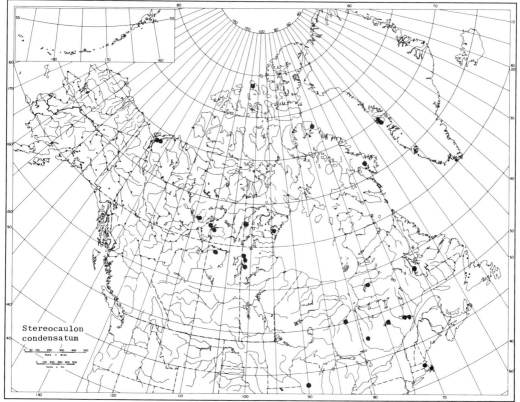

Stereocaulon
condensatum

8 spored; spores 4-celled, cylindric, one end narrower, $22-40 \times 2.5-3$ μ. Pycnidia immersed in the basal phyllocladia, conidia straight or slightly bent, $5-6.5 \times 0.8$ μ.

Reactions: asci I+ blue; central medulla K−, cortex K+ yellow.

Contents: atranorin and lobaric acid or only atranorin.

This species grows on sandy soil. It is classed as probably amphi-Atlantic boreal by Lamb (1951), is common in central Europe, and occurs in Kamtchatka and North America. In the latter it ranges trans-Canada and Alaska south to Connecticut, New Hampshire, Massachusetts, and Wisconsin.

8. Stereocaulon coniophyllum Lamb
Bot. Not. 114:266. 1961.

Primary thallus disappearing; pseudopodetia erect, firmly attached to the substratum by expanded holdfasts, 2.5–4.0 cm tall, robust, rigid, woody, irregularly branched, decorticate, glabrous; phyllocladia lacking, replaced by whitish, granular soredia, effuse or in spathulate apical expansions of the branches; cephalodia conspicuous, lateral, subglobose to becoming convolute-tuberculate or divided into verrucose masses, 1–2 mm in diameter; brown or glaucous-gray, sacculate with a solid core and well-developed cortical layer 35–75 μ thick, containing *Nostoc* or occasionally *Scytonema*.

Apothecia terminal, concave or flat to becoming convex, large, 1.5–4 mm broad; disk dark brown to blackish brown; exciple 75–100 μ thick of hyaline to partly brownish gelatinized radiating structure; the fistulose hyphal lumina 1–2 μ wide; hypothecium 45–80 μ, brownish; hymenium 80–90 μ, hyaline, upper part brown; paraphyses discrete, 1–2 μ thick, simple or sparingly branched, tips brown, capitate, 3–4 μ; asci 70–80 $\times$ 12–15 μ, the walls I+ blue; spores 8, elongate fusiform, straight or slightly flexuose, (3–) 5–7 septate, 40–65 $\times$ 3.5–4 μ. Conidia rod-shaped, straight, 4–5 $\times$ 0.8 μ.

Reactions: K+ yellow, P−; soredia P+ sulphur-yellow.

Contents: atranorin and lobaric acid.

This species grows on rocks, rarely on soil, apparently in rather moist places as in the spray of waterfalls. It is widely circumpolar, disjunct in the northern hemisphere and in Uganda, Africa. In North America it is so far known from Alaska and Baffin Island, N.W.T.

9. Stereocaulon dactylophyllum Flörke
Deutsch. Lich. 4:13. 1819. *Stereocaulon coralloides* Fr., Sched. Crit. Lich. Exsic. Suec. 3:24. 1825. *Stereocaulon dactylophyllum* f. *major* Sommerf., Fl. Lapp. Wahlenb. 125. 1826. *Stereocaulon corallinum* Laur. in Fries, Lich. Europ. 101. 1831 (non Schrad. 1794 = *Pertusaria*). *Stereocaulon paschale* var. *corallinum* Schaer., Lich. Helv. Spicil. 6:273. 1833. *Stereocaulon coralloides* var. *dactylophyllum* (Flörke) Th. Fr., De Stereoc. et Piloph. Comment. 16. 1857. *Stereocaulon paschale* var. *dactylophyllum* (Flörke) Branth. & Rostr., Bot. Tids. 3:162. 1869.

Primary thallus disappearing; pseudopodetia erect, 2–8 cm tall, firmly attached to rocks; simple at the base, branching upwards with spreading long and short branches, the base 1–3 mm thick, anisotomic with the main branches distinct, naked or white to roseate or gray, partly blackish, tomentose, especially at the base; several pseudopodetia forming tufts or cushions; phyllocladia originating as short side branches which then branch repeatedly mostly more or less dactyliform or as hands with many fingers, coralline, cylindrical, to 1 mm long, 0.15–0.3 mm thick, usually tapered toward the apices, occasionally forming into crust-like masses; cephalodia rare, immersed in the dark tomentum, warty granular, 0.1–0.3 mm broad, gray or brownish-gray, containing *Stigonema*.

Central cylinder with parallel, lightly conglutinate, somewhat loose hyphae, 2.5–5 μ thick, pachydermatous or leptodermatous intermixed, KOH+ yellow in the central part, KOH− in the outer part; phyllocladia with a uniform 15–25 μ thick opaque cortex of gelatinous hyphae 3–4 μ thick; medullary hyphae 4–6 μ.

Apothecia common, terminal, 1–2 mm broad, flat with a pale margin, becoming convex with a revolute margin; disk light to dark brown, older apothecia segmenting into smaller swollen units; underside irregularly tomentose, exteriorly with a palisade-like cortex; hypothecium indistinct; hymenium 50–75 μ high, iodine reaction variable, blue or not or only the

Stereocaulon coniophyllum Lamb

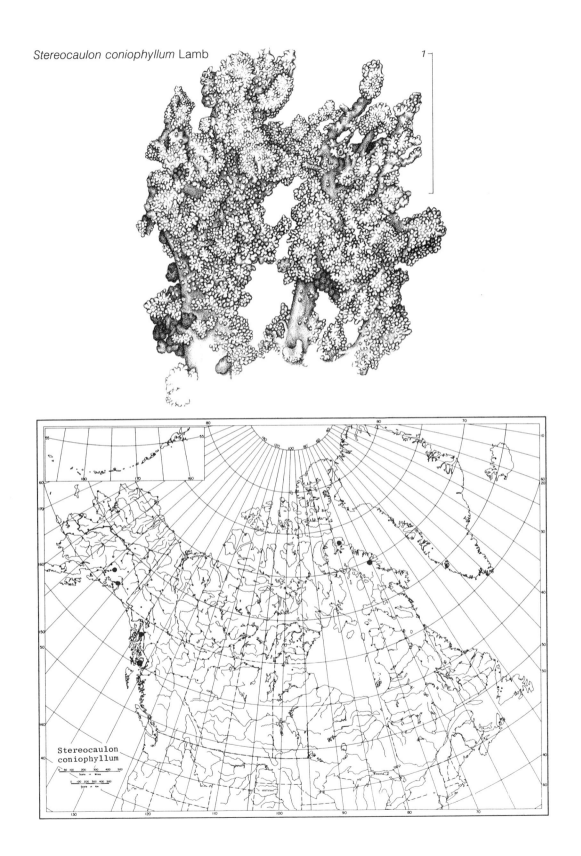

Stereocaulon
coniophyllum

Stereocaulon dactylophyllum Flörke

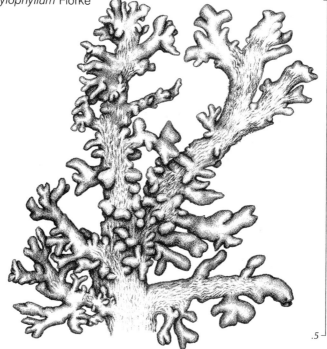

.5

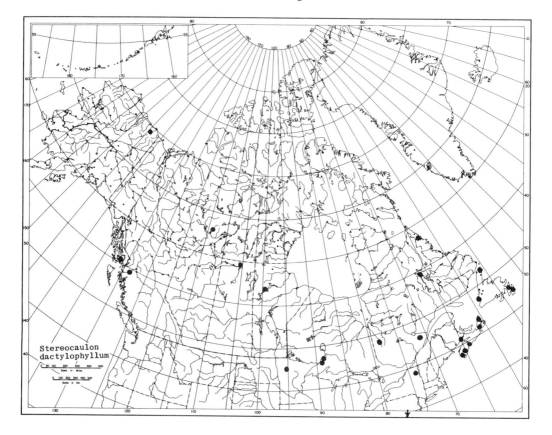

Stereocaulon
dactylophyllum

asci colored; paraphyses weakly gelatinous, 1.4–1.8 μ thick, the tips brown capitate to 4 μ; asci 40–60 × 9–11 μ; spores 8, to 6 celled, 40–72 × 8–14 μ. Conidia rod-like, 3.5–5 × 1 μ.

Reactions: central cylinder K+ yellow; phyllocladia K+ yellow, P+ orange.

Contents: atranorin and stictic acid plus small amounts of norstictic acid (Lamb 1977).

This species grows on acid rocks. It is a disjunct amphi-Atlantic boreal species, occurring in western boreal Europe and in the eastern part of North America from southern Greenland and Labrador south to North Carolina, and inland in Ungava and Ontario to Michigan.

When this species forms small tight cushions on the rocks with densely packed pseudopodetia it is called var. *occidentale* (Magn.) Grumm. (in some records as *S. spissum* (Nyl.) ex Hue).

10. Stereocaulon glareosum (Savicz) Magn.
Göteborgs K. Vetensk. Samh. Handl. 30:60. 1926. *Stereocaulon tomentosum* f. *glareosum* Savicz, Bull. Jardin Imp. Pierre le Grand 14:121. 1914.

Primary thallus composed of dense clusters of phyllocladia over the ground; pseudopodetia firmly attached to the ground, slightly tufted or more or less dispersed, pale, 1–2.5 cm tall, the base to 1 mm thick, tapering to 0.5 mm at the apices, entirely white or grayish white tomentose, sparingly branched; phyllocladia abundant at the base as well as on the substratum, more sparse above, rarely granular, usually cylindrical or papilliform 0.3–1 mm long, 0.2–0.3 mm thick; cortex continuous for long stretches on the pseudopodetia; cephalodia very characteristic, abundant among the phyllocladia on the branches, verruciform or globose, pale brownish-violet. or rose-whitish, containing *Nostoc.*

Central medullary hyphae parallel, conglutinate, leptodermatous, 3.5–6 μ in diameter, exterior hyphae yellowish, densely conglutinate, axis KOH−; cortex of phyllocladia 15–20 μ thick, hyphae intricate, not very gelatinous.

Apothecia rare, at the apices or lateral on upper branches, plane with thin white margin, soon convex, sometimes dividing, 0.7–1.5 mm, or in clusters to 4 mm; disk dark brown;

lower side with tomentum; hyphae of extreme margin perpendicular to the surface; hypothecium hyaline, 20–50 μ; hymenium grayish, 45–50 μ, only asci I+ blue; paraphyses somewhat free, 1.5 μ, brownish, clavate, tips 4.5 μ; asci 48–54 × 9–10 μ; spores 8, 3 septate, cylindrical-fusiform, slightly bent, 22–42 × 2.5–3.5 μ. Conidia unknown.

Reactions: K+ yellow, P−, KC−.

Contents: atranorin + lobaric acid.

This species grows on bare soil, on frost boils, or among mosses on acid soils. It is circumpolar, alpine, boreal-arctic in the northern hemisphere, ranging in North America south to Mexico, Colorado, and Wyoming. In South America Lamb (1977) lists it from the northern Andes.

11. Stereocaulon grande (Magn.) Magn.
ex Nilss., Ark. f. Bot. 24A (3):71. 1931. *Stereocaulon paschale* var. *grande* Magn., Göteborgs K. Vetensk. Vitterh. Samh. Handl. 30:49. 1926.

Primary thallus disappearing; pseudopodetia 4–8 cm tall, in loose growths, the base to 2 mm thick, slowly tapering to the apices, with few and short branches, the main axis dominant to the tip, pale roseate tomentose; phyllocladia distributed in groups along the pseudopodetia, clearly coralline, larger than in *S. paschale,* 0.4–0.8 mm long, 0.2 mm in diameter, not as coalescent as in *S. alpinum;* cephalodia common, small, low, pale brown, tomentose, containing *Nostoc.*

Central strand of the pseudopodetium quite loose; hyphae 3–4.5 μ, thinner walled than in *S. alpinum* and thicker walled than in *S. paschale;* cortex of phyllocladia thin, 15–20 μ, hyphae 3–4 μ thick.

Apothecia 2 mm broad, flat and margined then slightly swollen and with reflexed margin; disk blackish-brown, underside tomentose; hypothecium 40–50 μ thick, I+ blue; hymenium 50–60 μ, I blue; paraphyses 1–1.5 μ, capitate, tips to 4.5 μ; asci 8 spored, 42–52 × 8–10 μ; spores 4–5 celled, one end thicker, often slightly bent, 20–44 × 2–3 μ. Conidia 5–6 × 0.6 μ.

Reactions: KOH+ yellow, P−.

Contents: atranorin and lobaric acid.

Stereocaulon glareosum (Sav.) Magn.

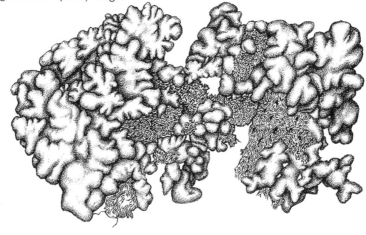

.5

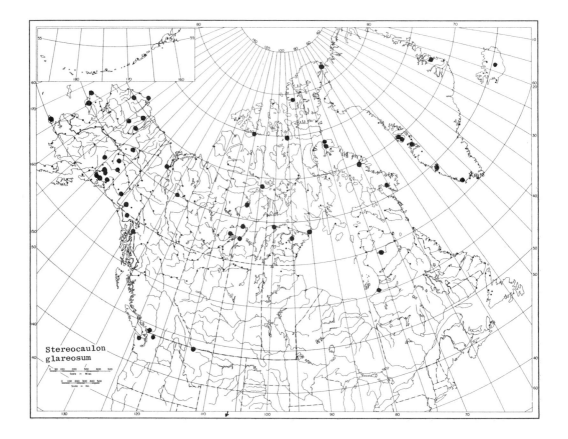

Stereocaulon
glareosum

Stereocaulon grande (Magn.) Magn.

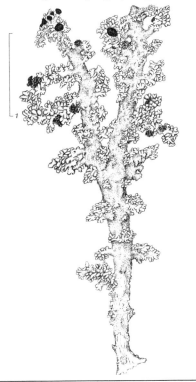

This species grows among mosses and in tundras over gravel soils. It is low arctic and circumpolar boreal, in North America occurring trans-Canada and in Alaska, south to Maine, New Hampshire and Oregon (Lamb 1977).

12. **Stereocaulon groenlandicum** (Dahl) Lamb

Occas. Papers Farlow Herb. 5:10. 1973. *Stereocaulon rivulorum* var. *groenlandicum* Dahl, Medd. om Grønland 150(2):117. 1950.

Pseudopodetia erect or prostrate, firmly attached to rocks or soil, rigid and woody (like *S. botryosum*) 1.5–5 cm tall, to 2 mm thick, terete, sparingly branched, sometimes more or less corymbose; whitish, dull, decorticate, smooth to longitudinally striate, with pale tomentum which may be browner toward the base; phyllocladia whitish, grain-like, becoming thickly crenate or digitate squamuliform, occasionally lobed and branching like those of *S. rivulorum,* lacking soredia; cephalodia incon-

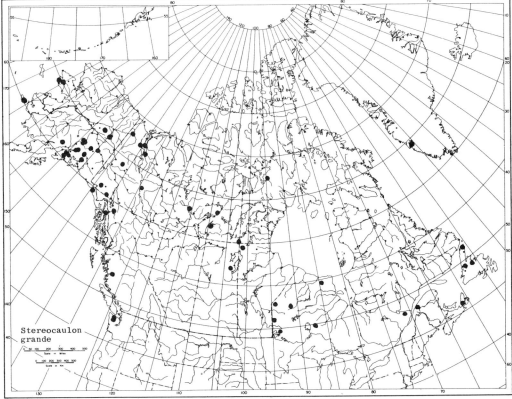

Stereocaulon grande

spicuous, sometimes lacking, among the phyllocladia, subglobular, usually pale brownish or yellowish, when well developed with slightly tuberculate appearance, the units with dark central spots, containing *Nostoc*.

Apothecia large, to 3–6 mm, becoming compound, dark brown to black brown, terminal on branches, exciple soon disappearing, palisade-radiate; hypothecium hyaline to cloudy; hymenium 50–75 μ; asci 50–60 μ; spores 4–8, elongate fusiform, straight, 3-septate, 20–40 × 3–4 μ.

Reactions: K+ yellow, PD+ pale yellow.

Contents: atranorin, miriquidic acid, perlatolic acid, and anziaic acid (Lamb 1973).

This species is known from rocks and gravels in Scandinavia, Greenland, Alaska, and an uncertain record from Spitzbergen. Lamb (1977) states that it occurs also in the Northwest Territories but cites no localities in that paper. This species is very close to *S. rivulorum*, differing in the more woody pseudo-

podetium and the presence of miriquidic acid. The other contents are similar to those of strain I of *S. rivulorum*.

13. **Stereocaulon incrustatum** Flörke

Deutschl. Lich. 4:12. 1819. *Stereocaulon tomentosum* var. *incrustatum* Schaer. Lich. Helvet. Spicil. 6:276. 1833. *Stereocaulon tomentosum* * *incrustatum* Th. Fr., Lich. Scand. 50. 1871.

Primary thallus disappearing, of small granules; pseudopodetia in thick or loose tufts, ascending at the circumference, erect centrally, simple or with few branches, 1.5–3 cm tall, 0.8–1.2 mm thick, with thick whitish-gray, ashy or roseate tomentum; for long stretches without phyllocladia, the phyllocladia irregularly distributed along the pseudopodetia, a little thicker toward the apices, whitish-gray or bluish-gray, verruciform, very small, 0.08–0.15 mm thick, often concealed by the tomentum or in var. *abduanum* larger and projecting above the tomentum in warty groups; soredia rare; cephalodia

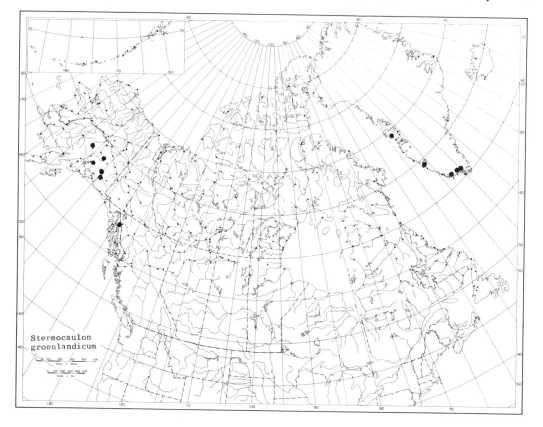

Stereocaulon groenlandicum

Stereocaulon incrustatum Flörke

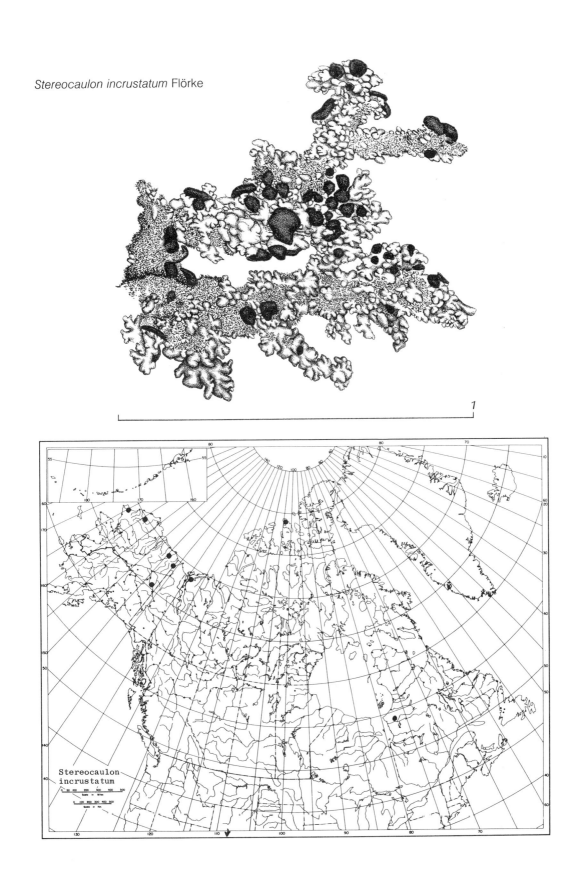

1

Stereocaulon
incrustatum

mainly where the phyllocladia are sparse, covered by the tomentum or larger and to 2 mm. broad, dark brown, mainly containing *Nostoc*.

Central medullary hyphae not firmly conglutinate, pachydermatous, 4–5 μ thick, KOH−; exterior hyphae gelatinous, not sharply distinguished, 6–8 μ thick; cortex of phyllocladia 15–30 μ thick; hyaline or yellowish, very gelatinous.

Apothecia common, terminal, single or confluent, immarginate; disk dark brown; hypothecium 30–45 μ, prosenplectenchymatous, I−; hymenium 60–65 (Magn. −50) μ, I+ blue; paraphyses somewhat coherent, 1–1.5 μ, tips capitate, brown, 2–3.5 μ; asci 52–60 × 9–12 μ; spores 4–6, cylindrical-fusiform, straight, 3 septate, 35–44 × 2.5–3 μ. Conidia bacilliform, 4.5–5.5 × 1.2 μ (Magn.) or 5–7 × 0.8–1 (Frey).

Reactions: K+y, P−, containing only atranorin (Lamb 1951).

Contents: atranorin ± lobaric acid.

This species grows on sandy soils. It is distributed from the Alps in Europe to Kamtchatka and in Alaska, across extreme northern Canada to the Arctic Archipelago (map in Lamb 1961) and in Colorado.

Var. *abduanum* (Anzi) Grey in Rabenh., Krypt. Fl. Deutschl. 9(4):176. 1933 (*Stereocaulon abduanum* Anzi, Comment. Soc. Crittogamol. Ital. 2(1):5. 1864) has larger phyllocladia more piled up, over the tomentum, in part coralline.

14. Stereocaulon intermedium (Sav.) Magn.

Göteborgs Kgl. Vetens. Vitterh. Samh. Handl. 30:23. 1926. *Stereocaulon coralloides* f. *intermedium* Savicz, Notul. Syst. Inst. Crypog. Horti. Bot. Petrop. 2–163. 1923.

Primary thallus disappearing; pseudopodetia densely caespitose, attached to rocks, 5–8 cm tall, 1–3 mm thick, with scattered long branches at acute angles, whitish, thinly tomentose; phyllocladia evenly distributed on the podetia, verrucose to verrucose-coralloid, 0.15–0.18 mm thick; cephalodia inconspicuous among the phyllocladia, to 1 mm, irregularly tuberculate, grayish to brownish gray, containing *Nostoc*.

Central medullary hyphae parallel, very

loosely united, pachydermatous, 2–3 μ thick, the exterior of the central cylinder of yellowish, gelatinous, very pachydermatous, hyphae 6–10 μ in diameter, KOH−; cortex of phyllocladia 16–20 μ thick, grayish, hyphae firmly conglutinate, unclear.

Apothecia numerous, terminal, margin soon disappearing; disk convex, dark-brown, the margin reflexed and the center dividing into several convex parts; underside of perpendicular, gelatinous, pachydermatous hyphae; hypothecium yellowish, 35–40 μ; hymenium 55–65 μ, pale, I+ blue; paraphyses not firmly conglutinate, 1.8 μ thick, the apices capitate, 3.5–4.5 yellowish; asci narrowly clavate, 50 × 8–10 μ; spores 6–8, usually 3-septate, rarely 4–5 septate, bacilliform, with one end thicker, the immature spirally twisted in the ascus, 35–48 × 3–3.5 μ. Conidia unknown.

Reactions: K+ yellow, P−.

Contents: atranorin and lobaric acid.

This species grows on stones in areas with masses of rocks. It is a Beringian bilateral radiant, reported from Japan and Kamtchatka in Asia and in Alaska and British Columbia in North America (Lamb 1951) and an unspecified disjunct occurrence in Newfoundland (Lamb 1977).

According to Magnusson it is very like *S. grande* but differs in the high hymenium, the blunt spores, the free hyphae of the main axis, the thick cortex of the lower side of the apothecium and the thin excipulum with thin hyphae.

15. Stereocaulon paschale (L.) Hoffm.

Deutschl. Flora 2:130. 1796. *Lichen paschalis* L., Sp. Pl. 1153. 1753. *Cladonia paschalis* (L.) Wigg., Primit. Florae Holsat. 90. 1780. *Baeomyces paschalis* (L.) Wahlenb., Flora Lapponica 450. 1812.

Primary thallus disappearing; pseudopodetia very loosely attached to the substratum or becoming free as the base dies off, crowded or dispersed among mosses, 2–6 cm tall, 0.6–2 mm thick at the base, naked below with a white to roseate tomentum above; phyllocladia few below, crowded toward the apices, whitish or grayish white, sometimes rose tinged, crowded in small stalked clusters, verruciform coralloid or rarely verruciform squamulose; cephalodia

Stereocaulon intermedium (Sav.) Magn.

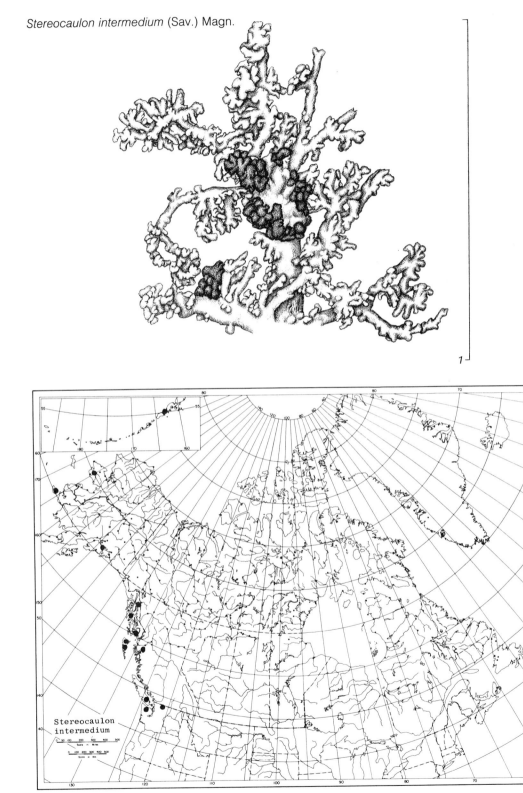

1

Stereocaulon paschale (L.) Hoffm.

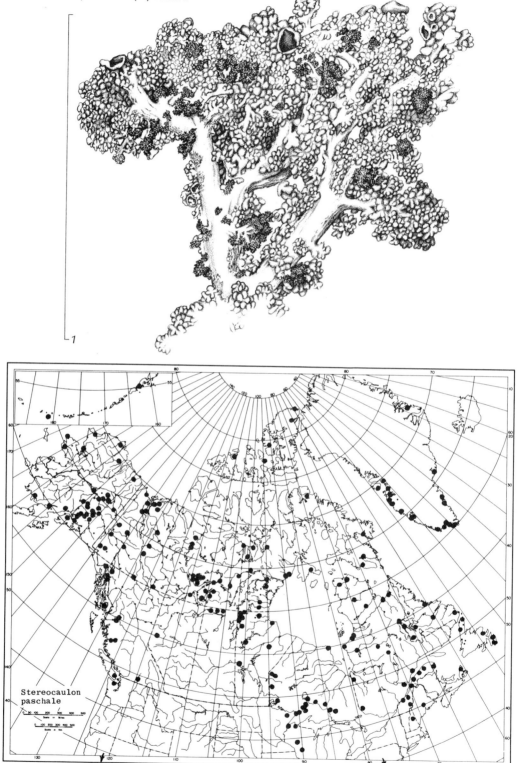

Stereocaulon
paschale

scabrid, dark olive brown, 0.3–1 mm, containing *Stigonema* or rarely *Nostoc*.

Central medullary hyphae fairly loose; hyphae leptodermatous, 3–4.5 μ thick, those of external cylinder more conglutinate, 6–8 μ thick; cortex of phyllocladia 20–25 μ thick, slightly gelatinous.

Apothecia somewhat rare, terminal on main or short side branches, 1–3 mm broad, disk flat or becoming slightly convex, dark brown, lower side smooth or slightly tomentose, of partly parallel, pachydermatous, gelatinous hyphae 7–9 μ thick; hypothecium 50 μ, hyaline; hymenium 45–50 μ, I+ dark blue; paraphyses 1.5–2 μ, loosely coherent, the upper part occasionally branched, tips yellow-brown, capitate, 3–4 μ; asci clavate, 35–40 × 8–10 μ according to Magnusson, 55–60 × 8–10 μ according to Frey; spores 6–8 per ascus, 3-septate, cylindrical, 25–30 × 2.5–3 μ, often slightly bent. Conidia straight or slightly bent, 5–12 × 0.8–1.2 μ.

Reactions: K+ yellow.

Contents: atranorin and lobaric acid.

This species grows on earth, mainly among mosses, when on rocks it is on an accumulation of humus. It is a circumpolar, arctic, and montane to boreal species, in North America ranging south to Virginia, Wisconsin, California.

The dark, scabrid, abundant cephalodia distinguish this species from *S. alpinum,* which has pale, spherical uncommon cephalodia. *S. tomentosum* differs in containing stictic acid.

16. **Stereocaulon pileatum** Ach.

Lich. Univ. 582. 1810. *Stereocaulon cereolinum* var. *pileatum* Th. Fr., De Stereoc. et Pilophor. Comment. 19. 1857.

Primary thallus persistent, crustose, grayish-white with white apices, of verruciform to more or less coralloid granules, occasionally squamulose, chinky areolate when in thick crusts. Pseudopodetia short, 2–5 (–20) mm tall, firmly attached to the substratum, sterile ones short,

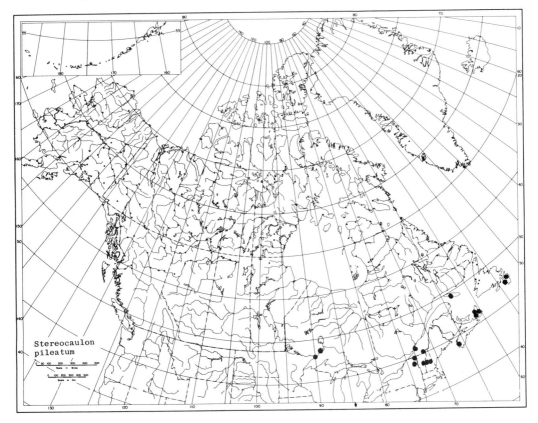

Stereocaulon pileatum

simple, with capitate white sorediate apex, fertile with terminal pileate apothecia, lacking tomentum; phyllocladia scattered or abundant, grayish-white, granulose or somewhat lengthened, short-stalked, to 0.3 mm long; cephalodia small, flattened, tuberculate, grayish brown or with a violet tinge, on the primary thallus or the pseudopodetia, containing *Stigonema.*

Central medullary hyphae densely conglutinate, the interstices very small, hyphae pachydermatous, 2.5–3.5 μ, outer hyphae yellowish gray, very thick walled, 6–10 μ in diameter; phyllocladia with cortex 15–20 μ, with shallow pits to 50 μ deep into the thallus, the hyphae 3–4 μ in diameter, indistinct; medullary hyphae looser.

Apothecia pileate, 1–1.5 mm broad, disk flat or becoming convex, sometimes fissuring, dark red-brown, the underside with a yellowish palisade-like cortex 60–85 μ thick, the central cone of dense conglutinate hyphae; hypothecium 20–30 μ , grayish-yellow; hymenium 50–65 μ, upper part yellow-brown, I+ blue; paraphyses loose, 1.5 μ thick, the tips brown capitate, 3 μ; asci clavate, 45–50 × 10–14 μ, 6–8 spored; spores blunt, one end thicker, 3-septate, 20–30 × 4–5 μ. Conidia cylindrical, straight or slightly bent, 6–8 × 1 μ.

Reactions: phyllocladia K+ yellow, P–.
Contents: atranorin and lobaric acid.

This species grows on granites or sandstones in very moist usually more or less shaded places.

This species is circumpolar, boreal. In North America it has the appearance of an amphi-Atlantic species, ranging from Quebec and Newfoundland south to Minnesota and in the eastern mountains to North Carolina. Yet I have collected it at Kirovsk on the Kolskiy Peninsula, in tundra, and Lamb (1977) reported an unchecked specimen from Siberia in the Irkutsk region, and hence I have included it in the possibility that it may later be collected farther north on the American continent.

17. Stereocaulon rivulorum Magn.
Göteborgs Kgl. Vetens. Vitterh. Samh. Handl. 30:63. 1926.

Primary thallus disappearing. Pseudopodetia loosely or firmly adhering to the ground, erect or decumbent, forming low lax tufts, whitish with a bluish tinge 0.5–1.2 mm thick, bare or with a whitish to roseate thin tomentum, *very fragile,* sparingly branched with no distinct main axis; phyllocladia glaucous-white, granular or somewhat elongated, broader toward the apices, dispersed in groups along the pseudopodetia, leaving spaces showing on the axis, conglomerate and angular crustose on some decumbent pseudopodetia; cephalodia infrequent, on the underside of dorsiventral pseudopodetia, inconspicuous, tuberculiform with granular surface, of the same pale color as the tomentum or brownish violet, containing *Nostoc.*

Central medullary hyphae parallel, not firmly conglutinate, leptodermatous, 2.5–3.5 μ thick, the lumina 1.3–1.7 μ; the exterior hyphae forming a pale yellowish cylinder of gelatinous, more conglutinate, hyphae 6–10 μ thick, colored pale yellow by KOH; phyllocladia with cortex 25–45 μ thick, pale yellowish, of firmly conglutinate hyphae, not easily discernible, partly parallel to the outer surface.

Apothecia usually numerous, terminal or sublateral, 1–3 mm broad; disk dark brown, with long persistent pale margin, lower surface plane, white, tomentose; hypothecium hyaline, 35–50 μ thick; hymenium 60–70 μ thick; paraphyses contiguous, 1.5 μ thick, the tips capitate to 4.5 μ brown; asci cylindrico-clavate, 50–60 × 10–12 μ, I+ pale blue, hymenial gelatine not colored; spores 8, cylindrico-fusiform with one end narrower, 3-septate, 25–37 × 3–3.5 μ. Conidia straight, cylindrical 5–6 × 1–1.2 μ.

Reactions: K+ yellow, P–.
Contents:
Strain 1: atranorin, perlatolic acid, and anziatic acid. This strain occurs in Scandinavia, Iceland, Greenland, and Alaska. The similar appearing *S. groenlandicum* has the same content as this strain but with the addition of miriquidic acid.
Strain 2: atranorin and lobaric acid. This is the common arctic-alpine strain.
Strain 3: atranorin, stictic and norstictic acids. Lamb (1977) cites only Greenland and Jan Mayen Island for this strain.
Strain 4: only atranorin present. A widely distributed strain.

Stereocaulon rivulorum Magn.

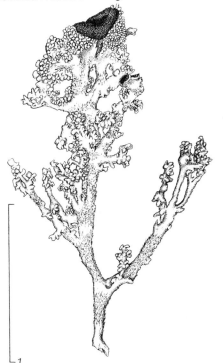

1

This species grows in low places below permanent snowbanks and along streams where it is occasionally inundated. It is circumpolar, arctic, and alpine, in our sector distributed from Greenland, trans-Canada to Alaska and south to Colorado and Washington.

18. **Stereocaulon saviczii** Du Rietz
Ark. f. Bot. 22A(13):13. 1929.

Primary thallus persistent, closely attached to the substratum, of more or less globose granules; pseudopodetia short, to 1.5 cm tall, 1–3 mm thick, caespitose or simple, subterete to flattened, decorticate, more or less sulcate or split into fibrils, tomentum thin, roseate white, thicker above than below; phyllocladia dense on the upper portion of the pseudopodetium, sparser toward the center, at first subglobose, becoming cylindrical to more or less coralloid, with darker spot-like depressed tips, simple to sparsely branched, robust, to 2 mm long and 0.5 mm diameter, tubulose, whitish ashy, subshiny, lacking isidia or soredia; ce-

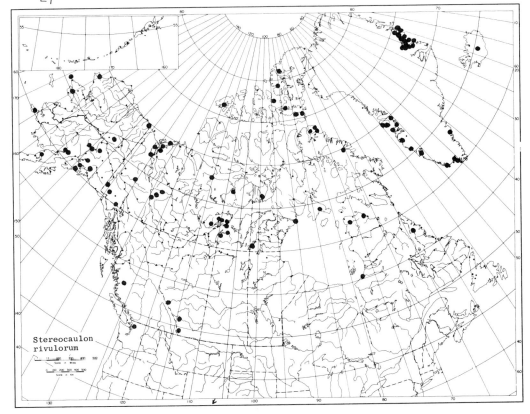

Stereocaulon
rivulorum

Stereocaulon saviczii Du Rietz

phalodia on the upper part of the pseudopode-tia, large, to 5 mm broad, granulose-glomerate, dark bluish-ashy to olive brownish, containing *Scytonema*.

Apothecia terminal, solitary, large, to 7 mm broad, rounded; disk flat to slightly convex, immarginate, black, dull; hypothecium dark brown; paraphyses slender, distinct, the apices capitate, dark brown; asci subclavate, 8 spored; spores hyaline, subfusiform, blunt or with one tip blunt, straight, $15-37 \times 3-6.5 \mu$.

Reactions: K+ yellow, P−.

Contents: atranorin and lobaric acid.

This species grows on rocks. It is an amphi-Beringian species known from Japan, Kamtchatka, and Alaska.

19. **Stereocaulon saxatile** Magn.

Göteborgs K. Vetens. Vitterh. Samh. Handl. 30:41. 1926. *Stereocaulon paschale* var. *evolutoides* Magn., ibid. 30:50. 1926. *Stereocaulon evolutoides* (Magn.) Frey, Rabenhorsts Krypt. Flora 9, Abt. 4(1):145. 1932.

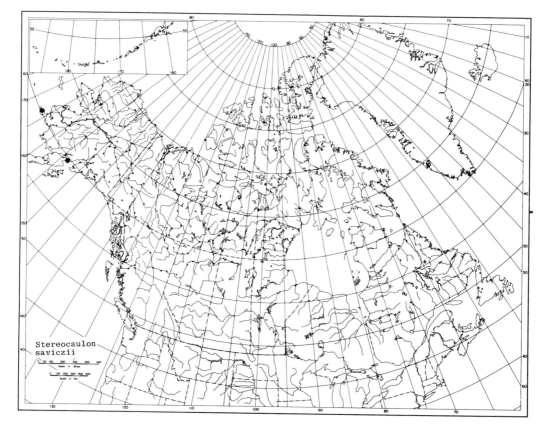

Stereocaulon saviczii

Stereocaulon saxatile Magn. central pseudopodetia

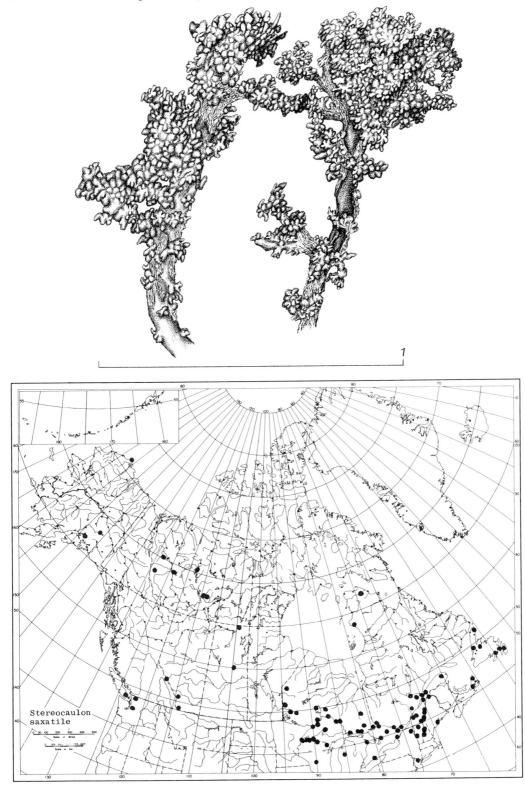

1

Stereocaulon
saxatile

Stereocaulon saxatile Magn. border pseudopodetia

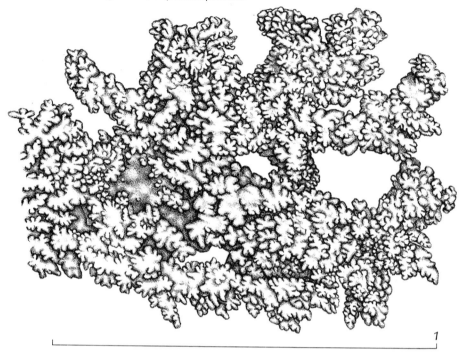

1

Primary thallus disappearing; pseudopodetia firmly attached to rocks, forming dense tufts to 7 cm across, stout, 0.5–1.5 mm thick, naked or grayish tomentose toward the apices; phyllocladia numerous, almost completely covering the pseudopodetia, squamuliform and more or less incised becoming coralloid squamuliform, grayish white; cephalodia rare, inconspicuous, containing *Stigonema*.

Central medulla loosely conglutinate, faintly KOH+ yellow; hyphae pachydermatous, 3–4 μ in diameter, the outer cylinder indistinct, hyphae 7–9 μ thick, darkened.

Apothecia rare, terminal on marginal pseudopodetia, convex, dark brown, less than 1 mm broad; the underside with a palisade-like cortex of perpendicular hyphae 6–10 μ thick; hypothecium 35–50 μ, grayish; hymenium 55–65 μ, I+ dark blue; paraphyses slightly coherent, 1.2–2 μ thick, the tips irregularly swollen clavate to 3–4 μ; asci cuneiform to cylindrico-clavate, 50×9–12 μ, the apices I+ darker blue spores 6, cylindrical, 4-celled, 20–40×3.5–4 μ. Conidia not known.

Reactions K+ yellow, P+ faint yellow. Contents: atranorin and lobaric acid.

This species grows on rocks and outcrops. It is boreal, amphi-Atlantic, occurring from the northern forest edge south to New Jersey, Minnesota, and Washington.

This species is very close to *S. paschale*, differing according to Magnusson (1926) in habitat, greater height of hymenium, and breadth of spores. Cephalodia are rare in this as compared with the abundant dark cephalodia of *S. paschale*.

20. Stereocaulon spathuliferum Vain.
Ark. f. Bot. 8(4):36. 1909. *Stereocaulon fastigiatum* f. *spathuliferum* (Vain.) Magn., Göteborgs K. Vetens. Vitterh. Samh. Handl. 30(7):37. 1926. *Stereocaulon botryosum* f. *spathuliferum* (Vain.) Frey, Rabenh. Krypt.-Fl. 9(4):125. 1932.

Primary thallus disappearing; pseudopodetia 30–50 mm tall, repeatedly sympodially branched; the branches spreading, thickened basally, terete or more or less flattened, mainly decorticate, pale white, K+ yellow, the interior K−; phyllocladia verrucose, with spatulate apical expansions, 0.3–3 mm broad, the upper side bluish gray, the underside commonly sorediate; the apices often recurved and round, entire or crenulate; in f. *dissolutum* (Magn.) Lamb the spatulate apical soralia are replaced by effuse soredia but the tips of the branches are still flattened somewhat; cephalodia usually stipitate, densely branched, tuberculate verrucose to botryose, bluish ashy to brownish, containing *Nostoc*.

Apothecia at the tips of branches, to 3 mm broad, convex, the details of spores not given by any authors and no material seen by me.

Reactions: K+ red, P+ red.

Contents: atranorin, stictic acid, norstictic acid and several unidentified substances (Lamb 1977).

This species, reported as growing on rocks, is classed by Lamb as boreal-arctic, amphi-Atlantic bicentric or possibly circumpolar. In the Arctic, Lamb reports it from Europe, Iceland, Greenland and in North America from Alaska, British Columbia, Oregon, and Washington but does not cite exact localities. A specimen of f. *pygmaeum* (Magn.) Lamb from Minnesota determined by Lamb is in my herbarium.

21. Stereocaulon subcoralloides (Nyl.) Nyl.
Flora 57:6. 1874. *Stereocaulon paschale* f. *subcoralloides* Nyl., Notis. Sällsk. pro. Fauna Flora Fenn. 5:64. 1861.

Pseudodetia loosely attached to stone, loosely caespitose or dense, erect, 6–15 mm tall, 0.5–1 mm thick, base dark, upper part pale, lacking tomentum, few branches on lower part of axis, upper part with numerous small branches; phyllocladia few in lower part, very abundant in the upper part, sometimes almost confluent, from grain-like to coralloid, 0.1–0.3 mm thick, 0.3–0.6 mm long, rarely incised squamuliform; cephalodia numerous, among the phyllocladia on the upper branches, olive-brown, tuberculate, containing *Stigonema*.

Apothecia few, terminal, 1.5–2 mm broad; disk flat, dark brown with pale margin, becoming convex or divided and with margins reflexed; hypothecium 35–45 μ, hyaline; hy-

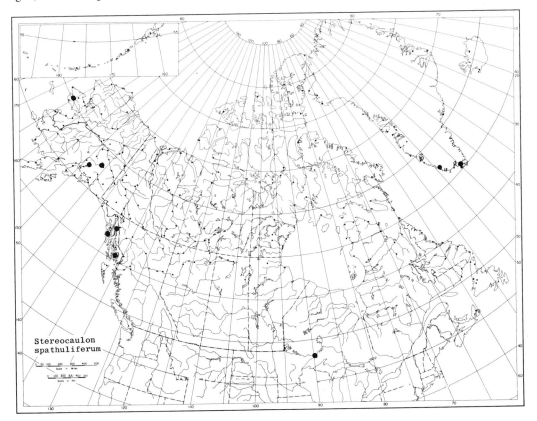

Stereocaulon
spathuliferum

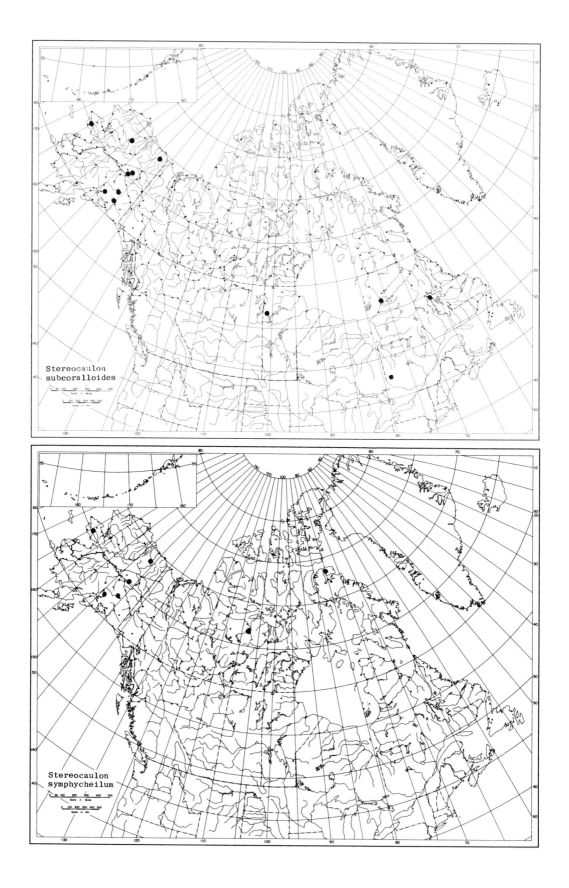

Stereocaulon
subcoralloides

Stereocaulon
symphycheilum

menium 45–50 μ, I+ pale blue, upper part dark brown; paraphyses coherent, 1.5–1.8 μ, tips brown, capitate 3–3.5 μ; asci broadly clavate; spores 8, narrowly ellipsoid with blunt ends, 3–6 septate, mainly 4 septate, 27–37 × 3–4.5 μ.

Reactions: K+ yellow, inner part K−.

Contents: atranorin and lobaric acid.

This species grows directly on rocks and outcrops. It is circumpolar, boreal, in North America distributed south to Maine, New York, Ontario and Alaska (Lamb, 1977).

22. Stereocaulon symphycheilum Lamb
Bot. Not. 114:271. 1961.

Primary thallus disappearing, of minute peltate squamules with pale margins; pseudopodetia firmly attached to rock, prostrate and strongly dorsiventral, short, 3–15 mm tall, simple or sparingly branched, the dorsal surface more or less continuously corticate, the ventral ecorticate, white to pale ochraceous, smooth and glabrous to finely subpubescent but not tomentose, mainly tipped by conspicuous globose capitate soralia which are pulverulent or granulose and greenish white to gray; phyllocladia beginning as nodules more or less crenate and becoming peltate as in *S. vesuvianum;* cephalodia more or less abundant, verruculose tuberculate or scabrid, dark brown to brown-black, lacking cortex, containing *Stigonema.*

Apothecia rare, usually terminal, quite large, 1–2.5 mm broad, flat to becoming pileate-convex, margin disappearing; disk brown-blackish; exciple 50–90 μ thick, colorless to pale brown, of conglutinous, radiating, pachydermatous hyphae, occasionally with granules in striae between the hyphae; hymenium 50–60 μ, pale to dark brown in upper part; paraphyses discrete, 1–2 μ thick, simple or branched, tips capitate, brown, 3–4 μ; asci clavate, 45–50 × 10–14 μ, I+ blue; spores (4–) 6–8 per ascus, elongate fusiform, straight, 3-septate, 20–35 × 3–4.5 μ. Conidia rare, straight or slightly curved, 4.5–0.7 μ.

Reactions: K+ yellow, P+ sulphur-yellow.

Contents: atranorin and lobaric acid.

This species grows on rocks and outcrops. It is probably circumpolar boreal to arctic. Lamb (1977) has reported it from Scandinavia, Kamtchatka, Alaska, and Quebec.

23. Stereocaulon tomentosum E. Fries
Sched. Crit. 20. 1825. *Stereocaulon paschale* var. *tomentosum* (Fr.) Duby, Bot. Gall. 2:618. 1850. *Stereocaulon tomentosum a campestre* Körb., Syst. Lich. Germ. 11. 1855.

Primary thallus disappearing; pseudopodetia loosely attached to the substratum, caespitose, erect or decumbent, clearly dorsiventral, 3–8 cm. tall, 1–2 mm thick, little branched at the base, the upper part with many recurved short branches, more or less anisotomic, the underside thickly tomentose, the upper side with abundant phyllocladia which are crenate squamulose, ashy gray, overlapping and thickly covering the pseudopodetia; cephalodia inconspicuous, pale, 0.3–0.8 mm broad, concealed in the tomentum of the underside, more or less spherical, containing *Nostoc.*

Central medullary hyphae densely conglutinate, KOH−, pachydermatous, 3.5–4.5 μ thick, exterior hyphae gelatinous, with more or less distinct inner limit, yellowish, 5–8 μ thick.

Apothecia usually abundant toward the ends of the branches but mainly lateral, 0.3–0.6 (−1.2) mm broad, the pale margin soon disappearing; disk dark brown, slightly convex, smooth, becoming more convex; lower side of gelatinous yellowish pachydermatous hyphae, 8–11 μ thick, the interior part intricate, the outer parallel, perpendicular to the surface; hypothecium 30–50 μ, grayish; hymenium 45–50 μ, I+ dark blue; paraphyses coherent, 1.5–1.8 μ, the tips brown, capitate, 3–4 μ; asci cylindricoclavate, 35–50 × 8–10 μ; spores 8, cylindricofusiform, somewhat spirally twisted in the ascus, usually 4-celled, rarely to 8-celled, 20–35 × 2.5–3 μ. Conidia rare, straight, 4–5 × 0.8 μ (Harmand 1907).

Reactions: KOH+ yellow, P+ yellow or orange red.

Contents: atranorin with stictic acid, norstictic acid and partly constictic and consalazinic acids (Lamb 1977).

Stereocaulon tomentosum E. Fries

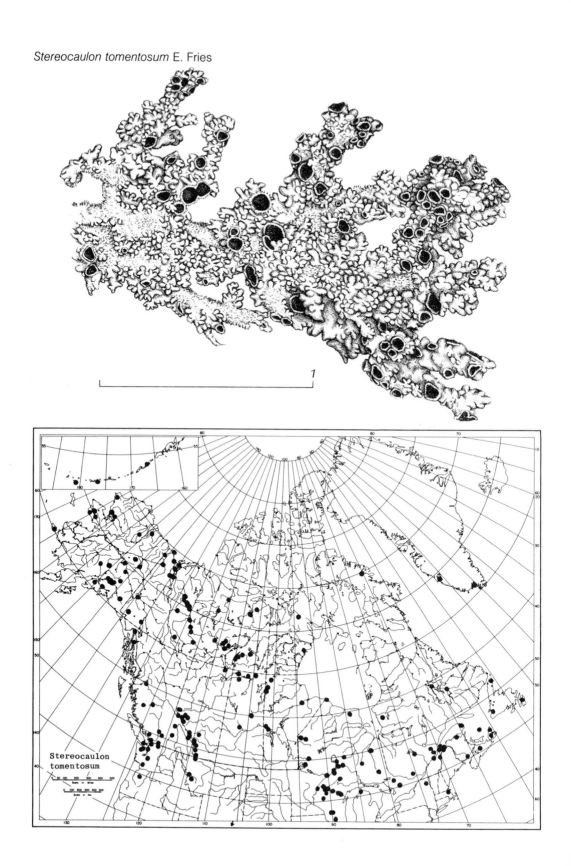

Stereocaulon
tomentosum

This species grows on earth in open places, occasionally on gravels. It is circumpolar, boreal-montane in the northern hemisphere, extending in South America to the tip of the continent.

S. tomentosum and *S. paschale* give much trouble in identifications. In *S. tomentosum* the apothecia are numerous, lateral, small, and convex, the phyllocladia more squamuliform, appressed to the pseudopodetia, the pseudopodetia dorsiventral with the phyllocladia on the upper side, the cephalodia pale, inconspicuous and containing *Nostoc;* in *S. paschale* the apothecia are few, terminal, larger and flatter, the phyllocladia spreading, digitately divided and radiating to all sides of the pseudopodetia; the cephalodia dark, obvious, and containing *Stigonema*. In western North America, ranging inland to Michigan, is a morphologically similar counterpart to *S. tomentosum* but containing lobaric acid. This species, *S. sasakii* Zahlbr. var. *tomentosoides* Lamb, however, does not reach north into the tundras. Also troublesome may be the distinction between *S. paschale* and *S. alpinum*. Both have the same chemistry. However the latter has gray-white spherical cephalodia instead of the conspicuous blackish brown scabrid cephalodia of *S. paschale*.

24. Stereocaulon uliginosum Lamb
J. Hattori Bot. Lab. 43:240. 1977.

Pseudopodetia erect, caespitose or dispersed, fragile, irregularly branched, the base dying, 2.5–4.0 mm tall, 1.0 mm thick, white, lacking tomentum, the apices with soralia 0.5–1.3 mm broad; phyllocladia sparse, mainly in the lower part of the pseudopodetia, broadly squamulose and digitate branched, white, unlike those of *S. rivulorum* (another fragile species), the soralia abundant, terminal, white, densely grouped, globose, the soredia farinose; cephalodia sparse, minute, 0.5 mm broad, inconspicuous, lateral on the pseudopodetia, subglobose, slightly verruculose to nearly smooth, pale brownish, containing *Nostoc* or *Stigonema*.

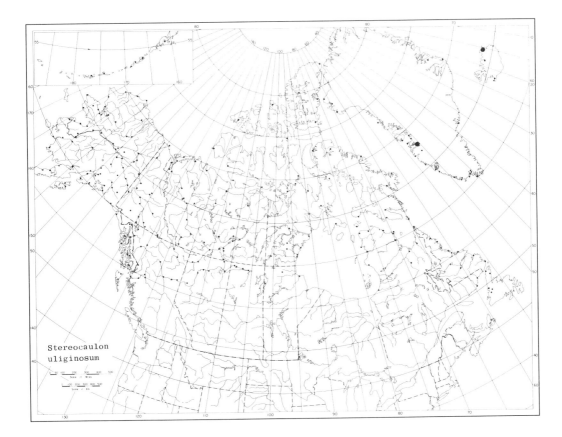

Stereocaulon
uliginosum

Stereocaulon vesuvianum Pers.

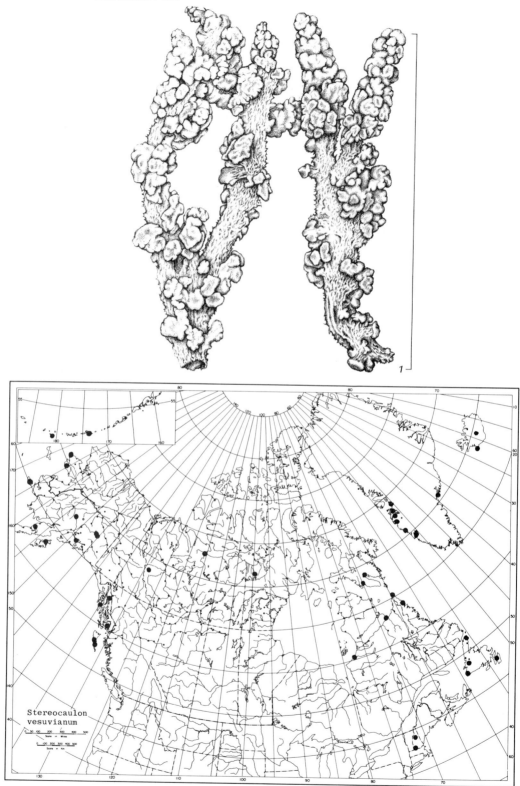

Apothecia not known.

Reactions: K+ yellow, P+ pale yellow.

Contents: atranorin and lobaric acid.

This species is reported as growing in heath or grassland and is known with certainty only from west Greenland and Iceland. A poorly developed uncertain specimen from Sweden was also seen by Lamb (1977).

This species is similar to *S. capitellatum* but with fragile pseudopodetia with dying bases and a very different chemistry.

25. Stereocaulon vesuvianum Pers.

Ann. Wetterau. Ges. 2:19. 1810. *Stereocaulon botryosum β vesuvianum* (Pers.) Ach., Synops. Meth. Lich. 285. 1814. *Stereocaulon denudatum* Flörke, Deutsch. Lich. 4:13. 1819. *Stereocaulon denudatum* var *vesuvianum* Laur. in Hepp, Lich. Europ. 2. 1853.

Primary thallus disappearing; pseudopodetia caespitose, erect, woody, cylindrical or compressed, 1–7 cm tall, 0.5–1 mm thick, lacking tomentum, whitish or grayish, the base dark, unbranched or the upper part branched; phyllocladia sparse or crowded, the small ones verruciform, the larger flattened to concave, peltate, the margins pale, crenulate, the centers darker olive, occasionally breaking up into granular soredia; cephalodia rare, verruculose, dark olive brown to blackish, containing *Stigonema*.

Cortex of the phyllocladia 20–25 μ thick, grayish, the hyphae indistinct; medulla 35–70 μ of gelatinous hyphae 5–6 μ in diameter, central medullary hyphae more or less parallel, densely contiguous, 2.5–3 μ, apparently leptodermatous but actually with scarcely discernible lumina; outer hyphae yellowish, 6–10 μ thick.

Apothecia uncommon but occasionally abundant, lateral, appressed to the pseudopodetia, 0.5–1 mm broad, flat or slightly convex, immarginate, pale or dark brown; lower side smooth, formed by the tips of densely conglutinate gelatinous pachydermatous hyphae oriented radially, the central part about 100 μ thick

of distinct pachydermatous hyphae like those of the main axis, brownish yellow; paraphyses 1.5–1.8 μ thick, the tips thickened to 4.5 μ, brownish; asci cylindrico-clavate, 35–45 × 11–14 μ, the apices I+ blue; spores 8, cylindrico-fusiform, one end thicker, 3–6 septate, 26–38 × 3–3.5 μ. Condia straight or bent, 8–12 × 0.6–0.8 μ.

Reactions: axis K−, phyllocladia K+ yellow, P+ orange.

Contents: atranorin, stictic acid, usually norstictic acid and in one variety (var. *umbonatum* (Wallr.) Lamb) dendroidin (= porphyrilic acid) (Lamb 1951 but not included in Lamb 1977). An acid deficient strain lacking the stictic acid is reported by Krog (1968).

This species grows on rocks, usually in more or less wet places. It has a reputation for colonizing lava flows. It is in both northern and southern hemispheres, practically cosmopolitan and circumpolar. Rather variable, a number of subspecific categories are recognized by Lamb (1977).

26. Stereocaulon leprocephalum Vain.

Ark. f. Bot. 8(4):35. 1909.

This species is rare, known only from the type specimen from northwest Siberia and a report from Baffin Island, Hale no. 289. *S. leprocephalum* is reported by Lamb to have a distinguishing content of atranorin plus porphyrilic acid. The Hale specimen however contains lobaric acid plus the atranorin. The specimen was first determined by Lamb as *S. leprocephalum,* then in a later annotation as *S. coniophyllum* and yet apparently retained again in *S. leprocephalum* in the Lamb 1977 Conspectus. In view of the discrepant chemistry and the confusing annotation I have not accepted this species as being in the American Arctic but have retained it in the key in case of future discovery. It would be more expected in the Alaskan sector than so disjunct in the Canadian eastern Arctic.

STICTA Schreb.

Genera Plant. 768. 1791.

Thallus foliose, dorsiventral, erect in some non-arctic species; upper side smooth to tomentose; underside with a fine nap of tomentum, cyphellate below; upper and lower cortex paraplectenchymatous; attached by rhizinae.

Apothecia adnate or marginal, with or without algae in the margin, the margin paraplectenchymatous; subhymenium hyaline or brownish; upper part of hymenium colored; paraphyses unbranched, septate, coherent; spores 8, hyaline to brown, spindle-shaped to acicular, transversely 1–7 septate. Pycnidia immersed, marginal or laminal; conidia short, straight, cylindrical or with the ends thickened.

Algae: *Palmella* (green) or *Nostoc* (blue green).

Type species: *Sticta filix* (Sw.) Nyl.

1. Thallus lacking soredia or isidia. . . . . 1. *S. arctica*
1. Thallus especially on the margins, with soredia or isidia. . . . . 2. *S. weigelii*

1. Sticta arctica Degel.
Medd. Göteborgs Bot. Trädg. 12:108. 1937.

Thallus foliose, the lobes small, to 30 mm long, 12 mm broad, the edges somewhat crisped and turned up, upper side smooth, brown; underside pale at the edges, dark centrally, covered with a fine tomentum and with scattered cyphellae, attached by rhizinae.

Apothecia and pycnidia not known.

Upper and lower cortices paraplectenchymatous, 25–40 μ thick; medulla lax, algae *Nostoc;* rhizinae simple or branched.

Reactions: K–, C–, KC–, P–, I–.

Contents: not known.

This species grows among mosses and on hummocks in both dry and moist types of tundras. It is an amphi-Beringian species, known from Siberia, Kamchatka, and in North America from Alaska east as far as Baffin Island.

2. Sticta weigelii (Isert ex Ach.) Vain.
Acta Soc. Fauna Flora Fenn. 7:189. 1890. *Sticta damaecornis* b. S. *weigelii* Isert ex Ach., Lich. Univ. 446. 1810.

Thallus brown, foliose, the lobes narrow and elongate, to 1.5 cm broad; the edges crisped and with an abundance of black soredia and isidia; the isidia may also line the edges of cuts and holes in the upper side; underside covered with a mat of fine tomentum, lighter brown toward the lobe tips, blackening centrally, with an abundance of scattered impressed cyphellae; upper cortex 30–40 μ, paraplectenchymatous; algae *Nostoc.*

Apothecia not seen, described by Vainio from tropical material as up to 3 mm broad, the margin thin; the disk red to brown; the exciple sometimes tomentose; hypothecium hyaline; hymenium 120 μ, hyaline, I+ blue; paraphyses coherent, septate, the tips slightly capitate; spores hyaline, 8, biseriate, fusiform, 1–3 septate, 30–35 × 7–9 μ. Conidia subcylindrical or the tips slightly thickened, 3 × 0.5 μ.

Usual reactions all negative and no substances described from this species.

This species grows on trees and on rocks. It is tropical to temperate, Asian and North and South American but with occurrences north into Alaska with a disjunct occurrence at Anaktuvuk Pass which correlates with the occurrence of other Appalachian lichens as *Umbilicaria caroliniana* and *Baeomyces roseus*.

Sticta arctica Degel.

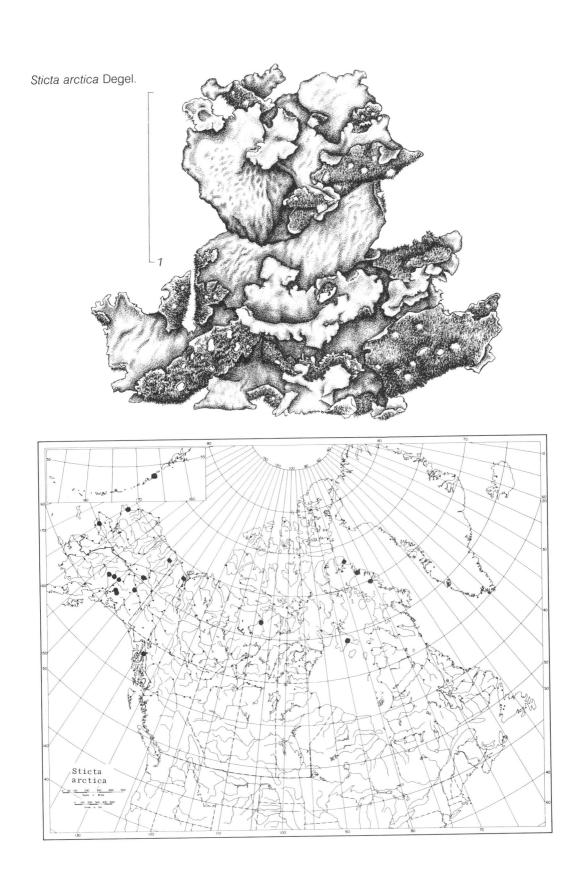

1

Sticta weigelii (Isert ex Ach.) Vain.

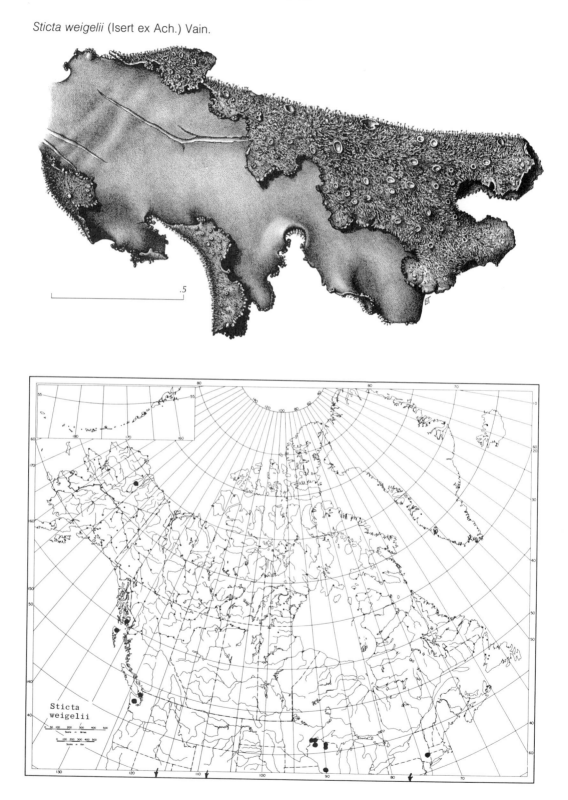

TELOSCHISTES Norm.

Nyt. Mag. Naturvid. 7:228. 1853. *Tornabenia* Mass., Mem. Lichenogr. 41. 1853. *Parmelia* D. *Platisma* + +*xanthophanae* Wallr., Flora Crypt. Germ. 1:531. 1831.

Thallus more or less fruticose, lacking rhizinae, attached by a holdfast, cylindrical or more or less flattened, entirely corticate; the cortex of vertical, dense, not paraplectenchymatous hyphae; medulla of more or less parallel vertical hyphae, occasionally partly fistulose.

Apothecia peltate, circular, on the surface or tips of the lobes, the margin lecanoroid; epithecium granulose, yellow, KOH+ red-violet; hymenium and hypothecium hyaline; paraphyses septate, simple or the upper part branched; the apices usually capitate; asci clavate; spores 8, hyaline, 2-celled polaribilocular. Pycnidia round, orange-red, immersed or with the mouth projecting; fulcra endobasidial; conidia short, straight.

Algae: *Trebouxia*.

Type species: *Teloschistes flavicans* (Swartz) Norm.

REFERENCE
Lambert, J. D. H. 1966. *Teloschistes arcticus* Zahlbr. reaffirmed in the Canadian arctic. Bryologist 69:358–361.

Only one species is found in the Arctic, the majority are southern in distribution.

1. Teloschistes arcticus A. Zahlbr.

Cat. Lich. Univ. 7:311. 1931. *Borrera* (?) *aurantiaca* R. Br. in a Suppl. to Appendix of Capt. Parry's Voyage Nat. Hist. 35. 1824 (non *Teloschistes aurantiaca* Norm., Nyt. Mag. Naturvid. 7:229. 1853).

Thallus fruticose, 1–2 cm tall, the tips cylindrical, becoming flattened toward the base, the tips nodulose, reddish orange, branching dichotomous to fastigiate-flattened; a few very short hyaline fibrillae on the nodulose tips.

Outer cortex discontinuous, of loose, irregular, more or less radiating leptodermatous hyphae 5–7 μ in diameter; within is a continuous layer of more or less pachydermatous vertical hyphae incrusted yellow with parietin, the layer about 25 μ thick; medulla of loose, anastomosing pachydermatous hyphae 5–7 μ in diameter; the lumen 1–2 μ, more or less radiate, and with the algae in scattered glomerules, the center more or less fistulose.

Apothecia not known.

Reactions: KOH+ red-violet.

Contents: parietin.

This species grows on gravelly and sandy soils in exposed places along old beach lines. It is known only from a few sites in the central northern Arctic, one on Cape Parry, the other sites on the Arctic islands.

THAMNOLIA Ach.

ex Schaer., Enum. Crit. Lich. Europ. 243. 1850.

Thallus fruticose, erect or decumbent, simple or little branched, the decumbent podetia forming erect branches which multiply the thallus; the base usually dying and growth continued from the apices, cylindrical, the tips pointed, bone-white or cream colored; the cortex paraplectenchymatous, of more or less longitudinally oriented hyphae; medulla thin, of longitudinally oriented hyphae, the interior hollow.

Apothecia and pycnidia are not known with certainty, reports seem to be based on lichen parasites.

Algae: *Trebouxia*.

Type species: *Thamnolia vermicularis* (Sw.) Ach. ex Schaer.

Teloschistes arcticus A. Zahlbr.

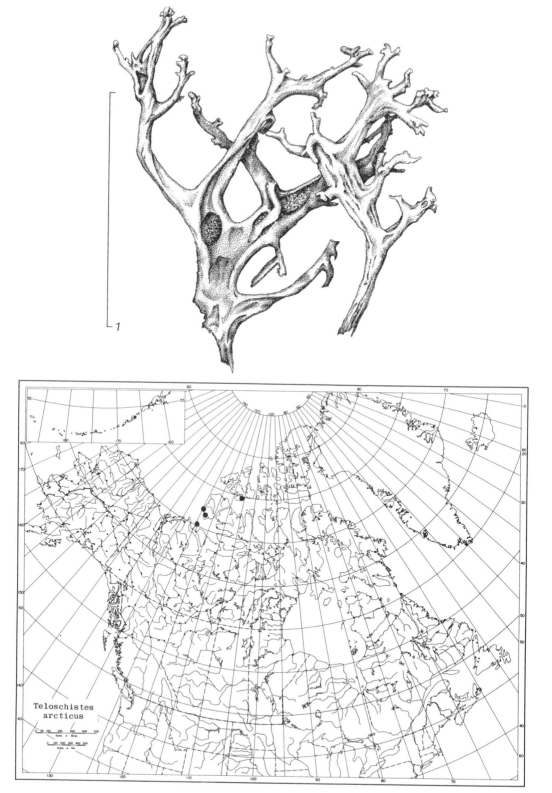

REFERENCES

Choisy, M. 1955. Sur le *Thamnolia vermicularis* et la systematique de ce genre. Bull. Soc. Linn. Lyon 10:245–266.

Culberson, W. L. 1963. The lichen genus *Thamnolia*. Brittonia 15:140–144.

Filson, R. B. 1972. Studies in Australian Lichens. 2. The alpine lichen *Thamnolia vermicularis* (Sw) Schaer. in Australia. Muelleria 2:180–187.

Hakulinen, R. 1962. Über *Thamnolia vermicularis* (Sw.) Schaer. in ostfennoskandien. Arch. Soc. Zool. Bot. Fenn. 'Vanamo' 17:134–138.

Merrill, E. D. 1900. The occurrence of *Thamnolia* in Maine. Rhodora 2:155.

Motyka, J. 1960. De variabilitate *Thamnoliae vermicularis* (Sw.) Schaer. Frag. Flor. Geobot. 6:627–635.

Sato, M. 1959. Mixture ratio of the lichen genus *Thamnolia* collected in Japan and the adjacent regions. Miscell. Bryol. Lich. 22:1–6.

—— 1962. Additional notes to "Mixture ratio of the lichen genus *Thamnolia* collected in Japan and the adjacent regions." Miscell. Bryol. Lich. 2:141–144.

—— 1963. Mixture ratio of the lichen genus *Thamnolia*. Nova Hedw. 5:149–155.

—— 1965. Distribution and ecology of the lichen genus *Thamnolia*. Bull. Fac. Arts & Sci., Ibaraki Univ. Nat. Sci. 16:25–35.

—— 1965. The mixture ratio of the lichen genus *Thamnolia* in New Zealand. Bryologist 68:320–324.

—— 1968. Two new varieties of *Thamnolia* from South America. Bryologist 71:49–50.

Sauer, E. G. F. 1962. Ethology and ecology of Golden Plovers on St. Lawrence Island, Bering Sea. Psychol. Forsch. 26:399–470.

Sheard, J. W. 1977. Palaeogeography, chemistry, and taxonomy of the lichenized Ascomycetes *Dimelaena* and *Thamnolia*. Bryologist 80:100–118.

Steiner, M. 1960. Zur unterscheidung von *Thamnolia vermicularis* und *subvermicularis* sensu Asahina. Osterr. Bot. Zeit. 107:113–114.

Weber, W. A. & S. Shushan 1955. The lichen flora of Colorado: *Cetraria, Cornicularia, Dactylina* and *Thamnolia*. Univ. of Colo. Studies 3:115–134.

1. Thallus UV−, K+ yellow, P+ orange to red, containing thamnolic acid . . . . 2. *T. vermicularis*
1. Thallus UV+, K+ weakly yellowish, P+ yellow, containing squamatic and baeomycesic acids
. . . . 1. *T. subuliformis*

1. Thamnolia subuliformis (Ehrh.) W. Culb.

Brittonia 15:144. 1963. *Lichen subuliformis* Ehrh., Beitr. 3:82. 1788. *Thamnolia subvermicularis* Asah., J. Jap. Bot. 13:317. 1937.

Thallus as in the following species, of erect, simple to sparingly branched, pointed, cylindrical podetia which are white but do not change color or stain herbarium paper on long standing.

Reactions: UV+ yellow, K− or pale yellow, P+ yellow.

Contents: squamatic and baeomycesic acids.

This species, like the following, grows on many types of tundras but is more common than it in the more sheltered localities. It is used by the golden plover in decorating its nest sites (Sauer 1962). It is both more abundant and more wide ranging than *T. vermicularis* in the northern hemisphere where it is circumpolar. In the southern hemisphere it is rare.

2. Thamnolia vermicularis (Sw.) Ach. ex Schaer.

Enum. Crit. Lich. Europ. 243. 1850. *Lichen vermicularis* Sw., Meth. Musc. Illus. 37. 1781. *Lichen tauricus* Wulf. ex Jac., Coll. Bot. 2:177. 1788.

Thallus erect or decumbent, occasionally forming tufts by branching from the base or along the decumbent podetia, cylindrical, the tips pointed, white or cream white; in the herbarium becoming pinkish and staining paper brown on long standing; smooth, seldom with short lateral pointed branches, very variable in size, usually 1–2 mm broad but reaching 8 mm broad and 12 cm tall.

Reactions: UV−, K+ yellow, P+ orange to red.

Contents: thamnolic acid.

This species grows on many types of tundras, from bare, open gravels and frost boils to rich moist, mossy thickets among the willows and heaths. It is circumpolar, arctic-alpine, the studies by Sato demonstrating clearly that it is more abundant than *T. subuliformis* in collections from the southward and into the southern hemisphere. Even in the northern

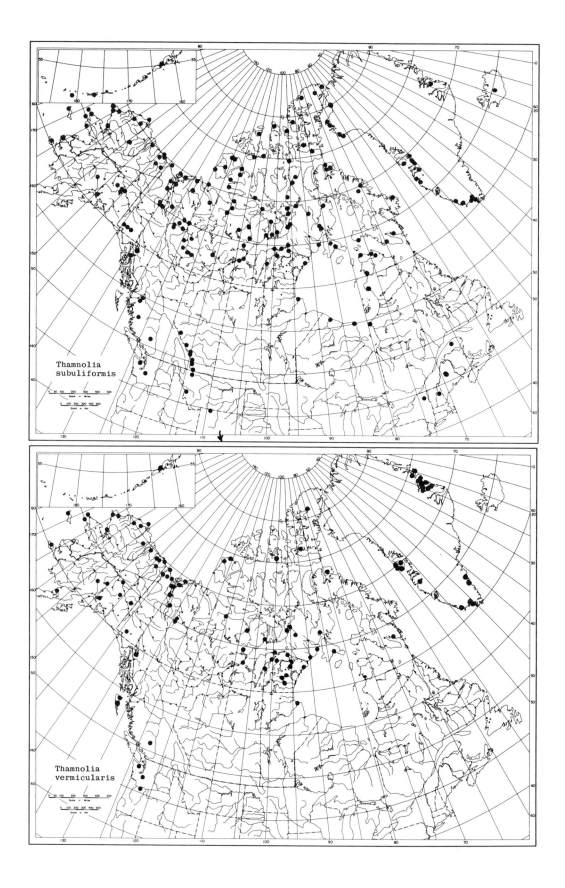

Thamnolia
subuliformis

Thamnolia
vermicularis

Thamnolia vermicularis (Sw.) Ach.

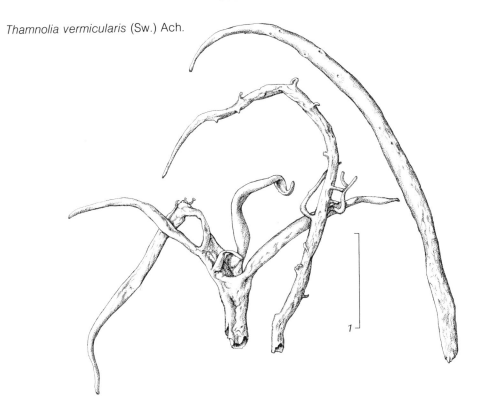

hemisphere the collections showed increasing pro-portions of *T. vermicularis* over *T. subuliformis* as Sato studied the populations which had been collected. In North America and Greenland its range appears to be more restricted than that of *T. subuliformis,* neither ranging as far north nor as far south as the latter. The extent of range of the *Thamnolia* species as bipolar and in such widely disjunct spots as New Guinea, New Zealand, Australia, and the Andes of South America is difficult to account for in the light of the apparent lack of easily dispersed diaspores. The comment of Sato (1965) that it occurs only on nonvolcanic mountains and dead volcanoes, never on resting or active volcanoes, suggests that dispersal is of great antiquity and not currently occurring.

UMBILICARIA Hoffm.

Descr. Adumbr. Pl. Lich. 1(1):8. 1789.

Foliose, umbilicate lichens, mono- or polyphyllus; heteromerous with upper and lower cortices, the upper cortex palisade- or scleroplectenchymatous; the medulla loose, sometimes not clearly differentiated from the lower cortex, the lower cortex usually scleroplectenchymatous, algal layer continuous.

Apothecia adnate to stipitate, lecideine with a proper margin which is continuous, gyrate, the young apothecia with a simple sterile column or of small gyri, the gyri splitting dichotomously and becoming concentric; asci with 8 spores, spores simple and hyaline or in some species becoming brown and muriform. Reagent color reactions in the Umbilicarias are best seen with acetone extracts on slides and over white paper.

Alga: *Trebouxia.*

Type species: *Umbilicaria proboscidea* (L.) Schrad.

REFERENCE
Llano, G. A. 1950. A monograph of the lichen family Umbilicariaceae in the Western Hemisphere. Navexos P-831:1–281.

1. Upper cortex rugose with vermiform or reticulate ridges.
  2. Lower side lacking rhizinae.
    3. Upper side with vermform ridges which do not form a network.
      4. Underside with black area around umbilicus, rest of underside dove-gray . . . . . 3. *U. arctica*
      4. Underside entirely brown to clove-brown or black . . . . 10. *U. hyperborea*
    3. Upper side reticulate ridged.
      5. Underside dark gray, umbo high with prominent ridges, margins entire, apothecia common; medulla C+ red . . . . 14. *U. proboscidea*
      5. Underside dark sooty black, with rare sooty black fibrils; margins perforate and incised; umbo low and eroded; apothecia rare; medulla C− . . . . 8. *U. havaasii*
  2. Lower surface with rhizinae.
      6. Lower surface with sparse, short cylindrical rhizinae; underside entirely pale gray . . . . 10. *U. hyperborea* var. *radicicula*
      6. Lower cortex with cylindrical to flattened, coarse, pale rhizinae in patches or over lower surface; underside sooty black with scattered gray patches . . . . 2. *U. aprina*
1. Upper cortex lacking raised ridges, smooth, chinky, areolate or sometimes folded.
  7. Underside lacking rhizinae, trabeculae, or lamellae.
    8. Upper side with papillate, cylindrical, or flattened isidia . . . . 7. *U. deusta*
    8. Upper side lacking isidia.
      9. Thallus monophyllous; upper side smooth or chinky, light brown; underside dove gray or dark gray, verrucose; apothecia common. . . . . 12. *U. phaea*
      9. Thallus polyphyllous; upper side smooth, blackish brown; underside black to black-brown, smooth; apothecia rare . . . . 13. *U. polyphylla*
  7. Underside with rhizinae, trabeculae, or lamellae.
    10. Underside with rhizinae only, lacking trabeculae or lamellae.
      11. Upper side pitted, gray to light brown, sometimes pruinose; underside light tan or pink, smooth or bullate; apothecia common, stipitate . . . . 6. *U. cylindrica*
      11. Upper side not pitted, smooth, brown or olive-brown; underside black or reddish black, lacunose, verrucose; apothecia rare, sessile to substipitate . . . . 4. *U. caroliniana*
    10. Underside with rhizinae and/or trabeculae or lamellae.
      12. Underside with a tangled mat of black, branched, granulose rhizinae but no ball-tipped short rhizinae, the under surface black verrucose, with lamellae and trabeculae near the umbilicus . . . . 1. *U. angulata*
      12. Underside with shorter, not tangled rhizinae but in some species forming a nap; or else with trabeculae or lamellae.
        13. Underside with trabeculae or lamellae, no rhizinae; thallus thin (250–400 μ), usually perforate . . . . 15. *U. torrefacta*
        13. Underside with rhizinae plus trabeculae or lamellae, thallus thicker (300–600 μ); usually not perforate.
          14. Underside covered with a thick black nap of longer attenuate cylindrical rhizinae interspersed with shorter ball-tipped ones.
            15. Upper side white to violet-brown, sometimes stained reddish, often pruinose; underside with both cylindrical and ball-tipped rhizinae; spores small, $8-14 \times 7-10\ \mu$ . . . . 16. *U. vellea*
            15. Upper side olive-brown to darker brown; underside with only short ball-tipped rhizinae; spores larger, $16-26 \times 8-15\ \mu$ . . . . 11. *U. mammulata*
          14. Underside with pale rhizinae or few short rhizinae.

16. Upper side esorediate toward the margins; underside with short ball-tipped rhizinae; lower cortex black and verrucose; upper side dark gray, reddish or violet . . . . 5. *U. cinereorufescens*
16. Upper side sorediate toward the margins; underside with numerous pale rhizinae; lower cortex brown, areolate; upper side ashy to ashy brown. . . . . 9. *U. hirsuta*

1. Umbilicaria angulata Tuck.

Proc. Amer. Acad. Arts. & Sci. 1:266. 1848. *Umbilicaria semitensis* Tuck., Gen. Lich. 31. 1872. *Umbilicaria angulata* var. *semitensis* (Tuck.) Tuck., Syn. North Amer. Lich. 1:88. 1882. *Gyrophora angulata* (Tuck.) Herre, Contr. U.S. Nat'l. Herb. 13:318. 1911. *Gyrophora semitensis* (Tuck.) Schol., Nyt. Mag. Naturvid. 75:28. 1934.

Thallus to 9 cm, monophyllus, thick, rigid, round to irregular often divided into overlapping lobes, the umbo puckered in between creases and folds; upper side smooth to obscurely wrinkled and cracks with occasional coarse black cracks, dull or subshiny, dark brown or blotched with grayish olive, with thin pruinosity toward the margins; underside black or dark brown, usually covered with a tangled mat of coarse cylindrical or flattened granulose rhizinae, the lower cortex black and coarsely papillate-verrucose.

Thallus 200–425 μ, upper cortex 13–23 μ, paraplectenchymatous; algal layer continuous; medulla not well differentiated from the lower cortex; the lower cortex with black zone about 20 μ.

Apothecia common, adnate or partly immersed more or less angular; disk black with a few gyri, margin thick, the gyri sometimes becoming beaded, the surface dull or shiny; hypothecium brown; hymenium 75 μ; paraphyses unbranched, septate, 2.4 μ; spores 8 in ascus, simple and hyaline but becoming reticulated and then brown and muriform according to Llano (1950); 12–29 × 7–18 μ.

Reactions: K−, C+ red, KC+ red, P−.

Contents: gyrophoric acid ± umbilicaric acid (Krog 1968).

Growing on very exposed, dry rocks, this is a species endemic to western North America from southern California to Alaska, reaching into the Arctic only on Great Sitkin Island in the Aleutian Islands.

2. Umbilicaria aprina Nyl.

Syn. Lich. 2:12. 1863. *Umbilicaria syagroderma* Nyl., Flora 13:418. 1860. *nom. nud. Gyrophora aprina* (Nyl.) Hue, Archiv. Mus. 3:36. 1891.

Thallus to 3 cm broad, monophyllus, rigid, deeply incised; umbo slightly raised, smooth to slightly vermiform or areolate chinky, partly coarsely granulose; upper side dull, powdery, dark gray to pale olive-buff, vermiform ridged to chinky-areolate; underside black, peripherally or spottedly pale gray, with few to many irregularly distributed cylindrical or flattened branched rhizinae.

Thallus 275–325 μ thick; upper cortex 20–30 μ, palisadeplectenchymatous; algal layer continuous; medulla loose; lower cortex 50–66 μ, scleroplectenchymatous.

Apothecia not seen, described by Llano from Frey (1936); to 1.5 mm broad, of both omphalodisc and gyrodisc types, with 1–2 fissures, no central sterile column; spores simple, hyaline, 18–20 × 7–8 μ.

Reactions: K−, C+ red, KC+ red, P−.

Contents: gyrophoric acid and umbilicaric acid (small rectangular crystals in GAQ test) were present in Hale 450 from Baffin Island.

Growing on exposed gneiss rocks, this is a peculiarly disjunct species known from collections in Ethiopia, Scandinavia, and a collection by Mason E. Hale, Jr., at Clyde Inlet, Baffin Island, N.W.T. (Hale 1954).

3. Umbilicaria arctica (Ach.) Nyl.

Herb. Mus. Fen. 84. 1859. *Gyrophora arctica* Ach., Method. Lich. 106. 1803.

Thallus to 20 cm broad, usually much less, monophyllus, leathery, upper side with vermiform ridges which do not form a net, more

Umbilicaria angulata Tuck.

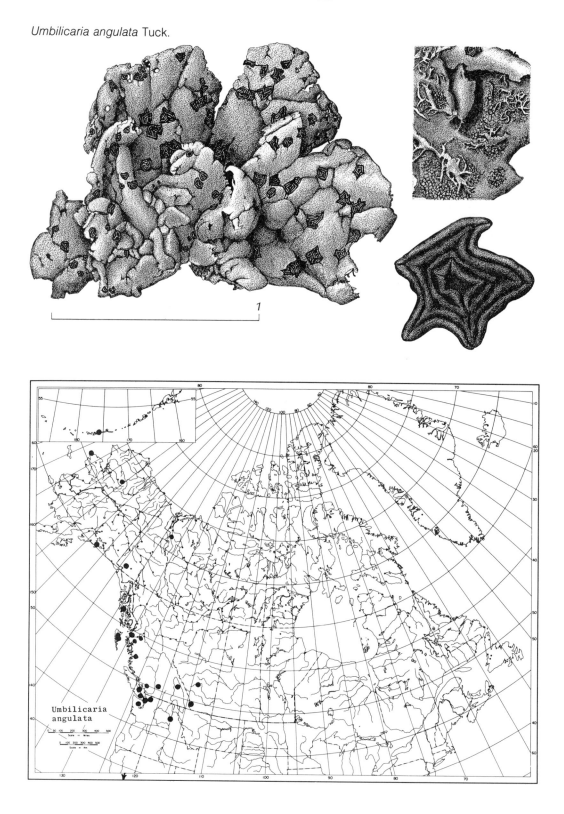

1

Umbilicaria
angulata

or less brainlike convoluted, the umbo moderately raised, occasionally with marginal perforations dull, brown, brown-black or tawny; underside bare, with few soft folds, black around the umbilicus, becoming pinkish brown marginally, slightly papillate-areolate.

Thallus 1100 μ thick, upper cortex 115–130 μ; scleroplectenchymatous; algal layer continuous; medulla very loose; lower cortex 210 μ, becoming scleroplectenchymatous.

Apothecia to 2.5 mm broad, adnate, gyrose, strongly convex; hypothecium black, irregular; hymenium 66 μ; paraphyses simple or branched, septate, 1.7 μ; spores 8, hyaline, simple, ovoid, 8.5–15.3 × 4.2–8.5 μ.

Reactions: K−, C+ red, KC+ red, P−.
Contents: gyrophoric acid.

Growing on acid rocks, this is a circumpolar high arctic-alpine species ranging south to California in the west and Minnesota in the east.

Var. *diomedensis* Merrill ex Llano was described by Llano from material collected by L. J.

Palmer on Diomede Island and differs in that the underside is mostly dark and the apothecia more irregular in shape.

4. Umbilicaria caroliniana Tuck.

Proc. Amer. Acad. Arts. & Sci. 12:167. 1877. *Umbilicaria mammulata* Tuck., Proc. Amer. Acad. Arts. & Sci. 1:261. 1848 *non* Ach. *Gyrophoropsis caroliniana* (Tuck.) Elenk. & Sav., Trav. Mus. Bot. Acad. Petersbourg 8:34. 1910. *Gyrophora caroliniana* (Tuck.) Schol., Nyt. Mag. Naturvid 75: 28. 1936.

Thallus to 6 cm broad, mono- or polyphyllus, thin and fragile, irregular, the surface strongly undulating, margins curled; upper side dull or shining, smooth, brown to olive-brown; underside black, verrucose or granular, with scattered black, simple to cribrose or split rhizinae.

Thallus 250–350 μ; upper cortex 17–33 μ, scleroplectenchymatous; algal layer continuous, thin; medulla loose; lower cortex ca. 33 μ, paraplectenchymatous with hyphae becoming parallel, lower portion black.

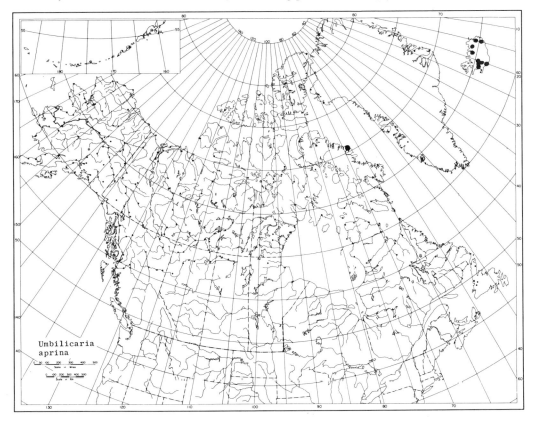

Umbilicaria
aprina

Umbilicaria arctica (Ach.) Nyl.

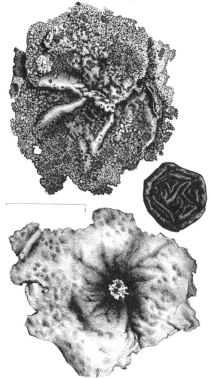

Apothecia rare, sessile to subpedicellate, with few gyri and thick margin; disk black, convex; hypothecium brown; hymenium 85 μ; paraphyses branched, septate, capitate to 4.8 μ; spores 6–8 per ascus, brown, muriform with many cells, 30–50 × 20–27 μ.

Reactions: K−, C+ red, KC+ red, P−.
Contents: gyrophoric acid.

Growing on exposed, vertical, acidic rocks. this species has a very disjunct range, occurring in North Carolina, Tennessee, Alaska, and in Asia in Japan and eastern Siberia. Llano discussed its distribution in terms of unglaciated areas but Hildur Krog (1968) has since reported it from glaciated terrain in interior Alaska.

5. **Umbilicaria cinereorufescens** (Schaer.) Frey

Hedwigia 71:109. 1931. *Umbilicaria vellea* γ *spadochroa* ε *cinereorufescens* Schaer., Enum. Crit. Lich. Eur. 25. 1850. *Gyrophora cinereorufescens* (Schaer.) Schol., Nyt. Mag. Naturvid. 75:28. 1936.

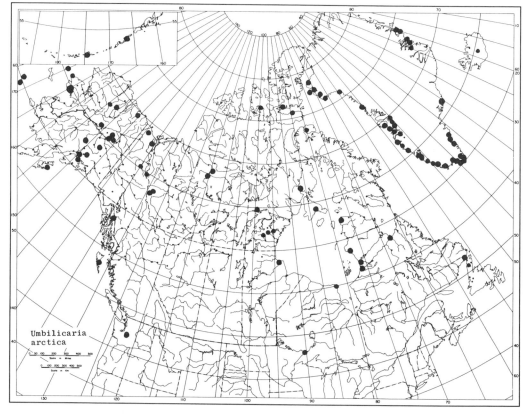

Umbilicaria arctica

Umbilicaria caroliniana Tuck.

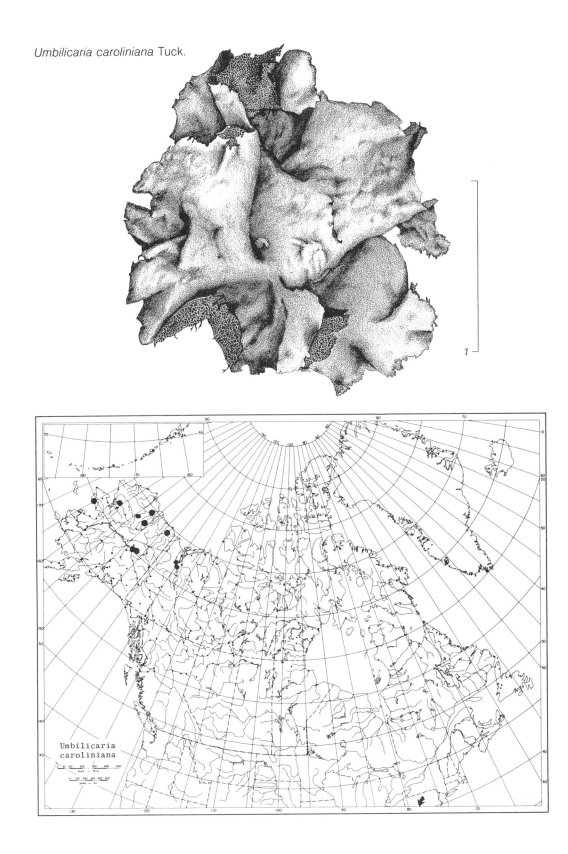

1

Umbilicaria
caroliniana

Umbilicaria cinereorufescens (Schaer.) Frey

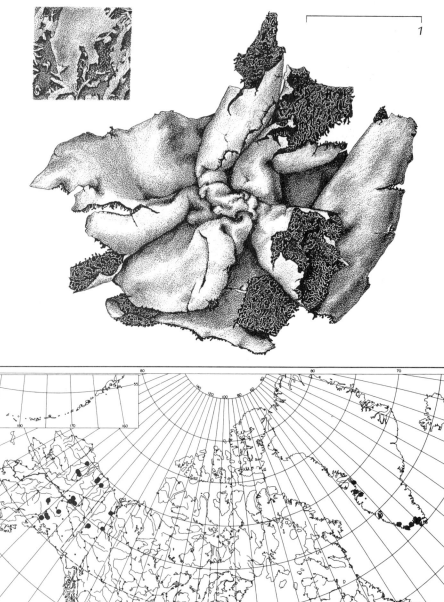

1

Umbilicaria
cinereorufescens

Thallus to 12 cm, usually less, mono-phyllus or polyphyllus, rigid, irregular in shape, umbo with folds or fading rugi; upper surface strongly undulating, rarely smooth, margins wavy, often reflexed; upper side dull, mouse-gray to blotched with cinnamon brown; under-side black, strongly verrucose, the umbilicus with ray-like coarse lamellae becoming fibrillose; rhizinae absent or very short, stubby, ball-tipped, marginally the underside slightly lighter and more verrucose.

Thallus 200–300 μ, upper cortex para-plectenchymatous, hyaline with light brown up-per zone, 13–20 μ; algal layer continuous; medulla and lower cortex not clearly differen-tiated; lower cortex scleroplectenchymatous, about 16 μ.

Apothecia rare, adnate to partly im-mersed; irregular, convex, gyrose, the margin becoming cracked; hypothecium black; hymen-ium 80–90 μ; paraphyses simple to branched, septate, slender, to 2 μ; spores brown, muri-form, 12–20 × 7–14 μ.

Reactions: K−, C+ red, KC+ red, P−.
Contents: gyrophoric acid.

This species grows on acid rocks, especially sandstones. It is known from Europe (Scandinavia and the Alps), Greenland, and North America where it is known from Alaska, the Yukon, and Arizona.

6. Umbilicaria cylindrica (L.) Del. in Duby
Bot. Gall. 2:595. 1830. *Lichen cylindricus* L., Sp. Pl. 1144. 1753. *Gyrophora cylindrica* Ach., Meth. Lich. 107. 1803.

Thallus to 26 cm broad but usually in range of 2–10 cm., polyphyllus, raised and curled, rigid, more or less perforated, with oc-casional rhizinoid growths; upper side smooth or obscurely vermiform or areolate-pruinose, light brown to dark brown, dull or shiny; under-side with well-developed long cylindrical or flat branched rhizinae, the lower cortex smooth or finely granular, light brown or tan to pink, sometimes darker toward the umbilicus, some-times with a gray pruinose marginal zone.

Thallus 450 μ thick; upper cortex ca. 80 μ; algal layer more or less continuous; medulla loose; lower cortex 100–115 μ, scleroplecten-chymatous.

Apothecia to 4 mm broad, markedly stipitate, the stipe and underside of apothecia of same color as the thallus; disk black, convex with few gyri, the gyri deep; hypothecium brown; hymenium 66 μ; paraphyes unbranched or slightly so, septate, 1.7 μ; spores simple, hya-line, ellipsoid, 8–15 × 3–9 μ.

Reactions: K−, C−, KC−, P−.
Contents: no lichen substances demon-strated.

Growing on acid rocks, this is a circumpolar arctic-alpine species which ranges south in the moun-tains to the White Mountains in New England to Or-egon and Colorado in the west.

7. Umbilicaria deusta (L.) Baumg.
Flora Lips. 571. 1790. *Lichen deustus* L., Sp. Pl. 1150. 1753.

Thallus to 9 cm broad, mono- or poly-phyllus, paper thin and fragile when dry, some-what gelatinous when moist, undulating, the margins lobed and crenate, the edges often curled under, umbo weakly formed, edges often per-forate; upper side dull, brownish black or dark brown, more brown when moist, usually with abundant isidia which vary from fine to coarse and lobed or tiny squamulose; underside smooth to areolate-papillate, weakly or strongly veined with papules, lacking rhizinae, black.

Thallus 200–300 μ, upper cortex 16 μ, algal layer continuous, medulla loose, lower cortex 50 μ, scleroplectenchymatous.

Apothecia rare, to 1.5 mm broad, ad-nate to partly depressed; disk flat to convex, gy-rose; hypothecium light brown; hymenium 85 μ; paraphyses unbranched, septate, 2.5 μ; spores simple, hyaline, 18.27 × 7–12 μ (material in North America has not been found with spores, data is after Frey 1933).

Reactions: K−, C+, KC+, P−.
Contents: gyrophoric acid and umbili-caric acid.

This species grows on acid rocks particularly in rills or water channels on the rock face. It may grow in the open or in partial shade. It is a circum-

Umbilicaria cylindrica (L.) Del. in Duby

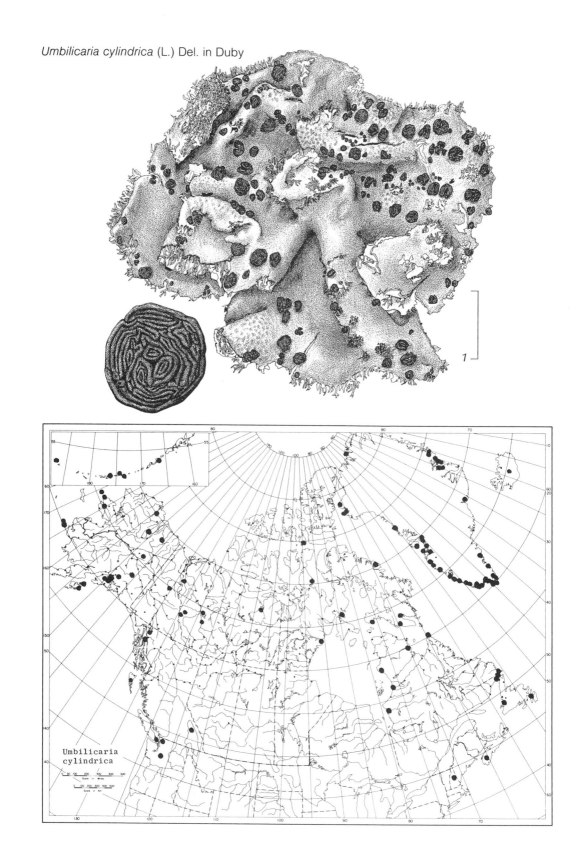

Umbilicaria deusta (L.) Baumg.

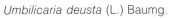

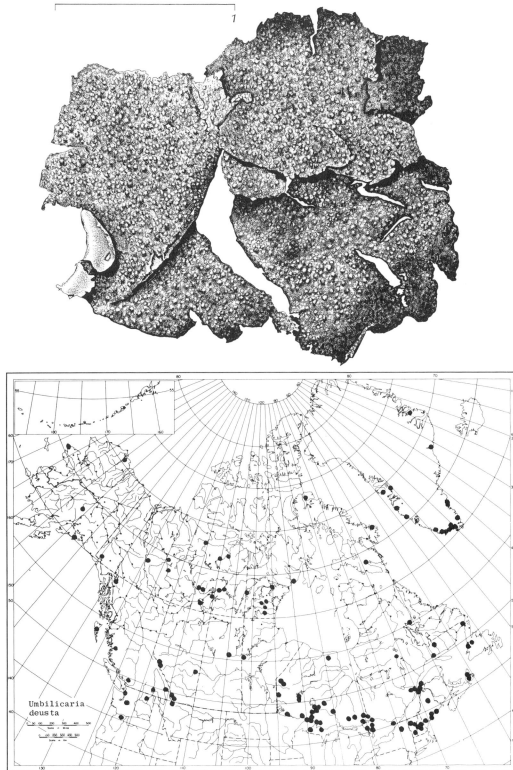

Umbilicaria
deusta

polar boreal and arctic species which in North America ranges south in the eastern states to the Adirondack and White Mountains, in the west to Arizona and is also in the Great Lakes region.

8. Umbilicaria havaasii Llano

Monog. of the Umbilicariaceae 136. 1950. *Gyrophora wenckii* Müll. Arg., Flora 50:533. 1867 pr. p. *Gyrophora fuliginosa* Havaas in Lynge, Videnskap. Skrift. Mat. Naturvid. Klasse 7:96. 1921. *Umbilicaria fuliginosa* (Havaas) Zahlbr., Cat. Lich. Univ. 8:491. 1932 (non Pers., Ann. Wetter. Ges. 2:19. 1810).

Thallus to 13 cm broad, monophyllus, thin to fragile; the umbo only slightly elevated, chinky, obscurely wrinkled, granulose; the margins weakly vermiform, wrinkled, and irregular, becoming perforate; upper surface dull, gray, dark mouse-gray, strongly pruinose-white over the umbo; underside unevenly folded, verrucose-granulose, centrally dove-gray to black streaked, mainly sooty black, bare or with sooty fibrils toward the margins, sometimes with scattered branching fibrils like rhizines erect from the upper side.

Thallus 120–225 μ thick; upper cortex 13–26 μ, scleroplectenchymatous; algal layer discontinuous; medulla not sharply differentiated; lower cortex 46 μ, paraplectenchymatous to scleroplectenchymatous, dark brown with imbedded black granules.

Apothecia rare, to 2 mm broad; sessile to subpedicellate; disk black, at first flat then becoming strongly gyrose; hypothecium brown; hymenium 60–80 μ; paraphyses simple, septate, hyaline, to 2 μ; spores simple, hyaline, ellipsoid or spherical, $10.2–20.4 \times 6.3–8.5\ \mu$.

Reactions: K–, C–, KC–, P–.

Contents: most specimens showed no lichen substances in GAQ test but Lich. Groenlandici Exs. 4 showed traces of gyrophoric acid.

Growing on acid rocks, this is an amphi-Atlantic species with broad disjunctions, occurring in Norway, Greenland, the eastern American Arctic, and in the west in British Columbia and Washington.

9. Umbilicaria hirsuta (Sw.) Ach.

Svensk. Vetensk.-Akad. Nya Handl. 15:97. 1794. *Lichen hirsutus* Sw. in Westr., Vetensk.-Akad. Nya Handl. 14:47. 1793. *Gyrophora hirsuta* (Sw.) Ach., Method. Lich. 109. 1803.

Thallus to 8 (–14) cm, monophyllus or polyphyllus, thin but rigid, the umbo only slightly elevated, upper cortex becoming powdery sorediate toward the margins; upper surface with fine network of fissures, margins torn-lacerate, often reflexed upwards; surface dull, light to dark brown; underside with lamellae extending toward edges or lamellae absent, the lamellae areolate-papillate, rhizinae poorly developed, cylindrical, short, simple or branched, sparse to fairly dense, of same color as underside or shaded darker, brown to dove-gray or black.

Thallus 140–225 μ, upper cortex 33 μ, paraplectenchymatous; algal layer discontinuous; medulla and lower cortex not clearly differentiated, lower part scleroplectenchymatous.

Apothecia rare, adnate and slightly immersed to sessile, to 2 mm broad; disk black, convex with narrow gyri; hypothecium brown; hymenium 60–70 μ; paraphyses unbranched, septate, the tips thickened to 3.4 μ; spores simple, hyaline, ellipsoid, $10–14 \times 5–7\ \mu$.

Reactions: K–, C+ red, KC+ red, P–.

Contents: gyrophoric acid.

This species grows on acid rocks in the open. It is a rare species known from Europe, Greenland, Ross Lake, N.W.T., Canada, and the United States where it occurs in Alaska, California, Montana, Colorado, Vermont, and New Hampshire.

10. Umbilicaria hyperborea (Ach.) Hoffm.

Descr. Adumb. Plant. Lich. 3:9. 1800. *Lichen hyperboreus* Ach., Vetensk.-Akad. Nya Handl. 15:89. 1794. *Gyrophora hyperborea* (Ach.) Ach., Meth. Lich. 104. 1803. *Umbilicaria papillosa* DC. in Lam. & DC., Flore Franc. 3d ed. 3:411. 1805. *Umbilicaria corrugata* Nyl., Lich. Scand. 119. 1861. *Gyrophora nylanderiana* Zahlbr., Cat. Lich. Univ. 4:720. 1927. *Umbilicaria intermedia* Frey in Zahlbr., Lich. Rar. Exsic. 300. 1933.

Thallus to 10 cm broad, monophyllus, rigid or thin and pliable, the umbo raised, sometimes with lobed excrescences over it; upper side with raised irregular ridges, vermiform but not forming a network, the ridging extending to the margins; upper side dull to slightly shiny, olive-brown to dark brown or gray-brown, smooth other than the bullations; lower cortex smooth to granulose, brown to clove-brown or

Umbilicaria havaasii Llano

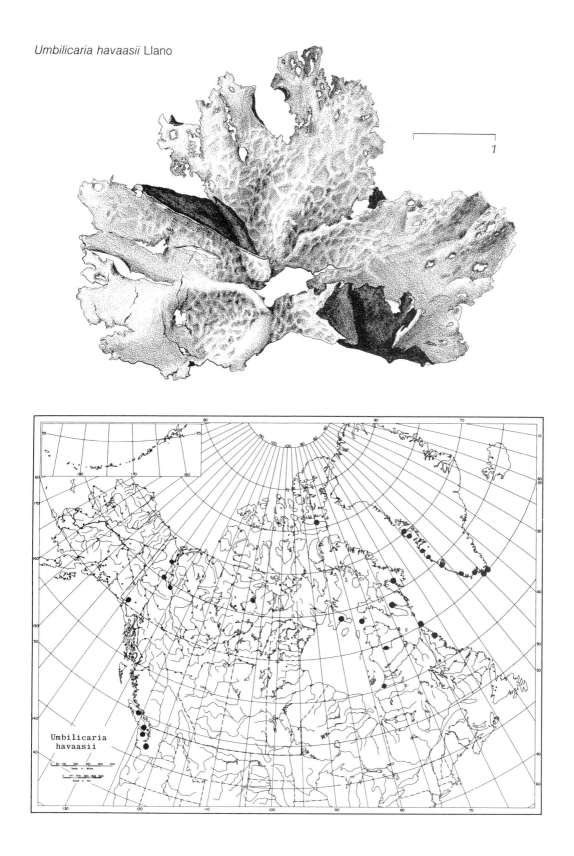

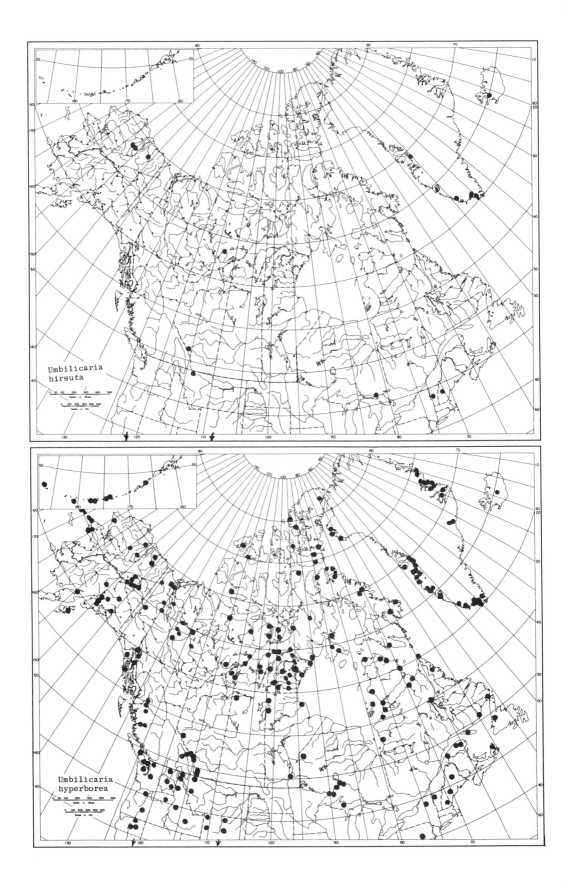

Umbilicaria
hirsuta

Umbilicaria
hyperborea

Umbilicaria hyperborea (Ach.) Hoffm.

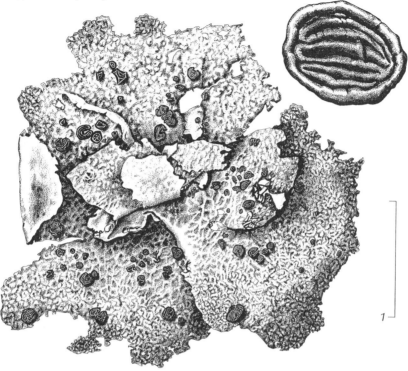

1

black, rarely pruinose marginally, lacking rhizinae.

Thallus 190–230 μ thick, upper cortex 16–20 μ thick, algal layer continuous, medulla very loose; lower cortex 66 μ, scleroplectenchymatous.

Apothecia common, to 2 mm broad, sessile, usually between the ridges, circular or irregular in shape; disk convex, black, gyrose, the gyri regularly dichotomous, hypothecium brown; hymenium 85 μ; paraphyses simple, septate, rarely branched, tips to 3 μ; spores simple, hyaline, ovoid, 6.6–20 × 3.3–8 μ.

Reactions: K−, C+ red, KC+ red, P−.

Contents: gyrophoric acid with accessory umbilicaric acid (Krog 1968).

Growing on acid rocks, this is a circumpolar arctic, boreal, and alpine species which also ranges in the southern hemisphere in South America and New Zealand. A variety with some rhizinae on the underside, var. *radicicula* Zett. is known from the western United States.

11. Umbilicaria mammulata (Ach.) Tuck.

Proc. Amer. Acad. Arts & Sci. 1:261. 1848. *Gyrophora mammulata* Ach., Syn. Lich. 67. 1814. *Umbilicaria dillenii* Tuck., Proc. Amer. Acad. Arts & Sci. 1: 264. 1848. *Gyrophora dillenii* (Tuck.) Müll. Arg., Flora 72:364. 1889. *Umbilicaria vellea* var. *dillenii* (Tuck.) Nyl., Flora 52:389. 1869. *Gyrophora vellea* var. *mammulata* (Ach.) Arn., Verhandl. Zool.-Bot. Ges. Wien 28:266. 1878.

Thallus to 30 cm broad, monophyllus, leathery, the umbo not raised, surface flat to undulating or folded, margins entire, rarely torn or perforate; upper surface dull or shiny, smooth to slightly papillose and pruinose, brown, olive-green, to greenish brown, greener when wet; underside black, smooth to coarsely papillate, covered with a short nap of fine, ball-topped rhizinae.

Thallus 300–600 μ, upper cortex 33 μ, scleroplectenchymatous; algal layer layer thin, continuous; medulla not clearly differentiated from lower cortex; lower cortex about 20 μ, black, irregular.

Apothecia rare, to 4 mm broad, sessile

Umbilicaria mammulata (Ach.) Tuck.

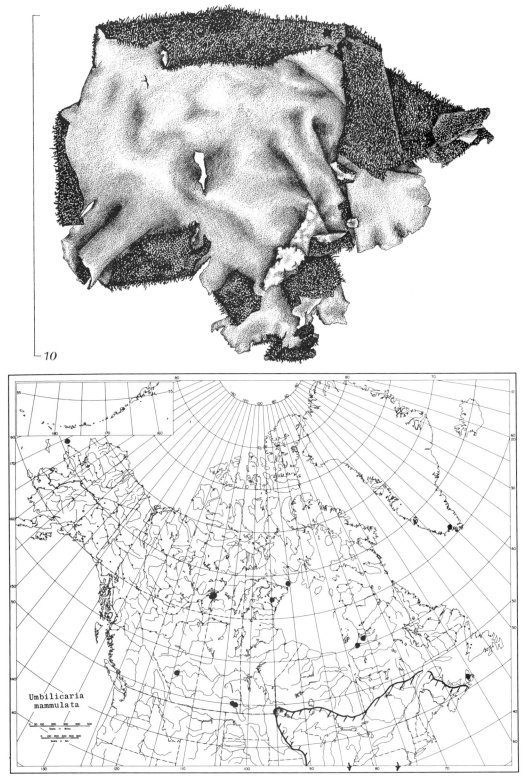

10

to stipitate, becoming convex, irregular, the gyri becoming separate, common margin persistent or broken; hypothecium thick, dark; hymenium 66 μ; paraphyses rarely branched, septate, tips slightly thickened, hyaline to brown, 2.4–3.3 μ; spores simple, hyaline, ellipsoid, 16–25×8–15 μ.

> Reactions: K−, C+ red, KC+ red, P−.
> Contents: gyrophoric acid.

This species grows on shaded rocks, sandstones, granites, slates, etc. It is a North American endemic species, very common in the eastern United States and Canada, rather sparsely distributed in the subarctic but distributed into the Bering Straits region on Diomede Island. The very similar *U. esculenta* is in Japan. This is the largest of the rock tripes and Llano reported a specimen of 63 cm diameter collected by A. J. Sharp in Tennessee.

12. Umbilicaria phaea Tuck.
Lich. Calif. 15. 1866. *Gyrophora phaea* (Tuck.) Nyl. in Hue, Nouv. Arch. Mus. 3:37. 1891.

Thallus to 6 cm broad, monophyllus, sometimes appearing polyphyllus; rigid with thinner margins, umbo raised, sometimes pruinose; upper surface smooth, chinky between the apothecia and marginally, the margins entire, torn, or perforated, dark brown to grayish; underside coarsely papillate to verrucose, dark all over or blotched and brown to dove-gray; sometimes pruinose peripherally, rhizinae rare, short, in patches.

Thallus 150–200 μ thick, upper cortex 16–23 μ, algal layer continuous, medulla not well differentiated from the lower cortex; lower cortex 66–75 μ, scleroplectenchymatous.

Apothecia common, to 2.2. mm broad, immersed to adnate, round to irregular with thin exciple, the gyri regular or irregular; disk black, convex; hypothecium brown; hymenium 80–100 μ; paraphyses septate, branched or unbranched, 1.7 μ with the tips to 5 μ; spores simple, hyaline, ellipsoid, 6.8–17×3.5–8.5 μ.

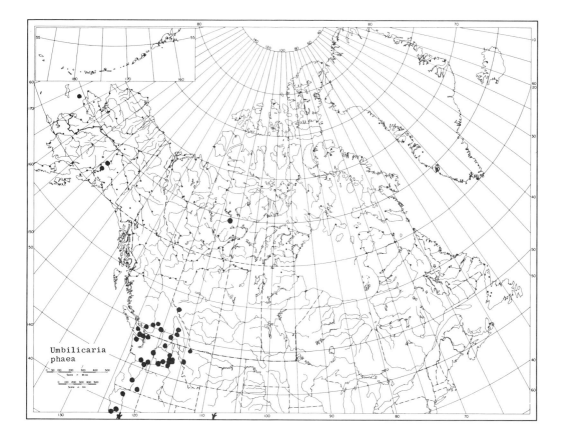

Umbilicaria
phaea

Reactions: K−, C+ red, KC+ red, P−.
Contents: gyrophoric acid.

A xerophytic species which grows on exposed, hot, dry rocks, it is endemic in western North America from Alaska to California and Mexico. A variation is also described by Llano (1950) from Chile. The farthest northeast collection was made by John Richardson on the Franklin Expedition in 1823 at Fort Enterprise between Great Slave Lake and Great Bear Lake (64° 28′ N, 113° 11′ W.).

13. Umbilicaria polyphylla (L.) Baumg.

Flora Lips. 571. 1790. *Lichen polyphyllus* L., Sp. Pl. 1150. 1753. *Gyrophora polyphylla* Funck, Cryptog.-Gewächse 4:4. 1804. *Gyrophora stygia* Hook. f. & Taylor, London Jour. Bot. 3:638. 1844.

Thallus to 6 cm broad, mono- or polyphyllus, thin, crisp, and fragile; very irregular in shape, lobes rounded and overlapping; partly lacerate and perforated, reflexed; upper side smooth to roughened, weakly ridged to squamulose, dull or shiny, shades of dark brown; underside smooth to finely areolate, sooty black or blotched black and dark brown.

Thallus 155–350 μ; upper cortex 10–23 μ, palisadeplectenchymatous; algal layer continuous; medulla loose; lower cortex 60 μ, scleroplectenchymatous.

Apothecia rare, sessile, to 1.5 mm broad; disk convex, with few gyri and thick margin, black; hypothecium brown-black; paraphyses much branched, septate, 2 μ; spores simple, hyaline, ellipsoid, 39–58 × 13–16 μ.

Reactions: K−, C+ red, KC+ red, P−.
Contents: gyrophoric acid, umbilicaric acid.

Growing on acid rocks, preferably along rills such as the habitat of *U. deusta*, this species is circumpolar, low arctic, and alpine, ranging south into the eastern mountains in New England and into California in western North America.

14. Umbilicaria proboscidea (L.) Schrad.

Spicil. Fl. Germ. 1:103. 1794. *Lichen proboscideus* L., Sp. Pl. 1150. 1753. *Gyrophora proboscidea* (L.) Ach., Meth. Lich. 105. 1803.

Thallus to 10 cm broad, usually less, monophyllus, rigid to thin and fragile, the umbo raised; upper side with a reticulating pattern of ridges, or nearer the umbo arranged circularly, white pruinose, fading to vermiform ridges toward the edges; margins perforate or whole or ragged; granulose over the umbo, surface dull, clove-brown to dark brown; underside smooth to areolate, usually without rhizinae; when present the rhizinae are simple or branched, buff or dove-gray to dark brown, sometimes with a pruinose peripheral zone.

Thallus 120–230 μ thick; upper cortex 24–56 μ, algal layer irregular, medulla very loose; lower cortex 60–70 μ, paraplectenchymatous.

Apothecia common, sessile to subpedicellate, circular; disk black, gyrose, the gyri regularly dichotomous; hypothecium 210 μ, brown; hymenium 78 μ; paraphyses simple or branched, septate, to 1.7 μ; spores simple, hyaline, ovoid, 10.17 × 3.5–6.8 μ.

Reactions: K−, C+ red, KC+ red, P− (Krog 1968, reported some Alaskan specimens P+ with norstictic acid).

Contents: gyrophoric acid with accessory norstictic acid.

Growing on acid rocks, this is a circumpolar arctic-alpine species, which in North America ranges south to Washington in the west and the White Mountains in the east.

15. Umbilicaria torrefacta (Lightf.) Schrad.

Spicil. Fl. Germ. 1:104. 1794. *Lichen torrefactus* Lightf., Flora Scotica 2:862. 1777. *Lichen erosus* Web., Spicil. Fl. Götingens 259. 1778. *Umbilicaria erosa* (Web.) Ach., Vetensk.-Akad. Nya Handl. 15:87. 1794. *Gyrophora erosa* var. *torrida* Ach., Meth. Lich. 104. 1803. *Gyrophora torrida* (Ach.) Lam., Encyl. Method. Bot. Suppl. 3:423. 1813. *Gyrophora koldeweyii* Körb., Zweite Deutsch. Nordpolfahrt 2:77. 1874. *Umbilicaria sclerophylla* Nyl., Flora 55:82. 1862. *Gyrophora sclerophylla* (Nyl.) Nyl., in Hue Nouv. Archiv. Mus. 3:37. 1891. *Gyrophora torrefacta* (Lightf.) Cromb., Monogr. British Lich. 1:329. 1894.

Thallus to 6 cm broad, monophyllus, thin or thick, rigid or membranaceous and fragile, umbo not much developed, center folded or puckered, margins always perforate; upper side dull or shining, dark brown to olive-brown; underside smooth or finely granulose-papillose with well-developed trabeculae over the entire underside or becoming more or less fimbriate, the

Umbilicaria polyphylla (L.) Baumg.

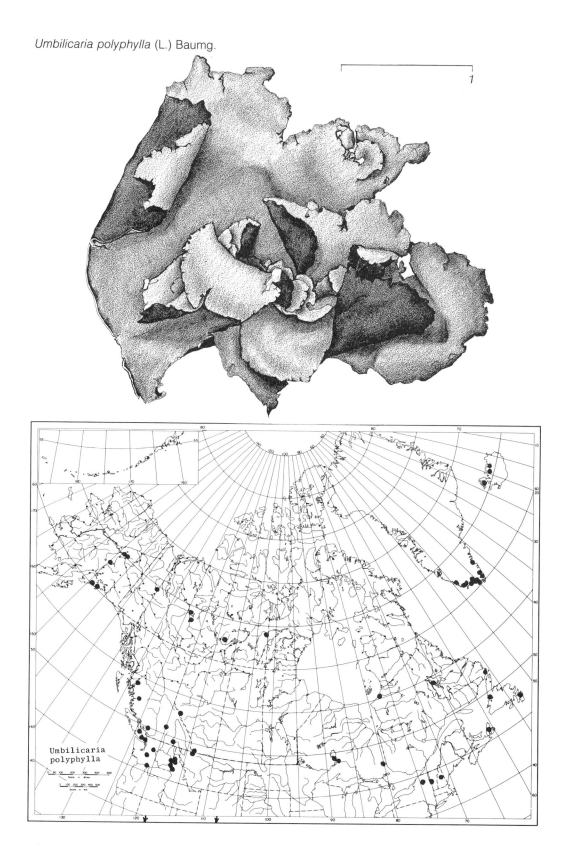

1

Umbilicaria
polyphylla

Umbilicaria proboscidea (L.) Schrad.

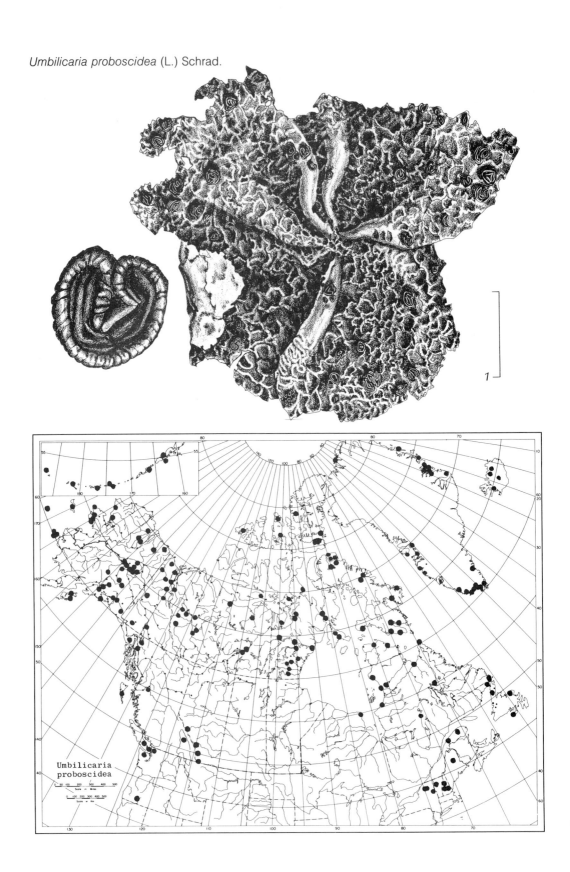

1

Umbilicaria
proboscidea

Umbilicaria torrefacta (Lightf.) Schrad.

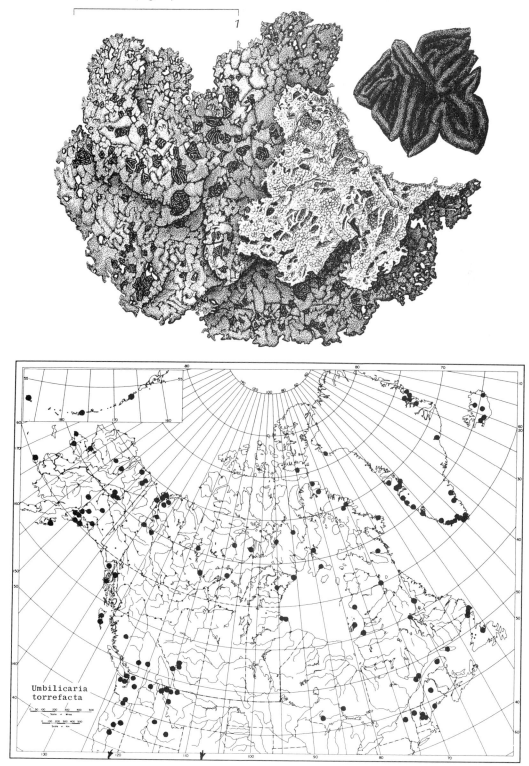

1

Umbilicaria
torrefacta

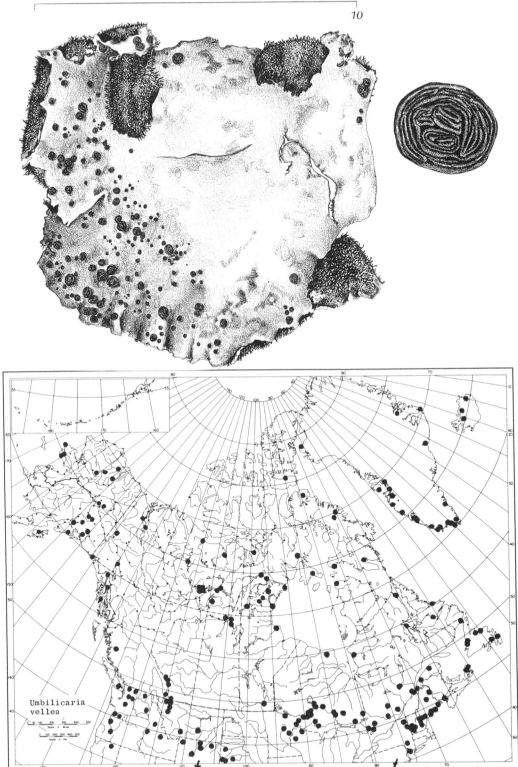

Umbilicaria
vellea

trabeculae occasionally poorly developed, buff to dark brown.

Thallus 250–400 μ, upper cortex 17–25 μ, algal layer continuous; medulla and lower cortex not clearly differentiated; lower cortex scleroplectenchymatous.

Apothecia common, to 2 mm, adnate to subimmersed in depressions, round or irregular; disk black, convex, irregular, margins soon disappearing; hypothecium dark brown; hymenium 70–80 μ; paraphyses unbranched, septate, tips brown to 2 μ; spores 8, simple, hyaline, ellipsoid, 7–12 × 5–7 μ.

Reactions: K–, C+, KC+, P–; or K+ reddish, C–, KC–, P–.

Contents: there appear to be three chemical strains of this species, one with stictic acid, one with gyrophoric acid, and one with gyrophoric and umbilicaric acids (Krog 1968). Those which I tested were mainly with the gyrophoric acid content; the specimens with stictic acid were from western North America. A more thorough investigation seems necessary for this species.

This species grows on acid rocks in exposed places. It is circumpolar arctic-alpine, in North America ranging south into the New England mountains, the Lake states, and in the west into Mexico.

16. Umbilicaria vellea (L.) Ach.

Vetenskap.-Akad. Nya Handl. 15:101. 1794. *Lichen velleus* L., Sp. Pl. 1150. 1753. *Gyrophora vellea* (L.) Ach., Meth. Lich. 109. 1803.

Thallus to 25 cm broad, usually less, monophyllus, rigid, umbo only slightly raised or level, with or without broad folds in the thallus, the margins entire to incised; upper surface dull, smooth to roughened, areolate-cracked, occasionally with rhizinae protruding through cracks or splits; pale gray to violet-gray or slaty with occasional reddish staining, often pruinose; underside black with lower cortex coarsely papillate-verrucose, covered with dense mat of rhizinae which are of longer cylindrical dark brown rhizinae mixed with more abundant shorter ball-tipped black and papillose ones.

Thallus 300–500 μ, upper cortex 8–20 μ, of palisadeplectenchymata; algal layer continuous; medulla dense; lower cortex 83–132 μ, inner part hyaline, outer brown, scleroplectenchymatous.

Apothecia rare, to 3 mm broad, sessile; disk black, convex, with well-developed gyri, hypothecium thick, brown; hymenium 75–80 μ; paraphyses simple or branched, septate, to 2 μ; spores 8, simple, hyaline, ellipsoid, 8–14 × 7–10 μ.

Reactions: K–, C+ red, KC+ red, P–.

Contents: gyrophoric and umbilicaric acids.

This species grows on either open or shaded cliffs, often near shores. It is circumpolar and wide-ranging, from arctic to temperate, in eastern North America south into Georgia and in the west into Mexico.

USNEA Dill. ex Adans.

Familles de Plantes 2:7. 1763. *Usnea* Dill., Hist. Muscor. 56. 1741.

Thallus fruticose, erect or hanging, usually attached to the substratum by a holdfast; branching sometimes dichotomous, more commonly with a central axis from which longer or shorter side branches project; the main axis often dark at the base, smooth, papillate, tuberculate, or with soralia or isidia; the lateral or peripheral branches often differing from the main axis, sometimes the main branches are foveolate; corticate, the medulla with an outer layer which may be more or less solid or loosely arachnoid, and an inner central cartilaginous strand that is very diagnostic.

Apothecia lateral or appearing terminal, discoid, with thalloid margin concolorous with the thallus, usually fibrillate; the disk pale, sometimes pruinose; epithecium pale; hypothecium hyaline; paraphyses coherent, branched, septate; asci clavate; spores 8, hyaline, simple, thin walled, ellipsoid or almost spherical. Pycnidia lateral, immersed or slightly projecting, pale or dark; fulcra exobasidial; conidia spindle-shaped, acicular, or cylindrical, straight.

Algae: *Trebouxia*.

Type species: *Usnea florida* (L.) Wigg.

REFERENCES

Asahina, Y. 1956. Lichens of Japan. Vol. 3 Genus *Usnea*. Res. Inst. Nat. Sci., Tokyo. 1–129.
Asahina, Y. 1954. Lich. Not. 99. A new method in describing the relation between cortex, medulla and axis of Usneae. J. Jap. Bot. 29:11–17.
Carlin, G. & U. Swahn. 1977. De svenska Usnea-arterna (skägglavar). Svensk Bot. Tidskr. 71:89–100.
Keissler, K. 1960. Usneaceae in Rabenh. Kryptogamen-Flora. 1–755.
Klingstedt, F. W. 1965. Über farbenreaktionen von flechten der gattung *Usnea*. Acta Bot. Fenn. 68:3–23.
Motyka, J. 1936–1938. Lichenum generis *Usnea* studium monographicum. Leopoli. 1–651.
Tallis, J. H. 1959. The British species of the genus *Usnea*. Lichenologist 1:49–88.
Watson, W. The species of *Usnea* in Great Britain and Ireland. Trans. Brit. Mycol. Soc. 34:368–375.

Usnea is a difficult genus which is much in need of further monographic work on the species occurring in North America. The treatment given here must be considered tentative in the light of the confusions attending this genus.

1. Thallus elongate, pendent.
  2. Thallus foveolate, epapillose, lacking short fibrils or tubercles or isidia . . . .   1. *U. cavernosa*
  2. Thallus not foveolate but papillose and tuberculate, medulla lax.
    3. The primary branches minutely papillate, the secondary tuberculate and often with isidiose soredia, thallus dark green; K+ red, containing salazinic acid; CMA 60–75:150–180:250
      . . . . 2. *U. dasypoga*
    3. The branches all with a mixture of obvious papillae and tubercles or entirely tuberculate sometimes also sorediate or isidiate, thallus light ashy to straw-colored, K–; CMA 40–50:165–200:200–250 . . . . 5. *U. scabrata* ssp. *nylanderiana*
1. Thallus caespitose, short.
    4. Thallus with all branches constricted and articulated basally, with dense short spinules and soredia, base pale, epapillose; medulla K–; CMA 60:300:250–300 . . . . 4. *U. hirta*
    4. Thallus branches not articulated basally, not densely short fibrillose-spinulose.
      5. Lacking papillae, short, less than 7 cm long, whitish to pale green or milky straw, becoming darker yellow to tan or brown to dark reddish in herbarium; soredia in cuff-like or excavate soralia; medulla lax and wider than the axis, K–; CMA 35:250:200 . . . . 3. *U. glabrata*
      5. Thallus with papillae.
        6. Soralia isidiose; medulla dense to sublax; thallus ashy to yellowish green, rather stiff; soralia initiated as tubercles on the lobes; K+ red or K–; CMA 60–100:120–200:300–380 . . . . 6. *U. subfloridana*
        6. Soralia not isidiate; medulla lax; thallus greenish straw-colored to very yellow; soralia cuff-like at the joints, excavate on the sides of the lobes; K–; CMA 65–80:270–280:270–330 . . . . 7. *U. substerilis*

1. Usnea cavernosa Tuck.

In Agassiz, Lake Superior. 171. 1850.

Thallus elongate, pendent, soft, with few branchlets, esorediate, straw yellow to pale green, darkening in the herbarium, irregularly dichotomously branching, the branches to 1 mm diameter, the thicker branches distinctively foveolate pitted, sometimes bluntly angled, more or less articulate fractured. CMA 30–50:180–200:240–275; medulla lax.

Apothecia rare, small, to 5 mm broad; exciple thin, marginal fibrils few, short or long;

disk flat to convex, bare or slightly pruinose.

Reactions: K−, P−, or P+ orange, K+ red.

Contents: usnic acid with or without salazinic acid.

This species grows pendent, sometimes in large masses, on mainly conifers but also on deciduous trees. It is expecially well developed under conditions with high atmospheric humidity.

It is circumpolar, boreal to temperate, ranging south to Mexico and Cuba in the New World. The specimens which contain salazinic acid turn red with KOH and have been designated ssp. *sibirica* (Räs.) Mot. The specimen from Alaska collected by Hildur Krog represents this latter taxon.

2. Usnea dasypoga (Ach.) Röhl.

Deutschl. Flora 2:144. 1813. *Usnea plicata* var. *dasypoga* Ach., Meth. Lich. 312. 1803.

Thallus elongate, hanging, usually 200–300 mm long but reaching 750 mm on the Pacific Coast, more or less compressed basally with few fibrils; ashy to dirty green, in the herbarium becoming yellow-brown to brown, the base often blackened; primary branches to 1 mm thick, minutely papillate, the secondary branches very long and hanging alongside the main branches, limp, more or less tuberculate; the tips mainly sorediate, the branches more or less distinctly articulate; isidia usually present among the soredia on the upper part of the branches.

Apothecia rare, not seen in American material.

CMA: 60–75:150–180:250; medulla lax to very lax, K+ yellow turning red, containing salazinic acid.

This species grows on conifers and on deciduous trees. It is very common in the western coastal states, rarer eastward. It is a circumpolar boreal to northern temperate species.

Variation in the color of the thallus from very yellow to the more characteristic dark greens, and the

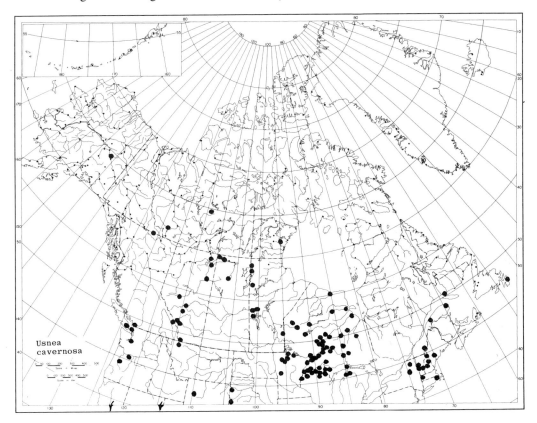

Usnea
cavernosa

Usnea cavernosa Tuck.

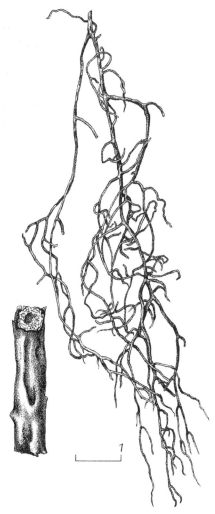

amount of black at the base, have been used as the basis for subspecific units but these are too intergrading and are probably under environmental control rather than genetic. In the British checklist this species is listed as *U. filipendula* Stirt.

3. Usnea glabrata (Ach.) Vain.
Ann. Acad. Sci. Fenn. Ser. A, 6:7. 1915. *Usnea plicata* var. *glabrata* Ach. Lich. Univ. 624. 1810.

Thallus erect to partially hanging, small, 30–70 mm long and wide but often very tiny; fruticose to caespitose; pale green in the field, in the herbarium turning yellow to reddish brown

or dark brown-red; the main base and the branches constricted and the thallus with marked inflated sections jointed and with the white medulla showing between the segments, to 1.5 mm thick, with no fibrils and with infrequent branching so rather coarse appearing; the surface smooth and polished or with very low bumps, not quite forming papillae; the tips attenuate, sorediate with farinose soredia in open soralia forming cuff-like edges of the smaller segments or more excavate soralia similar to those of *U. substerilis*.

Apothecia reported only once from Finland.

CMA: 35–75:210–250:190–375: medulla very lax, K−, P−.

This species grows on conifers, broad-leaved trees, and on old wood and fenceposts. It rarely reaches the tundra at the northern edge of the boreal forest. It is reported by Herre (in ms.) to occur from eastern United States to Alaska south to Mexico but some of his reports may be misdeterminations for *U. substerilis*. Although its range is circumpolar, boreal to temperate, its North American range is not clear.

4. Usnea hirta (L.) Wigg.
Prim. Flora. Holsat. 91: 1780. *Lichen hirtus* L., Sp. Pl. 1155. 1753.

Thallus varying from tiny to 80 mm long, rarely more, pendent, caespitose, usually with many fibrils and dense short spinules; the branching more or less dichotomous, the larger branches sometimes foveolate; the bases of the branches narrowed and jointed, the finer branches with an abundance of soredia which mainly become short isidia, the whole giving a dense, rough, dropping characteristic appearance; pale to darkening olive green, the base usually not blackened, epapillate; the soralia only on the outer ends of the plants.

Apothecia not seen, described by Motyka as terminal, to 7 mm broad; the exciple flat, smooth with short marginal fibrils; disk flat to slightly wrinkled, dark flesh-colored.

CMA 60:250–300; medulla lax to very lax, K− or K+ yellow to red.

Usnea dasypoga (Ach.) Röhl.

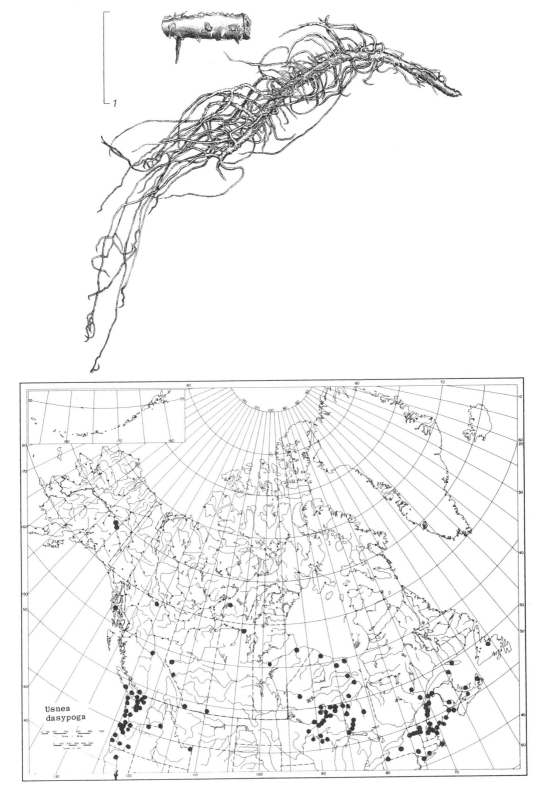

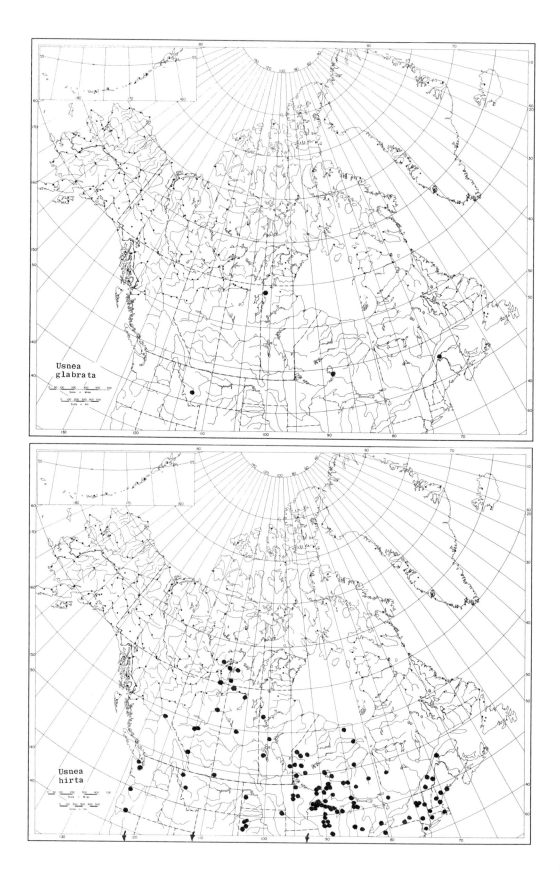

Usnea
glabrata

Usnea
hirta

Usnea glabrata (Ach.) Vain.

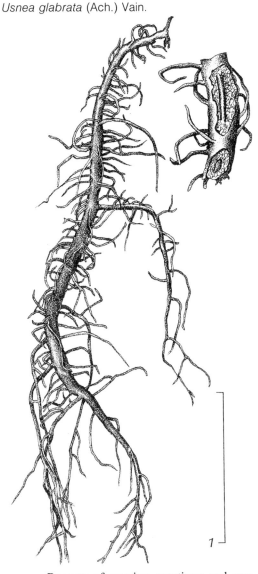

Usnea hirta (L.) Wigg.

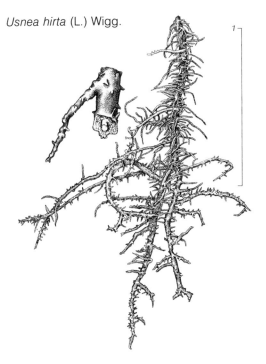

boreal to temperate, in North America ranging south to New England and in the west to Mexico.

Reports of varying reactions and contents for this species including not only the ubiquitous usnic acid but a number of others suggest the need for investigation of strains and their distribution. Klingstedt (1965) examining 2600 specimens from Finland found 80 contained salazinic acid and the papers listed by C. Culberson (1969) in the Chemical Guide include several other lichen substances.

U. hirta occurs on a wide variety of substrates, deciduous trees, conifers, old wood, boards, wooden fences, and rarely rocks. It is circumpolar,

5. Usnea scabrata Nyl.
Flora 58:103. 1873

ssp. nylanderiana Mot.
Usnea Stud. Monogr. 1:148. 1936.

Thallus pendulous, to 300 mm or more, pale green, becoming pale dusky straw in the herbarium; base short, distinct, to 2 mm thick; primary branches freely subdichotomously to subsympodially divided; the branches straight or slightly flexuous, continuous or slightly fractured, attenuate, the tips thread-like; primary branches with abundant thin pointed or cylindrical papillae and sometimes short branching spinules, the smaller branches abundantly wrinkled tuberculate; the tubercles more or less elongate, the wrinkles sometimes confluent, rarely intermingled with cylindrical papillae; lateral fibrils rare; soredia with isidia also sometimes present on the tubercles.

Apothecia rare, small, to 5 mm broad;

margin smooth, fibrils few, limp, thin; disk flat, pale.

CMA 40–50:165–200:200–250; very lax, K− in this ssp., K+ red in ssp. *scabrata*.

Contents: only usnic acid found in this subspecies.

This species commonly grows on *Picea* and *Pinus* but also on deciduous trees as *Betula*. This range of this species includes Europe, western Siberia, and western North America, barely reaching the margin of the tundras.

6. Usnea subfloridana Stirt

Scott. Nat. 6:294. 1882. *Usnea comosa* (Ach.) Vain. non Pers.

Thallus usually 50–70 mm but may reach 100 mm or more, fruticose, erect, fairly stiff, pale ashy green or ashy straw, base ca 1 mm, dusky to black, not constricted; primary branches 1–1.3 mm, soon branched, with many short cylindrical papillae with globose tips; lateral branches rather numerous, elongate attenuate, continuous or rarely with fractures; the upper elongate branchlets with many small soralia starting as rounded small tubercles which burst with farinose soredia soon intermixed with short isidia which are easily dislodged and drop off in the herbarium, leaving the farinose soredia; pale flesh-colored "pseudocephalodia" caused by *Abrothallus parmeliarum* Sommerf. may often be present according to Keissler (1960).

Apothecia described as rare.

CMA: 60–100:120–200:300–380: cortex hard and stiff; medulla more or less dense reacting K− in ssp. *comosa* but K+ red in ssp. *similis* Mot. (of. *U. comosa*).

The chemistry and taxonomy of this species is in considerable need of intensive study. There appear to be four chemical strains, plus considerable morphological variability. The four strains have been separated by Asahina in his treatment of *U. comosa* as: 1) with squamatic acid as *U. comosa* ssp. *eucomosa* Mot.; 2) with thamnolic acid as ssp. *colorans*

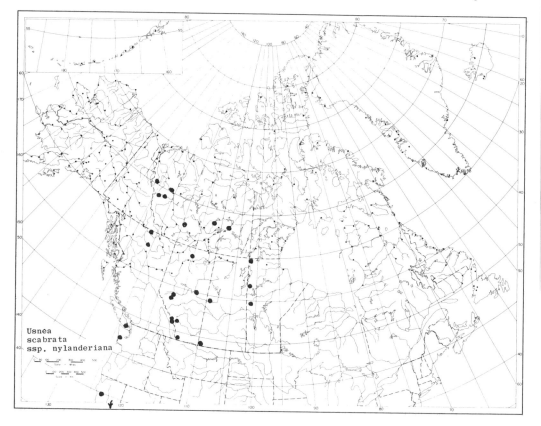

Usnea scabrata ssp. nylanderiana

Usnea subfloridana Stirt.

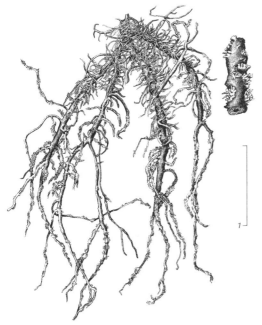

Asah. As the type specimen of *U. subfloridana* contains thamnolic acid according to C. Culberson, this species name should be tied with this content (Hawksworth, Field Stud. 3:535–578. 1972); 3) with salazinic acid as *C. comosa* ssp. *melanopoda* Asah. = *C. subfloridana* var. *melanopoda* (Asah.) Hawksw.; and 4) with norstictic acid as *U. comosa* ssp. *praetervisa* Asah. The chemistry of the three K+ red ssp. of *U. comosa*, ssp. *similis* Mot., ssp. *glaucina* Mot., and ssp. *gorganensis* Mot. as well as the K+ red vars. *cumulata* Mot. and *scabriuscula* Mot. should be investigated as to the type specimens to know how they fit into the Asahina scheme. It will be noted that the major part of the North American material fits into the squamatic acid strain which does not agree with the type of *U. subfloridana* and actually there may be no epithet for this strain at the species level; that is if one separates the four strains into species. *U. comosa* (Ach.) Vain. is apparently preempted by *U. comosa* Pers., *U. hirsutula* Stirt. is embarrassed by a lack of a type specimen, and *U. subcomosa* (Vain.) Vain. refers to material with a dense medulla and a K+ red reaction provenant in the West Indies and Brazil.

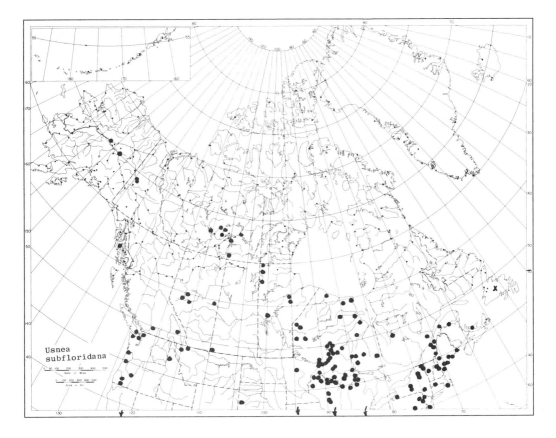

Usnea subfloridana

Usnea substerilis Mot.

1

7. **Usnea substerilis** Mot.
Wydawn. Mus. Slaskiego Katowicach 3:24. 1930.

Thallus usually very small, less than 65 mm, usually 25 mm or less, the width about the same as the length, in erect tufts on the substratum; greenish straw-colored, sometimes quite yellowish; the base with a narrow dark zone, sympodially branched, the branches terete, the base 1.5 mm or usually less, thickly and distinctly papillate, the papillae more or less cylindrical; lateral branches epapillate, nearly at right angles to the main stems, smooth or slightly tuberculate sorediate with the soredia at the joints or in enlarged soralia which expand the branchlets, the soredia farinose, whitish, soon lost, and the soralia often appearing excavate.

Apothecia lacking.

CMA: 65–80:270–280:270–330; medulla lax, K−.

This species is abundant toward the timberline on spruce and on birch and other trees and shrubs.

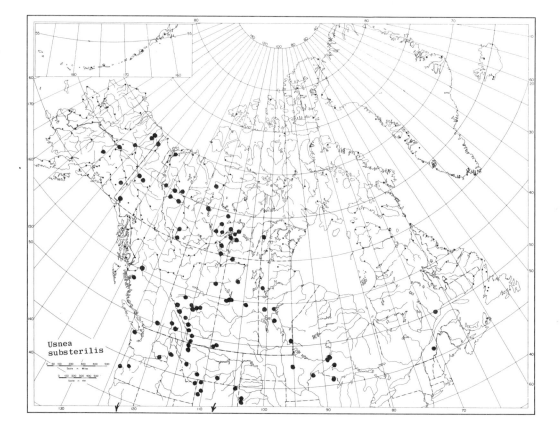

Usnea
substerilis

It is very variable, paler at the higher latitudes and in stronger insolation, darker green in greater shade, the branch length short in the more xeric conditions, the tips more elongate in greater moisture. These variations give rise to some of the confusion of names in this complex. See the following discussion.

This species is circumpolar boreal, ranging south in the mountains to Arizona and Mexico.

It has been very difficult to sort out the entities involved in the *U. comosa* group of species and determine what is represented in the northern material. There is a group of species with small thalli, erect habit, papillate main branches and especially characteristic soralia which burst from the interior at joints and particularly as well defined roundish soralia which bear farinose, not isidiate, soredia. The specimens may vary somewhat in color from very pale straw-colored to dark green, probably this determined by the density of the algae in the thallus and by the amount of insolation in the habitat. They may also vary in the length of the tips of the branches and this has been used in separating groups of species but seems a difficult character as it appears as if it is an environmentally induced character.

Just as in *U. subfloridana* there is a representation of material which reacts K+ red or red-orange but which really represents specimens which may contain salazinic acid, norstictic acid or stictic acid. This may include a considerable number of species recognized by Motyka but the chemistry of the type specimens needs to be determined and the recognition of the species adjusted accordingly. Those which are thus K+ include *U. perplexans* Stirt. (1881) (*U. perplectans* Stirt. according to Motyka), *U. glabrescens* (Nyl. ex Vain.) Vain. (1924), *U. extensa* Vain.

(1928), *U. wasmuthii* Räs. (1931), *U. fulvoreagens* (Räs.) Räs. (1935), and *U. betulina* Mot. (1936). *U. perplexans* has been reported to contain salazinic acid in Asiatic material (Asahina and Tukomoto 1933) but the type has not been checked, *U. glabrescens* contains norstictic acid according to several reports but not including the type in the reports. *U. wasmuthii* contains salazinic acid (Duvigneaud). Hale (1969) reported three strains in *U. fulvoreagens:* salazinic acid, norstictic acid, and stictic acid strains.

Of those which are reported to be K− there are contradictory reports, Klingstedt summarized his observations of the color reactions of these. The earliest name in the purportedly K− series is *U. lapponica* Vain. (1924). Then follow *U. laricina* Vain. (1931), *U. substerilis* Mot. (1936), but not *U. sorediifera* (Arn.) Lynge (1921), which belongs to the K+ red series of epithets. The type specimen of *U. lapponica* at Turku was not examined by Klingstedt, but isotypes at H gave K+ red reactions, hence that material may also belong in the K+ red series. *U. arnoldii* material determined by Motyka and supposedly K− gave Klingstedt K+, not K−, reactions and so also belongs to the K+ series. *U. substerilis* Mot. (1930) appears to be the earliest almost unequivocal name for the K− material although Keissler threw some doubt into this in saying K+ or −. *U. compacta* Mot. (1936) and *U. glabrella* (Mot.) Räs. (1940) are names which also might apply to this type of material as they are K−. Of all of these chemical strains or species only K−, P− material reaches the northern edge of the forest at its contact with the tundra and therefore is the only one represented here. In this tentative treatment I have adopted the name *Usnea substerilis* for the reasons mentioned above.

VESTERGRENOPSIS Gyeln.

in Rabenh., Krypt.-Fl. Deutschl. u. d. Schweiz 9(2):265–266. 1940.

Thallus of radiating small lobes closely attached to each other and to the substratum, isidiose or not, esorediate; medullary hyphae mainly longitudinally extended.

Apothecia hemiangiocarpic with thalloid margin; hymenium brown above, I+ blue; paraphyses septate, simple or sparingly branched, tips thickened and moniliform; asci cylindrical or clavate with apically thickened wall; spores 12–16, hyaline, simple, ellipsoid.

Algae: *Scytonema*.

Type species: *Vestergrenopsis elaeina* (Wahlenb.) Gyeln.

REFERENCES
Gyelnik, V. 1940. in Rabenh., Krypt. Fl. Deutsch l.u.d. Schweitz 9(2):265–266, 1940.
Henssen, Aino. 1963. A study of the genus *Vestergrenopsis*. Can. J. Bot. 41:1359–1366.

Vestergrenopsis elaeina (Wahlenb.) Gyeln.

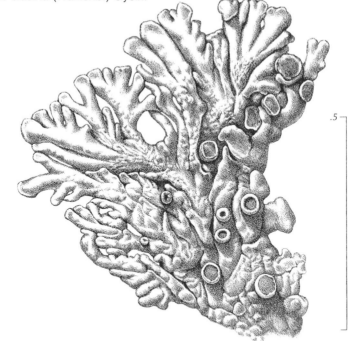

.5

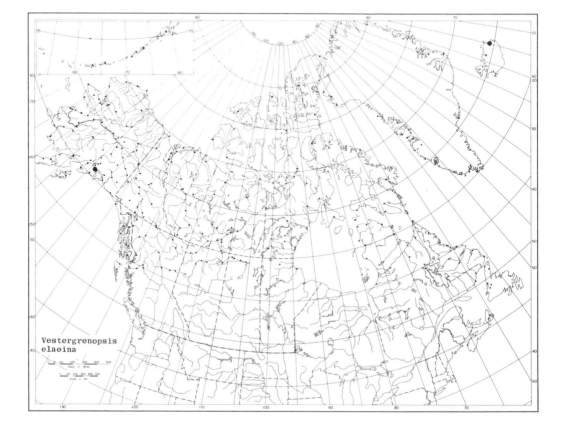

Vestergrenopsis
elaeina

Vestergrenopsis isidiata (Degel.) Dahl

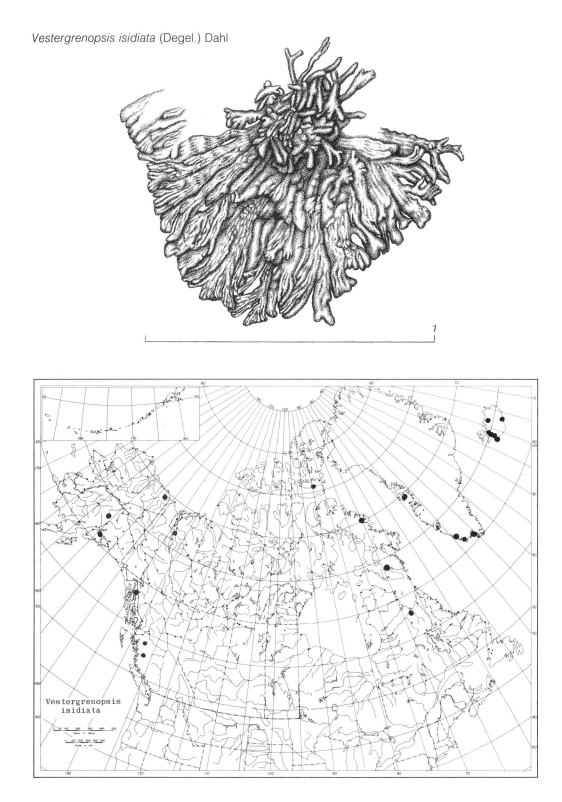

1

Vestergrenopsis isidiata

1. Vestergrenopsis elaeina (Wahlenb.) Gyeln.

in Rabenh., Kryptog.-Fl. Deutsch. u.d. Schweiz 9(2):266. 1940. *Parmelia elaeina* Wahlenb. in Ach., Meth. Lich. Suppl. 45. 1803.

Thallus of tiny radiating lobes, flat, longitudinally striate, closely attached to the substratum, the lobes 0.2–0.5 mm broad, the tips fan-shaped and flattened to 1–2 mm, the interior of the thallus more areolate; lacking isidia or soredia; upper side olive brown to dark brown; underside pale, rhizinae short, pale; medullary hyphae oriented longitudinally.

Apothecia to 1 mm broad; margin thalloid; disk brown to brownish black; exciple hyphae radially oriented; hypothecium dense, 40–70 μ; epithecium brown; hymenium 85–125 μ; paraphyses 1.5 μ, the tips thickened to 8.5 μ and moniliform; asci clavate, 37–50 × 7.5–10 μ; spores 12–16, hyaline, 1-celled but with cytoplasmic bridges making them appear 2-celled, 8–10 × 4–6 μ.

This species grows on siliceous rocks which are frequently wet by seepage. It is a rare species known from alpine Scandinavia, the Faeroe Islands, and Iceland. A recent collection established this species as new to North America. Alaska: on rocks on moraine at Portage Glacier campgrounds, 1967, J. W. Thomson (17797) and T. Ahti.

2. Vestergrenopsis isidiata (Degel.) Dahl

Medd. om Grønland 150:55. 1950. *Pannaria isidiata* Degel., Bot. Not. 90–91. 1943.

Thallus of tiny radiating lobes, flat, longitudinally grooved and striated, rarely smooth; 2–6 mm long, 0.2–0.4 mm broad, the tips broadened to 0.8 mm; isidia on the upper surface globose or cylindrical, becoming flattened; underside pale; upper side dark olive brown, rhizinae pale; interior hyphae longitudinally extended, those in the center cylindrical, those toward the margins ellipsoid, loosely arranged.

Apothecia to 1 mm diameter with thalloid margin; disk brown; hymenium 80–125 μ thick, hyaline, the upper part brown; subhymenium hyaline, 40–70 μ, of ellipsoid cells in radiating rows; pseudoexcipulum with large, broadly ellipsoid cells; paraphyses septate, simple or sparingly branched, 2 μ thick, the tips broadened to 7 μ, often moniliform; asci 55–58 × 11.5–13 μ; spores (8–) 12–16, hyaline, 1-celled but frequently with cytoplasmic bridges and appearing 2-celled, 7–10 × 4–6 μ.

This species grows on rock faces which are temporarily covered with running water, sometimes on vertical cliffs. It is known from Sweden, Greenland, and arctic Canada. It ranges south to British Columbia.

XANTHOPARMELIA (Vain.) Hale

Phytologia 28:485. 1974. *Parmelia* section *Xanthoparmelia* Vain., Acta Soc. Faun. Fl. Fenn. 7(7):60. 1890.

Thallus foliose with narrow lobes, yellow; upper surface with pored epicortex; upper cortex of palisadeplectenchyma; underside with simple rhizines; cortex containing usnic acid.

Apothecia adnate; margin persistent, concolorous with the thallus; disk brown to chestnut brown; hypothecium hyaline; hymenium hyaline with brown epithecium; paraphyses branched, slender; asci clavate; spores 8, hyaline, simple, ellipsoid.

Alga: *Trebouxia*.

Type species: *Xanthoparmelia conspersa* (Ach.) Hale.

REFERENCES

Hale, Mason E., Jr., 1974. *Bulbothrix, Parmelina, Relicina,* and *Xanthoparmelia,* four new genera in the Parmeliaceae (Lichenes). Phytologia 28:479–490.

Nash, T. H., III. 1974. Chemotaxonomy of Arizonan lichens of the genus *Parmelia* subgen. *Xanthoparmelia.* Bull. Torrey Bot. Club. 101:317–325.

1. Thallus with capitate soralia on upper side; medulla UV+, KC+ red, K+ yellow; containing alectoronic acid and atranorin... 3. *X. incurva*
1. Thallus lacking soredia.
   2. Medulla UV−, K+ red, containing salazinic acid.
      3. On rocks, lobe margins not turning under.
         4. Underside pale, tan to brown.................................... 5. *X. taractica*
         4. Underside black.. 6. *X. tasmanica*
      3. On soil, unattached, lobe margins turning under...................... 2. *X. chlorochroa*
   2. Medulla UV+, K+ yellow, KC+ red; containing alectoronic acid and atranorin.
      5. Underside white, medulla I−; lobes short and wide..................... 1. *X. centrifuga*
      5. Underside dark, mouse gray to black; medulla I+ blue; lobes elongate .... 4. *X. separata*

1. Xanthoparmelia centrifuga (L.) Hale

Phytologia 28:486. 1974. *Lichen centrifugus* L., Sp. Pl. 1142. 1753. *Parmelia centrifuga* (L.) Ach., Meth. Lich. 206. 1803.

Thallus foliose, closely adnate to the substratum, commonly dying in the center and forming concentric rings, the lobes irregularly branching, the tips pointed, flat to slightly convex, 2 mm broad; upper side greenish yellow, partly whitish green, darkening toward the center; dull, sometimes chinky, lacking soredia or isidia; underside whitish or pale brownish with brown to blackish rhizines; medulla white.

Apothecia adnate, to 5 mm broad, the margin entire to crenulate or inflexed, concolorous with the thallus; disk reddish or dark brown, dull; epithecium brownish; hypothecium and hymenium hyaline; spores ellipsoid, 8–14 × 4.5–6 μ.

Reactions: hymenium I+ blue; cortex K+ yellow; medulla I−, K−, KC+ red, P−.

Contents: usnic acid, alectoronic acid, atranorin.

This species grows on rocks, preferably acid, usually in strong light but in more shady spots becoming softer in texture. It is circumpolar, arctic-alpine, ranging south in North America into the Appalachian Mountains and near the northern shores of the Great Lakes.

2. Xanthoparmelia chlorochroa (Tuck.) Hale

Phytologia 28:486. 1974. *Parmelia chlorochroa* Tuck., Proc. Amer. Acad. Arts Sci. 4:383. 1860.

Thallus loose and blown by the wind, on arid soil; lobes long, narrow, irregularly to di-chotomously branched, convex, the upper parts of the lobes reflexed; upper side smooth, whitish gray to yellowish gray, convex; underside dark brown to black, with small black, simple rhizinae; cortex of palisade plectenchyma; medulla white, of loosely interwoven hyphae.

Apothecia rare, adnate, to 6 mm broad; margin smooth or crenulate, concolorous with the thallus; disk dark brown, concave to flat; spores 8, ellipsoid, simple, hyaline, 6–8 × 3–4 μ.

Reactions: cortex K+ yellow, medulla K+, yellow to red, KC+ yellow; P+ orange.

Contents: usnic acid and salazinic acid.

This species grows in the open on arid soils, loose and moving with the winds as a migratory lichen. It is a lichen of the Great Plains, its range being from Alberta and Saskatchewan south to New Mexico and Arizona. It has been shown to have an outlying population in the Arctic in the Reindeer Preserve with a range resembling that of *Buellia elegans*.

3. Xanthoparmelia incurva (Pers.) Hale

Phytologia 28:488. 1974. '*Lichen incurvus* Pers. in Usteri, Neue Ann. Bot. part 3:24. 1794. *Parmelia incurva* (Pers.) Fr., Nov. Schedul. Crit. 31. 1826.

Thallus forming small rosettes, closely attached to the substrate; the lobes narrow, to 1 mm broad, convex, closely set to each other, with capitate soralia scattered over the surface of the thallus; upper side grayish green, yellow-green, or grayish yellow, darkening to the center, dull; the soralia at the tips of short lobes and to 4 mm broad, usually less; underside brown at the edges, black centrally with blackish simple rhizinae; medulla white.

Xanthoparmelia centrifuga (L.) Hale

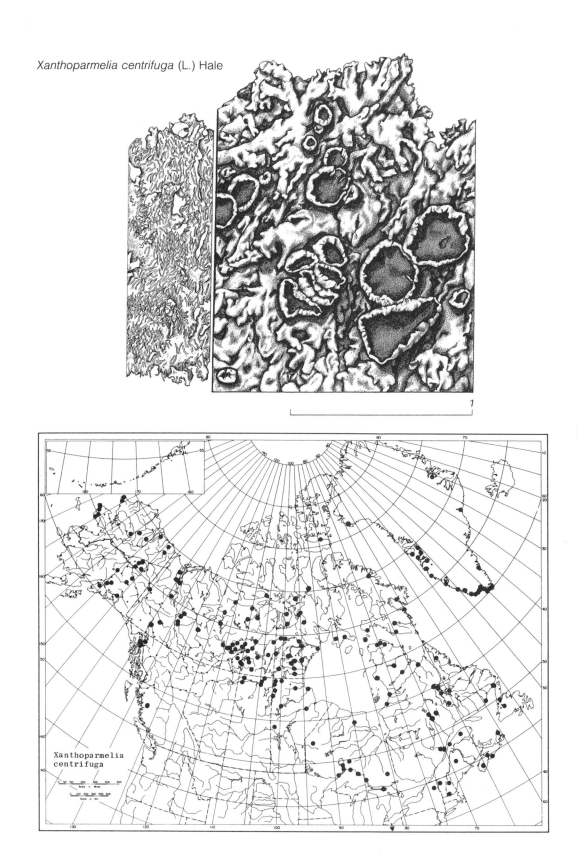

1

Xanthoparmelia
centrifuga

Apothecia uncommon, small, 2 (–4) mm broad, adnate; margin inflexed, entire to crenulate, concolorous with the thallus; disk chestnut brown, concave or flat, dull to shining; epithecium yellowish; hypothecium hyaline; spores ellipsoid, thick-walled, 9–10.5 × 6.8 μ.

Reactions: medulla K+ yellow, KC+ red, P–, UV+.

Contents: usnic acid, alectoronic acid, and atranorin.

This species grows on acid rocks usually in full sun. It is circumpolar, arctic, and alpine. In eastern North America it ranges south to the Adirondack Mountains and to a disjunct occurrence on Grandfather Mountain, N.C. (Dey 1973). In the west it appears to be southward only in Alaska and the Yukon.

4. Xanthoparmelia separata (Th. Fr.) Hale

Phytologia 28:489. 1974. *Parmelia separata* Th. Fr., J. Linn. Soc. Bot. 17:353. 1880. *Parmelia birulae* Elenk., Ann. Mycologici 4:36. 1906.

Thallus foliose, loosely attached to the substratum, dichotomously to irregularly branched, the lobes elongate as compared to *Xanthoparmelia centrifuga;* upper side yellow, brownish in spots, often chinky separated crosswise, smooth to wrinkled; underside mouse gray or dove gray to black, smooth with sparse simple rhizinae, medulla white.

Apothecia adnate, slightly pedicillate; margin entire to lobulate, concolorous with the thallus; disk concave, to 8 mm broad, bright chestnut brown, shining, epruinose; cortex of amphithecium poorly differentiated, heavily interspersed with yellow granules and opaque, the surface rough; epithecium and upper 12 μ of hymenium clear brown; lower part of hymenium hyaline; hymenium 65 μ, I+ blue; hypothecium 90 μ, hyaline to yellowish or grayish yellow, the upper 25 μ I+ violet, no reaction in the lower part; paraphyses conglutinate, thick-walled, little branched, septate, the tip cells enlarged, brown to 4 μ; asci clavate, 30 × 12 μ;

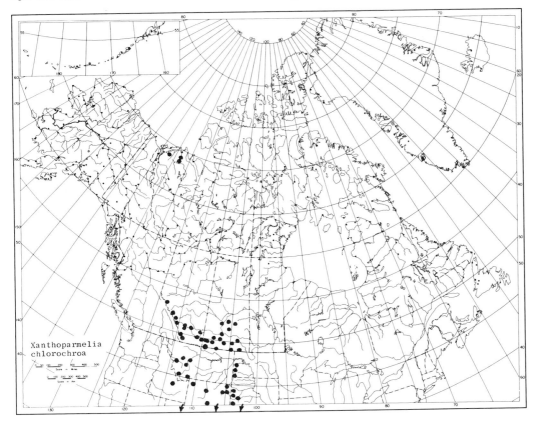

Xanthoparmelia
chlorochroa

Xanthoparmelia incurva (Pers.) Hale

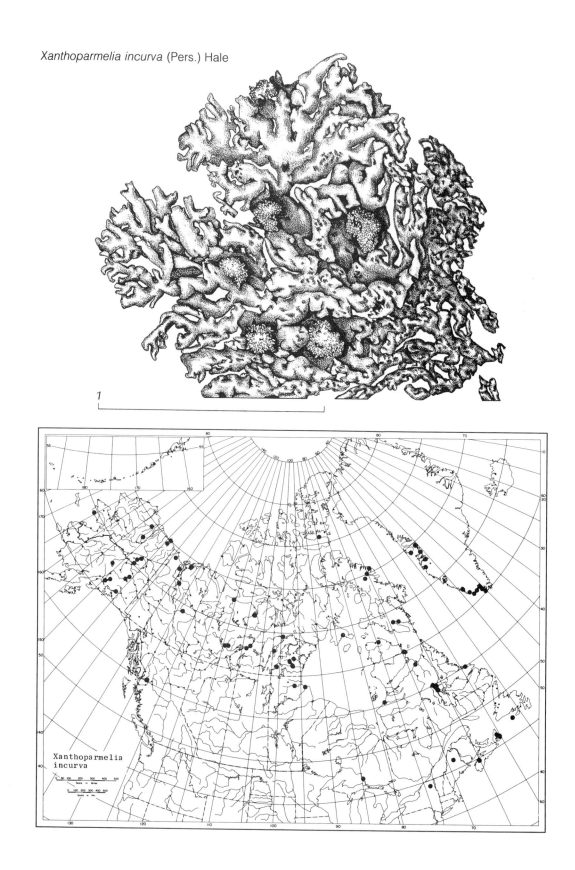

1

Xanthoparmelia
incurva

Xanthoparmelia separata (Th. Fr.) Hale

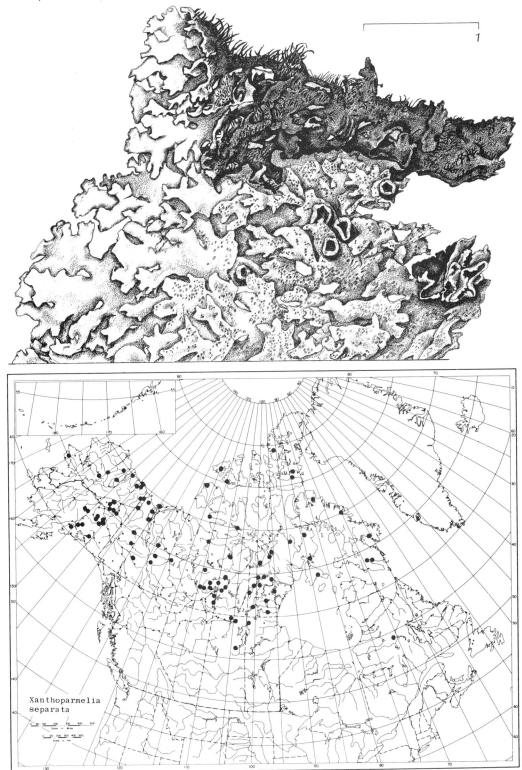

1

Xanthoparmelia
separata

Xanthoparmelia taractica (Krempelh.) Hale

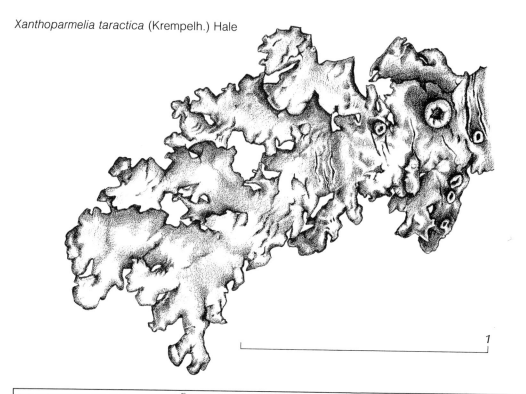

1

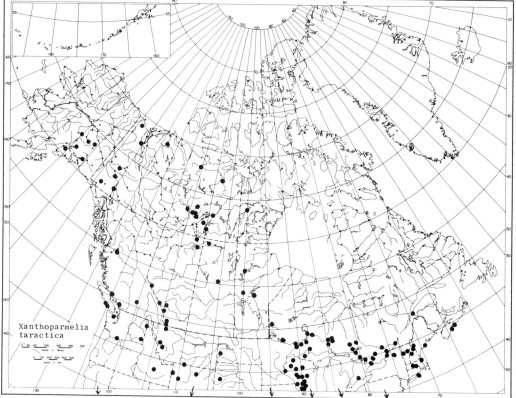

Xanthoparmelia
taractica

spores 8, hyaline, elongate ellipsoid, simple, 10–12.5 × 4.5–6.5 μ.

Reactions: cortex K+ yellow; medulla K–, KC+ red, P–, I+ blue.

Contents: usnic acid, alectoronic acid, and atranorin.

This species grows over mosses, on humus, and on rocks. It is probably circumpolar, arctic, ranging from Novaya Zemlya, across Siberia and northern North America, but has not yet been found in Greenland.

5. Xanthoparmelia taractica (Kremph.) Hale
Phytologia 28:489. 1974. *Parmelia taractica* Kremph., Flora 61:439. 1878.

Thallus foliose, irregularly branched, the lobes loosely attached to the substratum, 1–6 mm broad, overlapping, flat to slightly rounded; upper side yellow, smooth, lacking isidia or soredia; underside brown, tan at the margins, with simple unbranched rhizinae; medulla white.

Apothecia concave or deeply concave, to 8 mm broad; margin concolorous with the thallus, irregularly wavy to lobulate; disk chestnut brown to dark brown, shiny or dull; epithecium brown; hypothecium and hymenium hya-

line; spores 8, hyaline, ellipsoid to oval or slightly curved, 8–11 × 5–6 μ.

Reactions: medulla K+ yellow turning red, KC+ yellow, P+ orange.

Contents: usnic and salazinic acids.

This species grows on acid rocks in full sun. It is circumpolar, arctic to temperate, with very wide distribution, and it ranges over most of North America. Only the lighter color of the underside separates this from *P. tasmanica* which occupies the same range but is less common.

6. Xanthoparmelia tasmanica (Hook. & Tayl.) Hale
Phytologia 28:489. 1974. *Parmelia tasmanica* Hook. & Tayl., London J. Bot. 3:644. 1844.

Other than the black color of the underside of the thallus in this species instead of the tan to brown color of the underside of *P. taractica* there are no differences between these two species and the description and contents of the two are congruent. This species with the black underside occurs through the range of *P. taractica* but is less common. It is quite possible that the color of the underside of the thallus is not a very valid species criterion in this case. If these are to be combined then it should be noted that *P. tasmanica* is the older epithet.

XANTHORIA (FR.) Th. Fr. em Arn.
Ber. Kgl. Bayr. Bot. Ges. 1:40. 1891. *Parmelia* D. *Xanthoria* E. Fr., Syst. Orb. Veget. 1:243. 1825. *Xanthoria* Th. Fr., Lich. Arct. 66. 1860.

Thallus foliose, dorsiventral, erect or spreading; underside with rhizinae, heteromerous; both surfaces corticate, the upper cortex paraplectenchymatous, yellow to orange; the medulla loose, of thin-walled anastomosing hyphae.

Apothecia peltate, adnate or on short stipes, on the upper surface of the lobes; disk round, yellow or orange-red, with an entire, crenulate, or sorediate margin which is lecanoroid; epithecium violet with KOH; hymenium and hypothecium hyaline; paraphyses loose, simple or branched above, septate, the apices often thickened; asci clavate; spores 8, hyaline, polaribilocular. Pycnidia sunken in small warts in the thallus, fulcra endobasidial; conidia short, straight, the center slightly larger than the ends.

Alga: *Trebouxia*.

Type species: *Xanthoria parietina* (L.) Th. Fr.

The presence of parietin in the cortex and epithecium gives the KOH purple-red reactions.

REFERENCES

Hillmann, J. 1922. Übersicht über die Arten der Flechtengattung *Xanthoria*. Hedwigia 63:198–208.

—— 1935. Teloschistaceae. In Rabenhorst, Kryptogamenflora. 9(6):1–36.

Rudolph, E. D. 1955. Studies in the lichen family Blasteniaceae in North America north of Mexico. Unpubl. thesis, Washington Univ., Univ. Microfilms, Ann Arbor.

1. Thallus lacking isidia or soredia.
  2. Thallus with well-defined thread-like rhizines; growing on trees and shrubs, marginal lobes flat . . . . 4. *X. polycarpa*
  2. Thallus attached by broad hapters rather than thread-like rhizines; growing on rocks; marginal lobes rounded . . . . 2. *X. elegans*
1. Thallus with either isidia or soredia or both.
  3. Thallus with soredia only.
    4. Soredia in labriform marginal soralia, lobe tips round. . . . . 3. *X. fallax*
    4. Soredia bordering the lobes which break down irregularly to form soredia from the edges (soredia also on underside of lobe tips but not in well-defined soralia) lobe tips dividing into narrow segments . . . . 1. *X. candelaria*
  3. Thallus forming pustular isidia which may break down at the tips to form soredia as well . . .
    . . . . 5. *X. sorediata*

1. Xanthoria candelaria (L.) Arn.

Flora 62:364. 1879. *Lichen candelaris* L., Sp. Pl. 1141. 1753.

Thallus foliose, small, irregular to rosette-shaped, the margins ascending to erect, the lobe tips sorediate to isidia-like divided, yellow to chrome yellow or orange, K+ red; upper cortex paraplectenchymatous, becoming yellow granular, 25–35 μ; lower cortex also paraplectenchymatous, pale, 25–35 μ; medulla loose.

Apothecia rare, sessile to slightly stipitate, to 1.3 mm, margin concolorous with thallus; disk flat, orange, K+ red; epithecium orange granular; hypothecium hyaline; paraphyses slender, 2 μ, branching toward the tips; asci clavate, 45–56 × 14–18 μ; spores 8, polaribilocular, 10.4–13.0 × 6.1–7.0 μ, the isthmus 4 μ.

This species grows on trees, wood, rocks, and over mosses. It is a circumpolar arctic to temperate species which ranges over much of North America.

Xanthoria candelaria and *X. fallax* are sometimes difficult to distinguish, particularly in depauperate material from severe arctic conditions. The former tends to make the lobes more erect and the soralia more spreading over the underside of the tips of the lobes as well as having the lobes at least in part more finely divided. The latter has the lobes lying closer to the substratum and the soralia take on a more labriform definite shape while the lobing is much less divided. Some of the reports of *X. fallax* in the Arctic

(and for which I take responsibility) have turned out on closer study to be based on poorly defined material of *X. candelaria*. *X. fallax* is much more southern in distribution than *X. candelaria* but does reach into the Arctic along with other southern species as *Physcia aipolia* to the Firth River on the Alaskan-Yukon boundary.

2. Xanthoria elegans (Link.) Th. Fr.

Nova Acta Reg. Soc. Sci. Upsala Ser. 3, 3:169. 1861. *Lichen elegans* Link., Ann. Naturges. I Stück 37. 1791. *Caloplaca elegans* (Link.) Th. Fr., Lich. Scand. 1:168. 1871.

Thallus crustose to subfoliose, radiate lobed at margins, orange to reddish-orange, K+ purple (parietin), the lobes irregularly branched, to 1 mm broad; upper cortex fastigiate or paraplectenchymatous, to 40 μ thick; lower cortex pale with scattered pale rhizinae, similar to upper cortex; medulla of loose hyphae.

Apothecia sessile to slightly stalked, to 2 mm broad; margin of same color as disk or lighter; disk flat or convex, yellow to orange; epithecium yellow or orange granular, K+ purple; hypothecium hyaline; paraphyses slender, septate, 3.5 μ, the tips capitate; asci clavate, 35–56 × 9–17 μ; spores 8, polaribilocular, 8.7–15 × 3.8–8 μ; isthmus 5 μ.

This species grows on limestones and on enriched bird-perch rocks, on old bones, and occasionally on old wood. Nesting cliffs of birds in the Arctic

Xanthoria candelaria (L.) Arn.

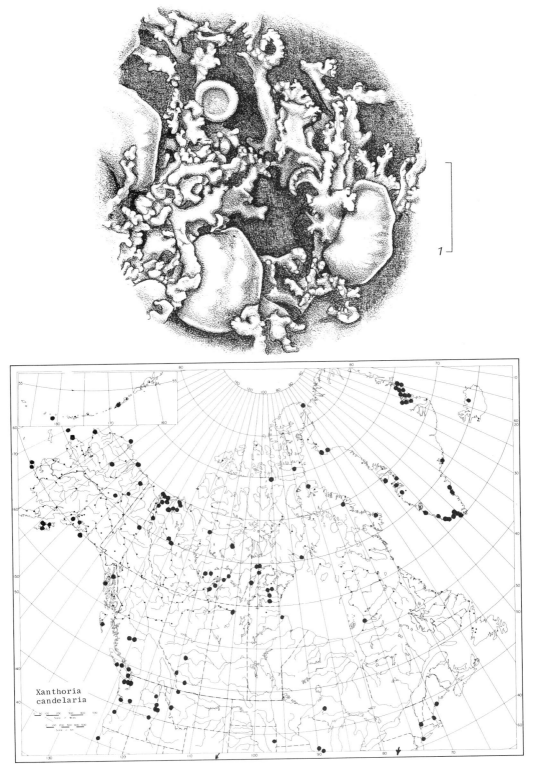

1

Xanthoria
candelaria

Xanthoria elegans (Link.) Th. Fr.

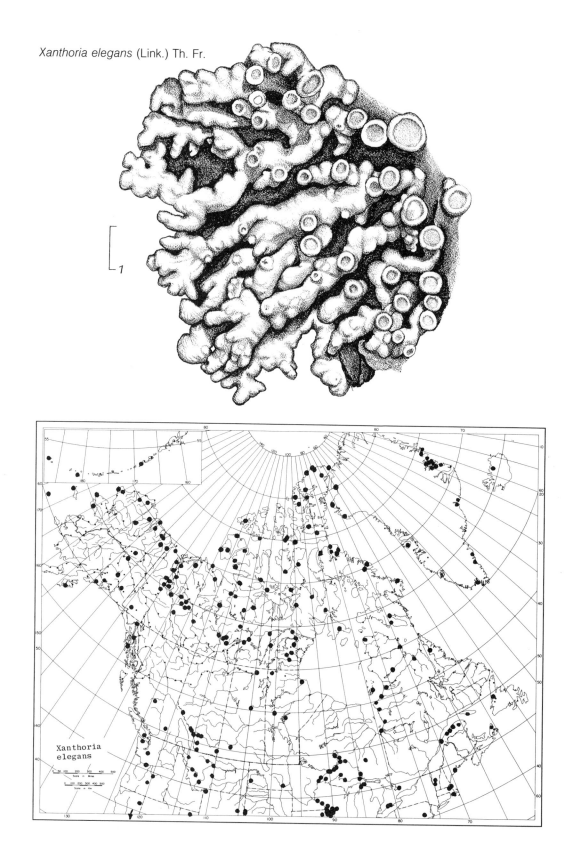

1

Xanthoria
elegans

are often distinguished from afar by the splashes of orange color of this lichen. This species is circumpolar, arctic to temperate, one of our commonest lichens. *Caloplaca splendens* (Darb.) Zahlbr., described from the Arctic, appears to be only a very luxuriant form growing over mosses in heavily manured places.

3. Xanthoria fallax (Hepp in Arn.) Arn.

Verhandl. zool. Bot. Ges. Wien 30:121. 1880. *Physcia fallax* Hepp in Arn., Flora 41:307. 1858.

Thallus foliose, lobes rounded, small, to 2 mm broad, the tips only slightly ascending as compared with *X. candelaria;* yellow to orange, K+ red-purple, margins and underside of margins with mainly distinctly limited soralia; upper cortex paraplectenchymatous, 17–50 μ; lower cortex pale, fastigiate to paraplectenchymatous, 25–35 μ, with scattered pale rhizinae; medulla loose.

Apothecia sessile, to 2 mm, margin concolorous with the thallus, thick, becoming crenulate or sorediate; disk orange, flat, dull; epithecium yellow granular; hypothecium hyaline; paraphyses to 3 μ, septate, branching toward apices; asci clavate, 35–42 × 10–14 μ; spores 8, polaribilocular, 10–14 × 5–7 μ, isthmus 3.5 μ.

This species grows on trees and rocks, occasionally on moss and old wood. It is circumpolar, boreal to temperate, reaching the edge of the tundra in the boreal forest at its northern border.

4. Xanthoria polycarpa (Ehrh.) Rieber

Jahres. Ver. vaterl. Naturk. Württ. 47:252. 1891. *Lichen polycarpus* Ehrh., Plant. Crypt. Exs. 136. 1789.

Thallus yellow to orange-yellow, seldom gray-yellow, K+ red, forming small tufts on the substrate, usually with abundant apothecia; lacking soredia or isidia; the lobes irregularly branching, narrow, to 0.5 mm broad, flat, more or less ascendent and overlapping; underside pale, with scattered white rhizines; with upper and lower cortex, the center with scattered algae and no definite medullary layer.

Apothecia adnate or with short stalks, to 4 mm broad; the disk dull, dark or yellowish orange, the margin thick, thalloid, often crenulate; epithecium yellow or greenish yellow; K+ purple; hymenium and hypothecium hyaline; paraphyses capitate; asci cylindrical to narrowly clavate, spores 8, hyaline, polarilocular, 11–16 × 6–8.5.

This species grows on the stems and branches of conifers and broad-leaved plants. It is circumpolar, boreal to subtropical, with very wide distribution in North America.

5. Xanthoria sorediata (Vain.) Poelt

Mitt. Bot. Staatssamml. München 11:29. 1954. *Lecanora elegans* var. *sorediata* Vain., Medd. Soc. Fauna Flora Fenn. 6:143. 1881. *Caloplaca sorediata* (Vain.) DuRietz, Svensk Bot. Tids. 10:477. 1916.

Thallus crustose centrally, foliose marginally; the marginal lobes branching irregularly, expanded and flat at the tips, granular and isidiose sorediate toward the center; the coarse soredia breaking out of pustular isidia, orange, K+ purplish-red; upper cortex paraplectenchymatous, 10–20 μ; lower cortex pale, similar to the upper; medulla loose.

Apothecia rare, similar to those of *X. elegans.*

This species grows on calcareous rocks. It is circumpolar, arctic-alpine, and boreal. It ranges south in the mountain ranges in the west to Mexico and in the east to the Great Lakes and eastern Canada.

Although *Xanthoria parietina* has been reported in the American Arctic by many authors these are undoubtedly based upon misidentifications. That species is maritime and ranges only as far north as Gaspé in the east and California in the west (Hale 1955). Reports of *Xanthoria etesiae* in the American Arctic are based upon small specimens of *X. elegans* with very dense medulla instead of the usually excavate medulla.

Xanthoria fallax (Hepp in Arn.) Arn.

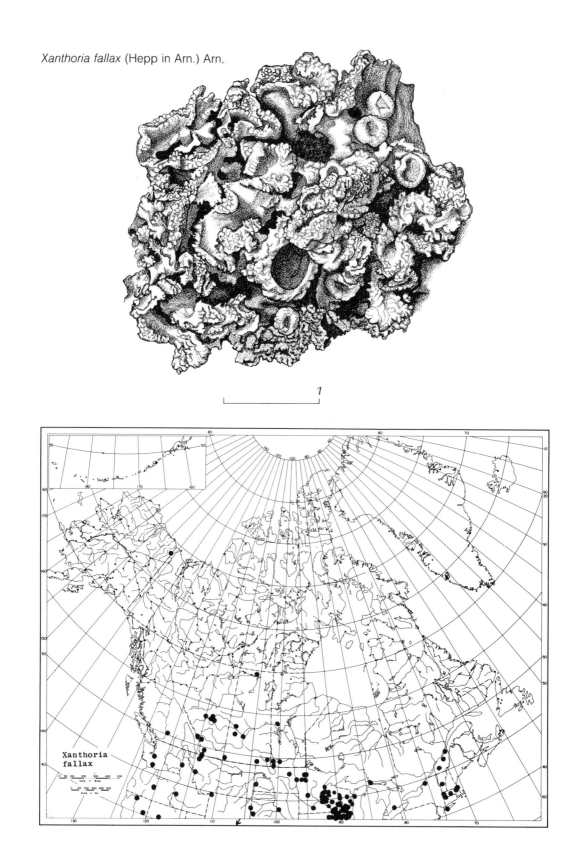

1

Xanthoria
fallax

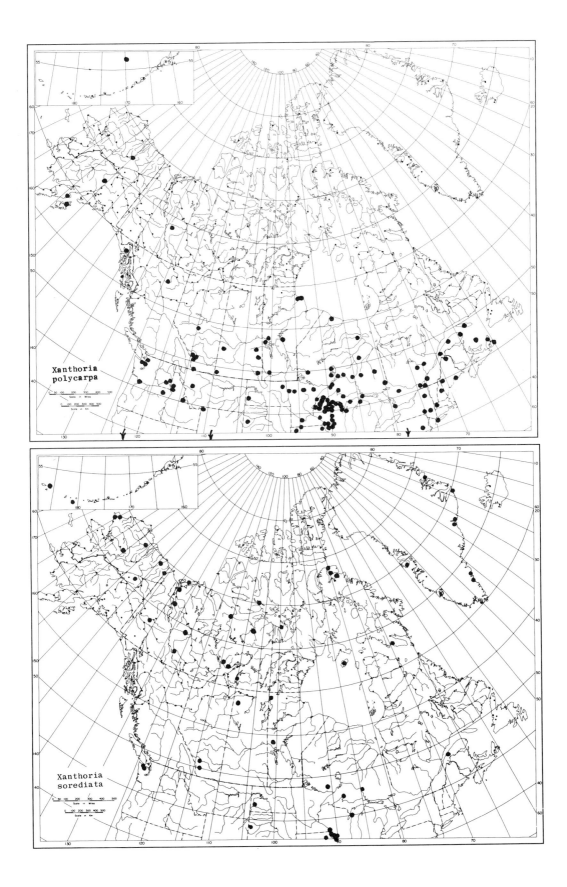

Xanthoria
polycarpa

Xanthoria
sorediata

ZAHLBRUCKNERELLA Herre

J. Wash. Acad. Sci. 2:384. 1912. *Zahlbrucknera* Herre, Proc. Wash. Acad. Sci. 12:129. 1910 (non Reichenb., 1832).

Thallus minutely fruticose, thread-like, olive-colored to blackish, dichotomously branched, lacking rhizinae, attached by holdfasts; hyphae either with round cells and parallel or forming irregular nets of angular cells.

Apothecia lateral with thalloid margin, disk green or brown; hymenium gelatinous; paraphyses net-like branched; asci cylindrical to subclavate; spores numerous, simple, hyaline. Pycnidia lateral, conidia terminal, rod-shaped.

Alga: *Scytonema.*

Type species: *Zahlbrucknerella calcarea* (Herre) Herre

REFERENCES

Henssen, A. 1963.. Eine revision der Flechtenfamilien Lichenaceae und Ephebaceae. Symbol. Bot. Upsal. 18(1):1–123.

——1977. The genus *Zahlbrucknerella*. Lichenologist 9:17–46.

Only one species reaches the American Arctic.

1. Zahlbrucknerella calcarea (Herre) Herre

J. Wash. Acad. Sci. 2:384. 1912. *Zahlbrucknera calcarea* Herre Proc. Wash. Acad. Sci. 12:129. 1910.

Thallus of minute thread-like filaments, olive to brownish black, forming clumps or mats, the filaments in the center more or less erect, the lateral ones decumbent, the tips attenuated; the hyphal cells more or less angular and in rows, toward the base with 2–4 rows externally.

Apothecia lateral, with thalloid margin, to 0.4 mm broad; disk red-brown; hymenium 100–130 μ, I+ greenish blue; hypothecium 10–50 μ; paraphyses net-like branched, becoming septate, 1.5–2.5 μ, the tips becoming capitate; asci subclavate, 70–90 × 10–20 μ; spores (16–)24, simple, hyaline, oval, 6–9 × 5.5–6 μ. Conidia cylindrical, 3–4.5 × 1 μ.

Reactions: K– , C– , P– .

Contents: not known.

This species grows over calcareous rocks in moist sites and among mosses over the rocks. It is circumpolar in the northern hemisphere (Henssen 1963). In North America it is known in the west in California (type locality), Colorado, Wyoming, in Canada in British Columbia and Alberta (Henssen 1977), reaching to near Inuvik in the American Arctic (Ahti et al. 1973).

Henssen suggests that *Z. fabispora* Henss. with 8 bean-shaped spores and slender filaments should be expected in the Canadian Arctic as its distribution seems to be circumpolar.

Zahlbrucknerella calcarea (Herre) Herre

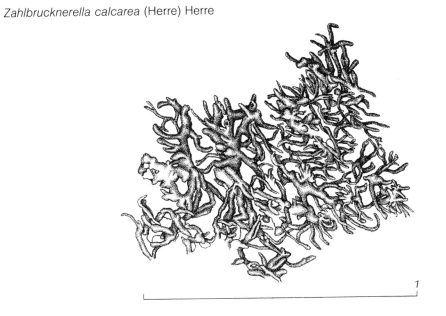

1

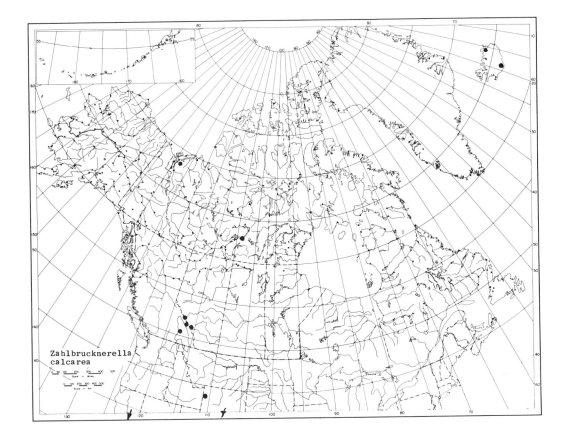

Zahlbrucknerella
calcarea

Index

Accepted names are in roman type, synonyms are in italics. Treatments of the species are indicated in bold-face, drawings and/or maps are in italics, other references are in roman type.